Single Neuron Studies of the Human Brain

Single Neuron Studies of the Human Brain

Probing Cognition

edited by Itzhak Fried, Ueli Rutishauser, Moran Cerf, and Gabriel Kreiman

The MIT Press
Cambridge, Massachusetts
London, England

MIT Press books may be purchased at special quantity discounts for business or sales promotional use. For information, please email special_sales@mitpress.mit.edu.

This book was set in Times by Toppan Best-set Premedia Limited, Hong Kong. Printed and bound in the United States of America.

Library of Congress Cataloging-in-Publication Data
Single neuron studies of the human brain : probing cognition / edited by Itzhak Fried, Ueli Rutishauser, Moran Cerf, and Gabriel Kreiman.
     p. ; cm.
Includes bibliographical references and index.
ISBN 978-0-262-02720-5 (hardcover : alk. paper)
I. Fried, Itzhak, editor of compilation.   II. Rutishauser, Ueli, editor of compilation.   III. Cerf, Moran, editor of compilation.   IV. Kreiman, Gabriel, 1971– editor of compilation.
[DNLM: 1. Neurons—physiology.   2. Brain—physiology.   3. Brain Mapping.   4. Cognition—physiology.
5. Synaptic Transmission.   WL 102.5]
QP360.5
612.8'2—dc23
2013041744

10   9   8   7   6   5   4   3   2   1

Contents

Open Letter to a Beginning Researcher in the Field of Human Single Neuron Investigations

From your vantage point as a researcher just entering the field of human single neuron research, you may already experience the excitement of potential contributions to human knowledge inherent in directly working with the human brain. However, the neurosurgical theatre of the mind may often look intimidating and complex, a foreign environment without the reassurance of complete experimental control. Being in the operating room is an intense experience. Individuals new to this field of research may easily find themselves overwhelmed by the cast of characters, doctors, nurses, and other ancillary personnel; by a great variety of instruments, the life support and anesthetic machinery, and the boundaries of a sterile field; and by the overwhelming presence of a patient with an exposed brain, sometimes awake during procedures performed under local anesthesia. On the hospital ward, the situation may be daunting as well, with an abundance of health care personnel, visitors in the patient's room, the constant possibility of imminent epileptic seizures, and a myriad of noise sources, electrical and psychological. What advice, then, can help you as a scientist entering this complex field of single neuron recordings in humans?

The first step in an organized research project would entail *choosing the right question.* However, contrary to the tradition of carefully preconceived lines of scientific investigation, as a researcher in *this* field, you need to be an experimental opportunist. You cannot choose just any question and hope to record from the relevant neurons. The sites of recordings will always be completely determined by the clinical imperative and thus will be fixed in locations that cannot be altered. The question you elect to explore has to be grounded in animal physiology and has to build on this knowledge. Yet, the question must also be relevant and unique to the human condition. In particular, you need to take what we know from nonhuman primate neurophysiology to the next level, the human level. Single neuron human neurophysiology is a small field between animal neurophysiology and human functional neuroimaging and other noninvasive methods customarily used in cognitive neuroscience. However, it is not enough to confirm findings from these areas. To merely confirm results obtained with other methods is to fail to take advantage of the unique opportunities for advancing knowledge afforded by the methodology of human single neuron research. You need to ask the next question, the one which can only be answered using *this* technique.

When you come to address your question with a designed experiment, you must keep in mind the most important tenet in this field: It is a privilege to work with patients. Always remember, patients come first. This might be difficult to keep in mind when all experimental control is lost because, for example, in the middle of a recording session that was laboriously set up, your patient has a pressing need. This need comes first. Naturally, individuals who must have total control of the experimental situation will not thrive in this environment. You need to be able to listen, to observe, and to not lose a rare moment of insight which may fleet by as your subjects, patients who can declare their thoughts and wishes, may have an illuminating comment. Indeed, Penfield was able to listen to his patients as they were lying awake under the surgical drapes and was able to correlate his stimulation of the temporal lobe with past recollections. In my own research, it was a particularly verbally gifted and insightful patient who once told me that she was feeling an urge to move her hand when I was stimulating a site in her supplementary motor area at a low current, thus providing much welcomed evidence of the importance of this brain region in volition. However, while having the flexibility to take advantage of research opportunities that present themselves in the moment is crucial to success in our field, at the same time you will find that, as a researcher, you need to stick to a paradigm and not change paradigms too often. It takes a few years to gather a sufficient number of neurons in a region to make a meaningful statement. Investigators also need to be technically savvy with data analysis (e.g., facile with computer programming) and competitive yet collaborative. Above all, achievement in this field requires that a researcher be someone who is passionate about working with the living human brain and yet always remembers that at the center of this unique situation is a courageous patient who is indeed the focus of all our efforts. Finally, the successful human single neuron researcher is someone who maintains a long-term vision guided by recognition that this is probably the only opportunity in neuroscience to access, at the most basic level, the substance that makes us human in patients who can declare and share with us their memories, perceptions, emotions, and wishes.

Bridging single neurons and human behavior is at the core of the mind–body problem. Our field, then, is challenging but also very rewarding. I hope you will find your journey in these new territories as meaningful and inspiring as I have.

—Itzhak Fried

1 Introduction

Itzhak Fried, Ueli Rutishauser, Moran Cerf, and Gabriel Kreiman

There has been tremendous progress in our understanding of faraway galaxies, probing the rules that govern the function of subatomic particles, elucidating the basic principles of life and describing the biochemical reactions inside cells. Yet, our own brains remain a major frontier for scientific investigation. Neurons and their interactions must give rise to the bewildering complexity that underlies our perceptions, memories, intentions and emotions. In a large number of unfortunate circumstances, malfunctioning of these circuits can give rise to some of the most devastating disorders. The magic behind the function of the human brain defies our intuitions. Elucidating this mystery can lead to profound transformations in how we understand ourselves, how we treat brain disorders, and how we build more intelligent machines.

The last century has seen major strides toward understanding neurons and how they interact, particularly through the examination of diverse animal models. Pioneers such as Adrian, Sherrington, Hodgkin, Huxley, Hubel, Wiesel, and many others paved the way to listen to the activity of neurons and correlate this activity with sensory, motor, and cognitive phenomena. Through their work, we have begun to understand specialized subcircuits that represent visual information, brain nuclei that can help consolidate experiences into long-term memories, and how circuits of neurons can orchestrate behavioral output.

While studying many animal species continues to illuminate the road ahead, there remain multiple phenomena that are difficult to examine outside the human brain. To cite a few, the nature of language, imagery, subjective feelings, free will, and consciousness are not easy to rigorously examine in animal models. Multiple techniques exist to study the human brain in a noninvasive manner including scalp electroencephalography (EEG), magnetoencephalography (MEG), functional magnetic resonance imaging (fMRI), and transcranial magnetic or electrical stimulation (TMS, TES). While these techniques have provided important correlations of cognitive phenomena, their spatial and/or temporal resolution is quite limited (see figure 1.1). For example, fMRI and EEG provide information on scales of millimeters to centimeters. A cubic millimeter contains on the order of 100,000 neurons. In the temporal domain, we know that the dynamics of brain computations change rapidly on scales of milliseconds whereas fMRI provides average measurements over scales of seconds. Furthermore, the biophysics underlying such

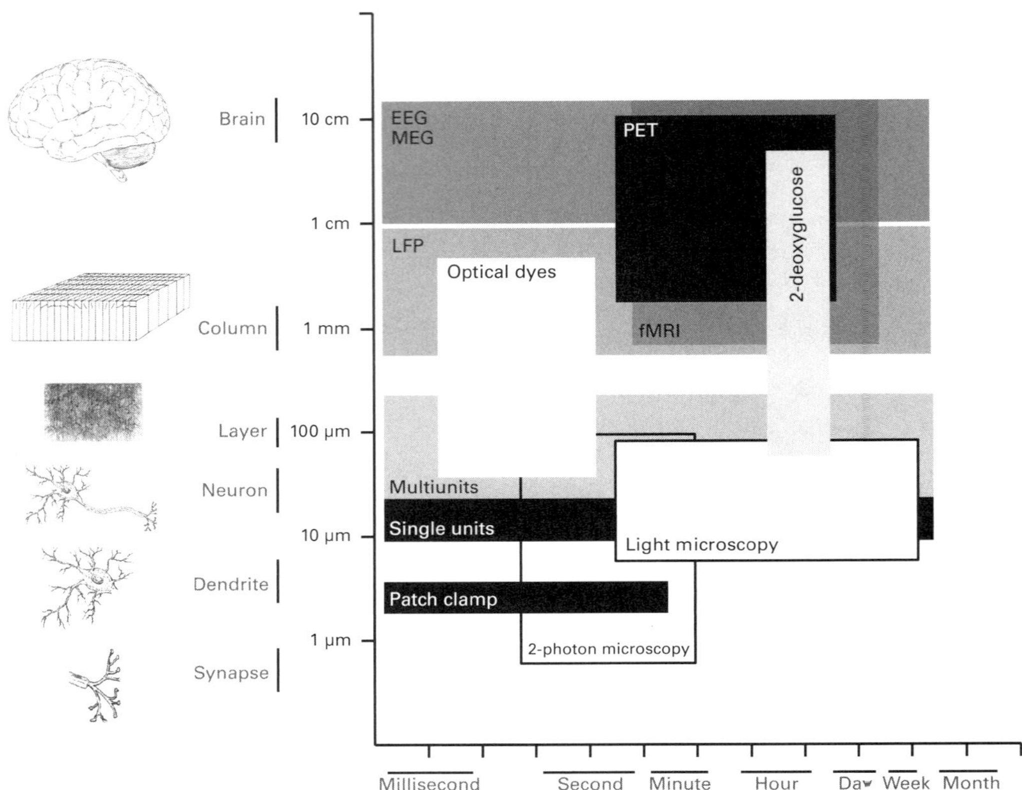

Figure 1.1
Schematic rendering of the spatial and temporal resolution of multiple different techniques to study brains. The x-axis describes temporal resolutions from milliseconds to months, and the y-axis describes spatial resolutions from micrometers to centimeters. EEG, electroencephalography; MEG, magnetoencephalography; PET, positron emission tomography; LFP, local field potential; fMRI, functional magnetic resonance imaging.

noninvasive signals and their relationship to neuronal function is very complex, is indirect, and remains poorly understood (Logothetis, 2002; Buzsáki et al., 2012).

To our rescue, in some rare circumstances, it is possible to directly interrogate the activity of spiking neurons in the awake, behaving human brain. The techniques for such recordings are similar to those used for extracellular recordings in animals such as nonhuman primates but differ in several critical aspects. The invasive nature of the procedures require that such cases always accompany the treatment of clinical conditions such as Parkinson's disease, brain tumors, and epilepsy (Engel et al., 2005). These rare opportunities have provided major clinical insights and at the same time have opened the doors to examine cognitive phenomena at unprecedented resolution (Crick et al., 2004). In this book, the world leaders in invasive studies of the human brain take us through the adventures, successes, failures, challenges, and opportunities in the field.

We hope that the book will provide a solid foundation to this emerging field and inspire the next generation of investigators coming from a variety of disciplines including neurosurgeons, neurologists, psychiatrists, physicists, engineers, neuroscientists, and cognitive neuroscientists among others.

Some readers may want to read through the entire book in order. Chapters are approximately grouped into the following clusters. The first chapters provide a historical and methodological reference and background (chapters 2–6). The next section includes a variety of cognitive science domains where single neuron studies have shed novel insights (chapters 7–14). The last section documents the importance of examining single neuron activity in studies of several clinical conditions (chapters 15–18).

Other readers may be looking for information on methods, advances in cognitive science, or clinical translational insights and may want to focus on specific chapters. For those readers, we provide an approximate road map and description of the chapters here. A historical overview is provided by George Ojemann, one of the pioneers in the exploration of the human brain through single electrodes (chapter 2). He recounts how neurosurgeons began examining the activity of individual neurons. Success in any effort to interrogate the activity of individual electrodes depends strongly on the surgical implantation of the electrodes and the clinical environment. Itzhak Fried, who has perfected the techniques and has performed hundreds of such operations, takes the reader into the operating theater and to the clinical environment. Fried discusses the methodological and clinical aspects of surgery and electrode implantation and the approach to basic research in this clinical environment, providing advice to the young researcher entering the field of single neuron recordings in neurosurgical patients (chapter 3). At the center of our impetus to push the frontiers of clinical practice and scientific investigation is the patient. In chapter 4, Adam Mamelak provides a thorough and lucid discussion of the ethical and practical considerations in this type of effort. Another important methodological consideration concerns the electrodes themselves, how to build them, how they interact with the tissue, and how they impact the types of signals that we record. Richard Staba and colleagues provide a thorough documentation of the properties of different electrodes and hardware commonly used to interrogate neural function (chapter 5). When conditions are right, neurophysiological recordings can yield a wealth of data, and multiple quantitative procedures and algorithms are commonly used to analyze and interpret the recordings. These algorithms range from how to filter the signals, how to cluster the multiunit recordings to obtain putative single units, and how to decode the signals from neural ensembles among other topics. Some of these data analyses techniques are described in chapter 6 by Ueli Rutishauser and colleagues.

Following the methodological considerations are several chapters that provide an enticing overview of achievements and contributions to elucidating different aspects of cognitive function (chapters 7–14). Many electrodes in epileptic patients are implanted in areas of the medial temporal lobe (MTL), which are known to play a critical role for memory consolidation. Rutishauser and colleagues describe the roles of neurons in the hippocampus and surrounding structures in memory formation as previously unfamiliar stimuli become familiar. They further describe how neural activity at the time of encoding predicts which stimuli will later be remembered and the critical role of theta oscillations in this process (chapter 7). Mormann and colleagues summarize and discuss several efforts to describe the highly selective responses of MTL neurons in response to visual stimuli as well as the relationship between activity in the MTL and conscious perception (chapter 8). Suthana and Fried describe the insights into spatial navigation that have resulted from recording from the human MTL. This work builds a bridge to the extensively studied area of hippocampal place fields in rodents by demonstrating that cells with similar properties can also be found in the human MTL during virtual spatial navigation (chapter 9).

The function of sleep has remained largely mysterious, but it is clear that the brain is not simply suspended or shut off during sleep. On the contrary, the human brain is highly active during sleep, as evidenced by prominent oscillatory patterns. Recent work utilizing intracranial recordings during sleep has started to reveal critical new insights that are reviewed by Nir and colleagues (chapter 10). Another situation in which the brain shows strong activation patterns even in the absence of input involves internal mental processing of the type encountered during visual imagery. Cerf and colleagues describe the specific patterns of neuronal responses encountered in the human brain during imagery and how those signals could be utilized in real-time experiments and procedures (chapter 11).

The continuous anticipation, prediction, and evaluation of rewards constitute a crucial role of the human brain. Areas such as the basal ganglia, prefrontal cortex, and anterior cingulate cortex are known to be crucial for such processes. Neuronal activity from these areas can be studied during intraoperative procedures for treatment of certain neurological conditions. Such recordings have yielded the first evidence so far on how these areas contribute to reward-driven behavior in humans as summarized by Patel and colleagues (chapter 12).

Neuroimaging work shows that the human brain contains areas highly responsive to the processing of faces and the emotions conveyed by face stimuli. Lesion studies have shown that the amygdala is necessary for processing emotions in faces, but the neuronal mechanisms by which the amygdala does so remain unknown. Adolphs and colleagues review what single unit recordings from the amygdala have revealed on the neuronal mechanisms underlying processing of faces (chapter 13).

One of the most difficult aspects of cognition to study in animal models is language. Further, in many cases language is of particular clinical importance given that electrodes may target areas in the vicinity of so-called "eloquent cortex." Critical to the clinical procedure is to eliminate seizures without interfering with language and other cognitive functions. In chapter 14, Ojemann describes decades of work studying the responses of single neurons during a variety of tasks that involve language. In contrast to chapters 7–11, the recordings presented in chapters 12 and 14 take place during surgery performed under local anesthesia.

The ability to directly interrogate neuronal function in the human brain has important implications for translational work to help alleviate multiple brain disorders including epilepsy, Parkinson's disease, brain tumors, motor disorders, and many others. The next section of the book focuses on clinical neuroscience insights derived from neurophysiological recordings from the human (chapters 15–18). The implantation of deep brain stimulation (DBS) electrodes into deep nuclei of the basal ganglia frequently requires localization using microelectrode recordings. This procedure typically requires that the patient be awake during the recordings, thus presenting a unique opportunity for behavioral experiments during such procedures. The techniques and challenges of such recordings are discussed by Goodwin and colleagues (chapter 15). While highly effective, the mechanisms by which DBS affects stimulated neuronal tissue remain poorly understood. Recording single neurons during DBS implantation allows the direct investigation of the effect of stimulation on neuronal activity, as described by Patra and colleagues (chapter 16). There has been substantial progress toward the development of a motor prosthesis device to help patients with paralysis due to spinal cord injury or strokes. Bansal describes how such prostheses are driven by directly reading out spiking activity from motor cortex or related areas and the resulting clinical trials (chapter 17).

The majority of human intracranial recordings stem from patients with epilepsy, and the primary reason for implanting the electrodes in the first place is to record seizures. A few groups have managed to record robust single unit activity directly before and during seizures, yielding insights into mechanisms of seizure initiation and propagation as summarized by Schulze-Bonhage and colleagues in chapter 18.

We conclude the book with an outlook perspective on where the field is headed in the next ten years and beyond, highlighting some of the challenges that need to be overcome, some of the fundamental basic science questions that are ripe for investigation, and some of the exciting avenues for translational efforts (chapter 19). Neurons constitute the "atoms" of cognition. Francis Crick simply stated, "You are nothing but a pack of neurons" (Crick, 1994). Nothing more. Nothing less. While such a reductionist statement may invite vigorous objection, it is still the prevailing view of modern neuroscience that the neurons and their complex connections, what has recently been referred to as the *connectome*, are the substrate of mental life. A few atom types and their interactions can lead to the rich variety of molecules that form all matter, including brains. In a similar vein, it is conceivable that various neuronal types and their interactions may lead to the rich variety of cognitive phenomena and behavior that we are fascinated with and constitute the basic fabric of our minds. We have a tremendous opportunity, and hence a significant duty, to probe the inner workings of the human brain. Many of the questions that have puzzled philosophers and scientists for millennia can now begin to be investigated in a rigorous, systematic, and mechanistic manner. Capitalizing on this unique opportunity holds the potential to radically transform how we think about brain function, how we interact with brains in clinical and scientific environments, and how the magic of cognition is orchestrated by the bewildering and fascinating interactions of neurons.

Acknowledgments

The authors want to thank all the patients that have participated in the research efforts described in this work. As described in chapter 6, the patients are the heroes of this story. The authors would also like to thank the talented, industrious, and radiant people that have made this work possible, including Eric Behnke, Salaz Brooke, Jessica Capucci, Jack Connolly, Paul Dionne, Tony Fields, Vanessa Isiaka, Lixia Gao, Mariana Holliday, Eve Isham, Vanessa Isiaka, Sheryl Manganaro, Anna Pastolova, Jane Tingley, Michelle Tran, Irene Wainwright, Nanon Winslow, Stephen McAllister, Melissa Murphy, and Karen Walters. Financial support from NIH, NSF, Dana Foundation, Simmons Foundation, and the Swartz Foundation has played a critical role in the completion of this book and the work described here.

References

Buzsáki, G., Anastassiou, C. A., & Koch, C. (2012). The origin of extracellular fields and currents—EEG, ECoG, LFP and spikes. *Nature Reviews. Neuroscience, 13*, 407–420.

Crick, F. (1994). *The astonishing hypothesis*. New York: Simon & Schuster.

Crick, F., Koch, C., Kreiman, G., & Fried, I. (2004). Consciousness and neurosurgery. *Neurosurgery, 55*, 273–282.

Engel, A. K., Moll, C. K., Fried, I., & Ojemann, G. A. (2005). Invasive recordings from the human brain: Clinical insights and beyond. *Nature Reviews. Neuroscience, 6*, 35–47.

Logothetis, N. K. (2002). The neural basis of the blood-oxygen-level-dependent functional magnetic resonance imaging signal. *Philosophical Transactions of the Royal Society of London. Series B, Biological Sciences, 357*, 1003–1037.

2 Fifty-plus Years of Human Single Neuron Recordings: A Personal Perspective

George Ojemann

Human single neuron recording began with the work of Arthur Ward, Jr. (see figure 2.1), first Professor and Chairman of Neurological Surgery at the University of Washington in Seattle. Dr. Ward's interest in the electrophysiology of the human nervous system was a product of his undergraduate experience at Yale University in the late 1930s, working in the distinguished physiology department directed by Prof. John Fulton. As he told the story, the career that eventually led to human single neuron recordings began during overnight duty in prolonged acute primate experiments, a duty shared between undergraduates and visiting scientists. During one such overnight shift, the visiting scientist was Percival Bailey, on sabbatical from his position as head of neurosurgery at the University of Chicago, then arguably the most research oriented neurosurgery department in the country. Bailey convinced the undergraduate Ward that he should be trained as a neurosurgeon so that he could undertake neurophysiological observations in humans, using clinical settings. This is what Ward did, receiving his M.D. from Yale medical school and neurosurgical training at the Montreal Neurological Institute with Wilder Penfield. Once situated in Seattle, Dr. Ward established an epilepsy surgery program, utilizing the awake, electrocorticographically guided technique of Penfield. It was in this setting, in the mid-1950s, that with the assistance of the neurologist–electroencephalographer Dr. Louis B. Thomas he performed the first human microelectrode recordings (Ward & Thomas, 1955; see figure 2.2).

These initial intraoperative recordings utilized technology from nonhuman experimental studies: glass micropipettes in a hydraulic micromanipulator without a footplate, inserted obliquely into posterior temporal cortex. Activity was recorded photographically from an oscilloscope. The much larger pulsations of exposed human cortex represented a major technical problem. The author's first experiences with human microelectrode recording began in this setting when he joined Dr. Ward's program in 1960 initially as a neurosurgical resident.

Dr. Ward's major research interest was in the electrophysiology of experimental epileptic foci, with particular interest in the experimental focus produced by the injection of alumina cream into sensorimotor cortex of monkey (Schmidt, Thomas, & Ward, 1959). However, he was always concerned that the experimental epilepsy models had parallels in human epilepsy, so that the focus of his human single neuron studies was on the similarities in the electrophysiology of human and experimental epilepsy. He demonstrated similarities in "burst" activity (Ward and

Figure 2.1
Arthur A. Ward, Jr. (1916–1997).

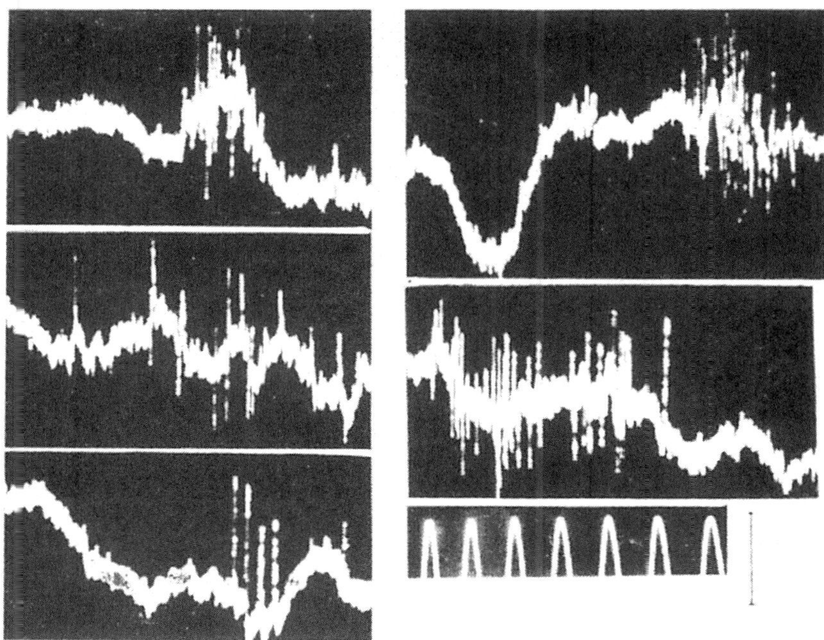

Figure 2.2
First reported human single neuron recordings. From Ward and Thomas (1955).

Schmidt, 1961), including the "structured" burst that he thought particularly identified the experimental epileptic focus (Calvin, Ojemann, & Ward, 1973). These human single neuron studies in epilepsy were then extended to the relation between interictal electrocortical "spikes" and single neuron activity (Wyler, Ojemann, & Ward, 1982; Creutzfeldt et al., 1993). The findings from even the earliest of these studies are similar to those of a recent report based on a much larger number of neurons analyzed with modern techniques (Keller et al., 2010).

During the early experience with human microelectrode recording, we also experimented with different electrodes and different microdrive arrangements, in an effort to achieve greater stability and reliability of recording, finally settling on electrolytically sharpened tungsten mounted in hydraulic drives with transparent footplates to damp cortical pulsations. Better methods of separating the extracellularly recorded activity into that of single neurons were developed, beginning with visual identification of action potential amplitudes from movie strips, to the beginnings of computer-based amplitude discrimination with the aid of storage oscilloscopes. With that shift, recordings were increasingly considered to represent activity of single neurons rather than multiunit activity. Other centers also begin human microelectrode recording during procedures for epilepsy both in cortex (Rayport & Waller, 1967) and medial temporal lobe (Verzeano et al., 1971), including the development of the recording technique utilizing microwires through depth electrodes that has proved so useful subsequently (Babb et al., 1973). Additionally, there were early reports of human microelectrode investigations of activity related to movement and its disorders, both from premotor cortex (Li & Tew, 1964) and thalamus (Albe-Fessard et al., 1963; Jasper & Bertrand, 1966), investigations conducted during stereotaxic thalamotomies.

Human Microelectrode Recording during Cognition

Reports of human neuronal activity during cognitive measures began appearing in the 1970s. Halgren et al. (1978), in recordings from medial temporal lobe through chronically implanted microwires, reported units responding to visual input, choices, and recall from recent memory. Bechtereva et al. (1971, 1979), from the Institute for Experimental Medicine in what was then called Leningrad, reported differences in the frequency of thalamic and inferior frontal activity during language measures, including proposing codes for words in thalamic recordings. In the mid-1980s, our group at the University of Washington shifted the focus of our intraoperative single neuron recording from lateral temporal cortex during awake surgery for epilepsy to changes during cognitive measures, particularly language and recent verbal memory (Creutzfeldt et al., 1987, 1989a, 1989b; Creutzfeldt & Ojemann, 1989; Ojemann et al., 1988). Subsequently, as indicated in the remainder of this book, assessing changes during measures of many cognitive processes during human single neuron recordings has been the focus of investigations by a number of groups, sampling several different brain regions. The current findings from our lateral temporal cortex studies are reviewed in another chapter.

Utilizing human single neuron studies to assess cognitive process rather than the pathophysiology of epilepsy raises a number of specific issues. The first is in what brain areas to record from. For studies directed at epilepsy this is rarely an issue. Those recordings have generally been obtained from the regions of brain thought, either during diagnostic evaluation or therapeutic procedures, to be involved in the epileptic process. Surgical access to these regions is part of the process of treating this patient's epilepsy, and the recordings are in tissue that it is thought (at least at the time the recordings are made) likely to be part of the resection directed at treating the epilepsy. That is not necessarily the situation with recording directed at assessing changes with cognition. The ideal recording sites based on current models of the brain regions involved in different cognitive processes are not necessarily the areas that would be surgically exposed for purely diagnostic or therapeutic reasons and, indeed, might be areas that would generally be spared in any resection, such as sensorimotor cortex or areas considered "eloquent" for language. While recordings from such areas have been reported (Goldring & Ratcheson, 1972), it is the author's view that since microelectrode recording is invasive, and associated with some risk of significant damage to the region of recording, it should be restricted to tissue that would be subject to surgical injury independent of any microelectrode recording. Thus recording should be restricted to tissue that would be resected as part of a therapeutic procedure independent of any considerations of microelectrode recording or along the track or at the target of electrodes placed for clinical reasons independent of microelectrode recording. Moreover, the extent of surgical exposure and anesthetic techniques utilized for an operation should be based only on clinical considerations and not microelectrode recording.

However, when opportunities for microelectrode recording that meet these conditions exist, every effort should be made to utilize them, with the patient's informed consent that includes issues of risk and comfort, and after Institutional Review Board approval. During the author's 50-year neurosurgical career, some of those opportunities have come and gone, with the most dramatic example being the fluctuations in the performance of awake thalamic stereotaxic surgery with changes in the available medical management of movement disorders. It is quite conceivable that advances in noninvasive imaging techniques may substantially reduce the use of intracranial electrodes in the evaluation of patients for epilepsy surgery (and the opportunities for microelectrode recording that occur with their use). As is evident from the remainder of this book, microelectrode recording can provide unique insights into the neural mechanisms of human cognition, findings that are essentially unavailable with any other technique. It is thus a tragedy for human knowledge when an effort is not made to utilize the available opportunities. A current example is the limited number of microelectrode investigations of cognition using the opportunities presented by placement of deep brain electrodes for dyskinesias, even though microelectrode recording is currently part of one standard technique for localizing the target in those operations.

All human single neuron investigations involve collaboration between several disciplines: neurophysiology, neuropsychology, and neurosurgery. However, in the author's experience, active commitment of the neurosurgeon responsible for the patient to these investigations is essential to their success. It also provides additional protection to the patient. It is to that neurosurgeon that the patient is most likely to turn with concerns about comfort. And it is that neurosurgeon who will have to deal with any consequences of complications. In modern medicine, there are many pressures neurosurgeons must deal with that encourage them to limit or omit investigative studies: pressures to reduce operative time and length of hospital stay, pressures (largely financial) to "do more cases," and pressures of competition from groups that do no research and thus are considered more "cost-effective" by insurance payers and perceived by patients to have better clinical outcomes. Without the incentive of an active interest and role by the neurosurgeon in the research, those pressures result in the research's not occurring. Many neurosurgical trainees enter residency programs interested in neuroscience. And there are a number of training programs, including the one at the University of Washington, that try to actively involve trainees in research. It has been the author's privilege to participate in the training of several neurosurgeons who have made significant contributions to human single neuron recording, including Itzhak Fried and Matthew Howard. However, the long period of neurosurgical training, which entails the acquisition of clinical and technical skills with little relation to neuroscience, often blunts any neuroscience interests. And the current practice climate makes active participation in research even more difficult. Finding neurosurgeons who also have ongoing research interests is likely to be a major future limitation on successful human single neuron studies.

Another set of issues in microelectrode recording during cognition involve what behaviors to assess and how. Two different philosophies have guided different investigators. One approach has been to investigate behaviors that have been well studied in nonhumans and are related to the function of the brain region that will be investigated. An example is the investigation of hippocampal neuronal activity during route finding using behavioral measures similar to those extensively studied in rodent hippocampus (Ekstrom et al., 2003). The other approach has been to assess uniquely human cognitive processes likely to be represented in the brain region to be sampled, based on findings with other methods of relating brain to cognition, especially findings from the literature on the effects of focal brain lesions. Behaviors likely to be represented have also been derived from the theoretical models of cognitive processes derived from neuropsychology, as have many of the assessment tools. Examples of this approach are to be found in the author's studies of lateral temporal cortical activity.

Utilizing these approaches in human single neuron studies requires comparing activity during a behavior to control conditions. Comparisons have been made to "baseline" activity immediately before the behavior or to average activity during the experiment. This approach biases the findings toward finding "phasic" activity time-locked to some aspect of the behavior, though of course one does not know what nonovert behavior the subject may be engaged in during the "baseline" period. An alternative is to compare two different defined behavioral conditions. In the author's studies, comparisons have often been made between activity with identification of common verbal items and recent memory encoding that also requires identifying the same or similar verbal items but, additionally, with the instruction to retain the item in memory. Such comparisons often identify "tonic" as well as "phasic" changes with one behavior (Ojemann & Schoenfield-McNeill, 1999). If the changes in single neuron activity are to be related to a specific behavior, control measures for effects of perception, motor output, and attention would be necessary also. Additionally, over the half century of human neuron recording there has been an evolution of the analytic strategy for establishing a relation between a behavior and changes in activity from a description of observed changes to a variety of statistical approaches for establishing that any correlations between changed activity and a behavior are "nonrandom," an evolution that makes comparison of findings from earlier and more recent studies often difficult.

There are a number of factors that affect all findings from human single neuron studies that are little discussed but affect the ability to generalize findings. All recordings come from brains of patients with some disease, even recordings from the "good" hemisphere. Many of the human single neuron studies of cognition have been in patients with medically refractory epilepsy. There is evidence that even at the level of gross anatomy there are widespread differences between these patients and other populations, changes that extend beyond any gross pathology (McDonald et al., 2008), though whether these changes alter single neuron activity is unknown. Similar evidence exists for most other patient populations available for single neuron recording. Epilepsy effects in single neuron recordings can be mitigated to some extent by limiting recording to tissue that does not have electrocorticographic evidence of epileptiform activity, and neurons that do not show "bursting" activity, but that does not mitigate any unknown effects of the more widespread "reorganization" of cognitive processes. The sample of single neuron activity recorded in all studies is almost ridiculously small compared to the number of neurons actually present in the sampled region. Moreover, that sample is clearly biased to recording from larger pyramidal neurons, with little or no information on activity of other neuron types. All micro-electrode recording techniques are invasive and, at least at a microscopic level, produce injury. Although these effects also can be mitigated some by avoiding recording from neurons with "injury discharge" firing, more widespread alterations in the extent or nature of participation of an individual neuron in the network for a behavior remain. To some extent this may be even more of a problem for chronic compared to acute recording, secondary to the development of local swelling about the chronic electrodes. Finally, there are consistent biases in the activity in our small sample from only one neuron type, most notably the brain bias to perceive novelty, so that the response to the initial presentation of an item will differ consistently from that to repetitions of the item, though without repetitions, statistical analysis of any changes is impossible.

Overall, the current situation in single neuron studies of human cognition is one of marked fragmentation of findings. The various groups that have reported findings in this area (and in this book) have sampled different recording sites, using different microelectrode techniques, in different patient populations, using different behavioral paradigms and with different control measures. About the only area of some consistency now is in the analytic techniques used to separate activity into that of individual neurons. The overall result is that outside of some of the changes with epilepsy, essentially none of the very interesting changes reported with different cognitive processes have been replicated by any other completely independent group, simply because no one has attempted to do so. A major contributing factor to this is that for much of the past 50+ years, the actual recording of human single neuron activity during cognitive mea-sures was limited to a very few programs. The reports of recordings by several additional pro-grams in the last decade are thus an encouraging development for the future of human single neuron recording.

There are presently many different methods for investigating the relation between brain func-tion and a specific cognitive process. Each method provides a somewhat different perspective on that relationship. One group of methods relates brain function to cognition by identifying the functional deficits associated with inactivation of specific brain regions. Those methods thus identify the areas of brain that are *crucial* for that function, at least at that time. They include the effects of spontaneous brain lesions, most often strokes, and the basis for most of the clas-s cal models of brain–cognition relationships. Also included are methods of temporary inactiva-tion of portions of the brain, such as the intracarotid amobarbital perfusion test (ICAP or Wada test; Wada & Rasmussen, 1960), and electrical stimulation mapping (Ojemann et al., 1989). These methods differ in the time course of the inactivation (months to indefinite for lesions, seconds for stimulation mapping), the extent of inactivation (usually large for lesions, small for electrical stimulation mapping), and application selectively to special populations (elderly with vascular disease for strokes, refractory epilepsy for ICAP). Each has issues in determining the exact extent of the inactivation. And each has somewhat different findings: for example, more focality and individual variability for stimulation mapping compared to lesions. The other methods for establishing brain–cognition relationships do not necessarily identify brain areas that are crucial for a function but rather show only brain areas that *participate in* the function, by correlating a change in the measure of brain function between a particular cognitive process and a "control" measure. These methods include functional magnetic resonance imaging (fMRI), electroencephalography (EEG) and its intracranial components, electrocorticography (ECoG) and local field potentials (LFPs), and the single neuron recording that is the topic of this book. Each method reflects some different aspect of brain function: fMRI, the change in oxygen extraction and blood flow, and thus only indirectly a change in neuronal activity; EEG, ECoG and LFP, predominately the graded activity associated with synaptic input in neuronal aggregates of varying sizes (very large for EEG, very small for LFP); and single neuron activity, the output of an individual neuron. The interrelationships between these different methods for identifying participation of brain areas in a cognitive process have been the subject of much recent investigation (e.g., fMRI–LFP–single neurons, Ojemann et al., 2010; LFP–single neurons, Zanos et al., 2012). The perspectives provided by these different methods can be quite different. In assessing the relation between language and cerebral cortex, the methods that identify crucial areas show strong lateralization to one hemisphere, usually left. The methods that identify participating areas show much less or, in the case of the proportion of single neurons showing changes, no evidence of lateralization (Schwartz et al., 1996). Thus, our models for brain mechanisms of cognitive process need to integrate findings from all these methods, including the changes in single neuron activity described in this book.

References

Albe-Fessard, D., Arfel, G., & Guiot, G. (1963). Activités électriques caractéristiques de quelques structures cérébrales chez l'homme. *Annales de Chirurgie, 17*, 1185–1214.

Babb, T. L., Carr, E., & Crandall, P. H. (1973). Analysis of extracellular firing patterns of deep temporal structures in man. *Electroencephalography and Clinical Neurophysiology, 34*, 247–257.

Bechtereva, N. P., Bundzen, P.V., Gogolitsin, Y. L., Malyshev, V. N., & Perepelkin, P. D. (1979). Neurophysiological codes of words in subcortical structures of the human brain. *Brain and Language, 7*, 145–163.

Bechtereva, N. P., Bundzen, P. V., Matveev, Y. K., & Kaplunovsky, A. S. (1971). The functional reorganization of activity in neuronal populations of the human brain in short-term verbal memory. *Fiziologia Z'iurnal SSSR, 12*, 1745–1761.

Calvin, W. H., Ojemann, G. A., & Ward, A. A., Jr. (1973). Human cortical neurons in epileptogenic foci: Comparison of inter-ictal firing patterns to those of "epileptic" neurons in animals. *Electroencephalography and Clinical Neurophysiology, 34*, 337–351.

Creutzfeldt, O., & Ojemann, G. (1989). Neuronal activity in the human lateral temporal lobe: III. Activity changes during music. *Experimental Brain Research, 77*, 490–498.

Creutzfeldt, O., Ojemann, G., & Chatrian, G. (1993). Activity of single neurons and their relationship to normal EEG waves and interictal epilepsy potentials in humans. In W. Haschke, A. Roitbak, & E. Speckmann (Eds.), *Slow potential changes in the brain* (pp. 21–42). Boston: Birkhauser.

Creutzfeldt, O., Ojemann, G., & Lettich, E. (1987). Single neuron activity in the right and left human temporal lobe during listening and speaking. In D. Ottoson (Ed.), *Duality and unity of the brain. Wenner-Gren International Symposium Series, Vol. 47* (pp. 295–310). Hampshire: MacMillan Press.

Creutzfeldt, O., Ojemann, G., & Lettich, E. (1989a). Neuronal activity in the human lateral temporal lobe: I. Responses to speech. *Experimental Brain Research, 77*, 451–475.

Creutzfeldt, O., Ojemann, G., & Lettich, E. (1989b). Neuronal activity in the human lateral temporal lobe: II. Responses to the subjects own voice. *Experimental Brain Research, 77*, 476–489.

Ekstrom, A. D., Kahana, M. J., Caplan, J. B., Fields, T. A., Isham, E. A., Newman, E., et al. (2003). Cellular networks underlying human spatial navigation. *Nature, 425*, 184–188.

Goldring, S., & Ratcheson, R. (1972). Human motor cortex: Sensory input data from single neuron recordings. *Science, 175*, 1493–1495.

Halgren, E., Babb, T., & Crandall, P. H. (1978). Activity of human hippocampal formation and amygdala neurons during memory testing. *Electroencephalography and Clinical Neurophysiology, 45*, 585–601.

Jasper, H. H., & Bertrand, G. (1966). Thalamic units involved in somatic sensation and voluntary and involuntary movememts in man. In D. P. Purpura & M. D. Yahr (Eds.), *The thalamus* (pp. 365–390). New York: Columbia University Press.

Keller, C. J., Truccolo, W., Gale, J. T., Eskander, E., Thesen, T., Carlson, C., et al. (2010). Heterogeneous neuronal firing patterns during interictal epileptiform discharges in the human cortex. *Brain, 133*, 1668–1681.

Li, C.-L., & Tew, J., Jr. (1964). Reciprocal activation and inhibition of cortical neurons and voluntary movements in man: Cortical cell activity and muscle movement. *Nature, 203*, 264–265.

McDonald, C. R., Hagler, D. J., Jr., Ahmadi, M. E., Tecoma, E., Iraqui, V., Gharapetian, L., et al. (2008). Regional neocortical thinning in mesial temporal lobe epilepsy. *Epilepsia, 49*, 794–803.

Ojemann, G. A., Corina, D. P., Corrigan, N., Schoenfield-McNeill, J., Poliakov, A., Zamora, L., et al. (2010). Neuronal correlates of functional magnetic resonance imaging in human temporal cortex. *Brain, 133*, 46–59.

Ojemann, G. A., Creutzfeldt, O. D., Lettich, E., & Haglund, M. M. (1988). Neuronal activity in human lateral temporal cortex related to short-term verbal memory, naming and reading. *Brain, 111*, 1383–1403.

Ojemann, G., Ojemann, J., Lettich, E., & Berger, M. (1989). Cortical language localization in left, dominant hemisphere: An electrical stimulation mapping investigation in 117 patients. *Journal of Neurosurgery, 71*, 316–326.

Ojemann, G., & Schoenfield-McNeill, J. (1999). Activity of neurons in human temporal cortex during identification and memory for names and words. *Journal of Neuroscience, 19*, 5674–5682.

Rayport, M., & Waller, H. J. (1967). Technique and results of micro-electrode recording in human epileptogenic foci. *Electroencephalography and Clinical Neurophysiology, 25*(Suppl), 143–151.

Schmidt, R. P., Thomas, L. B., & Ward, A. A., Jr. (1959). The hyperexcitable neuron: Microelectrode studies of chronic epileptic foci in monkey. *Journal of Neurophysiology, 22*, 285–296.

Schwartz, T., Ojemann, G., Haglund, M., & Lettich, E. (1996). Cerebral lateralization of neuronal activity during naming, reading and line-matching. *Brain Research. Cognitive Brain Research, 4*, 263–273.

Verzeano, M., Crandall, P. H., & Dymond, A. (1971). Neuronal activity of the amygdala in patients with psychomotor epilepsy. *Neuropsychologia, 9*, 331–344.

Wada, J., & Rasmussen, T. (1960). Intracarotid injection of sodium amytal for the lateralization of cerebral speech dominance: Experimental and clinical observations. *Journal of Neurosurgery, 17*, 266–282.

Ward, A. A., Jr., & Schmidt, R. P. (1961). Some properties of single epileptic neurons. *Archives of Neurology, 5*, 308–313.

Ward, A. A., Jr., & Thomas, L. B. (1955). The electrical activity of single units in the cerebral cortex of man. *Electroencephalography and Clinical Neurophysiology, 7*, 135–136.

Wyler, A. R., Ojemann, G. A., & Ward, A. A., Jr. (1982). Neurons in human epileptic cortex: Correlation between unit and EEG activity. *Annals of Neurology, 11*, 301–308.

Zanos, S., Zanos, T., Marmarellis, V., Ojemann, G., & Fetz, E. (2012). Relationship between spike-free local field potentials and spike timing in human temporal cortex. *Journal of Neurophysiology, 107*, 1808–1821.

3 The Neurosurgical Theater of the Mind

Itzhak Fried

Neurosurgery in the operating theater requires a series of carefully executed steps involving manipulation of brain parenchyma, its vasculature and surrounding membranous and bony structures, and the various spaces encasing the brain and embedded in it including the extradural, subdural, subarachnoid, and ventricular spaces. In the course of this highly focused, somewhat mechanical activity one never forgets the subject of this effort, the human brain, and the patient under the surgical drapes. However, it is often quite difficult to direct one's attention to the opportunities that the rare situation of brain surgery pose for learning and better understanding of the brain in ways that are not available in any other field in neuroscience. While such opportunities, when seized, may not benefit the patient directly, they may hold unusual promise for future patients and for development of better tools to meet the challenges posed by neurological afflictions.

Electrical Stimulation of Human Cortex and Subcortical Structures

In the history of neurosurgery extraordinary individuals have opened up brain surgery to investigations that provided rare insights into brain functions. These were individuals who, in addition to tending to patient clinical needs and developing their surgical skills, applied methods and knowledge from neuroscience in the operating room.

Foremost among these pioneers was Wilder Penfield, who introduced electrical stimulation to map the cortex in awake patients in the operating room. Penfield knew of the pioneering studies of Fritz and Hitzig mapping motor cortex of the dog and finding somatotopic organization there. He was also exposed to the prominent neurophysiologist of the era, Sherrington, working in his laboratory at Oxford. However, Penfield was able to take the additional and bold step of applying this knowledge in humans. He saw the clinical opportunity to provide a functional road map during surgical resections for epilepsy. Penfield went on to describe the human homunculus and, by keeping his patients awake, was able to probe language cortex (Penfield & Roberts, 1959). Performing operations under local anesthesia was no small feat, yet Penfield's stimulation produced unexpected findings. Here is how Penfield described it years later:

Twenty six years ago I was operating on a patient under local anesthesia…. She told me suddenly that she seemed to be living over a previous experience. She seemed to see herself giving birth to her baby girl…. This I thought was a strange moment for her to talk of that previous experience…. Nevertheless I noted the fact that it was while my stimulating electrode was applied to the left temporal lobe that this woman has had this unrelated and vivid recollection. This was in 1931. (Penfield, 1958)

Penfield had discovered the "experiential responses" (Penfield & Perot, 1963). These were subjective experiences which were remarkable in their vividness as if the patient was reliving a past memory, yet there was also a "doubling of consciousness," what Jackson had once termed "mental diplopia"; concurrent with the vivid experience the patients never failed to realize that they were in the operating room. Intrigued by these responses, often elicited with electrical stimulation of the temporal lobe, Penfield (1958) postulated, "There is a permanent record of the stream of consciousness within the brain. … hidden in the interpretive areas of the temporal lobes, there is a key mechanism that unlocks the past."

Electrical stimulation mapping at discrete sites of the brain has, since Penfield, evolved into a complex tool used by many neurosurgeons and gradually yielding a wealth of information enriching cognitive neuroscience. Indeed, electrical stimulation has evolved from a tool of brain mapping to a therapeutic role in the form of a field known as deep brain stimulation (DBS), where application of electrical stimulation at specific brain sites is employed to manipulate neural circuits in neurological diseases such as Parkinson's disease (PD), dystonia, depression, and other disorders (also see chapter 16).

Various neurosurgical procedures provide an opportunity to eavesdrop on the electrical signals produced by the human brain. These can be classified according to (1) the type of signal recorded (electroencephalography [EEG], local field potentials [LFPs], multiunit activity, and single unit activity), (2) the clinical condition for which recording is employed (viz., movement disorder such as PD or dystonia, epilepsy, brain tumor, etc.), and (3) the clinical setting of recording (i.e., whether these signals are recorded acutely in the operating room or chronically in patients implanted with indwelling brain electrodes placed for specific clinical reasons).

Penfield, together with Herbert Jasper, used electrocorticography—that is, recording of EEG from the surface of the exposed cortex during surgery—to identify those abnormalities associated with the chronic state of epilepsy. This "functional map" of epilepsy, together with the information on brain function gleaned from electrical stimulation, provided guidance to Penfield as to the extent of resection required to treat his patients' epilepsy while preserving their various cognitive functions, such as language, movement, sensation, vision, and so on. Later, coupling recordings of surface or depth EEG with specific cognitive paradigms yielded a wealth of information which is still gathered nowadays using increasingly more elaborate recording electrodes and tools of imaging, recording, and data analysis. These studies are beyond the scope of this book.

Unique among the signals recorded from the brain are the action potentials of individual neurons. These indeed are the building blocks of the human mind, and the opportunity to capture these signals in awake human patients is the pinnacle of what neurosurgery can offer to neuroscience. Contemporary neurosurgery has several clinical conditions providing the opportunity to record single unit activity.

Procedures for Epilepsy Surgery

Neurosurgical procedures for treatment of severe intractable epilepsy have provided the most opportunities to record single neuron activity over relatively long periods of time in awake conscious patients. Epilepsy surgery underwent major advancement in the last few decades with the advent of modern neuroimaging and stereotactic and other neurosurgical techniques and with the refinement of diagnostic tools enabling better identification of neuronal networks responsible for the seizures.

Candidates for epilepsy surgery are a subset of epilepsy patients in whom pharmacological treatment does not yield seizure control (also see chapter 18). In some of these patients the neuronal network causing the seizure can be identified in sufficient detail to enable focal surgical intervention in the form of surgical resection, disconnection, or chronic stimulation to neutralize the seizures. These surgical procedures, and in particular precise focal surgical resections, can often produce a complete cure or considerable improvement in seizure frequency or severity with very few, if any, consequences in terms of neurological or cognitive function. In the majority of candidates for these procedures, noninvasive tools are sufficient to localize the epileptogenic network. These include video-EEG monitoring of the patients to capture seizures, detailed high-resolution magnetic resonance imaging (MRI), positron emission tomography scanning, single photon emission computed tomography, magnetic source imaging (MSI), neuropsychological tests, functional magnetic resonance imaging (fMRI), and fMRI combined with EEG.

However, in a small subset of these patients placement of intracranial electrodes is needed to further pinpoint the epileptogenic network. Some of these are electrodes inserted into brain parenchyma and thus termed "depth electrodes." Advancement in electrode technology has enabled the introduction of microwires or microcontacts, which enable recordings of not only LFPs but also extracellular multiunit and single unit activity (see chapter 5). The clinical requirement is that following placement of these electrodes performed in the operating room, the patients need to be monitored for several days and sometimes for up to two or three weeks until a sufficient number of spontaneous seizures are captured, thus enabling the correlation of brain activity with the actual seizures captured on the video monitor in the hope of then identifying the network underlying the seizures. If this monitoring is successful, later surgery may be offered for resection is likely to prove curative.

The technique of depth electrode placement has been described in detail (Fried et al., 1999). In general the electrodes are placed stereotactically using detailed imaging data. Since this procedure involves placement of many electrodes, often between 8 and 14 electrodes, a clear depiction of the vasculature is essential. In addition to MRI we have used digital subtraction angiography in the surgical planning. Magnetic resonance angiographic techniques as well as CT angiography have provided adequate tools, which may replace traditional angiography. Robotic techniques are under development to enable the rapid use of these imaging data in the placement of a large number of electrodes.

Since 1993, with the development of a special flexible electrode with microwires inserted through the electrode lumen (the Behnke–Fried electrode; see chapter 5) and specialized software for implantation, a strict protocol has been used to ensure the safest and most precise placement of multiple electrodes (Fried et al., 1999). On the morning of surgery at UCLA, the patient is affixed with a Leksel stereotactic frame and then undergoes a detailed MRI. The patient then undergoes a cerebral angiogram. Based on these studies, targets and trajectories are planned. We customarily use orthogonal placement. The electrodes are inserted in the operating room under general anesthesia. At each entry site a small 2-mm skin incision is made and a twist drill is used to make a 2-mm hole in the skull. The dura is coagulated and punctured, and a guide screw is placed. We first place the guide screws for all the electrodes and subsequently insert the electrodes, usually of 1.3-mm diameter each and with seven to eight contacts along the shaft. Through the lumen of each electrode the microelectrodes made of platinum iridium alloy are inserted to the correct depth so that they protrude about 4 mm beyond the electrode distal end. Usually eight microwires 40 micra in diameter each with one reference microwire are inserted. A critical element is the guide screw cap, which is secured over every guide screw to ensure no cerebrospinal fluid (CSF) leak. In some of the electrodes we have introduced a modification enabling insertion of a tiny microdialysis catheter (along with about four microwires) enabling the in- and outflow of artificial CSF fluid. This technique affords the monitoring of neuroactive substances in the extracellular space across a membrane, which can later be correlated with significant events such as a seizure, sleep (Zeitzer et al., 2006), a cognitive event such as learning (Fried et al., 2001), or social interactions (Blouin et al., 2013). Electrode technology will likely develop rapidly, and we will see depth electrodes with many more microcontacts, so that recordings from hundreds and maybe a thousand neurons simultaneously may become a reality.

Following surgery, the patient is placed in the epilepsy monitoring unit for video-EEG monitoring. While the contacts along the electrode shaft enable recording of intracranial EEG signals, the microelectrodes afford recordings of LFP as well as multiunit and single unit activity. Microwires have proven advantageous in recording of epileptogenic activity where a series of studies have demonstrated high-frequency pathological oscillations associated with the seizure focus which had not been previously recognized because of the limited sampling frequency offered by traditional EEG recording filters (Bragin et al., 2002; Engel et al., 2005).

At the center of this unique clinical situation is the patient who is awake for most of the monitoring and thus able to participate in specific paradigms where neuronal activity can be correlated not only with epileptic seizures but also with various cognitive functions and tests. Obviously, these studies of cognitive functions are performed under strict Institutional Review Board regulations, and patients provide informed consent (also see chapter 4). We find that often patients are eager to participate in these studies as they provide a relief of the monotonous wait for spontaneous seizures to occur.

It is important to understand the unique setting in which these studies occur. First, the placement of the electrodes is based solely on clinical requirements as decided by the group of clinicians caring for the patient. Thus, the experimental questions are always surrogate to the particular array of implantation required for the individual patients. However, since many of the epilepsies and the suspected foci involve the medial temporal lobe and particular frontal lobe sites such as the supplementary motor area, the anterior cingulate, or the orbitofrontal cortex, these sites are frequent targets for depth electrode placement. Second, the placement of the electrode is fixed—that is, one cannot change location of the electrodes once they are placed. While minute movements of the microwires are possible, we have occasionally observed relatively stable units over a period of several days. The yield of the microwires diminishes with time, perhaps because of some mini-scar formation. However, we often get good recordings over a period of ten days.

It is also important to realize that the fact that the recording setting is, of necessity, a hospital often makes the conduction of carefully controlled studies difficult. A hospital is an electrically noisy environment, and ingenuity and patience are required to eliminate possible sources of electrical noise. More importantly, the patient's comfort and well-being are always the dominant and foremost consideration, and thus situations may arise where less than optimal control of the behavioral paradigm can be achieved.

A frequent criticism is that these recordings are obtained in patients with neurological disease, and thus the ability to generalize these findings to normal brain physiology may be limited. While this criticism is valid, it should be borne in mind that the epileptogenic focus is not known a priori; thus, several sites need to be implanted, and eventually many of these sites will prove unrelated to the seizure network or focus. Additionally, we see that decades of neurophysiological investigations in these patients have yielded useful insight into normal brain function, such as in Penfield's discovery of the Rolandic homunculi or Penfield's, and later Ojemann's, detailed description of language cortex (see chapter 14). Finally, the neurophysiological data obtained from these patients should always be evaluated in the context of what is known from animal and nonhuman primate neurophysiology and from other techniques in humans, especially fMRI and other neuroimaging techniques, MSI, and so on. For instance, the finding of place cells in rodents has been a major achievement in animal neurophysiology, providing a striking correlation between behavior and hippocampal neurophysiology. Indeed, we have found similar cells in human hippocampus (Ekstrom et al., 2003) (see also chapter 9), yet place cells in rodents have probably evolved in humans to encode a much more complex "map" of concepts (Quian Quiroga et al., 2005) (see also chapter 8). In the supplementary motor area, major advances in knowledge gained by nonhuman primate neurophysiology (Kurata & Tanji, 1986) led the way to probing this region in humans with particular emphasis on the additional question of a single neuron basis for volition studies in subjects who can report their wishes and intention (Amador & Fried, 2004; Fried et al., 2011).

Deep Brain Stimulation Procedures

Over the last two decades DBS of specific sites in brain parenchyma has gained significant clinical use, and various potential clinical indications for DBS are under preliminary or advanced clinical trials. The widest clinical use has been for movement disorders, particularly PD, dystonia, and essential tremor, but emerging indications include depression, obsessive–compulsive disorder, epilepsy, pain, Tourette's syndrome, and dementia (see Lozano & Lipsman, 2013, for a review).

In some of these procedures multiunit activity is recorded during the implantation procedure in order to arguably identify in more precise physiological detail the anatomic target for the implantation. This has been particularly common for implantation of DBS in the subthalamic nucleus for PD, in the globus pallidum interna for PD and dystonia, and in the ventral intermediate nucleus of the thalamus for essential tremor. Recordings have also been performed in the cingulate gyrus for procedures related to pain and depression.

The opportunity to record unit activity in this setting is limited by the acute nature of the procedure and the limited time allotted to it. Recordings here are usually performed via a single microelectrode along a single or several tracks, so that nuclei en route to target can be sampled, albeit over a limited period of time. This, then, is an altogether different setting compared to the fixed microelectrodes in chronic monitoring. However, in most of these procedures the patient is awake and thus experimental questions can be addressed (Engel et al., 2005) (see also chapters 15 and 16).

Brain–Machine Interfaces (BMI)

The emerging field of brain–machine interfaces and neuroprosthetic devices is likely to provide another clinical setting for microelectrode recordings (see chapter 17). The decoding of volitional motor acts from direct neural recordings has been most efficient using single and multiunit recordings from multiple channel recordings via microelectrode arrays such as the Utah array. While considerable work has been done in nonhuman primates, initial human trials have been launched, some with promising results.

Surgery here involves implantation of the array into the cortical gray matter from the surface of the cortex over a small region. Electrodes here are implanted for long periods of time, essentially for chronic therapeutic use, and thus present an excellent opportunity to study neuronal activity with respect to specific cognitive and motor states (Carmena et al., 2003; Musallam et al., 2004; Hochberg et al., 2006; Collinger et al., 2013; Tankus et al., 2013).

References

Amador, N., & Fried, I. (2004). Single-neuron activity in the human supplementary motor area underlying preparation for action. *Journal of Neurosurgery, 100,* 250–259.

Blouin, A. M., Fried, I., Wilson, C. L., Staba, R. J., Behnke, E. J., Lam, H. A., et al. (2013). Human hypocretin and melanin-concentrating hormone levels are linked to emotion and social interaction. *Nature Communications, 4*, 1547.

Bragin, A., Wilson, C. L., Staba, R. J., Reddick, M., Fried, I., & Engel, J., Jr. (2002). Interictal high-frequency oscillations (80–500 Hz) in the human epileptic brain: Entorhinal cortex. *Annals of Neurology, 52*, 407–415.

Carmena, J. M., Lebedev, M. A., Crist, R. E., O'Doherty, J. E., Santucci, D. M., Dimitrov, D. F., et al. (2003). Learning to control a brain–machine interface for reaching and grasping by primates. *PLoS Biology, 1*, E42.

Collinger, J. L., Wodlinger, B., Downey, J. E., Wang, W., Tyler-Kabara, E. C., Weber, D. J., McMorland, A. J., Velliste, M., Boninger, M. L., & Schwartz, A. B. (2013). High-performance neuroprosthetic control by an individual with tetraplegia. *Lancet, 381*, 557–564.

Ekstrom, A. D., Kahana, M. J., Caplan, J. B., Fields, T. A., Isham, E. A., Newman, E. L., et al. (2003). Cellular networks underlying human spatial navigation. *Nature, 425*, 184–187.

Engel, A. K., Moll, C. K., Fried, I., & Ojemann, G. A. (2005). Invasive recordings from the human brain: Clinical insights and beyond. *Nature Reviews. Neuroscience, 6*, 35–47.

Fried, I., Mukamel, R., & Kreiman, G. (2011). Internally generated preactivation of single neurons in the human brain predicts volition. *Neuron, 69*, 548–562.

Fried, I., Wilson, C. L., Maidment, N. T., Engel, J., Behnke, E., Fields, T. A, et al. (1999). Cerebral microdialysis combined with single-neuron and electroencephalographic recording in neurosurgical patients. *Journal of Neurosurgery, 91*, 697–705.

Fried, I., Wilson, C. L., Morrow, J. W., Cameron, K. A., Behnke, E. D., Ackerson, L. C., et al. (2001). Increased dopamine release in the human amygdala during performance of memory tasks. *Nature Neuroscience, 4*, 201–206.

Hochberg, L. R., Serruya, M. D., Friehs, G. M., Mukand, J. A., Saleh, M., Caplan, A. H., et al. (2006). Neuronal ensemble control of prosthetic devices by a human with tetraplegia. *Nature, 442*, 164–171.

Kurata, K., & Tanji, J. (1986). Premotor cortex neurons in macaques: Activity before distal and proximal forelimb movements. *Journal of Neuroscience, 6*, 403–411.

Lozano, A. M., & Lipsman, N. (2013). Probing and regulating dysfunctional circuits using deep brain stimulation. *Neuron, 77*, 406–424.

Musallam, S., Corneil, B., Greger, B., Scherberger, H., & Andersen, R. (2004). Cognitive control signals for neural prosthetics. *Science, 305*, 258–261.

Penfield, W. (1958). Some mechanisms of consciousness discovered during electrical stimulation of the brain. *Proceedings of the National Academy of Sciences of the United States of America, 44*, 51–66.

Penfield, W., & Perot, P. (1963). The brain's record of auditory and visual experience: A final summary and discussion. *Brain: A Journal of Neurology, 86*, 595–696.

Penfield, W., & Roberts, L. (1959). *Speech and brain mechanisms*. Princeton, NJ: Princeton University Press.

Quian Quiroga, R., Reddy, L., Kreiman, G., Koch, C., & Fried, I. (2005). Invariant visual representation by single neurons in the human brain. *Nature, 435*, 1102–1107.

Tankus, A., Fried, I., & Shoham, S. (2013). Cognitive-motor brain-machine interfaces. *Journal of Physiology (Paris)* [Epub ahead of print].

Zeitzer, J. M., Morales-Villagran, A., Maidment, N. T., Behnke, E. J., Ackerson, L. C., Lopez-Rodriguez, F., et al. (2006). Extracellular adenosine in the human brain during sleep and sleep deprivation: An in vivo microdialysis study. *Sleep, 29*, 455–461.

4 Ethical and Practical Considerations for Human Microelectrode Recording Studies

Adam N. Mamelak

As methods to study the function of the human brain increase in sophistication, questions once relegated only to theoretical discussion, such as how human cognition works, how the brain computes reward-based choices, and even understanding the basis of human consciousness are now beginning to be tackled in rigorous fashion. While indirect means to measure human brain activity such as functional magnetic resonance imaging (fMRI), scalp-based electroencephalography (EEG), and magnetoencephalography (MEG) provide insight into function, simultaneous recording of neurons in a human and behavioral testing provides one of the most unique and powerful ways to study the mechanisms underlying fundamental brain function. The study of human neuronal activity in vivo has occurred in limited scope for many years. However, the opportunities for recording extracellular action potentials in humans have been limited by the clinical scenarios, the microelectrodes utilized to record, the hardware available for data acquisition, and the software often needed to decipher meaningful data from the raw data. In the past decade many of these once formidable barriers have been overcome largely by technological advances. The development of more flexible microwire bundles and tetrode arrays has permitted neurosurgeons to safely implant chronic microelectrodes in conjunction with clinical electrodes, thereby opening up a new era for chronic human recordings (Fried et al., 1999). Marked improvements in preamplifiers, head stages, and noise reduction circuitry, coupled with the increased affordability of high-capacity data storage systems and improved spike sorting algorithms, have led to a mini-explosion in the field of human microelectrode recording (MER) and the study of human brain activity (Aksenova et al., 2003; Rutishauser, Schuman, & Mamelak, 2006). The development of effective surgical strategies for diseases such as Parkinson's disease (PD) or essential tremor (ET) has also increased the clinical indications for microelectrode recordings in awake humans, in turn opening the door for more experimental investigations (Gale et al., 2011; Starr et al., 2004; Theodosopoulos et al., 2003).

While these advances provide a unique opportunity for scientists to learn how the human brain processes information, the experimentalist is also faced with a unique set of ethical and practical considerations that must be taken into account to ensure patient safety, minimize experimental risk, and protect the individual rights of patients. At the end of the day, regardless of the significance of the scientific question being investigated, the patient participating in these experiments is a volunteer, undergoing treatment for a real medical condition. The goal of this chapter is to outline the basic concepts underlying ethical experimentation for human single unit MER studies and to help the reader determine the best way to design experiments that are likely to yield positive results while preserving patient safety and integrity.

Background

While the present-day regulations and limitations imposed on most researchers by Institutional Review Boards (IRBs) may seem tedious and restrictive, the need for oversight represents a relatively new addition to the field of medical investigation. The very concept of ethical medical experimentation is less than a hundred years old (Steinberg, 2003). Physicians of antiquity did not generally perform experiments, and medical knowledge was primarily gained by observation alone. Over time, physicians such as Galen began to perform experiments in animals although progress was very slow. In the early period of modern medicine, physicians often experimented upon themselves, obtaining anecdotal support for new treatments. For example, John Hunter (1728–1793) injected himself with syphilis to define the etiology of the clinical illness it caused. James Simpson (1811–1870) inhaled chloroform in search of a new anesthetic agent. Some physicians experimented with relatives, such as Edward Jenner (1749–1823), who injected his son and his neighbor's children with a vaccine against smallpox. However, studies in unrelated volunteers or patients were not conducted. Despite the lack of formal investigative studies, the general concept held by most physicians prior to the 1940s was that human experimentation benefited medical science and, ultimately, patient care, and was therefore justifiable. No regulatory oversight other than the morality of the individual investigator existed. By the beginning of the 1900s physicians began to increasingly rely on cohorts of patients for human studies, often without their consent or knowledge of potential risks. Notable examples include Walter Reed's injection of the fly that transmits yellow fever into patients prior to the advent of any treatment for yellow fever. In the 1920s, the concept of eugenics (genetic selection of desirable and undesirable features in people) gained popularity, and tens of thousands of patients with epilepsy, mental disabilities, or other misunderstood mental conditions were sterilized (Gaudilliere, 2000; Jones, 1963). The majority of these patients did not have the capacity to understand what was being done to them, and their families were not asked to give informed consent In the United States, the poor, undereducated African Americans, prisoners, and other susceptible populations were often the subject of experimentation without consent. Perhaps the most notable example is the Tuskegee Syphilis Study. This study was designed to study the long-term and longitudinal effects of untreated syphilis. Several hundred black males with syphilis were enrolled in the study in 1932 and observed over many decades (1932–1972) without treatment to assess the long-term effect of syphilis. These patients were uninformed as to the nature of the study. They were offered free "medical care" in exchange for their participation but were not informed about the goals of the study or treatment options. Remarkably, even after a cure for syphilis was introduced and widely distributed in the 1940s, the subjects in this study were denied treatment, resulting in suffering, pain, and death in many cases (Schuman et al., 1955). Perhaps more significantly, the study was reviewed by a panel of the American Medical Association and found to be ethical and scientifically valid (Kampmeier, 1974)!

The most notorious examples of abusive human experimentation are the studies performed by Nazi Germany predominately on Jewish concentration camp prisoners and others during World War II (Angell, 1990). During World War II, several thousand Jews and other people were subjected to barbaric torture in the name of human experimentation. Experiments consisted of procedures such as placing people in ice water to document the effects of hypothermia, surgical mutilations, surgical procedures without anesthesia, and forced starvation (Mellanby, 1947). While one cannot be completely sure of the underlying motivations in these cases, the physicians performing these "experiments" did feel they were learning about the limits of human endurance and suffering. Careful documentation of their experimental results serves as a testimony of their belief in the scientific validity of the studies. Similar chemical warfare experiments conducted by the Japanese against Chinese prisoners of war reflected an equal disregard for human rights (Pappworth, 1990). Thus, while these studies may indeed have demonstrated scientific results, they were obtained at the expense of human torture and, as such, constitute crimes against humanity rather than valid scientific work.

With the end of World War II and the prosecution of war criminals, a new awareness of the possibilities for human abuse arose. The Nuremberg trials of Nazi war criminals laid out in detail many of the crimes conducted in the name of science, and for the first time established the concept that human experimentation needed to be conducted within ethical guidelines. Similar outrage over a number of studies in the United States, such as ones in which patients were injected with viable cancer cells, and many others (Pappworth, 1990), combined with the growth of the civil rights movement during the 1960s, led to a demand for the establishment of standards for human experimentation. In response, a government-appointed group was organized. In 1979 the Belmont Report was published, establishing for the first time a comprehensive set of guidelines for the ethical treatment of human participants in medical experimentation ("Protection of Human Subjects," 1979). The Belmont Report outlined three main principles to be adhered to in the establishment of human studies (see table 4.1): (1) Respect for Individuals, (2) Justice, and (3) Beneficence. Furthermore, the Belmont Report established the need for oversight and the establishment of IRBs to perform this oversight. Each of these concepts will be briefly discussed.

Respect for Persons   The concept of individual rights, freedoms, and the capacity to make independent decisions is embodied in this guideline. Respect for Persons indicates that each individual who participates in a clinical study must be fully informed of all the known risks and benefits, the reason for his or her participation, and the opportunity to accept or refuse participation without any impact on regular health care. Patients must have the study explained in detail in a language and at a level they understand. Further, investigators must respect the voluntary nature of human participation. Failure to adhere to this principle reflects involuntary participation in human experimentation and is forbidden in the present era. The practical application of this

Table 4.1
Principles of ethical research and their practical application

Ethical principles for research	Applications of ethical principles for research
Respect for Persons Individuals should be treated as autonomous agents Persons with diminished autonomy are entitled to protection	**Informed Consent** Volunteer research participants, to the degree that they are capable, must be given the opportunity to choose what shall or shall not happen to them The consent process must include: • Information, • Comprehension, and • Voluntary participation
Beneficence Human participants should not be harmed Research should maximize possible benefits and minimize possible risks	**Assessment of Risks and Benefits** The nature and scope of risks and benefits must be assessed in a systematic way
Justice The benefits and risks of research must be distributed fairly	**Selection of Participants** There must be fair procedures and outcomes in the selection of research participants

concept is the use of IRB-approved informed consents from patients. Signing of an informed consent that is written in plain language at a sixth-grade level or below is considered an indication that the individual has voluntarily agreed to participate. The informed consent will explain the protocol, its purpose, and all associated risks, as well as methods for protecting patient safety and confidentiality and means for compensation (if any) in the event of injury. All informed consent documents must be reviewed and approved for both language and content by the IRB before any study can be instituted. Informed consent is an absolute requirement for all modern human studies.

Justice   This concept indicates that studies performed on humans must demonstrate fairness in the selection of research participants. The participants in any study must be selected using criteria that allow broad participation for appropriate candidates, and the study must exclude inappropriate subjects based on similarly clear criteria. For example, a study that proposes testing a new cancer therapy should be made available to patients suffering with that type of cancer while potentially excluding those with other cancer types. Exclusion of minors might be appropriate and just for some studies but not others. Excluding non-English-speaking patients from a study that requires patients to fill out a questionnaire in English may be just, although if a translated version of the questionnaire could be easily obtained, exclusion may not be appropriate.

Beneficence   Beneficence is a simple concept yet one of particular importance to human MER and other neurophysiological studies of behavior. This concept states that there should be some benefit to the subject for participation in the study and that studies should be designed to maximize potential benefit while minimizing risk. In the case of a clinical trial of an investigational drug, surgical intervention, or other such medical therapy, the treatment should have the possibility of benefiting the specific disease being treated. In other studies, such as participation in a registry or database, the patient may not derive personal benefit, but the information learned by the study may be broadly applicable to the treatment of other patients with that medical issue. In the case of human MER studies, if the data are being collected to study neuronal behavior specific to the patient's disease, then it may be fairly easy to demonstrate benefit. However, if the data being collected are unrelated to the actual disease for which the patient is undergoing electrode testing, and the study questions are similarly unrelated, the benefit to patients is more difficult to demonstrate. For example, studying neuronal activity in the hippocampus during a learning task may increase general knowledge about how the brain encodes memories but has no direct or indirect benefit to the epilepsy patient. In contrast, studies that evaluate neuronal activity during and after a seizure may have more direct benefit. While increased general knowledge of how the brain works may potentially lead to better medical treatments, this benefit is far enough removed from the subject that beneficence is marginal. While this issue can be resolved, patients and investigators alike must be well aware of this and design experiments that reduce risk to negligible levels.

Institutional Review Boards

Oversight for research involving humans is performed by institutional review boards (IRBs). Many hospitals and research centers establish their own IRB while some facilities may contract with an external IRB that is paid to review research studies. The goal of the IRB is to make sure that all research conducted in humans abides by the main principles of ethical research.

An IRB is typically composed of both professionals and laypersons. Physicians, nurses, scientists, and other allied health professionals review and critique proposed studies and provide oversight for ongoing studies. Most IRBs include a statistician to evaluate the statistical validity of a trial, the number of subjects to be enrolled, and the likelihood that the proposed research can achieve its stated goals. This is an important role because if a study lacks the statistical power to demonstrate its desired goals, it is unethical and unjust to perform that study on humans in most cases. Ongoing review of data is often needed to ensure that statistical validity is maintained. Nonmedical persons such as community members, clergy, lawyers, and others comprise a critical component of the IRB. They provide the perspective of the patient, balance the views of the health professionals, help determine whether the research consent is understandable in plain English, and ensure that the research is perceived as safe and ethical. While most members of an IRB serve in a volunteer capacity, IRBs generally employ full-time staff who are responsible for prescreening protocols, ensuring regulatory compliance with state and national guidelines, and reporting deviations to appropriate governing bodies. IRB analysts provide reminders to investigators to submit progress reports, document adverse events, and make sure that an approved study actually does what it planned to do on paper. These IRB workers are trained to assist investigators in preparing appropriate consents, solving conflicts between institutions and research sponsors, and, if needed, halting a research protocol if deemed unsafe. When an IRB evaluates a study, they want to ensure that the research proposed (1) is achievable in a realistic time frame, (2) matches benefit with risk appropriately, and (3) makes every effort to respect the safety, privacy, and dignity of the humans who are volunteering for participation. When studies fail to convince the IRB that they can achieve these goals, a study is likely to be rejected or returned to the investigators for revision. When an IRB study requests permission to utilize tissues removed on the day of surgery, the stringencies are often much less. So long as appropriate informed consent is obtained, the use of the tissue is for appropriate scientific purposes, tissue is being removed for clinical reasons only, and patients are clearly told what the tissue is to be used for, most IRBs will approve such research.

Potential Patient Populations for Human Neurophysiological Studies

At present, the ability to record single units ("neurons") and intracranial local field potentials (LFPs) in humans is largely limited to a subset of patients who are undergoing implantation of electrodes for either seizure localization (epilepsy) or treatment of movement (deep brain stimulation; DBS) and patients undergoing removal of brain tissue for treatment of either a tumor or epilepsy in which the tissue to be removed is first recorded from and then removed. Each of these patient populations presents unique ethical and practical challenges that must be kept in mind when designing studies.

Epilepsy

Patients with medically refractory epilepsy in whom noninvasive studies suggest that the source of the seizure onset might be a discrete area of the brain are the main patients considered for further invasive electrographic studies. The goal of these invasive recordings is to either localize the seizure to a distinct area for subsequent resection or to assist with lateralizing the seizure focus to one side of the brain and/or lobe of the brain. Patients requiring monitoring with intracranial electrodes typically undergo either depth electrode or subdural grid electrode implantation surgery. Depth electrodes are placed into deep brain structures such as the amygdala and hippocampus. Depending on surgeon preference and indications for implantation, electrodes can be placed with orthogonal trajectories extending from the lateral cortical surface to deeper structures (Rutishauser, Schuman, & Mamelak, 2006) or along the length of the hippocampus and amygdala (Spencer et al., 1990). Depth electrodes record local field potentials (LFP) from the clinical electrode contacts. Alternatively, hybrid depth electrodes that contain clinical electrodes in conjunction with microwires that can record units can be implanted (Fried et al., 1999). The majority of published human MER studies in the past decade have utilized this hybrid microwire method although several other studies and electrode designs have been employed (Howard et al., 1996).

The most common means for intracranial seizure localization relies on the use of subdural grid electrodes. A subdural grid is a sheet of thin disk-like electrodes implanted in a silicone sheet that can flex and conform to the brain surface. The size, density, and number of electrode contacts can vary from small strips containing two electrodes to dense arrays containing over 256 electrodes. More recently, so-called microwire grids have been developed that allow recording of LFP from small wires placed microns apart (Kellis et al., 2011; Van Gompel et al., 2008). In general, these subdural electrodes record LFP and not units. Subdural grids are most often implanted along a large cortical surface via a craniotomy. The advantage of the grid is that it allows physicians to sample EEG from a large area of the cortex, thereby helping to discretely localize a specific seizure focus. Further, direct stimulation of the grid electrodes permits physicians to map cortical functions such as speech, movement, vision, and sensation to discrete areas of the cortex, thereby sparing these structures as needed during surgical resection. The decision of where to place subdural grid or depth electrodes and the number of electrodes to be placed is a clinical one and should not be influenced by research concerns.

Epilepsy patients represent perhaps the most valuable group of patients for potential MER and other behavior-related human neurophysiology experiments due to the nature of the monitoring and experimental setup. Once patients are implanted with electrodes, they recover from surgery and are transferred to an epilepsy-monitoring unit for video-EEG recording to capture seizure. The recording process is continuous and typically goes on for several days to weeks. Recording sessions of one to two weeks are most typical. During this time, patients are awake, interactive, in minimal pain, and generally quite willing to participate in experiments so long as those experiments do not interfere with clinical care or cause discomfort. This provides a relatively relaxed testing environment in which the experimenters can develop a rapport with the subjects, have time to test and verify the test equipment, and explain the testing paradigms repeatedly to subjects. The opportunity to repeat experiments from day to day can provide for increased trials. Further, the ability to perform a task at one time period and repeat or institute a similar task at a later time period provides a rare opportunity to study mechanisms of learning and memory for a variety of tasks. In general, human MER and LFP recordings in the epilepsy monitoring unit are the closest approximation to experimental designs routinely utilized in primates and nonhuman primates while at the same time providing the opportunity to study uniquely human aspects of cognition, memory, and other behaviors (Haglund & Hochman, 2004; Howard et al., 1996; Mellanby, 1947; Park et al., 2011; Quiroga et al., 2008; Rutishauser et al., 2010; Rutishauser, Schuman, & Mamelak, 2006; Rutishauser, Mamelak & Schuman 2008).

While these advantages are clear, there are several distinct limitations to performing human MER in epilepsy patients. First and foremost, some of the tissues in which electrodes are placed are presumed to be, at least in part, diseased and abnormal. Therefore, any data derived from these structures must be interpreted with caution, as they may not represent the behavior in "normal" tissues. Second, epilepsy patients take seizure medications that may interfere with both normal physiological functioning and cognitive behaviors. Anticonvulsant medications routinely slow reaction time, diminish memory, and alter decision making (Aldenkamp, De Krom, & Reijs, 2003). Therefore, results must be interpreted with caution. Similarly, patient recording and testing may occur in the setting of a patient's having had, having, or being about to have a clinical seizure. Pre- and postictal behavior and neuronal activity may be markedly different from normal activity, and therefore behavioral observations may not reflect typical brain responses (Aldenkamp, 1997). The behavioral tasks designed need to take these limitations into consideration.

Beyond these considerations, there are several ethical considerations that reflect the unique nature of this patient population. First and foremost, patients who agree to participant in human MER studies generally derive no direct clinical benefit from their participation. While the risk of neurological injury from the implantation of hybrid electrodes does not appear to be more than the risk of implantation of standard electrodes, patients must be aware of the lack of potential benefit. Furthermore, patients and/or their family members need to clearly understand the voluntary nature of their involvement in research, their rights to resign from the research, and the fact that they are unlikely to derive any benefit from their participation. Finally, experiments need to reflect the functional capabilities of patients. Experiments that are too complex or demanding for patients will likely not achieve their intended results and may even cause undo emotional stress in the patients.

The following guidelines should be employed when designing and performing tasks in this patient population:

1. *Keep the task simple and direct.* Avoid tasks that involve multiple steps or procedures. A good rule of thumb is that the task should be something an average sixth-grader could complete without difficulty.

2. *Pretest the task* with normal volunteers or, ideally, other epilepsy patients prior to using it in the clinical setting to make sure that the task can be completed. Patients in the hospital for noninvasive scalp EEG video telemetry may represent an excellent control group to test the utility of a particular behavioral task.

3. *Avoid tasks that may have a strong emotional component*, and if they do, make sure these tasks are consistent with ethical human studies.

4. *Make sure the tasks are short.* In general, patients will not tolerate more than about 1 hour of testing in any one session. Tasks that take a long time to set up or perform are likely to be aborted or not completed.

5. *Carefully obtain informed consent.* Patients participating in these studies have no direct benefit and perhaps not even a secondary societal benefit. This fact should be carefully explained to patients. If needed, obtain consent with family members present. For children under age 16 parental assent is always required. Provide a copy of the consent to the patients to read.

6. *Be aware that tasks that involve a reward or punishment* to the patient such as gambling tasks have the potential to cause distress. Therefore, financial rewards should be very small to avoid patient complaints or difficulty obtaining IRB approval.

Deep Brain Stimulation

Implantation of DBS electrodes most commonly is performed in patients with movement disorders such as ET, PD, or dystonia. These disorders typically cause profound loss of fine motor control and may also cause excessive postural problems as well as cognitive changes (PD in particular). Electrodes are typically placed into deep brain nuclei using a stereotactic frame and a microdrive. The globus pallidus interna, ventral intermediate nucleus of the thalamus, and subthalamic nucleus are the most common targets. Electrodes are advanced from a starting point in the frontal lobe to the end target. During the electrode placement process, neurophysiological recording of single units can be carried out with rigid microelectrodes that pass through various brain structures seeking a physiological signature most consistent with the nucleus to be targeted (Kellis et al., 2011; Spencer et al., 1990; Steinberg, 2003). Such recordings can be useful to fine-tune the final electrode placement. MER performed for purposes of target identification does not require IRB approval unless the data will be utilized for investigational purposes. However, recordings performed to evaluate any other aspects of behavior or those that will be used to report research results generally do require IRB approval. During these surgeries, most patients are awake and cooperate with physiological stimulation testing to ensure that electrode macrostimulation results in the desired effect on movement control while avoiding side effects due to electrodes placed inaccurately. Recordings of LFP can be obtained from the micro- and macroelectrodes.

A unique aspect of these surgeries is the extremely high-quality unit data that can be obtained. Because rigid tungsten or platinum iridium electrodes are used with positioning adjusted with a microdrive, investigators can search for large-amplitude high-quality units in the cortex, thalamus, basal ganglia, and brainstem. In contrast, electrodes placed for epilepsy patients are inserted to a fixed location with more flexible microwires. This tends to lead to far more variable recording of lesser quality. Because patients are awake and interactive, they can undergo experimental behavioral testing and simultaneous recordings. Furthermore, many of these patients do not often suffer from the same degree of cognitive impairment present in epilepsy patients. As such, they are likely to more quickly comprehend a task and perform it accurately. Finally, this patient population can be pretrained on tasks so that they are more readily able to perform testing in operating room settings.

Despite these advantages, there are several ethical and practical limitations that must be kept in mind when performing experiments in this patient population. First, MER passes and the insertion of multiple parallel electrodes have been associated with an increased risk for hemorrhage during DBS surgery (Ben-Haim et al., 2009; Binder, Rau, & Starr, 2005; Pappworth, 1990). This risk is not trivial and, in fact, many neurosurgeons do not perform MER at all due to this risk. In light of this possibility, it is critical to ensure that the surgeon performs MER as a routine part of his or her standard surgical procedure and not simply to obtain experimental data. Second, MER recordings and testing are being obtained during an ongoing surgery aimed at implanting the electrodes. Patients may have a heightened anxiety during surgery and therefore are less willing to participate in any experiments of an elective or uncertain nature and may also perform poorly due to this anxiety. This situation may be exacerbated by any surgical pain or symptoms due to their underlying disease. Along these lines, experiments need to be designed to be rapidly set up and rapidly administered with little opportunity to repeat tasks.

Experimenters must be prepared to abandon experiments depending on the clinical setting. Third, owing to the often-impaired movements of patients such as having severe tremor, slowness of movements, or spontaneous abnormal movements, tasks should be designed to avoid reliance on motoric response to obtain results. For example, a task that requires a patient to push a button when he or she responds to an image may give false results due to the patient's inability to press the button accurately or in a timely fashion. Despite these limitations, the high quality of DBS MER data combined with the potential ability to record from both cortex and deep structures simultaneously provides a uniquely important paradigm for human MER (Gale et al., 2011; Kellis et al., 2011; Steinberg, 2003).

In light of the potential limits of DBS surgery, the following guidelines should be kept in mind when designing experiments:

1. *Keep the task short.* Each experimental task should be able to be completed in 5 minutes or less. Due to the constraints of the operating room environment, experimenters will have no more than 30 minutes total to perform any experimental procedures.

2. *Whenever possible, pretest the task* with the patient to make sure he or she can perform it in the operating room environment. Attempting to explain a task to a patient in an operating room with his or her head open and undergoing surgery is likely to be frustrating to both the patient and the experimenter as well as unacceptable to the operating room staff and physicians. Careful preplanning and comfort with the operating room environment is critical to successful data collection.

3. *Make sure the surgeon uses MER as a routine part of his or her operative procedure.* If not, any MER experiments likely represent a major ethical breach due to increased risk.

4. *Avoid tasks that require complex or well-controlled motoric responses.* These are likely to be difficult for patients undergoing DBS surgery for movement disorders.

5. *Whenever possible, focus on task-relevant issues* concerning the basal ganglia rather than the cortex since these recordings will always be of the highest quality and most clearly able to be localized based on surgical planning.

6. *Be prepared for the reality that many patients may not be able to complete the tasks in the operating room.* The high volume of patients undergoing these types of procedures will likely more than compensate for the several patients who have consented to participate in research studies but are unable to successfully participate during surgery.

Tumor or Tissue Resection

This population represents the least frequently encountered and potentially most difficult group in whom to perform MER or other electrophysiological studies. Patients in this category typically have brain tumors, epileptic foci, or other structural lesions for which the patient is undergoing a craniotomy and subsequent removal of brain tissue. In these patients, MER is presently performed only for research purposes with essentially no novel clinically relevant information gained. In some of the original human MER studies, patients underwent physiological unit recording of epileptic regions to attempt to characterize neuronal activity (Van Gompel et al., 2008). In later studies, patients performed basic behavioral tasks such as reading, naming objects, spontaneous speech production, memory tasks, or simple motor responses while undergoing MER (Chan et al., 2011; Mormann et al., 2008; Ojemann et al., 1988; Ojemann & Schoenfield-McNeill, 1999; Schuman et al., 1955; Schwartz et al., 2000). These sorts of studies have provided neuroscientists with invaluable information as to the functional organization of the cortex and mesial temporal lobe. Recordings are generally obtained from rigid microelectrodes, and therefore unit quality can be quite high, but the limitations of recording under surgical conditions can be onerous. In general, patients are placed in a relatively uncomfortable position with limited field of vision and constrained mobility. Therefore, behavioral tasks tend to be restricted and limited to reading, speech production, or reporting responses to cortical stimulations. One particularly valuable aspect of these experiments is the opportunity to record from multiple trajectories simultaneously due to the fact that the tissue will be removed and therefore the risk of structural damage is of less significance. Since the tissues will be removed, the opportunity to combine intraoperative recordings with pathological evaluation of tissues, tissue slice physiology, or other in vitro experiments represents a unique feature of these studies. Studies of this sort also allow for other physiological studies such as optical imaging (Gaudilliere, 2000) or the opportunity to combine unit recordings with LFP recordings over broad cortical areas.

In light of the limitations posed by patients undergoing awake craniotomy, the following ethical and practical guidelines may prove useful:

1 *Make sure the testing paradigm can be comfortably accommodated* by a patient who will be lying down with the head turned to one side and essentially covered by surgical drapes. There will be a narrow working window and greatly restricted movement. Therefore, experiments involving passive responses such as squeezing a hand for a response or reading material are more likely to be successful.

2 *Experiments need to be coordinated not only with the surgeon but with the anesthesiologist* to make sure there is working room without interfering with clinical care of the patient. Further, anesthetic methods must be taken into account to make sure the patient is fully alert and not under the influence of significant narcotics or other sedating medications that may diminish both behavioral and neurophysiological responses.

3. *Many patients will likely be unable or unwilling to participate in such studies and at no time can they be coerced into participation.* As with other MER experiments, there is very likely no direct benefit to the patient while surgery is likely to be greatly extended in length.

4. *Recording must generally be limited to areas of brain tissue that will be resected.* Strict adherence to this policy is required to avoid IRB investigation and potential halting of the study. If recordings outside of this area are to be performed, expressed consent must be obtained.

Future Populations

Implantable electrode arrays for long-term recording of units via chronic indwelling electrode arrays have been the subject of great interest recently. The primary goal of these arrays is to record cortical neurons over periods of months to years and to utilize the neuronal signals as a means to drive visual, auditory, or motor prostheses. Due to their cost and method of implantation, they are generally not utilized for more acute studies. The "Utah Array" has been most extensively studied in both animals and humans (Maynard, Nordhausen, & Normann, 1997; Normann et al., 1999). Utilization of chronically implanted arrays represents a more significant risk to patients due to issues of infection, long-term reaction, and possibly the development of seizures or brain injury. Furthermore, these arrays tend to be implanted over viable, functional cortex and not tissue that will be resected. In light of these concerns, experiments involving chronic implants are likely to be more scrutinized by IRBs. In general, the use of these arrays will be restricted to patients with severe medical illnesses such as paralysis or blindness, in conjunction with studies aimed to drive or modify prosthetic devices. In these situations, ethical considerations such as respect of person, beneficence, and justice are far more evident, thereby justifying the potential for significantly greater risk. However, behavioral experiments not specifically aimed at the underlying disease in this patient population may still be feasible and provide unique data assuming patients are willing to cooperate in these tasks. The rarity of appropriate patients for these implants should serve as a caution to experimenters, but the trade-off of being able to record units over weeks to months or longer may justify pursuit of these studies in select instances.

General Considerations

While the opportunity to obtain human MER and LFP recordings during behavioral and cognitive testing paradigms offers a unique opportunity for scientists and clinicians to understand brain function, the ethical responsibilities that investigators must adhere to are substantial. Investigators must always remember that their subjects are people with serious neurological conditions who are undergoing surgery to treat their disease. Their interest in the experiments being performed is negligible, and their participation is offered due to generosity of spirit. However, in the absence of obvious clinical benefit the risks must be kept negligible. Furthermore, their privacy, the privacy of their families, and respect for their right to halt, withdraw, or refuse to participate must be adhered to at all times. The following guidelines are useful to keep in mind:

1. *Keep it simple.* Limit experiments to 10–20 minutes in the operating room setting from start to finish and 1 hour in the epilepsy unit. Furthermore, make sure the tasks are easy to understand. A good rule of thumb is that a 6th grader could perform the task with ease.
2. *Have a repertoire of tasks available*, especially when longer testing is feasible. Whenever possible, meet with the patient in advance to review these tasks. This will provide the opportunity

to take advantage of each testing session. Further, it allows the experimenter to develop a rapport with the patient prior to testing which will make the actual testing less anxiety-prone or difficult. This is especially true for intraoperative testing. An iterative trial and error process is often needed to define the best task, so be patient and be creative.

3. *Make sure there is IRB approval* and signed informed consent on every patient prior to any experiments. Failure to strictly adhere to this regulation can result in complete cessation of the research program. Strict adherence to IRB-established guidelines will generally avoid most problems.

4 *Be prepared to abandon experiments due to clinical concerns.* Remember essentially any clinical person ranging from nurse to technician to physician can halt experiments at any time due to clinical concerns. If clinical issues arise (e.g., patient has a seizure) during the experiment, the investigator on his own should stop the experiment and call for assistance immediately. Similarly, the patient or any family member has the right to stop the experiments at his wish and for almost any reason. Failure to respect this right represents a major breach of policy.

5 *Be prepared to work rapidly and troubleshoot quickly.* The clinical setting provides little opportunity for relaxed testing and troubleshooting. The slower an experimenter is with troubleshooting, the more likely it is that he or she will not be able to complete any experimental tasks.

Conclusion

Human MER offers a unique and unprecedented opportunity to study how the brain works. Research efforts along these lines should be applauded and encouraged as they are likely to lead to profound and potentially revolutionary insight into how the human brain functions and may indeed lead to novel treatments for a variety of brain diseases. However, due to the unique ethical and practical considerations in these studies, a strict adherence to ethical principles with a pragmatic approach to experiment design and execution are needed to ensure success in this endeavor.

References

Aksenova, T. I., Chibirova, O. K., Dryga, O. A., Tetko, I. V., Benabid, A. L., & Villa, A. E. (2003). An unsupervised automatic method for sorting neuronal spike waveforms in awake and freely moving animals. *Methods (San Diego, Calif.)*, *30*, 178–187.

Aldenkamp, A. P. (1997). Effect of seizures and epileptiform discharges on cognitive function. *Epilepsia*, *38*(Suppl 1), S52–S55.

Aldenkamp, A. P., De Krom, M., & Reijs, R. (2003). Newer antiepileptic drugs and cognitive issues. *Epilepsia*, *44*(Suppl 4), 21–29.

Angell, M. (1990). The Nazi hypothermia experiments and unethical research today. *New England Journal of Medicine*, *322*, 1462–1464.

Ben-Haim, S., Asaad, W. F., Gale, J. T., & Eskandar, E. N. (2009). Risk factors for hemorrhage during microelectrode-guided deep brain stimulation and the introduction of an improved microelectrode design. *Neurosurgery*, *64*, 754–762; discussion 762–763.

Binder, D. K., Rau, G. M., & Starr, P. A. (2005). Risk factors for hemorrhage during microelectrode-guided deep brain stimulator implantation for movement disorders. *Neurosurgery*, *56*, 722–732, discussion 722–732.

Chan, A. M., Baker, J. M., Eskandar, E., Schomer, D., Ulbert, I., Marinkovic, K., et al. (2011). First-pass selectivity for semantic categories in human anteroventral temporal lobe. *Journal of Neuroscience, 31*, 18119–18129.

Fried, I., Wilson, C. L., Maidment, N. T., Engel, J., Jr., Behnke, E., Fields, T. A., et al. (1999). Cerebral microdialysis combined with single-neuron and electroencephalographic recording in neurosurgical patients: Technical note. *Journal of Neurosurgery, 91*, 697–705.

Gale, J. T., Martinez-Rubio, C., Sheth, S. A., & Eskandar, E. N. (2011, January 6). Intra-operative behavioral tasks in awake humans undergoing deep brain stimulation surgery. *Journal of Visualized Experiments* (47), e2156.

Gaudilliere, J. P. (2000). Mendelism and medicine: Controlling human inheritance in local contexts, 1920–1960. *Comptes Rendus de l'Académie des Sciences. Série III, Sciences de la Vie, 323*, 1117–1126.

Haglund, M. M., & Hochman, D. W. (2004). Optical imaging of epileptiform activity in human neocortex. *Epilepsia, 45*(Suppl 4), 43–47.

Howard, M. A., III, Volkov, I. O., Abbas, P. J., Damasio, H., Ollendieck, M. C., & Granner, M. A. (1996). A chronic microelectrode investigation of the tonotopic organization of human auditory cortex. *Brain Research 724*, 260–264.

Jones, B. C. (1963). Prohibition and eugenics, 1920–1933. *Journal of the History of Medicine and Allied Sciences, 18*, 158–172.

Kampmeier, R. H. (1974). Final report on the "Tuskegee syphilis study." *Southern Medical Journal, 67*, 1349–1353.

Kellis, S., Greger, B., Hanrahan, S., House, P., & Brown, R. (2011). Platinum microwire for subdural electrocorticography over human neocortex: Millimeter-scale spatiotemporal dynamics. *Conference Proceedings; ... Annual International Conference of the IEEE Engineering in Medicine and Biology Society. IEEE Engineering in Medicine and Biology Society. Conference, 2011*, 4761–4765.

Maynard, E. M., Nordhausen, C. T., & Normann, R. A. (1997). The Utah Intracortical Electrode Array: A recording structure for potential brain–computer interfaces. *Electroencephalography and Clinical Neurophysiology, 3*, 228–239.

Mellanby, K. (1947). Medical experiments on human beings in concentration camps in Nazi Germany. *British Medical Journal, 1*, 148–150.

Mormann, F., Kornblith, S., Quiroga, R. Q., Kraskov, A., Cerf, M., Fried, I., et al. (2008). Latency and selectivity of single neurons indicate hierarchical processing in the human medial temporal lobe. *Journal of Neuroscience, 28*, 8865–8872.

Normann, R. A., Maynard, E. M., Rousche, P. J., & Warren, D. J. (1999). A neural interface for a cortical vision prosthesis. *Vision Research, 30*, 2577–2587.

Ojemann, G. A., Creutzfeldt, O., Lettich, E., & Haglund, M. M. (1988). Neuronal activity in human lateral temporal cortex related to short-term verbal memory, naming and reading. *Brain, 111*, 1383–1403.

Ojemann, G. A., & Schoenfield-McNeill, J. (1999). Activity of neurons in human temporal cortex during identification and memory for names and words. *Journal of Neuroscience, 19*, 5674–5682.

Pappworth, M. H. (1990). "Human guinea pigs"—A history. *BMJ (Clinical Research Ed.), 301*, 1455–1460.

Park, J. H., Chung, S. J., Lee, C. S., & Jeon, S. R. (2011). Analysis of hemorrhagic risk factors during deep brain stimulation surgery for movement disorders: Comparison of the circumferential paired and multiple electrode insertion methods. *Acta Neurochirurgica, 153*, 1573–1578.

Protection of human subjects; Belmont Report: Notice of report for public comment. (1979). *Federal Register, 44*, 23191–23197.

Quiroga, R. Q., Mukamel, R., Isham, E. A., Malach, R., & Fried, I. (2008). Human single-neuron responses at the threshold of conscious recognition. *Proceedings of the National Academy of Sciences of the United States of America, 105*, 3599–3604.

Rutishauser, U., Ross, I. B., Mamelak, A. N., & Schuman, E. M. (2010). Human memory strength is predicted by theta-frequency phase-locking of single neurons. *Nature, 464*, 903–907.

Rutishauser, U., Schuman, E. M., & Mamelak, A. N. (2006). Online detection and sorting of extracellularly recorded action potentials in human medial temporal lobe recordings, in vivo. *Journal of Neuroscience Methods, 154*, 204–224.

Rutishauser, U., Schuman, E. M., & Mamelak, A. N. (2008). Activity of human hippocampal and amygdala neurons during retrieval of declarative memories. *Proceedings of the National Academy of Sciences of the United States of America, 105*, 329–334.

Schuman, S. H., Olansky, S., Rivers, E., Smith, C. A., & Rambo, D. S. (1955). Untreated syphilis in the male Negro: Background and current status of patients in the Tuskegee study. *Journal of Chronic Diseases, 2*, 543–558.

Schwartz, T. H., Haglund, M. M., Lettich, E., & Ojemann, G. A. (2000). Asymmetry of neuronal activity during extracellular microelectrode recording from left and right human temporal lobe neocortex during rhyming and line-matching. *Journal of Cognitive Neuroscience, 12*, 803–812.

Spencer, S. S., Spencer, D. D., Williamson, P. D., & Mattson, R. (1990). Combined depth and subdural electrode investigation in uncontrolled epilepsy. *Neurology, 40*, 74–79.

Starr, P. A., Turner, R. S., Rau, G., Lindsey, N., Heath, S., Volz, M., et al. (2004). Microelectrode-guided implantation of deep brain stimulators into the globus pallidus internus for dystonia: Techniques, electrode locations, and outcomes. *Neurosurgical Focus, 17*, E4.

Steinberg, A. (2003). *Encyclopedia of Jewish medical ethics* (2nd ed., Vol. 2). Jerusalem, Israel: Feldheim.

Theodosopoulos, P. V., Marks, W. J., Jr., Christine, C., & Starr, P. A. (2003). Locations of movement-related cells in the human subthalamic nucleus in Parkinson's disease. *Movement Disorders, 18*, 791–798.

Van Gompel, J. J., Stead, S. M., Giannini, C., Meyer, F. B., Marsh, W. R., Fountain, T., et al. (2008). Phase I trial: Safety and feasibility of intracranial electroencephalography using hybrid subdural electrodes containing macro- and microelectrode arrays. *Neurosurgical Focus, 25*, E23.

5 Subchronic In Vivo Human Microelectrode Recording

Richard J. Staba, Tony A. Fields, Eric J. Behnke, and Charles L. Wilson

Overview of Clinical Electroencephalography

The fields of electrophysiology and electroencephalography (EEG) developed from numerous discoveries and scientific achievements that cannot be adequately describe here, and a detailed historical account can be gathered from several publications (Brazier, 1961; Grass, 1984; Niedermeyer, 1993). Nonetheless, it would be useful to first describe briefly the development of clinical EEG and its role in the presurgical evaluation of epilepsy where subchronic in vivo microelectrode studies are typically carried out. Two prominent scientists who made significant contributions to EEG were Richard Caton, who described brain electrical activity in animals (Caton, 1875), and Hans Berger, who is credited with recording the first EEG in humans (Berger, 1929). Berger also recorded EEG from patients with epilepsy and described abnormal EEG patterns during seizures (Berger, 1933). Prior to Berger's published human studies, Pavel Kaufman and Napoleon Cybuski had separately observed abnormal EEG discharges during electrically induced seizures in animals (Kaufman, 1912; Cybulski & Jelenska-Maciezyna, 1914). These earlier studies and work by Frederic Gibbs and colleagues, who provided a thorough description of EEG rhythms associated with different types of seizures, that is, 3-Hz spike-and-wave discharges in absence epilepsy (Gibbs et al., 1935; Gibbs et al., 1936), mark the beginning of the use of the EEG technique in clinical investigations of epilepsy and other neurological disease states.

Electroencephalography not only was important in the diagnosis and understanding of abnormal electrical activity associated with epilepsy but could also be used to help localize epileptogenic brain areas for surgical treatment (Jasper, 1941). Recordings from electrodes placed on the scalp or in some cases on the dura through small burr holes in the skull were used to localize epileptiform discharges to focal brain areas that could then be surgically removed to eliminate seizures (Penfield, 1939; Bailey & Gibbs, 1951). In the 1950s there were few active surgical epilepsy programs in North America, including the Montreal Neurological Institute, the National Institutes of Health Clinical Center in Bethesda, the University of Washington in Seattle, the Barrow Neurological Institute in Phoenix, and Illinois Neuropsychiatric Institute. It was during this time that Paul Crandall, Richard Walter, and colleagues at the University of California, Los

Angeles (UCLA) began using neurophysiology in the surgical treatment of epilepsy (Crandall & Babb, 1993). The UCLA Clinical Neurophysiology Research Laboratory opened in 1961 and included special operating rooms for teleradiography and neurophysiological procedures and a separate electrically shielded room for recording intracranial EEG from depth electrodes, which was a recording technique already being used in the neurosurgical intervention of movement disorders (Meyers & Hayne, 1948; Meyers et al., 1949).

UCLA depth electrode studies were carried out in patients who had clinical and pathological changes that indicated seizures arose from brain areas in one hemisphere—for example, mesial temporal lobe—but could not be confirmed on EEG recorded from the scalp electrodes. Initially, depth electrodes were surgically implanted using stereotactic procedures such as ventriculography in combination with the Talairach atlas (Talairach et al., 1958; Crandall et al., 1963). Each depth electrode consisted of a pair of small-diameter stainless steel wires each insulated except at the distal end that were affixed to a central strut (see figure 5.1A; Crandall et al., 1963). The bipolar electrodes were similar to those used in chronic animal studies that were proven safe

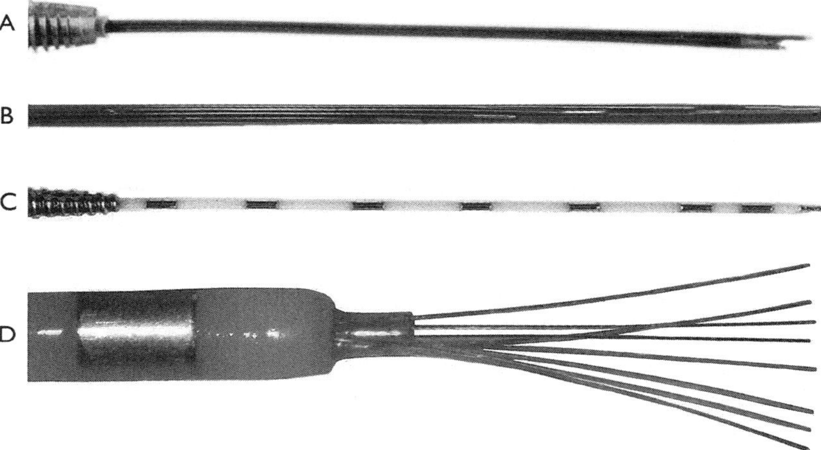

Figure 5.1
Photographs of clinical depth electrodes and microwire bundle. In order to provide appropriate visual detail of each electrode, photographs were taken at different levels of magnification. (A) Bipolar clinical depth electrode consisting of two stainless steel wires fused to central strut. Each wire is ~250 μm in diameter with 2-mm separation between wires and 0.5 to 1.0 mm of Epoxylite insulation removed at each tip for an impedance of 3 to 10 kΩ. On the left is a guide screw used to position and secure the electrode. (B) Epoxylite-coated 21-gauge stainless steel tube with 0.4-mm insulation removed from the tip, plus seven 200-μm wires laminated to the side with 2.5-mm insulation removed from the tip of each wire. The lumen of this tube contained the first bundles of microwires used for recording single unit activity in 1971 (Babb et al., 1973). (C) Clinical Behnke–Fried depth electrode (AD-TECH Medical Instrument Corp., Racine, WI) made from flexible polyurethane (outer diameter 1.25 mm) with seven platinum–iridium (Pt-Ir) contacts each 1.5 mm in length and spaced 5.5 mm center to center with the exception of two distal contacts that are separated by 3 mm for use in delivering biphasic stimulation. (D) Bundle of eight polyimide insulated Pt-Ir (80%/20%) microwires inserted through the lumen of AD-TECH depth electrode that can extend up to 5 mm beyond the distal tip. Each microwire is 40 μm in diameter with impedance between 100 and 300 kΩ (ninth uninsulated reference wire not shown).

and reliable for long-term recording (Adey et al., 1961). Subsequent depth electrodes were fabricated using multiple stainless steel wires laminated to a narrow, semi-rigid tube that could record from middle and inferior temporal gyri as well as mesial temporal structures (see figure 5.1B). This earlier multicontact electrode was similar to some current commercially manufactured clinical electrodes consisting of multiple platinum–iridium (Pt-Ir) alloy contacts spaced along the length of a narrow tube (see figure 5.1C). However, the central tube of the electrode is now manufactured from polyurethane that combines the durability of metal and the elasticity of rubber. An important feature shared by the electrodes shown in figure 5.1B and 5.1C was that the tube allowed the passage of very fine microwires or microelectrodes (see figure 5.1D). The size and impedance of microelectrode were capable of recording extracellular unit activity, that is, neuronal action potentials, at the distal end of depth electrode, which has been a technique used at UCLA for more than 40 years. In addition to microelectrodes, some depth electrodes were also adapted with a microdialysis probe to measure levels of biogenic amines within the extracellular space (Fried et al., 1999).

In Vivo Microelectrode Recording

Beginning in the 1940s extracellular unit recordings were being carried out in anesthetized or spinal transected animals using a single (or paired) glass micropipette pulled to a tip diameter of several microns (Renshaw et al., 1940; Schmidt et al., 1959). More durable metal microelectrodes were designed for long-term unit recordings in behaving animals that consisted of a small diameter (< 80 μm) tungsten or stainless steel wire that could be sharpened to a tip diameter < 10 μm (Hubel, 1957; Strumwasser, 1958). In some animal studies a bundle of flexible microwires was implanted in subcortical structures such as hippocampus or amygdala in order to record extracellular activity from many units simultaneously (O'Keefe & Bouma, 1969; Jacobs et al., 1970).

The development of clinical EEG and microelectrode design described in the preceding paragraphs was facilitated with many advances in the technology. Improvements in circuit design provided greater signal amplification and differential recording to remove common signals and noise between electrodes (Forbes & Thacher, 1920; Matthews, 1934). In addition, a critical design feature for microelectrode recording was the impedance buffer. The introduction of a buffer or head stage lowered the source impedance to levels at or below the amplifier impedance that preserved the relatively small amplitude unit potentials (tens to a few hundred microvolts) from electromagnetic noise and movement artifact picked up by the high-impedance microelectrode and cables.

Early in vivo microelectrode recordings of cortical unit activity from patients were carried out inside the operating room during epilepsy surgery using very fine tipped glass micropipettes (Ward & Thomas, 1955; Ward et al., 1956). At UCLA, interictal unit activity from the amygdala was recorded using a Pt-Ir microwire (5 to 10 μm diameter) inserted through the lumen of a depth electrode under aseptic conditions outside the operating room (Verzeano et al., 1971).

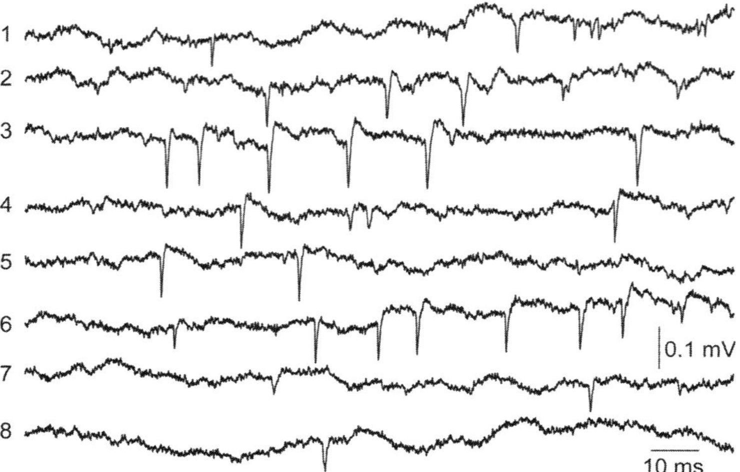

Figure 5.2
Example of spontaneous extracellular unit activity recorded from a microelectrode containing eight microwires (numbered 1–8; impedance 100–300 kHz measured at 1 kHz) with tips spaced at 500-μm intervals positioned in entorhinal cortex. Continuous data were sampled at 27.8 kHz per channel (16-bit resolution; 1 Hz to 6 kHz; Neuralynx, Inc., Bozeman, MT) and recorded with respect to adjacent uninsulated microwire used as a local reference (1–3 kHz measured at 1 kHz). Negative polarity of signals is oriented down.

Subsequent microelectrode studies involving presurgical patients were carried out with bundles of microwires at the distal tip of each depth electrode positioned in amygdala, uncus, hippocampus, and parahippocampal gyrus (Babb et al., 1973). The microwires illustrated in figure 5.1D were each cut to a similar length to extend the same distance from the depth electrode. Other microelectrodes were designed with microwires in a tetrode configuration to improve single unit isolation, or microwires could be cut to different lengths—for example, microwire bundles with 500-μm intertip spacing—to record units and local field potentials from different cell lamina (see figure 5.2) (Bragin et al., 2002; Le Van Quyen et al., 2008). The ability to record outside the operating room made it possible to study unit activity, as well as amino acid release using microdialysis (Wilson et al., 1996), during spontaneous seizures (Babb et al., 1973; Babb & Crandall, 1976; Babb et al., 1987), memory testing (Halgren et al., 1978; Cameron et al., 2001), sleep (Ravagnati et al., 1979; Staba et al., 2002a; Staba et al., 2002b), vision (Wilson et al., 1983; Wilson et al., 1984), and functional connectivity within the human mesial temporal lobe using stimulation-evoked responses (Wilson et al., 1990).

There are now a number of epilepsy centers with subchronic microelectrode recording capabilities, and the different designs reflect adaptations to the type of clinical electrode and neurosurgical approach used in the diagnostic evaluation of epilepsy. Some of these microelectrodes consist of multiple pairs of 50-μm Teflon-coated Pt-Ir microwires positioned between contacts and flush with the surface of the clinical depth electrode (Howard et al., 1996). Others include nonpenetrating 40-µm microwires with 1-mm spacing arranged in a 5 × 5 array embedded between contacts of subdural grid electrodes (Van Gompel et al., 2008). Similar to microwires at the distal end of the depth electrode, microwires positioned on the surface can record extracellular unit activity, while epipial microelectrodes can record local field potentials that are not captured on adjacent larger diameter clinical electrode contacts (Stead et al., 2010). In addition, there are several high-density intracerebral microelectrodes such as those with 24 Pt-Ir contacts (25-µm diameter) spaced at 100-µm intervals oriented linearly on a stainless steel shaft (Ulbert et al., 2001; Ulbert et al., 2004). Microelectrode arrays with up to 100 contacts have been fabricated from silicon and arranged in a 10 × 10 grid at 400-µm intervals (Campbell et al., 1991; Jones et al., 1992) or contain Pt-tipped contacts arranged in a 4 × 4 mm grid (Hochberg et al., 2006; House et al., 2006; Truccolo et al., 2008; Schevon et al., 2009).

Microelectrode Recording System

The conventional recording system illustrated in figure 5.3 begins with the microelectrodes that collect signals from the subject's brain. The electronic signals then proceed to a buffer or head stage followed by an amplification stage. The output of the amplifier is then sent to an analog-to-digital converter (ADC) and the digitized signals can then be manipulated by a digital signal processor (DSP) for real-time signal conditioning, display, and storage for offline analysis. Each aspect of the recording system beginning with recording environment and including cables that connect the microelectrodes with the buffers is discussed in the sections that follow.

The recording environment is an important factor since it can significantly affect the quality of signals. Currently at UCLA and likely most epilepsy centers, basic and clinical research

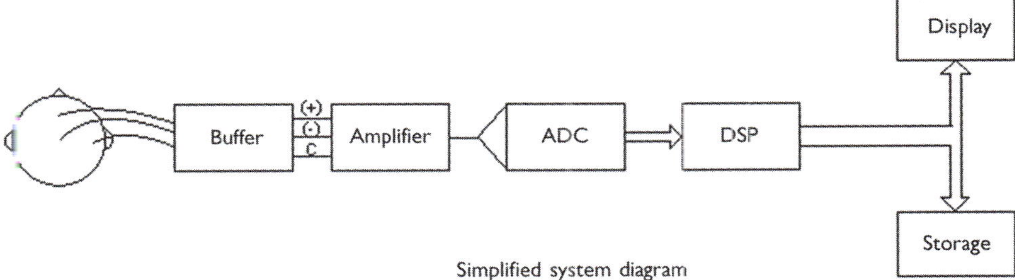

Figure 5.3
Block diagram of a conventional microelectrode recording setup. Beginning from the left the diagram illustrates signal collection from a single microelectrode that consists of an active electrode (indicated by +), reference electrode (indicated by −), and common (patient ground denoted by letter C) connection. The active and reference signals are connected to buffers or a head stage that performs impedance transformation and prevents loading of the bioelectric generating source. The outputs of the buffers are sent to a differential amplifier (instrumentation amplifier) where the signals are subtracted and amplified. The output of the amplifier is sent to the analog-to-digital converter (ADC) to be digitized by a digital signal processor (DSP). Note that anti-aliasing is implied between the amplifier output and the ADC input. The digitized output of the ADC is processed by a computer and streamed onto storage media and displayed.

studies are carried out in the patient's hospital room, an environment that was not specifically designed for high-impedance microelectrode recording. The largest sources of noise derive from overhead florescent lights, which can produce electromagnetic interference, and within hospital room walls that contain AC power lines and are often located in the wall directly behind the head of the patient's bed. Ideally, shielded hospital rooms could attenuate electromagnetic noise picked up by microelectrodes as well as clinical electrode recording systems, but few centers have this luxury. More practical solutions are simply to turn off overhead lights and adjust the patient's bed away from the head wall. In addition, reduced cable lengths and proper shielding can be used as is discussed below.

Human Microelectrode Materials and Fabrication

Microelectrode fabrication incorporates conducting and insulating materials that (1) are biocompatible, that is, nontoxic, (2) possess electrical properties for recording unit potentials, (3) are small in diameter to minimize tissue damage and increase proximity to neurons, and (4) contain several mechanical characteristics (e.g., stiffness, strength) to withstand sterilization and handling, particularly during insertion, to prevent buckling or breaking (Dymond et al., 1972). These are basic features of most electrodes designed for chronic recording applications, and even though the microelectrodes described here are used on average for two weeks, high levels of safety and reliability are required. Previous studies have evaluated metals and insulating materials for brain tissue toxicity and found metals such as gold, platinum (and Pt/Ir alloy), stainless steel, nickel–chromium, and tungsten and insulating materials of polyimide and Epoxylite are nontoxic. Metals such as silver, copper, and iron are toxic and associated with necrosis and chronic phagocytosis (Dymond et al., 1970; Babb & Kupfer, 1984). Each of these nontoxic metals recorded EEG of equivalent quality under different impedance, noise, and artifact conditions (Dymond, 1972). In addition, studies have found several of these metals, stainless steel, platinum, and nickel–chromium in particular, were safe in the magnetic resonance imaging (MRI) scanner (Zhang et al., 1993; Davis et al., 1999). Subchronic microelectrode studies will benefit from research that focuses on the design of chronic intracerebral electrodes that seek to minimize acute and long-term immune response, for example microgliosis, reactive astrocytosis and glial scar formation, and responses from neurons that are adjacent to the implanted electrode (Polikov et al., 2005; Winslow & Tresco, 2010).

As indicated in the preceding sections and illustrated in figure 5.1D, the microelectrodes fabricated at UCLA now were adapted from a design of flexible fine microwire bundles used in chronic animal studies and modified for subchronic applications in patients (Babb et al., 1973; Harper & McGinty, 1973). A typical microelectrode consists of a bundle of nine 40-μm microwires each 0.0015" diameter and composed of 80% platinum, 20% iridium alloy that is covered polyimide insulation (California Fine Wire[1]). The microwire is specified as stress relieved to unreel from a spool as straight as possible. For a single microelectrode, each microwire is cut to a length of 30 cm, and 2 mm of insulation is removed from one end using the flame from an alcohol burner. Insulation is also removed from the opposite end on one of the nine microwires that will be used as a local reference for the other eight recording microwires. Since the microwires are too fragile to extend at any great length to connect with recording amplifiers, a stronger ten-conductor flat ribbon cable is attached to the microelectrode.[2] To attach the ribbon cable to the microelectrode, 3 mm on the end of each ribbon cable wire one through nine is stripped and tinned and then made into a closed loop.[3] Note the tenth wire on the ribbon cable serves as the connection to common ground. The 2-mm flame-exposed end of each microwire is then soldered into a closed-loop wire on the ribbon cable using liquid flux and solid solder.[4,5] Each solder joint is carefully rinsed with ethanol to remove all flux residues before being covered with 10 mm of 1/16" diameter heat-shrink tubing.[6] The nine microwires are then carefully inserted down dual polyimide tubing. The inner polyimide tubing sleeve is cut to a length of 27 cm.[7] The outer polyimide tubing is cut to a length of 12 cm and acts as a sleeve to facilitate insertion into the clinical depth electrode by covering the microwire bundle.[8] The microwire-ribbon cable interface is covered with a 4-cm length 1/4" diameter heat-shrink tubing to contain potted urethane.[9,10] Potting with urethane epoxy consists of two steps. First, each microwire-ribbon cable connection is covered, and second, a custom socket Delrin interface is inserted and the exterior portion of Delrin is covered with epoxy. The Delrin socket mates with the clinical depth electrode.[11] An insulation-displacement connector is attached to the proximal end of the ribbon cable.[12] Microwires are then checked for continuity in a physiological saline bath using a low-current impedance meter.[13] Finally, each microelectrode is sterilized with ethylene oxide followed by an aeration period lasting at least seven days.

Cables and Shielding

Cost concerns often dictate that trade-offs must be made when setting up a recording system and environment to support recording from microelectrodes. Often the conductors that are used for carrying the signals from the electrodes to the head stage consist of off-the-shelf stranded plated copper. While this material is suitable to the task, there is a fair amount of inductance associated with this type of cabling and the magnitude of inductive pickup depends on the length and gauge of the wire. There are several commercial vendors that provide many low-inductance cables for connecting the electrodes to the head stage. From our experience, we have found that 23 American Wire Gauge (AWG) copper at lengths less than 30 cm sufficiently reduce noise produced by inductive pickup.

Electromagnetic fields in the hospital environment are difficult, if not impossible, to eliminate, but as mentioned already there are practical solutions to minimize noise. In addition, there are steps that can be taken to reduce the field intensities and that could further reduce noise in electrophysiology recordings. If a room is being designed, then the head wall should ideally have grounded Mu metal plating to provide a barrier to low-frequency electromagnetic fields such as 60-Hz line noise. If florescent lighting is utilized in the room and cannot be turned off, then ultrafine conductive metal mesh can be used to allow adequate lighting while attenuating

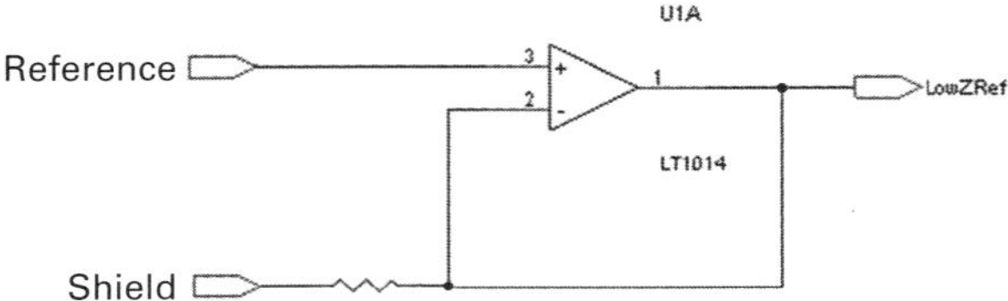

Figure 5.4
Schematic diagram of a buffer used on a reference electrode with the output of the buffer driving a shield. The resistor is used to limit the current drawn should the shield come in contact with ground. The reference electrode is relatively low impedance, and any line noise impressed on this signal leg should be expressed as a common mode signal in order to minimize noise in the output of the instrumentation amplifier.

the harmonically rich noise radiated by these lights. Moreover, ultrafine mesh can be used to make a loose-fitting helmet or drape that can be placed over the electrode–head stage connections that are atop the patient's head. This mesh can then be driven by the reference buffer output (see figure 5.4). At UCLA, we have used Flectron mesh (Monsanto) and found significantly lower levels of noise when this is placed over the electrode–head stage connection.

Grounds and Reference

The ground and reference scheme are two very important elements for high-fidelity microelectrode recordings. Careful electrode preparations, amplifier specifications, and noise source mitigation techniques can be rendered useless if system grounding and reference scheme are not sufficient. From the experience of the UCLA team, we have found that the ground impedance should be less than 500 Ω and the reference should be buffered (see figure 5.4). If there is more than one system connected at the same time (e.g., clinical EEG monitoring system), then the two systems must share an isolated patient ground and each system has a separate reference (ANSI/IEEE Standard 602–1986, 1986).

Buffering Stage

A critical component in any microelectrode recording system is the head stage or buffer. Neuron sources generating bioelectric signals are susceptible to loading effects from the amplifier that will reduce the amplitude of the signal (Geddes, 1972). In general, the buffer serves to transform the noise-susceptible high-impedance signal to a low-impedance signal prior to amplification (see figure 5.5). Buffers originally were designed using a vacuum tube configured as a cathode follower, but with respect to safety, the vacuum tube was not ideal. Failure of the vacuum tube such as a short circuit could generate large currents that could cause significant injury to the subject, although we were unable to find any studies that described such occurrences.

The introduction of miniaturized semiconductors with improved direct current (DC) characteristics produced signals with less noise and provides greater safety and subject comfort. Field

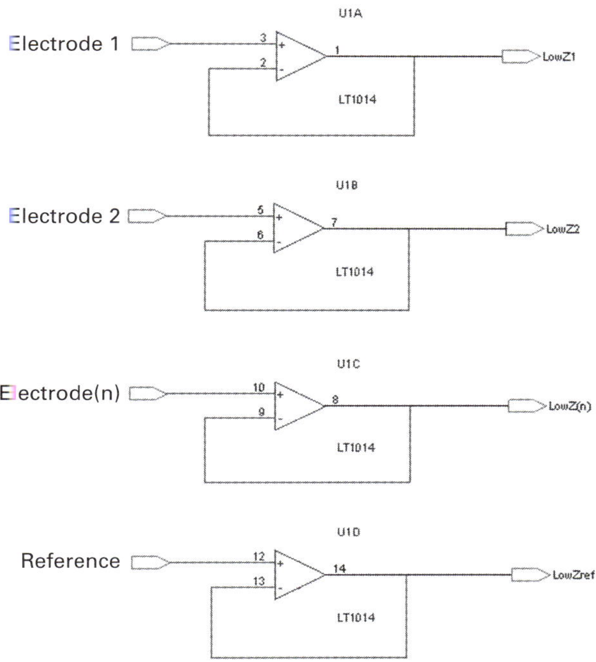

Figure 5.5
Schematic diagram of a standard unity gain voltage follower buffer configuration using a common operational amplifier (for example, LT1014). Note that the diagram only shows the general configuration of the voltage follower while a complete schematic would include a series current limiting resistor, compensation capacitor, and bleed impedance (resistor or back-to-back diodes from buffer input to ground that dissipates static charges) that should be included in any design. The part number for the operational amplifier is for illustration only, and custom designs should review part specifications for input noise, bias currents, and offsets.

effect transistors used as a voltage follower operated at lower voltages which increased subject safety in the event of a short circuit (Babb & Crandall, 1976). Technical advances in operational amplifier integrated circuits brought the ideal voltage follower closer to realization as far as input impedance, input bias current requirements, noise characteristics, and failure mode mitigation. The current state of the art has produced solutions that enable designers and therefore users of electrophysiology recording instruments that ensure fidelity of signals while addressing noise artifact and subject safety.

Since the cables that connect the microelectrodes to the amplifier can introduce movement-related noise, it is ideal to have the buffer positioned in close proximity to the microelectrode (Babb & Crandall, 1976). One advantage of some digital telemetry systems now used in animal studies is that buffers as well as amplification and digitization occur near the electrode, thus minimizing cable length and noise (Fenton et al., 2008; Davis et al., 2011). Generally, a buffer reduces signal distortion and provides safety by having the following characteristics:

- Input impedance equal to or greater than 100 MΩ
- Input bias current is at least < 10 pico amps
- The input to the buffer has a path to patient ground even when the input is open as this prevents charge build up on the input that can discharge through the microelectrode when the patient is connected
- The input to the buffer has a series current limiting resistor to prevent patient injury in the event that the buffer input shorts to either power rail
- The current limiting resistor has a compensation capacitor in parallel whose function is to mitigate cable inductance and increase recovery time
- The output impedance of the buffer should be less than 100 Ω, and a series resistor should be placed on the output to mitigate cable capacitance.

Amplification

Following head stage unity gain buffering is differential amplification (see figure 5.6). It should be noted here that due to the rapid development of converters and DSPs the amount of gain required for signal acquisition is determined by the type of ADC. In order to minimize or eliminate common mode signals such as power line noise, it is critical that the bioelectrical signal is amplified against a common buffered reference. Typically, the output of the buffer is AC coupled in order to remove any large DC offset voltages which, if amplified, would saturate the output of the amplifier.

Analog-to-Digital Conversion

Resolution

Most ADCs acquire at no more than 24 bits per sample and store data at 16 bits per sample. Two chief reasons are lower storage requirements and effective data resolution with respect to signal-to-noise ratio of the system. Thirty-two bits per sample and higher are conceivable although storage technology, in terms of data representation, would have to advance more rapidly to keep pace with data production rates as well as storage requirements.

Gain

System noise specifications are constantly improving; it is now possible to reliably acquire signals with amplitudes less than 10 μV, and new ADCs have increased the effective useful bit depths up to 20 bits per sample. Because of these improvements, amplification is not as critical to signal quality factor in overall system performance. However, it is important to have adequate differential amplification with a common mode rejection ratio (CMRR) of at least 100 dB, which would be the minimum requirement for clinical recordings or microelectrode recordings in a shielded environment, and microelectrode recordings in a clinical (unshielded) environment would benefit from higher CMRR values. The amount of gain/amplification as mentioned earlier

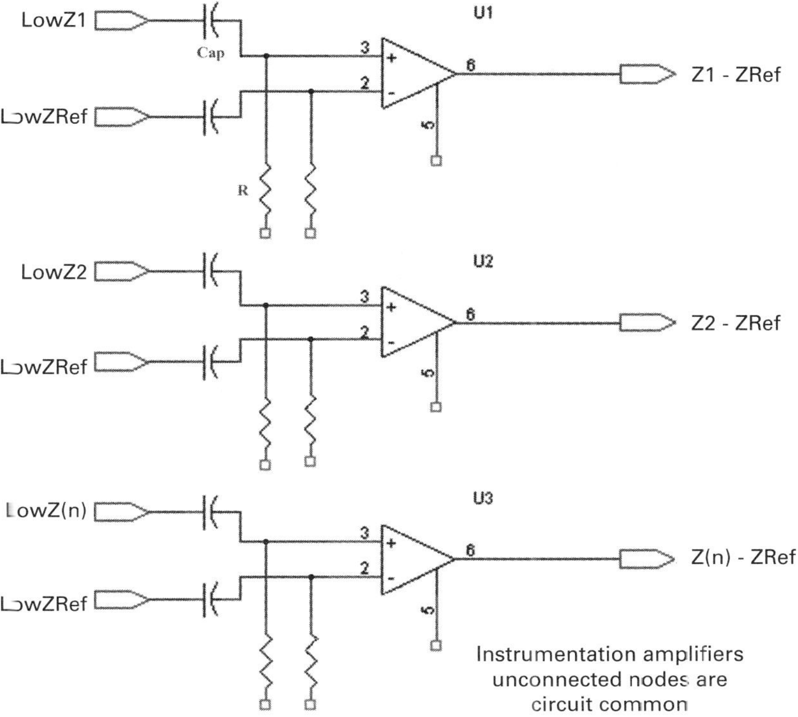

Figure 5.6
Schematic diagram of an instrumentation amplifier. Note that the resistors (R) and capacitors (Cap) on the input to the amplifier should be closely matched so that the common mode rejection ratio specifications of the selected instrumentation amplifier are not significantly degraded. The unconnected ends of the resistor and amplifier nodes are connected to common (patient) ground. The output impedance of an instrumentation amplifier is typically less than 600 Ω and should be taken into account when sampling the output since the charge time is affected by the output impedance of the amplifier and so the maximum sampling rate.

depends on the type of ADC employed in the signal path. Note that most ADCs provide a means of differential acquisition, and these devices have good CMRR specifications.

Sampling

Older systems typically sampled signals at rates of 10,000 samples per second, which was commonly used in single unit analysis. It was often difficult to perform post hoc cluster analysis on action potential waveforms that were sampled at low rates. Currently, ADC rates of 32,000 samples per second and higher are more common and have significantly improved single unit analysis and waveform classification to characterize different cell types (Viskontas et al., 2007; Le Van Quyen et al., 2008). In addition, higher order analysis such as phase analysis requires accurately sampled signals in order to reduce statistical errors. Fortunately, many recording systems support simultaneous channel sampling from subsets to individual microelectrodes that produce little to no sample jitter. One simple method to quantitatively evaluate sampling accuracy is to record a 1-kHz sinusoidal signal input to the recording system using equivalent sample rate and gain across all channels. Then select several distant channels, for example, channels 16, 48, 80, and 112 for a 128-channel system, and after subtracting a few minutes of the recorded signal between pairs of channels perform a fast Fourier transform (FFT). Peaks in FFT results will indicate the extent of spurious noise due to misalignment associated with sampling.

Digital Signal Processing

Signals of interest and the trade-offs that must be made in order to acquire them is a common dilemma that faces most researchers. Most useful neuronal information occurs between DC and 6 kHz, a frequency domain that encompasses ultra–slow changing charge gradients through very rapid unit discharge waveforms. Continuous wide bandwidth signals, however, increase data storage requirements and introduce issues such as choice of media and accessibility, duration of storage, data security, and cost. Digital signal processing modules on many recording systems support multiple sampling rates, signal conditioning, and real-time event detection algorithms (see figure 5.7). Custom configuration of DSPs can generate multiple data files that contain, for

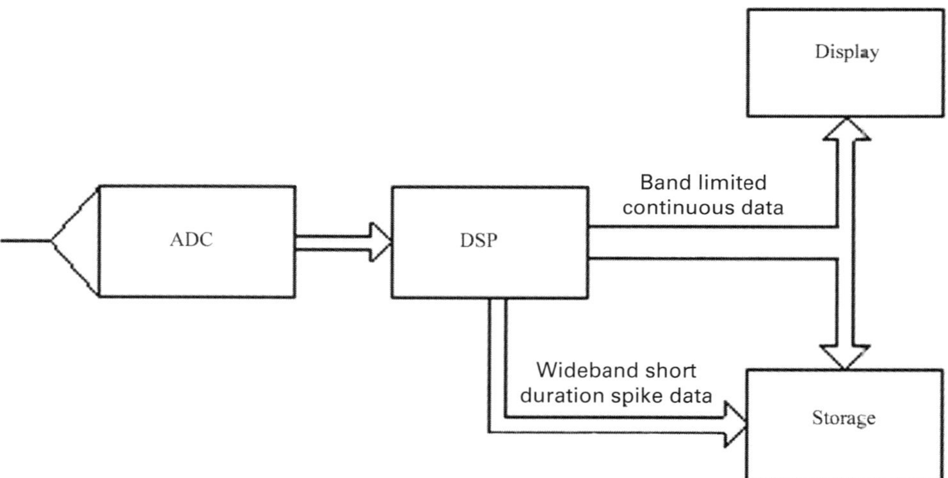

Figure 5.7
Block diagram of digital signal processing (DSP) that splits the digitized data into two alternate bandwidth signal paths in order to minimize storage requirement for multichannel continuous recordings. The DSP is optimized so that high-frequency signals, that is, neuronal action potentials, are stored as discrete events consisting of a time stamp that reflects the occurrence of maximum amplitude and action potential waveform. The lower frequency data is compressed using a standard compression algorithm such as run length limiting. The DSP also allows for selection of how the data are displayed, continuous, discrete, overlay, and so on. ADC, analog-to-digital converter.

example, continuous low-pass-filtered local field potentials and discrete high-pass-filtered (or wide bandwidth) unit waveforms and time stamps corresponding to their individual occurrence.

Summary

This chapter provides an abbreviated history of EEG in the clinical investigation of epilepsy. The clinical requirement for depth electrodes studies in some patients who are candidates for epilepsy surgery, along with development of electrodes and improvements in recording technology, played critical roles in extraoperative in vivo microelectrode recordings that are carried out today. The microelectrodes described here and technical aspects of recording derive from our experience at UCLA, which was greatly informed from prior animal studies and microelectrode work at UCLA and other programs. The microelectrode design and specific features of the recording system used in patient studies are strongly influenced by the type of clinical electrode, procedures required to implant the electrodes, recording environment, type of microelectrode study and signals of interest, and cost. Another defining factor and most important is safety. Patients who volunteer for research have the expectation that there is minimal risk and discomfort beyond that imposed by the clinical diagnostic evaluation. Therefore, materials used in the fabrication of microelectrodes must be nontoxic, minimize tissue trauma and foreign body response, and resist dissolution particularly if microwires will be used for stimulation. Since MRI is used regularly, microelectrodes cannot heat or induce currents during scanning and should not produce significant susceptibility artifact on MRI. Furthermore, microelectrodes must be robust to withstand sterilization, handling, and insertion and perform reliability for the duration of recording. Finally, recording system head stages and cables should be sturdy, yet lightweight, so that their use does not cause subject fatigue or restrict mobility. Most of these requirements are the focus of electrode research and digital telemetry systems intended for chronic recording in neural prostheses, and implementation of new microelectrode and recording technology will facilitate clinical and basic neurophysiological studies in human health and disease.

Notes

1   Microwire 0.0015"-diameter Platinum 20% Iridium Stress relieved H-ML enamel insulations CFW-400 spool from California Fine Wire (www.calfinewire.com).

2   Flat cable 3M—1700/10 (100')—Twist-to-Flat Cable, 10 COND, 100 ft, 28 AWG, Newark part number 87F4351 (www.newark.com).

3   Solder (RA Core Flux) Kester Solder "44" RA Core—24–6337–0027—Solder Wire, Sn63Pb37, 0.031" diameter, 133°C, 1 lb (www.newark.com).

4   Solder Flux Kester Solder Flux Formula 817.

5   Solder (Solid) Kester Solid Solder—14–6040–0031—Solder Wire, Sn60Pb40, 0.031" diameter, 183°C, 1 lb.

6   Heat Shrink (individual wire) Cole Flex STS-221–1/16 black HS Polyclefin.

7   Polyimide Tubing (inner) Polyimide Tubing Code 110-I × 32". Color: Natural (www.microlumen.com).

8   Polyimide Tubing (outer) Polyimide Tubing Code 140-I × 30". Color: Natural (www.microlumen.com).

9   Heat Shrink (bundle) Cole Flex STS-221–1/4 white HS Polyolefin.

10. Hardman urethane D-50 (#04022).

11. Ad-Tech Medical Instrument Corporation (www.adtechmedical.com/behnke-fried-depth-electrodes/).

12. Connector 3M—3473–6600—WIRE-BOARD CONN, SOCKET, 10POS, 2.54 mm, part number 46F4655 (www.newark.com).

13. Bak Electronics, IMP-1 (www.bakelectronicsinc.com).

References

Adey, W. R., French, J. D., Kado, R. T., Lindsley, D. F., Walter, D. O., Wendt, R., et al. (1961). EEG records from cortical and deep brain structures during centrifugal and vibrational accelerations in cats and monkeys. *IRE Transactions on Bio-medical Electronics, 8*, 182–188.

ANSI/IEEE 602–1986. (1986). Electrical safety and grounding. In *Recommended practice for electric systems in health care facilities* (p. 426). New York: Institute of Electrical and Electronics Engineers.

Babb, T. L., Carr, E., & Crandall, P. H. (1973). Analysis of extraceullular firing patterns of deep temporal lobe structures in man. *Electroencephalography and Clinical Neurophysiology, 24*, 247–257.

Babb, T. L., & Crandall, P. H. (1976). Epileptogenesis of human limbic neurons in psychomotor epileptics. *Electroencephalography and Clinical Neurophysiology, 40*, 225–243.

Babb, T. L., & Kupfer, W. R. (1984). Phagocytic and metabolic reactions to intracerebral electrical stimulation of rat brain. *Experimental Neurology, 86*, 183–197.

Babb, T. L., Wilson, C. L., & Isokawa-Akesson, M. (1987). Firing patterns of human limbic neurons during stereoencephalography (SEEG) and clinical temporal lobe seizures. *Electroencephalography and Clinical Neurophysiology, 66*, 467–482.

Bailey, P., & Gibbs, F. A. (1951). The surgical treatment of psychomotor epilepsy. *Journal of the American Medical Association, 145*, 365–370.

Berger, H. (1929). Uber das Electrenkephalogram des Menschen. *Archiv für Psychiatrie und Nervenkrankheiten, 87*, 527.

Berger, H. (1933). Uber das elektrenkephalogramm des menschen. *Archiv für Psychiatrie und Nervenkrankheiten, 100*, 301–320.

Bragin, A., Wilson, C. L., Staba, R. J., Reddick, M., Fried, I., & Engel, J., Jr. (2002). Interictal high-frequency oscillations (80–500 Hz) in the human epileptic brain: Entorhinal cortex. *Annals of Neurology, 52*, 407–415.

Brazier, M. A. B. (1961). *A history of the electrical activity of the brain: The first half-century.* London: Pitman.

Cameron, K. A., Yashar, S., Wilson, C. L., & Fried, I. (2001). Human hippocampal neurons predict how well word pairs will be remembered. *Neuron, 30*, 289–298.

Campbell, P. K., Jones, K. E., Huber, R. J., Horch, K. W., & Normann, R. A. (1991). A silicon-based, three-dimensional neural interface: Manufacturing processes for an intracortical electrode array. *IEEE Transactions on Bio-Medical Engineering, 38*, 758–768.

Caton, R. (1875). The electrical currents of the brain. *British Medical Journal, 2*, 278.

Crandall, P. H., & Babb, T. L. (1993). The UCLA Epilepsy Program: Historical review 1960–1992. *Journal of Clinical Neurophysiology, 10*, 226–238.

Crandall, P. H., Walter, R. D., & Rand, R. W. (1963). Clinical applications of studies on stereotactically implanted electrodes in temporal lobe epilepsy. *Journal of Neurosurgery, 20*, 827–840.

Cybulski, N., & Jelenska-Maciezyna, X. (1914). Action currents of the cerebral cortex [in Polish]. *Bulletin of the Academy of Science Krakov, Series B*, 776–791.

Davis, K. A., Sturges, B. K., Vite, C. H., Ruedebusch, V., Worrell, G., Gardner, A. B., et al. (2011). A novel implanted device to wirelessly record and analyze continuous intracranial canine EEG. *Epilepsy Research, 96*, 116–122.

Davis, L. M., Spencer, D. D., Spencer, S. S., & Bronen, R. A. (1999). MR imaging of implanted depth and subdural electrodes: Is it safe? *Epilepsy Research, 35*, 95–98.

Dymond, A. M. (1972). Comparison of metals for recording the EEG. *Medical & Biological Engineering, 10*, 393–401.

Dymond, A. M., Babb, T. L., Kaechele, L. E., & Crandall, P. H. (1972). Design considerations for the use of fine and ultrafine depth brain electrodes in man. *Biomedical Sciences Instrumentation, 9*, 1–5.

Dymond, A. M., Kaechele, L. E., Jurist, J. M., & Crandall, P. H. (1970). Brain tissue reaction to some chronically implanted metals. *Journal of Neurosurgery, 33*, 574–580.

Fenton, A. A., Jeffery, K. J., & Donnett, J. G. (2008). Neural recording using digital telemetry. In R. P. Vertes & R. W. Stackman (Eds.), *Electrophysiological recording techniques* (pp. 77–102). New York: Springer.

Forbes, A., & Thacher, C. (1920). Amplification of action currents with the electron tube in recording with the string galvanometer. *American Journal of Physiology, 52*, 409–471.

Fried, I., Wilson, C. L., Maidment, N. T., Engel, J., Jr., Behnke, E., Fields, T. A., et al. (1999). Cerebral microdialysis combined with single-neuron and electroencephalographic recording in neurosurgical patients. *Technical Note. Journal of Neurosurgery, 91*, 697–705.

Geddes, L. A. (1972). *Electrodes and the measurement of bioelectric events*. New York: Wiley.

Gibbs, F. A., Davis, H., & Lennox, W. G. (1935). The electroencephalogram in epilepsy and in conditions of impaired conciousness. *Archives of Neurology and Psychiatry, 34*, 1133–1148.

Gibbs, F. A., Lennox, W. G., & Gibbs, E. L. (1936). The electro-encephalogram in diagnosis and in localization of epileptic seizures. *Archives of Neurology and Psychiatry, 36*, 1225–1235.

Grass, A. M. (1984). The electroencephalographic heritage until 1960. *American Journal of EEG Technology, 24*, 133–173.

Halgren, E., Babb, T. L., & Crandall, P. H. (1978). Activity of human hippocampal formation and amygdala neurons during memory testing. *Electroencephalography and Clinical Neurophysiology, 45*, 585–601.

Harper, R. M., & McGinty, D. J. (1973). A technique for recording single neurons from unrestrained animals. In M. I. Phillips (Ed.), *Brain unit activity during behavior* (pp. 80–104). Springfield, IL: Charles C Thomas.

Hochberg, L. R., Serruya, M. D., Friehs, G. M., Mukand, J. A., Saleh, M., Caplan, A. H., et al. (2006). Neuronal ensemble control of prosthetic devices by a human with tetraplegia. *Nature, 442*, 164–171.

House, P. A., MacDonald, J. D., Tresco, P. A., & Normann, R. A. (2006). Acute microelectrode array implantation into human neocortex: Preliminary technique and histological considerations. *Neurosurgical Focus, 20*, E4.

Howard, M. A., III, Volkov, I. O., Abbas, P. J., Damasio, H., Ollendieck, M. C., & Granner, M. A. (1996). A chronic microelectrode investigation of the tonotopic organization of human auditory cortex. *Brain Research, 724*, 260–264.

Hubel, D. H. (1957). Tungsten microelectrode for recording from single units. *Science, 125*, 549–550.

Jacobs, B. L., Harper, R. M., & McGinty, D. J. (1970). Neuronal coding of motivational level during sleep. *Physiology & Behavior, 5*, 1139–1143.

Jasper, H. H. (1941). Electroencephalography. In W. Penfield & T. C. Erickson (Eds.), *Epilepsy and cerebral localization* (pp. 380–454). Springfield, IL: Charles C Thomas.

Jones, K. E., Campbell, P. K., & Normann, R. A. (1992). A glass/silicon composite intracortical electrode array. *Annals of Biomedical Engineering, 20*, 423–437.

Kaufman, P. Y. (1912). Electrical phenomena in the cerebral cortex (in Russian). *Oborz Psikhiat Nevrol eksper Psikol, 7–8*, 403–424, 513–535.

Le Van Quyen, M., Bragin, A., Staba, R., Crepon, B., Wilson, C. L., & Engel, J., Jr. (2008). Cell type-specific firing during ripple oscillations in the hippocampal formation of humans. *Journal of Neuroscience, 28*, 6104–6110.

Matthews, B. H. C. (1934). A special purpose amplifier. *Journal of Physiology, 81*, 28P–29P.

Meyers, R., & Hayne, R. (1948). Electrical potentials of the corpus striatum and cortex in parkinsonism and hemiballismus. *Transactions of the American Neurological Association, 73*, 10–14.

Meyers, R., Hayne, R., & Knott, J. (1949). Electrical activity of the neonstriatum, paleostriatum and neighbouring structures in parkinsonism and hemiballismus. *Journal of Neurology, Neurosurgery, and Psychiatry, 12*, 111–123.

Niedermeyer, E. (1993). Historical aspects. In E. Niedermeyer & F. Lopes da Silva (Eds.), *Electroencephalography: Basic principles, clinical applications, and related fields* (pp. 1–14). Baltimore: Williams & Wilkins.

O'Keefe, J., & Bouma, H. (1969). Complex sensory properties of certain amygadala units in the freely moving cat. *Experimental Neurology, 23*, 384–398.

Penfield, W. (1939). The epilepsies: With a note on radical therapy. *New England Journal of Medicine, 221*, 209–218.

Polikov, V. S., Tresco, P. A., & Reichert, W. M. (2005). Response of brain tissue to chronically implanted neural electrodes. *Journal of Neuroscience Methods, 148*, 1–18.

Ravagnati, L., Halgren, E., Babb, T. L., & Crandall, P. H. (1979). Activity of human hippocampal formation and amygdala neurons during sleep. *Sleep*, *2*, 161–173.

Renshaw, B., Forbes, A., & Morison, B. R. (1940). Activity of isocortex and hippocampus: Electrical studies with microelectrodes. *Journal of Neurophysiology*, *3*, 74–105.

Schevon, C. A., Trevelyan, A. J., Schroeder, C. E., Goodman, R. R., McKhann, G., Jr., & Emerson, R. G. (2009). Spatial characterization of interictal high frequency oscillations in epileptic neocortex. *Brain*, *132*, 3047–3059.

Schmidt, R. P., Thomas, L. B., & Ward, A. A., Jr. (1959). The hyper-excitable neurone: Microelectrode studies of chronic epileptic foci in monkey. *Journal of Neurophysiology*, *22*, 285–296.

Staba, R. J., Wilson, C. L., Bragin, A., Fried, I., & Engel, J., Jr. (2002a). Sleep states differentiate single neuron activity recorded from human epileptic hippocampus, entorhinal cortex, and subiculum. *Journal of Neuroscience*, *22*, 5694–5704.

Staba, R. J., Wilson, C. L., Fried, I., & Engel, J., Jr. (2002b). Single neuron burst firing in the human hippocampus during sleep. *Hippocampus*, *12*, 724–734.

Stead, M., Bower, M., Brinkmann, B. H., Lee, K., Marsh, W. R., Meyer, F. B., et al. (2010). Microseizures and the spatiotemporal scales of human partial epilepsy. *Brain*, *133*, 2789–2797.

Strumwasser, F. (1958). Long-term recording from single neurons in brain of unrestrained mammals. *Science*, *127*, 469–470.

Talairach, J., David, M., & Tournoux, P. (1958). *L'exploration chirurgicale stereotaxique du lobe temporal dans l'epilepsie temporale*. Paris: Mason.

Truccolo, W., Friehs, G. M., Donoghue, J. P., & Hochberg, L. R. (2008). Primary motor cortex tuning to intended movement kinematics in humans with tetraplegia. *Journal of Neuroscience*, *28*, 1163–1178.

Ulbert, I., Halgren, E., Heit, G., & Karmos, G. (2001). Multiple microelectrode-recording system for human intracortical applications. *Journal of Neuroscience Methods*, *106*, 69–79.

Ulbert, I., Heit, G., Madsen, J., Karmos, G., & Halgren, E. (2004). Laminar analysis of human neocortical interictal spike generation and propagation: Current source density and multiunit analysis in vivo. *Epilepsia*, *45*(Suppl 4), 48–56.

Van Gompel, J. J., Stead, S. M., Giannini, C., Meyer, F. B., Marsh, W. R., Fountain, T., et al. (2008). Phase I trial: Safety and feasibility of intracranial electroencephalography using hybrid subdural electrodes containing macro- and microelectrode arrays. *Journal of Neurosurgery*, *25*, E23–E29.

Verzeano, M., Crandall, P. H., & Dymond, A. (1971). Neuronal activity of the amygdala in patients with psychomotor epilepsy. *Neuropsychologia*, *9*, 331–344.

Viskontas, I. V., Ekstrom, A. D., Wilson, C. L., & Fried, I. (2007). Characterizing interneuron and pyramidal cells in the human medial temporal lobe in vivo using extracellular recordings. *Hippocampus*, *17*, 49–57.

Ward, A. A., & Thomas, L. B. (1955). The electrical activity of single units in the cerebral cortex of man. *Electroencephalography and Clinical Neurophysiology*, *7*, 135–136.

Ward, A. A., Jr., Thomas, L. B., & Schmidt, R. P. (1956). Some properties of single epileptic neurones. *Transactions of the American Neurological Association* (81st Meeting), 41–43.

Wilson, C. L., Babb, T. L., Halgren, E., & Crandall, P. H. (1983). Visual receptive fields and response properties of neurons in human temporal lobe and visual pathways. *Brain*, *106*, 473–502.

Wilson, C. L., Babb, T. L., Halgren, E., Wang, M. L., & Crandall, P. H. (1984). Habituation of human limbic neuronal response to sensory stimulation. *Experimental Neurology*, *84*, 74–97.

Wilson, C. L., Isokawa, M., Babb, T. L., & Crandall, P. H. (1990). Functional connections in the human temporal lobe. *Experimental Brain Research*, *82*, 279–292.

Wilson, C. L., Maidment, N. T., Shomer, M. H., Behnke, E. J., Ackerson, L., Fried, I., et al. (1996). Comparison of seizure related amino acid release in human epileptic hippocampus versus a chronic kainate rat model of hippocampal epilepsy. *Epilepsy Research*, *26*, 245–254.

Winslow, B. D., & Tresco, P. A. (2010). Quantitative analysis of the tissue response to chronically implanted microwire electrodes in rat cortex. *Biomaterials*, *31*, 1558–1567.

Zhang, J., Wilson, C. L., Levesque, M. F., Behnke, E. J., & Lufkin, R. B. (1993). Temperature changes in nickel-chromium intracranial depth electrodes during MR scanning. *AJNR. American Journal of Neuroradiology*, *14*, 497–500.

6 Data Analysis Techniques for Human Microwire Recordings: Spike Detection and Sorting, Decoding, Relation between Neurons and Local Field Potentials

Ueli Rutishauser, Moran Cerf, and Gabriel Kreiman

Data Acquisition and Processing

This chapter gives a technical perspective on the procedures for an experiment to proceed from the acquisition of continuously sampled data from microwires all the way to points of time a putative single neuron fired a spike, to local field potential (LFP) recordings, and to decoding neurophysiological signals in single trials and in real time.

In order to analyze extracellular waveforms, it is important to acquire data with high sampling rates. Sampling rates below 16 kHz can miss important aspects of the submillisecond structure of action potential waveforms. Current systems typically have sampling rates above 30 kHz. Another methodological consideration involves the use of 60 Hz notch filters. While purists will advocate examining raw data without any filters, the clinical environment often presents significant challenges and electrical artifact contaminations for neurophysiological recordings. Removing 60 Hz and harmonics with a notch filter can significantly enhance the signal-to-noise ratio (SNR) to discriminate action potentials. The focus in this chapter starts at the point of time the data has been stored by the acquisition system whereas chapter 5 in this volume describes the acquisition system and electrodes themselves.

Microwires record the extracellular voltage at a particular point in space. This signal, a voltage as a function of time, is the linear superposition of a great number of current sources generated by electrically active components of the brain (Buzsáki et al., 2012), including synaptic events and action potentials propagating down an axon or backpropagating inside dendrites. In general, it is extremely difficult to decompose the extracellular signal into the single events that give rise to it. An important aspect of the extracellular signals that can sometimes lead to a reasonable interpretation is the occurrence of action potentials in the near vicinity of the microwire tip. In the rat hippocampus CA1 region, it has been estimated that electrodes can distinguish extracellular spikes from neuronal processes located as far away as 140 μm from the tip of the electrode (Buzsáki, 2004). The peak amplitude depends on the physical size of the neurons under study, making it most likely that a majority of extracellular recordings involve pyramidal cells (Henze et al., 2000) (see also brief discussion of extracellular waveforms in chapter 8). To our knowledge, in human recordings, there are no quantitative estimates of the relationships between extracellular waveform amplitudes and shapes and neuronal types or distances to neighboring neurons. However, since the peak amplitude of extracellular spikes decreases rapidly as a function of distance and cell sizes are roughly comparable in rodents and humans, it is reasonable to assume that the basic properties of extracellular recordings summarized in Buzsáki et al. (2012) are also applicable in human neurophysiology. Thus, if action potential sources occur sufficiently close to the microwire and the sources are sparse in space and time, the extracellular signal can distinguish the shapes of the single waveforms sufficiently well to allow a clustering process that groups waveforms of sufficient similarity into clusters that originate from *putative* single neurons. Animal studies with moveable electrodes (such as implanted microdrives) allow experimenters to move the electrode in small steps to optimize the position till the waveforms show the desired properties (waveform, amplitude, number of clusters). This procedure is often followed in intraoperative recordings in humans, particularly during the implantation of deep brain stimulation devices in humans (see chapters 15–16). In contrast, during semi-chronic recordings in epileptic patients, microwires are implanted under anatomical guidance without simultaneous recordings and without moving the microwires to optimize the recordings.

For the reasons outlined above, the variety of extracellular waveforms encountered is large. In practice the question frequently arises as to whether a particular waveform could possibly be neuronal or rather some sort of artifact. To gain intuition into what kinds of waveforms can be obtained from extracellular recordings, it is instructive to consult computational studies simulating voltage in the extracellular milieu of reconstructed neurons from which both intra- and extracellular potentials were recorded simultaneously (Harris et al., 2000; Gold et al., 2006). These studies show that the extracellular waveform originating from a single neuron varies greatly as a function of the relative position of the electrode with respect to the neuron. Apart from the amplitude, other features that change systematically include the width, the number of peaks, as well as the sign (positive or negative) of the waveform (see figure 2 in Gold et al., 2006). The extracellular waveform results because of three different currents that flow in and out of cells: Na+ inflow, K+ outflow, and capacitive current. The different components are visible to different extents at different locations in the cell, which explains the great variability of waveforms. For example, positive-going waveforms can result from the capacitive current in distal dendrites. Usually, the simultaneously occurring Na+ current masks the capacitive current, but in distal dendrites the Na+ current is delayed. This results in bipolar or reverse-polarity waveforms. Situations where waveforms of different cells appear with different polarity on the same wire are thus possible. In our experience, this happens routinely in human recordings from the human medial temporal lobe. Another aspect that influences the shape of extracellular waveforms is the incidence of bursting. Subsequent spikes within a bust typically show distinct waveform properties. The discussion so far has focused on the extracellular waveform obtained from considering a single neuron. In actual recordings, the microwires capture the activity of multiple neurons and neighboring neurons can also affect the shape of the extracellular waveform. In particular, simultaneous (or nearly simultaneous) action potentials from nearby neurons can lead to action potential waveforms that do not resemble each individual waveform (but could perhaps be modeled as a linear superposition of such individual waveforms).

In addition to spike sorting to separate different units contributing to the extracellular recording, the shape of the extracellular waveform can be used to infer tentative information about the morphology and type of the neuron recorded from. For example, the width of the waveform can be used to distinguish between inhibitory and excitatory neurons, which has been done also in humans (Viskontas et al., 2007) (see also chapter 8). Also, there are significant correlations between the peak amplitude, spike width, and electrode distance that can be used to infer morphological features as well as the approximate electrode location (Gold et al., 2006). However, for such inferences to be accurate, the waveform has to be preserved as authentically as possible. This requires using software and hardware filters which do not distort the waveform. For example, filters that are used to discard the low-frequency components of the extracellular signal can greatly distort waveforms (Quiroga, 2009; Wiltschko et al., 2008). The distortions are caused by filters that have a phase lag that is a function of frequency, such as causal Butterworth filters (Quiroga, 2009). As a precaution, little or no filtering (if feasible) should be done in hardware, and all software filters should have zero-phase lag (which are noncausal). Also, real-time spike detection and sorting by necessity are based on causal filtering, which means that the waveforms produced by such methods are greatly distorted. It is thus advisable to post hoc redo all spike detection and sorting offline from the broadband signal. Another source of potential artificial waveforms is band-pass-filtered artifacts. Band-pass filtering almost any high-frequency signal (such as a static discharge) will result in a waveform which looks approximately like a spike. Therefore, great care has to be taken to distinguish between artifacts and neuronal spikes. One such approach is to use independent metrics such as statistics based on the distribution of inter-spike intervals (ISIs).

Spike Detection and Sorting

The first steps after continuously sampled raw signals have been acquired (see figure 6.1A, plate 1) are to (1) detect the spikes and (2) identify which putative neuron the spikes originated from ('spike sorting"). Inferring the number of putative neurons contributing to a collection of waveforms is an ill-posed inverse problem with no unique or "best" solution due to the sparseness of the available data. Recording the same spike simultaneously from multiple spatial locations greatly increases the ability to distinguish different neurons with similar waveforms. This can be achieved with tetrodes or silicon probes, but these techniques are not yet widely available for human recordings, and we thus focus on single wire recordings.

Various algorithms and procedures have been developed for manual, semi-automatic, or fully automatic spike sorting, and a number of software packages are available either as open source or commercially. These approaches have been reviewed and compared in the literature (Lewicki,

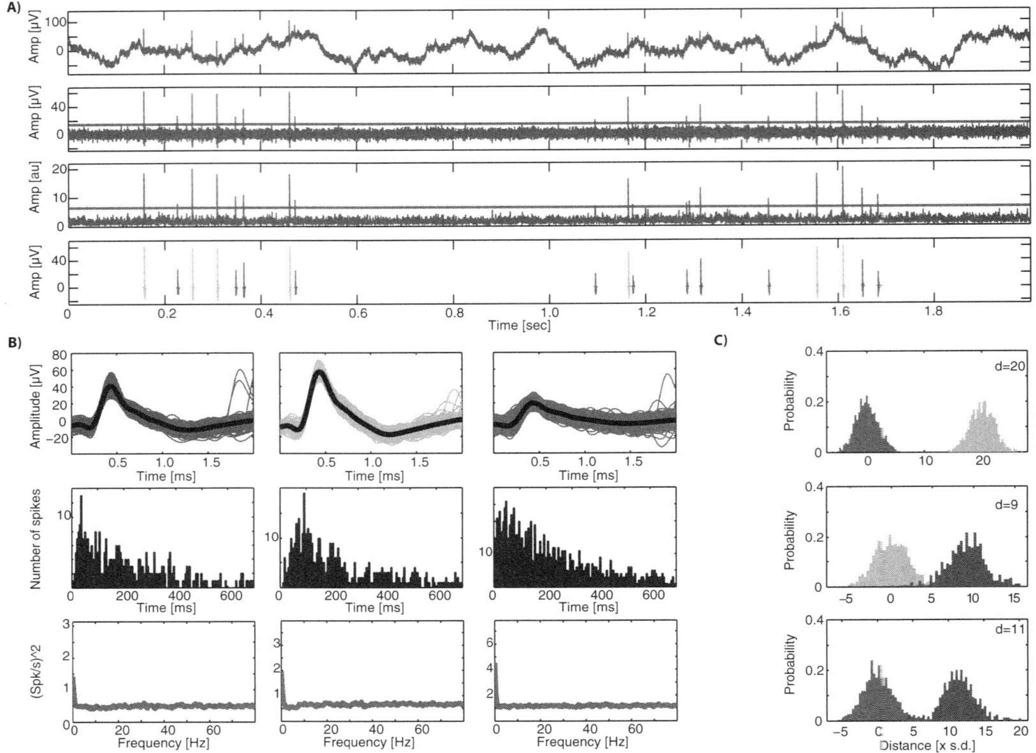

Figure 6.1 (plate 1)
Spike detection and sorting example. Shown are recordings obtained from a single wire implanted in the right anterior cingulate cortex of a human epileptic patient. The detection and sorting shown was done automatically using OSort. (A) From top to bottom: raw data (2 Hz high pass), band-pass filtered (300 Hz–3 kHz; red line is 4 times the estimated standard deviation; see text), energy-signal used for spike detection (line shows 5 × SD as used for spike detection in this example), and detected and sorted spikes (color indicates cluster identity, as computed by OSort). (B) Metrics of three of the identified clusters: raw waveforms (top), interspike interval (ISI; middle), and power spectrum of the spike train (bottom). All clusters were well separated, with the percentage of ISIs < 3 ms equal to 0%, 0%, and 0.19% respectively. (C) Pairwise projection test for all possible combinations shows that the clusters are well separated. The distance d is indicated (see text). Amp, amplitude; Nr, number; Spk/s, spikes per second.

1998; Pouzat et al., 2002; Quiroga et al., 2004; Rutishauser et al., 2006; Gibson et al., 2012). Procedures developed for animal models are generally applicable also for humans, but there are several aspects unique to human recordings that deserve discussion.

Spike Detection

The first step in the process of spike sorting is the identification of individual spikes and their waveforms (see figure 6.1A, plate 1). There is a critical dependency between sorting and detection: Better detection makes sorting easier. Detection involves several steps: (1) determining the point of time a spike occurred, (2) determining the waveform of this spike, and (3) realigning this waveform to a common reference frame that makes it comparable to all other collected waveforms. Here, we will highlight a few types of detection approaches that have been applied to human recordings.

Amplitude thresholding methods are the simplest: A spike is detected if the band-pass filtered raw signals (e.g., >300 Hz) crosses a predefined threshold such as a multiple of the standard deviation of the underlying signal. A threshold that is frequently used is

$$T = n \; median\left(\frac{|x|}{0.6748}\right),$$

where n is a constant (typically $n = 4$) and the second term is an estimate of the standard deviation of the noise in the voltage x (Donoho & Johnstone, 1994; Quiroga et al., 2004). An example of this threshold is shown in figure 6.1A (plate 1). Note that this method requires the user to choose a direction of spiking (positive, negative, or both). This choice is often performed manually on a channel-by-channel basis. If spikes are prominent in both directions, special care has to be taken to ensure the same spike is not detected twice as many waveforms have approximately equal amplitude in both directions.

More sophisticated spike detection methods rely on a derivate signal, which is then thresholded in a similar manner to amplitude thresholding. Such derivative signals have the advantage that their SNR can be higher due to selective amplification, a standard approach in signal processing (Kay, 1993). An example is an energy-type model that estimates the likelihood of a spike's being present as a function of time (Bankman et al., 1993; Kim and Kim, 2003). A method frequently used in human recordings (Rutishauser et al., 2006, 2008; Rutishauser et al., 2010) as well animal recordings (Csicsvari et al., 1998) is the following: Convolve the band-pass-filtered raw signal with a rectangular kernel that has the approximate width of a spike, that is, 1 ms (but note that for some recordings, this value will have to be adjusted accordingly depending on the neuron type, such as when recording dopaminergic neurons as in Zaghloul et al., 2009). The rectangular kernel is a matched filter that will amplify signals of appropriate width and suppress spike-like waveforms of widths that are smaller or larger. This can be computed very efficiently using a convolution kernel (see Appendix A.1. in Rutishauser et al., 2006, for implementation details). The resulting signal is strictly positive and can be thresholded at a multiple of its own standard deviation (typically in the range of 3.5–6 × SD). An example of this method is shown in figure 6.1A (plate 1) with a threshold of $T = 5$. There are other, more sophisticated, methods that compute a multiscale version of the energy signal (Choi et al., 2006) which are promising but have, to our knowledge, not been evaluated rigorously for use with human recordings.

A third class of detection methods that has been used is wavelet based. This is typically computationally more expensive but can lead to improved detection performance particularly in low SNR recordings (Nenadic & Burdick, 2005). There are well-studied wavelets such as the *bior* family, which resemble the waveforms of spikes and thus yield better detection compared to the simple rectangular kernel used by energy-signal methods.

Choosing an appropriate spike-detection method is based on carefully weighting several trade-offs, such as detection performance, computational cost and complexity, implementation complexity, real time versus offline performance, and for some applications power requirements. Investing more in detection makes sorting easier whereas investing more in sorting allows simpler detection methods. Our experience indicates that it pays to understand the behavior of the particular method used in detail by building intuition using simulated data or data recorded simultaneously intra- and extracellularly (Obeid, 2007). Only such data sets allow the rigorous determination of false-positive and true-positive rates as well as sweeps of thresholds to build detection receiver operating characteristic (ROC) curves (see figure 8 in Nenadic & Burdick, 2005, for an example). Performing an ROC-based performance evaluation using data sets simulated to resemble SNR ranges and waveform shapes in a particular data set can greatly improve the understanding of the parameters and trade-offs of a particular method.

A crucial step that follows detection is spike alignment. This can be an error-prone step, which can lead to substantial sorting quality problems or spurious clusters. Most sorting methods are based on distance metrics that assume that a common point of the waveform is at a fixed location in the matrix that holds all the detected waveforms. Common alignment points are the peak, trough, half-max amplitude, or the point of maximal slope. However, consistently identifying this point in waveforms is difficult. Note that the alignment point will typically be different from the point of threshold crossing, which shows considerable variability. Many waveforms are biphasic and have approximately equal amplitudes in both directions (on average). In such situations, using the location with the maximal absolute amplitude will lead to the creation of two clusters. This is because, due to variability, for some spikes the peak will be maximal whereas for others the trough will show a larger amplitude. A common solution to this problem is twofold: Either the peak or trough is always used, or a preference is enforced in which always the first significant (with respect to background noise) peak or trough is used as the alignment point (Rutishauser et al., 2006). The first case works well if spikes are dominant in one direction, which is often case. However, this requires a manual choice of alignment for each channel (usually by visual inspection). This is because in the bipolar recordings often used for chronically implanted microwires in humans, the direction of the spikes is arbitrary, depending on where the reference wire is located. An additional difficulty to consider is that the location of the peak is very sensitive to the sampling rate. Since the time spent at the peak is minimal, it is unlikely that the recording sampled the exact peak location. The probability of sampling the true peak increases with the sampling rate. For example, the peak location of a typical spike has less than 0.2 ms uncertainty, which for a sampling rate of 30 kHz is represented by eight data points. The accuracy of peak finding can be increased by upsampling the signal to a higher sampling rate (such as 100 kHz) before peak finding.

Spike Sorting

The goal of spike sorting is to assign each detected waveform to a putative single neuron that generated this spike (see figure 6.1A, plate 1, bottom). The number of possible unique neurons that could be present in a recording is unknown and also has to be estimated—that is, spike sorting is an unsupervised clustering problem where the number of the clusters is unknown. There are many unsupervised clustering approaches, and a number of such methods have been applied to spike sorting. These methods are reviewed extensively elsewhere (Lewicki, 1998; Pouzat et al., 2002; Quiroga et al., 2004; Rutishauser et al., 2006; Gibson et al., 2012). Rather than describe specific algorithms, we will review different spike-sorting approaches and discuss general issues that we found of relevance in our human recordings work.

Human recordings from semi-chronically implanted microwires are unique in that the microwires are not movable. Also, under most circumstances, no recordings take place during implantation, so microwire location cannot be optimized with respect to yield and signal quality. In contrast, electrode position can usually be optimized specifically for unit yield in animal models. In addition, experiment time with awake behaving humans is very limited. Thus, only a limited number of spikes is available for any given neuron, making spike sorting more difficult. Additionally, neurons in the brain areas typically covered by implanted electrodes tend to have very small baseline firing rates and sparse response properties. This further increases the demands on spike sorting as, in extreme cases, neurons might only respond to a few trials out of a long experiment (e.g., chapter 8). Not only will such neurons yield only a few spikes, but it is not possible to predict a priori what aspects of the task will elicit activity in those neurons. Detailed study of such neurons will thus require an adaptive experiment that shows stimuli which are chosen such that they are relevant for the neurons that are currently being recorded. This presents the challenge of requiring rapid spike sorting. Some experiments may also call for online sorting in order to achieve real-time decoding capacity (see this chapter and also chapter 17). Necessarily, such quick spike sorting, whether on- or offline, has to be semi- or fully automated. One criterion to consider is thus speed of sorting and possibilities to automate the process. Different approaches exist, starting with fully manual "cluster cutting" or window discriminators, semi-automatic cluster cutting where manual user interaction serves to refine automatic clustering, and automatic sorting either online or offline. Manual postprocessing is required even with so-called "fully automatic" spike-sorting approaches. This includes deciding between types of clusters, such as those that likely represent single units, those that are multiunits, and others that are noise. Clusters that represent noise should not be discarded but rather identified as such because these clusters will attract the noise spikes and prevent them from becoming intermingled with other clusters. Often, manual processing also includes deciding that some identified clusters have to be merged because they are too similar.

Examples of software packages and their mode of operation that have been used for human recordings are as follows:

1. MClust—manual; A.D. Redish et al.
2. KlustaKwik—semi-automatic with manual refinement (Harris et al., 2000)
3. OSort—automatic online sorting, manual cluster selection (Rutishauser et al., 2006)
4. Wave_clus—automatic offline sorting, manual cluster selection (Quiroga et al., 2004).

In addition, several commercial equipment manufacturers now offer integrated online spike detection and sorting. While these algorithms are necessarily less sophisticated, they can be good enough for quick online sorting during the experiment or to drive stimulus selection; however, offline reanalysis is usually required for more fine-grained analyses.

In addition to speed and accuracy, an important consideration in spike sorting is consistency. While some decisions made during this process are necessarily subjective, one would at least like to have these decisions always be made in a consistent, well-understood manner. Automatic spike sorting ensures this if it is combined with a systematic manner of implementing the manual postprocessing steps. This is of great importance to get consistent and comparable results across experimenters in a lab and, hopefully, across labs.

As with spike detection, it is important to understand and have an intuition for the spike-sorting approach utilized. Such an intuition can be built by using a data set of simulated neural recordings that is representative of the recording situation. Benchmark data sets of different difficulty levels are available for this purpose—for example, as part of OSort or Wave_clus (see above for references). These should be used to systematically evaluate the performance of spike sorting in terms of clusters identified, false positive/true positive/missed spikes.

Quality Metrics

Due to the uncertainty and subjectivity inherent in spike sorting, appropriate quantitative metrics should be applied and reported to allow comparisons between studies, groups, and investigators. For example, what exact definitions were used to classify a unit as "single unit" as opposed to "multiunit"? The exact parameters vary to some degree and are subjective, which requires that the definitions be consistent, quantitative, and clearly reported. Such metrics can be classified into two groups: (1) single unit measures and (2) comparisons between different units. The first includes metrics such as the mean waveform, its variance, and its SNR; properties of the ISI such as the proportion of ISIs below some minimal threshold or its distribution; or the autocorrelation of spike times. The second includes metrics to quantify the difference between two or more putative single units—for example, by using pairwise distance metrics (see below). Such metrics are instrumental in establishing consistent criteria to evaluate whether two units are distinct and whether such distinction remains stable over time.

The SNR of a spike quantifies to what extend the waveform is different from the "background." A common definition of the SNR is the root mean square of the individual or average waveform over some period of time (such as 2.5 ms), divided by the standard deviation of the noise (estimated from segments where no spikes were detected; Bankman et al., 1993). The SNR is positively correlated with the waveform amplitude and negatively correlated with the amplitude of the background noise. It is useful to compute the SNR for each individual waveform and then quantify the variance of the SNR for all spikes associated to a given cluster. Similarly, the SNR of the waveforms can be plotted as a function of time to evaluate the stability of the unit and its isolation over time. Generally, the higher the SNR of a set of waveforms, the higher the likelihood that a well-isolated single unit can be discriminated.

Typically, a collection of spikes that potentially originated from a single neuron is identified based entirely on the properties of the waveform (such as its shape, SNR, energy) alone. If this is the case, the points of time at which these spikes occurred is a statistically independent metric that can be used to evaluate the sorting result. Metrics that are useful for this purpose are either based on the ISI distribution or on the autocorrelation of the spike times. Due to the refractory period of neurons, very short ISIs (<3 ms) are highly unlikely and should therefore be rare in a well-isolated unit. A useful metric is thus the proportion of ISIs shorter than a few milliseconds apart (see figure 6.1B, plate 1, for an example). For example, in one study, we found that the average ISIs < 3 ms for all isolated units was 0.3% (Rutishauser et al., 2008). The refractory period also leads to a "dip" in the autocorrelation function of the spike train, which can be used to derive a similar test (Gabbiani & Koch, 1999). The autocorrelation and power spectrum of spike trains (figure 6.1B, plate 1, shows examples) is very sensitive to deduce whether a unit contains artificial spikes caused by highly regular sources such as line noise (leading to peaks at 16.6 ms for 60 Hz) or refresh-triggered noise of CRT or LCD screens (up to 240 Hz). In our experience, a combination of at least these three metrics is necessary to confidently declare a cluster to be a single unit. For example, a noise-corrupted unit can well have 0% ISIs below 3 ms and high SNR.

A second class of metrics serves to quantify how well a unit is separated from other isolated or unidentified units. Whether two or more units are truly distinct or not can be difficult to determine based on the waveforms alone, particularly because of the challenges in visualizing high-dimensional spaces. Statistical methods can be used both to help the sorting process itself and to quantify and document the results (Schmitzer-Torbert et al., 2005; Joshua et al., 2007; Hill et al., 2011). One metric that has been applied to human single unit data is the projection test (Pouzat et al., 2002). It is based on the observation that, given a particular level of background noise, two units need to be separated by a minimal distance to be statistically different. If the waveforms of two units are more similar than this minimal distance, they cannot be differentiated given the present noise level. This does not mean that the two units are not different but that they cannot be distinguished reliably given the recording conditions and thus have to be considered the same. The projection test is a pairwise metric that, for each pair of units, defines their distance as a multiple of this minimally distinguishable distance. The properties of the background noise are calculated from spike-free segments of the recording—that is, after all detected spikes have been removed. The background (neuronal and noise) in extracellular recordings typically has a significant negative or positive autocorrelation for periods of up to several milliseconds. For example, in one human data set the autocorrelation function was found to be significant up to 1.2 ms (figure 3 in Rutishauser et al., 2006). This means that the noise is colored—that is, the noise amplitude at successive data points is not independent. This violates the "white-noise" assumption of many statistical tests. The technique of "whitening" a signal, a standard tool in signal processing (Kay, 1993), removes correlated noise to make signals white. This reveals the actual waveforms without artificial smoothening due to correlated noise.

The projection test consists of the following steps: (1) estimation of the noise autocorrelation, (2) prewhitening of detected spikes and normalization of the standard deviation of the noise, (3) pairwise calculation of distance between two clusters. The distance between two clusters is the distance between the two mean waveforms at the center of the two clusters. It is strictly positive and is expressed in units of multiples of the standard deviation of the point around the cluster center. For example, a distance of 5 means that the two cluster centers are separated by 5 standard deviations whereas the points around each center have standard deviation 1 (by definition after the normalization). We recommend this metric as very convenient both for setting rigorous criteria for what minimal distances are acceptable and for documenting the population of neurons constituting a given study (figure 6.1C, plate 1, shows examples). For example, in one study (Rutishauser et al., 2008) we found that the mean pairwise distance between all possible pairs of units recorded on the same wire was 13.7 in units of standard deviations. A histogram of these pairwise distances is a convenient way to summarize the separation criteria of a study (see figure S12 in Rutishauser et al., 2008, for an example).

To what degree are the results of a particular study dependent on spike sorting? It will depend greatly on the experimental measure under investigation whether a result will depend greatly on separation criteria or not. To evaluate such concerns rigorously, it is useful to calculate the metric of interest (such as a response or selectivity index) for all neurons recorded followed by rank ordering the results according to spike-sorting/isolation metrics such as the SNR, the percentage of short ISIs (ISIs within the refractory period could reflect poor sorting), or the projection test distance. If there is a significant correlation between the metric under investigation and the quality metric, further steps have to be taken to avoid confounds. For example, the threshold for inclusion of neurons could be set appropriately based on examination of these correlations with isolation and recording quality (Schmitzer-Torbert et al., 2005).

The Relation between Spikes and the LFP

Signal Acquisition

The spikes originating from a single neuron are a local measure of neuronal spiking output. However, neurons are embedded in a multitude of networks. They receive inputs from and project to a large number of other neurons and can thus hardly be considered independent. Depending on brain state and the area the unit was recorded from, some neurons fire in synchrony with many others whereas in other instances their firing appears to be mostly independent of others. This changing extent of synchrony of firing relative to other neurons is of great interest for investigating phenomena such as functional connectivity, plasticity, and many aspects of cognition such as top-down attention. One approach to quantify such dependencies is to evaluate to what extent the spiking of a particular neuron correlates with aspects of the LFP recorded from the same or a different microelectrode.

Several different types of signals are often referred to as "LFPs." While investigators have used the term LFP to discuss scalp electroencephalographic recordings or intracranial recordings (sometimes referred to as "electrocorticography," or "ECoG"), those signals are less "local" than the ones we consider here. Here we focus on the low-pass band of the extracellular recordings from microwire electrodes. A critical variable that influences how *local* the field potentials are is the impedance of the electrode. We suggest restricting the term LFP to signals recorded from high-impedance (>100 kΩ) microwire electrodes. Typically, the low-pass corner frequency may be set at 100 Hz. Depending on the type of analyses, the exact corner frequency may be relevant since higher frequencies show a significant contribution from spiking activity in the same electrode (Logothetis, 2002; Zanos et al., 2011; Buzsáki et al., 2012). While there continues to be significant debate about how local LFPs really are (and the answer likely depends on the electrode diameter, impedance, recording area, relevant frequencies, and other variables), several studies have suggested that LFPs capture activity within a radius of a few hundred micrometers in the vicinity of the electrode (Buzsáki et al., 2012).

If spikes and LFPs are recorded from the same electrodes, it is necessary to consider whether the waveforms of the spikes could possibly "leak" into the LFP frequencies of interest and thus introduce artificial phase locking. Simulations indicate that this can be the case for frequencies above 50 Hz (Logothetis, 2002; Zanos et al., 2011; Buzsáki et al., 2012). For purposes of analysis of the LFP with respect to phase locking, such leakage of power into low frequencies is undesired. To prevent this, it is necessary to replace the samples around a detected spike such that the spike is removed from the LFP. This can be achieved by replacing spikes with a cubic spline interpolation between 1 ms before and 2 ms after the peak of the spike (in the high-resolution signal at full sampling rates). More sophisticated methods for spike removal have also been suggested (Zanos et al., 2011).

Of particular interest here is whether the probability of spiking correlates systematically with the power and/or phase of a specific oscillatory component (such as theta frequency oscillations) of the LFP. Other types of relationships have also been investigated, such as broadband power increases and their relationship to single unit firing (Miller et al., 2007; Manning et al., 2009). There are several methods for quantifying the spike–field relationship, of which we highlight two: direct estimation based on instantaneous power/phase and indirect estimation using spike–field coherence (SFC).

All methods require that the low-frequency (LFP) and high-frequency (spikes) parts of the signal are available simultaneously. This means that, if a spike occurs at t = 100.0 ms, we need to know the power and phase of the oscillations that are part of the LFP at this exact moment in time. While this might seem trivial, in reality this is sometimes not the case because of frequency-dependent time lags introduced by components of the acquisition system, the electrode, or the data processing pipeline. Such artifacts have to be corrected for, as otherwise they produce artificial spike–field relationships because systematic shifts that have a different (but fixed) lag as a function of frequency will make them appear systematically before/after the spike whereas in reality these two happened simultaneously. In practice, there are three principle sources of such artifacts: (1) filters used by the data acquisition system, (2) the head stage, and (3) filters in the postacquisition data processing.

Filters used by the data acquisition system are typically causal (as they are applied in real time), which means they have a phase lag dependent on frequency (as discussed above). Frequently, experimenters have the acquisition system apply two types of filters: LFP band (such as 0.5–100 Hz) and spikes (such as 300–3000 Hz). The system then processes and stores each component independently, such as in a continuous data file and in a file with detected spikes. In such a situation careful analysis is necessary to ensure that the filters used for the two bands have the same phase-lag response (which typically they do not). Some systems display the time lag introduced (which can reach several milliseconds) and allow switching on a correction mode. We advise avoiding this situation altogether by saving the data in as broadband a format as possible and then extracting spikes and the LFP from the very same file using noncausal zero-phase lag filters.

The second possible source of signal distortions is the head stage. Careful modeling and analysis studies have demonstrated that head stages with too low input impedances distort the LFP to such an extent that artificial spike–field correlations are induced and/or the spike waveforms are significantly distorted (Nelson et al., 2008; Nelson & Pouget, 2010). This can be avoided by using head stages with input impedances > 1 GΩ. This is necessary because of the relatively high impedance of the electrodes, typically in the range of 400 kΩ–1 MΩ. Some of the commercial head stages frequently used for human recordings are well characterized and have sufficiently high input impedances. One example is the HS-36 electrode we used in a recent study (Rutishauser et al., 2010) which has >1 TΩ input impedance. However, there are many other systems being used that have either unknown or too low input impedance for this type of recording. In such situations, manual corrections by deconvolution can be applied as has been done by some authors (Siegel et al., 2009), and some suppliers offer tools to achieve this (i.e., FPAlign by Plexon Inc.). An additional area of concern is filtering done directly in the head stage. The hardware filters in the head stage have to be known or experimentally measured to ensure that they do not introduce phase shifts in the range of interest. Such shifts can easily reach 90° for low frequencies (Nelson et al., 2008), meaning the measured signal will lead the actual signal by one fourth of the cycle length. This can lead to misinterpretations or null results that are due to these phase-distortion artifacts. The effective impedance of the head stage is reduced by the cabling between the head stage input and the electrode through capacitive coupling (Robinson, 1968), providing a further incentive to make these cables as short as possible.

Quantitative Measures

The aim of measures to quantify the relationship between spikes and the LFP is to quantify statistically whether the spike times of a particular neuron are related to the phase and/or power of an oscillatory component of the LFP. In this section, we focus on phase relationships, but similar methods can be applied for power. We summarize two methods: assessing phase locking using circular statistics and SFC.

Circular Statistics and Estimation of Instantaneous Phase   Circular statistics provide a method to assess the distribution of circular variables such as phase angles (Fisher, 1993). Examples are circular equivalents of the normal distribution (the von Mises distribution) and circular tests for uniform distribution of variables (the Rayleigh test, among many others). In the following we quantify phases θ in units of radians in the range of $-\pi,\dots,\pi$ where $\theta = 0$ is the peak and $\theta = \pm\pi$ is the trough (see figure 6.2). This notation is used in many analysis programs such as MATLAB.

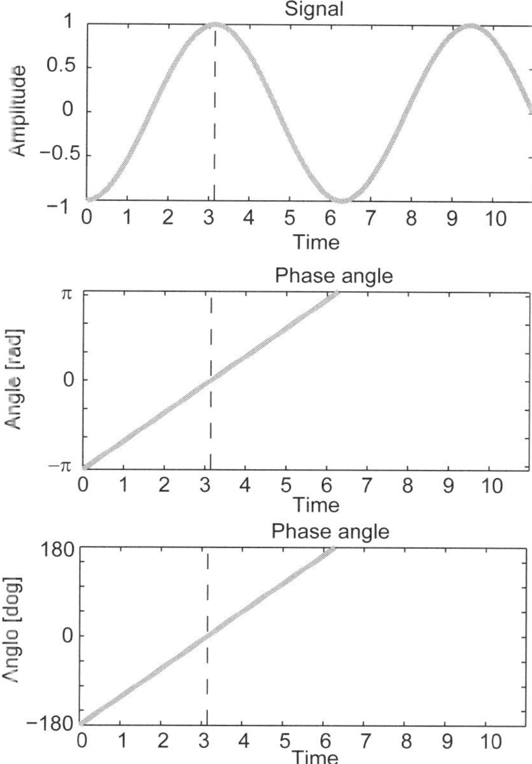

Figure 6.2
Illustration of the notation for phases. Shown are 1.5 cycles of an oscillation (arbitrary units). The corresponding angle is shown in the bottom two rows both in radians (rad; middle) and degrees (deg; bottom). In this notation, the peak of the oscillation corresponds to 0° and the trough to ±180°. Note the discontinuity at 180°, when the phase resets.

The free MATLAB toolbox CircStat contains implementations of many statistical tests and tools necessary for the analysis of circular data (Berens, 2009).

Applying circular statistical tools first requires that the phase of the underlying oscillation(s) be determined for every point of time where a spike occurred. There are many ways to achieve this, of which we outline two common methods that have been employed successfully to analyze human single neurons and their relationship to the LFP (Caplan et al., 2001; Jacobs et al., 2007; Rutishauser et al., 2010): (1) the Hilbert transform and (2) the wavelet transform (see also Le Van Quyen et al., 2001, for a comparison of these two methods). The Hilbert transform is particularly useful when there is a certain frequency of interest (such as 35 Hz) whereas wavelet methods are more natural when a range of frequencies is considered.

After the phases for all spikes fired by a neuron have been determined (see below), the next step is to determine whether the population of phase values θ_i is distributed significantly around their mean angle $\bar{\theta}$ or whether their distribution is, alternatively, uniform and thus random. The mean resultant vector $\bar{R}$ is the vector sum of all the phase angles:

$$C = \sum_{i=l}^{n} \cos(\theta_i), \ S = \sum_{i=l}^{n} \sin(\theta_i), \ R^2 = C^2 + S^2, \ \bar{R} = \frac{R}{n}. \tag{6.1}$$

In the above, C and S are the normalized sums of the sine and cosine of each phase angle, respectively, and n represents the total number of phase angles. The larger the length of the mean resultant $\bar{R}$ (range 0...1), the stronger the phase locking of the neuron is. The sample circular variance is $V = 1 - \bar{R}$. To test whether a neuron is significantly phase locked, the sample of all phase angles is compared against uniformity using a Rayleigh test. The Rayleigh test is based on a statistic Z:

$$Z = n\bar{R}^2, \ P = \exp(-Z)\left[1 + \frac{2Z - Z^2}{4n} - \frac{24Z - 132Z^2 + 76Z^3 - 9Z^4}{288n^2}\right]. \tag{6.2}$$

If P is sufficiently small, the null hypothesis of uniformity can be rejected. The alternative hypothesis is that the data are unimodal (one mean direction). Notice that the Rayleigh test is strictly a function of $\bar{R}$ as well as of n (number of spikes included). The same value of $\bar{R}$ thus leads to different significance values depending on the number of spikes that are included. In practice this leads to the problem that, given enough spikes, neurons that are not convincingly phase locked lead to statistical significance at $P < 0.05$. It is worthwhile to estimate a reasonable p-value cutoff for the numbers of spikes considered using simulations. Frequently, studies use cutoffs of $P < 0.01$ or $P < 0.001$ to avoid false positives. Due to these problems, we have found it advantageous to use summary metrics other than $\bar{R}$ and the Rayleigh-test significance value. These are based on the von Mises distribution, the circular equivalent of the normal distribution (Fisher, 1993):

$$f(\theta) = \frac{1}{2\pi I_0(\kappa)} \exp(\kappa \cos(\theta - \mu)). \tag{6.3}$$

The probability of observing a phase angle θ is a function of the mean angle μ and the concentration parameter κ. Both parameters can be determined easily from population data by maximum likelihood methods (see Fisher, 1993, p. 88). The concentration parameter is the analog to the standard deviation of a normal distribution, although of opposite direction: The larger κ, the more concentrated the distribution (the smaller its variance). For $\kappa = 0$, the von Mises distribution is equivalent to the uniform distribution on the circle. $I_0(\kappa)$ is the modified Bessel function of order zero. Plotting a histogram of the values of κ summarizes the phase-locking strength of a population of neurons, a form of visualization preferred by us over histograms of p values.

Hilbert transform To estimate the phase of an oscillation of a particular frequency, the signal is first narrowly band-pass filtered at the frequency of interest (such as 32–38 Hz if 35 Hz is of interest). The Hilbert transform can then be used to transform this signal $S(t)$ into a complex valued analytical signal $X(t) = S(t) + iS_H(t)$. The real part of the analytic signal equals the raw signal $S(t)$, and the complex part is the Hilbert transformed signal $S_H(t)$. The instantaneous phase $\phi(t)$ and power $R(t)$ of $S(t)$ can then be estimated based on $X(t)$ as follows:

$$R(t) = \Re\{X(t)\}^2 + \Im\{X(t)\}^2 \tag{6.4}$$

$$\phi(t) = \arg(X(t)) = \mathrm{atan2}(\Im\{X(t)\}, \Re\{X(t)\}),$$

where $\Re$ and $\Im$ refer to the real and imaginary part of $X(t)$, respectively.

Wavelet transform The raw signal $S(t)$ can be decomposed into a function of frequency and time using the continuous wavelet transform (CWT) (Torrence & Compo, 1998). While many different wavelet basis $\psi_0(\eta)$ could be utilized, we and others have used the complex Morlet wavelet, which has the two parameters: a center frequency f_0 and the number cycles. Typical values are $f_0 = 1$ and $\varpi = 4$ or 6 cycles. The resulting CWT is a function of both scale (frequency) and time: $W(t, s)$. It is computed by convolving the raw signal (of length N) with the wavelet function $\psi_0(\eta)$ for a number of different frequencies (scales) s. The effective resolution of the Morlet wavelet depends on the center frequency f_0 and the scale s. If δT is the spacing between two sampled points (sampling rate), the effective frequency of a Morlet wavelet at scale s is

$$f = \frac{f_0}{s\delta T}.$$

Thus, the higher the scale, the lower the frequency. The resolution is measured separately in terms of the standard deviation in time σ_t and frequency σ_f. Time resolution at scale s is $a\delta T$ and frequency resolution is σ_f / a. Thus, the better the resolution in time, the worse it is in frequency, and vice versa (uncertainty principle, a fundamental limit, dictates $\sigma_t\sigma_f \leq 1/(2\pi)$). The time width of a wavelet is defined as (Najmi & Sadowsky, 1997)

$$\sigma_t^2 = \frac{\displaystyle\int_{-\infty}^{\infty} t^2 \psi^2(t)\,dt}{\displaystyle\int_{-\infty}^{\infty} \psi^2(t)\,dt}. \tag{6.5}$$

Thus,

$$\sigma_f = \frac{1}{2\pi s \sigma_t}.$$

To illustrate this trade-off, figure 6.3 shows Morlet wavelets in both time and frequency space for different parameters together with their time and frequency resolution. Notice the trade-off between accuracy in time and frequency illustrated by the size of the error bars. Since the width in frequency space increases as a function of frequency, the frequencies at which the wavelets are calculated are typically logarithmically scaled. This leads to an even sampling in frequency space. Typically, we sample at frequencies of $f = 2^x$ with x linearly covering the frequencies of interest such as 1–50 Hz. The signal $W(t,s)$ for the frequency and time of interest can be used just as the Hilbert transformed signal $X(t)$ to estimate the instantaneous phase and power as shown above.

Spike–Field Coherence Determining the phase locking of neurons based on the instantaneous phase at the time of the spike provides a limited view of phase locking. While it allows an assessment of whether a neuron is locked or not, it is difficult to assess changes in the phase-locking strength across time or experimental conditions using this method. Also, it is based on a single value (the instantaneous phase) rather than the rich data that the LFP provides. Also, this approach cannot determine more complex relationships, such as complex oscillatory patterns happening some time before or after the spike that are in themselves not phase locked, including sharp waves (Buzsáki, 2006). Examples of these more complex scenarios include nesting of oscillatory power at higher frequencies coupled to phases of lower frequency oscillations, a phenomenon that is prominent in recordings in humans (Canolty et al., 2006). An alternative method that can detect such patterns is SFC, which has been used in a wide variety of studies to explore complex interactions between spikes and fields in the same and different areas in animals (Fries et al., 1997; Fries et al., 2001; Womelsdorf et al., 2006) and humans (Rutishauser et al., 2010). The SFC makes more efficient use of the data available because it considers many more data points for each spike. SFC-based estimates are thus more robust and can be made from fewer spikes compared to circular statistics–based measures.

The spike–field coherence $SFC(f)$ is a function of frequency f and takes values between 0 and 100%. The larger the SFC, the more accurately the spikes follow a particular phase of this frequency. The SFC is the ratio between the frequency spectrum of the spike-triggered average (the STA), divided by the average frequency spectrum of the LFP traces themselves (that were used to construct the STA). By design, the SFC is thus normalized for LFP power changes that co-occur with spikes. The average spectrum of the LFP traces themselves is referred to as the spike-triggered power (STP). Formally, the SFC is defined as follows:

$$SFC(f) = \frac{f STA(f)}{STP(f)} 100\%. \tag{6.6}$$

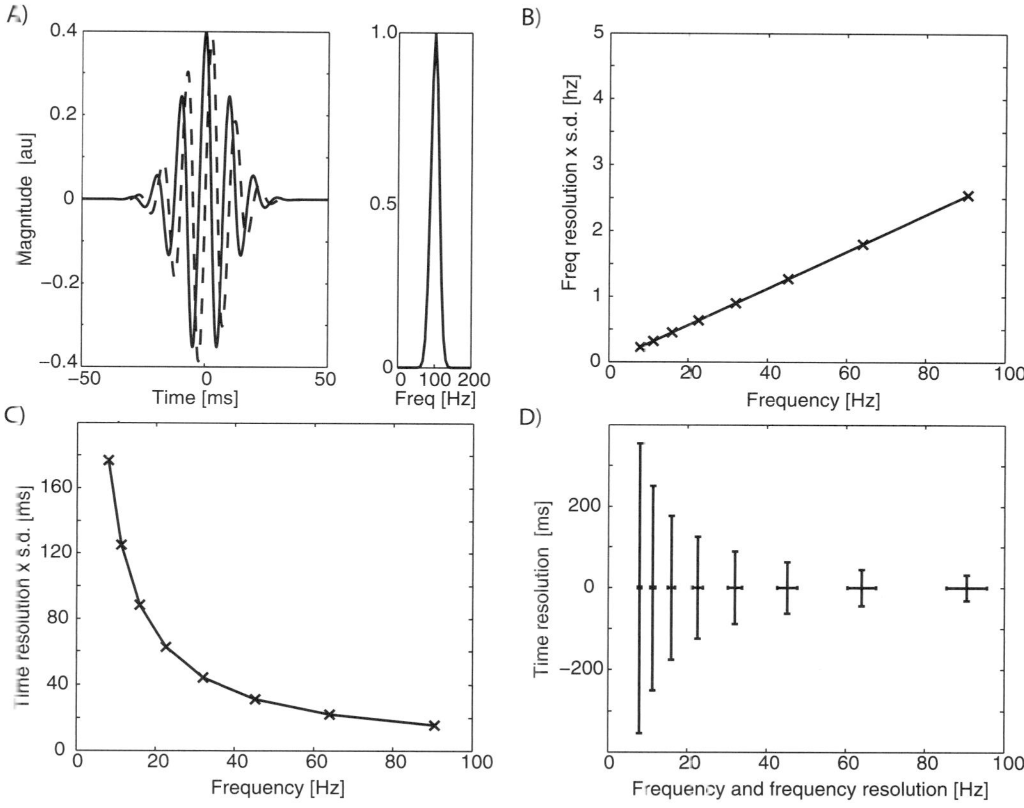

Figure 6.3

Illustration of the complex Morlet wavelet and trade-offs between frequency (Freq) and time resolution inherent in any form of local field potential analysis. All examples are for the complex Morlet wavelet with cycle number 2 and center frequency 1. (A) Wavelet in time (left; dashed lines show the complex part, straight lines the real part) and frequency (right) for one example scale (au, arbitrary units). (B–D) Illustration of the trade-off between specificity in time and frequency. (B) Frequency resolution as a function of frequency. Points on the y-axis are in units of SD. (C) Time resolution as a function of frequency. Points on the y-axis are in units of SD. (D) The 95% confidence intervals for time and frequency resolution plotted against each other.

The STA is the average of many small segments of the LFP, each of which is extracted by taking a small piece of LFP centered on every spike. The size of the window depends on the frequencies of interest. For the example of <10 Hz theta oscillations, we previously used ±480 ms. Averaging all such traces of LFPs results in the STA. We used multitaper analysis and its implementation in the Chronux Toolbox (Mitra & Bokil, 2008) to estimate the frequency spectra. Multitaper analysis is a powerful method to estimate robust single trial frequency spectra (Jarvis & Mitra, 2001). The choice of parameters (number of tapers and time–bandwidth product TW) for the multitaper analysis determines the time and frequency resolution of the *SFC*. For example, at 250-Hz sampling rate, the ±480-ms window consists of 240 data points, which result in a frequency resolution (half-width) of 4.2 Hz for seven tapers and TW = 4.

The *SFC* is a strictly positive measure that varies between 0 and 1 (or, sometimes 0 to 100%). It is a function of frequency but also the number of spikes used to calculate it. Consider the simulations shown in figure 6.4 (plate 2) to gain further insights into the properties of the *SFC*. We simulated a realistic LFP signal with a 1/f power structure and superimposed a simulated 25-Hz oscillation (see figure 6.4A, plate 2). We further simulated two neurons, one of which preferentially fires spikes at the trough of the simulated oscillation (red) whereas the other fires randomly (green) irrespective of the oscillation (see figure 6.4B, plate 2). The STA of the phase-locked neuron recovers the 25-Hz oscillation as expected, with the trough at t = 0 (see figure 6.4C, plate 2). Based on the STA, the spectrum of the STA and the SFC can be calculated (see figure 6.4D, plate 2). The peak of the SFC is around 25 Hz as expected, but shifted slightly to the right. This is due to the frequency resolution used for the multitaper estimates. Here the half-width was 10 Hz, as can be seen from the width of the bump. Notice that, despite perfect phase locking, the absolute value of the *SFC* reached is about 6%. To explore this apparent inconsistency, we calculated the amplitude of the peak SFC value for different numbers of simulated spikes (25–300 spikes) and for different phase-locking strengths (see figure 6.4F, plate 2). Three units were simulated: one with perfect phase locking (every spike follows the phase), one with 70% of the spikes phase locked, whereas the other was firing at random phases. As expected, the peak *SFC* value accurately indicated which of the units was most phase locked for any number of spikes used. But crucially this simulation also reveals that the peak SFC value decreases monotonically with the number of spikes used till it asymptotes after hundreds of spikes. This bias makes comparisons between groups (such as the 100%- and 70%-locked neuron) only valid if the same numbers of spikes is considered for each group. If, for example, unit 2 had 25 spikes and unit 1 had 300 spikes, the direct comparison of their SFC values would lead to the wrong conclusion that unit 2 was more phase locked whereas in fact the opposite is the case. This simulation illustrates the necessity of equalizing groups for sample sizes (e.g., by random subsampling of the bigger group) to allow a fair statistical comparison.

The reason for this bias is that, under the null hypothesis of no phase locking, the amplitude of the STA tends to zero only for large samples (the green line). This is because of the strictly positive nature of the *SFC* that does not allow noise to "average out" to zero. Also note that this explains the apparently low value of 6% above—for this figure, 1400 spikes had been used. Note

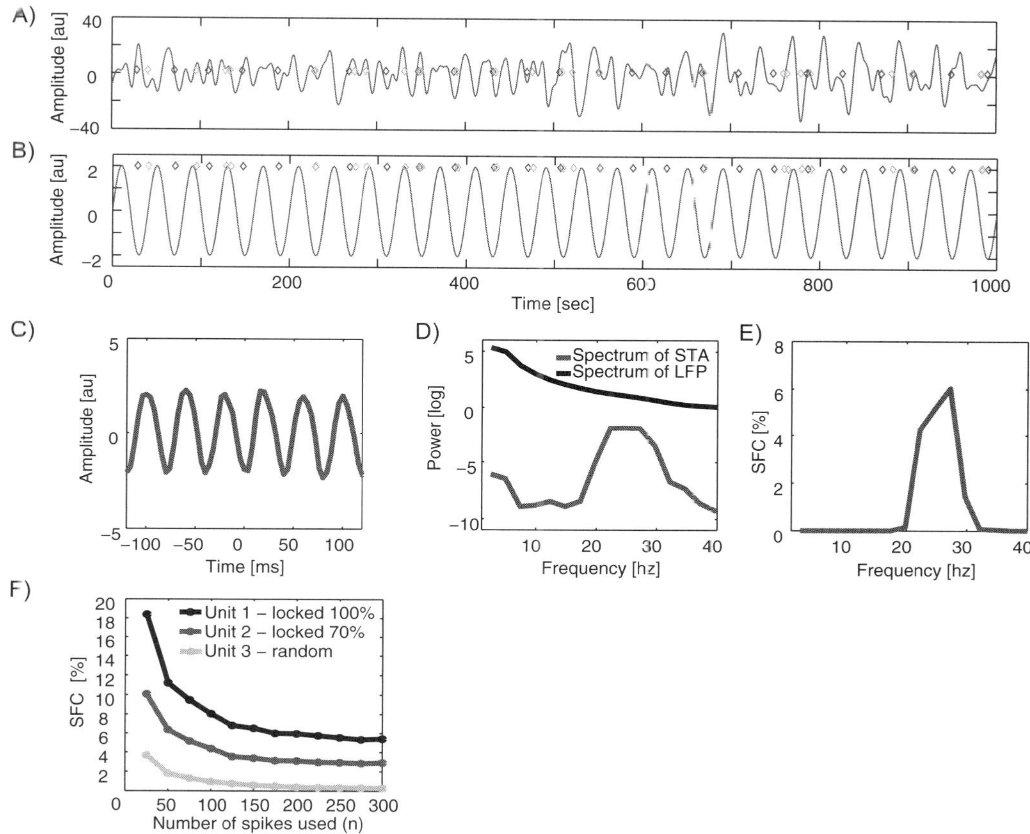

Figure 6.4 (plate 2)

The steps of calculating the spike–field coherence (SFC), illustrated using simulated data. (A) Simulated extracellular. Shown is a band-pass-filtered version of the simulated raw signal (10–300 Hz). (B) One part of (A) is a 25-Hz oscillation of amplitude 2. The red and green dots indicate the spikes of two simulated neurons. (C) The spike-triggered average (STA) for the red simulated neuron shown in (B). (D) The power spectrum of the STA (blue) shown in (C). (E) The SFC as a function of frequency. (F) The mean peak height of the SFC at 25 Hz as shown in (E) as a function of the number of spikes used for its calculation. Figure adapted from Rutishauser et al. (2010). LFP, local field potential; au, arbitrary units.

that this reinforces the point that the SFC is a relative measure—its actual amplitude is not meaningful, but what is meaningful is the proportional difference between two groups (such as the *SFC* is 50% bigger in group A compared to group B). Recent new methodological developments (Grasse & Moxon, 2010) have also led to the proposal of an analytical method to correct for this bias.

Decoding Brain Signals in Single Trials

Once we have detected spikes, separated them into clusters, and analyzed their relationship with LFPs, we are often interested in correlating the neurophysiological responses with aspects of the cognitive tasks under study (e.g., memory formation, stimulus presentation, attentional modulation, reaction times, etc.). The brain needs to be able to act upon the neurophysiological data in single trials. While averaging over trials is commonplace in an attempt to get rid of noise in many studies, neurons do not have that luxury.[1] Also, several practical applications such as driving prosthetic limbs based on the activity of a population of neurons (see chapter 17) require decoding responses in single trials. Here we provide an overview of the math, algorithms, and methodology that underlie decoding the activity of an ensemble of neurons in single trials. For a more detailed discussion of the mathematics of machine learning, we refer the reader to Bishop (1995), Vapnik (1995), and Poggio and Smale (2003). For a more extensive review of feature expression for machine learning and decoding approaches, see Meyers and Kreiman (2011) and Singer and Kreiman (2012).

We focus our discussion on decoding the activity of ensembles of spike trains or LFPs. To provide a concrete example, we imagine a scenario where the subject was presented with n stimuli labeled $S_1,...,S_n$ over multiple trials in pseudorandom order. The exact details of the experimental paradigm are not relevant here. $S_1,...,S_n$ could represent n different pictures, or n different motor commands, or behavioral measures such as correct recollection or not. In any given trial k, we want to examine the neurophysiological recordings and make an educated guess about which experimental condition was presented ($_k\lambda \in \{S_1,...,S_n\}$ denotes the condition for trial k).

Feature Extraction

The first question that arises in this procedure involves extracting relevant features from the input signals. This constitutes a critical step. If the features extracted from the data do not contain information about the experimental conditions, no machine-learning algorithm will be able to magically lead to decoding performance above chance levels.

We consider a set of spike trains $_ks_i(t) = \sum_{j=1} \delta(t -_k t_i^j)$, where $_k t_i^j$ denotes the timing of spike j from neuron i ($i = 1,...,n$) in trial k measured from trial onset. In the interest of simplicity, we consider a simple feature, namely, the spike count, which has historically been shown to convey interesting information: $_k x_i[w] = \int_w {}_k s_i(t)dt$, where w denotes a window of interest. The spike count defined here is by no means the only feature of interest. Other features that have provided interesting information include the number of spikes that are correlated across multiple neurons, the projections onto the first principal components, the coherence between spikes and LFPs, the power of the spike train in specific frequency bands, and so on. While in the definition above $_k x_i[w]$ is a scalar, one could also consider feature vectors such as $[_k x_i[w_1],\ldots,_k x_i[w_u]]$ where the feature of interest is computed in multiple temporal windows $w_1,\ldots,w_u$ within trial k.

It is also of interest to decode LFP signals. It should be noted that the brain does not have direct access to LFP signals. Except for small effects of extracellular fields (Anastassiou et al., 2011), postsynaptic neurons need to make their decisions (to spike or not to spike) based on the incoming set of spike trains from presynaptic neurons. Still, LFPs have often been shown to contain interesting information, their signals can be correlated with spike trains (see "The Relation between Spikes and the LFP"), they may be more readily accessible in situations where spike recordings are not possible and can also have interesting practical properties such as increased stability. In the interest of simplicity, we extract the total power in the LFP signal ($L(t)$) in a given window w: $_k x_i[w] = \int_w {}_k L_i^2(t)\,dt$. Again, there are multiple other possibilities including considering the power restricted to certain frequency bands, the phase of the signal, the coherence between LFPs recorded from different electrodes, and so on. As described above for spike trains, the feature of interest can be extracted in multiple windows $w_1,\ldots,w_u$.

It is worth mentioning several practical considerations. It is often of interest to compare decoding performance of multiunit activity versus single unit activity obtained after spike sorting (see the "Spike Detection" section in this chapter). Nonstationarities abound in neurophysiological recordings; it is a good idea to attempt to eliminate such nonstationarities before attempting to decode information (Meyers & Kreiman, 2011). In several cases, investigators have considered ensembles of neurons that are not recorded simultaneously (e.g., Hung et al., 2005, as opposed to simultaneous recordings in Wilson & McNaughton, 1993). These ensembles are referred to as *pseudo-populations* and assume that the responses are independent (i.e., decoding the activity of pseudo-populations does not take into account potential interactions across neurons; Meyers & Kreiman, 2011).

Learning from Examples

Decoding neural data using machine learning techniques constitutes a typical example of supervised learning from examples. We have a set of data $_k \mathbf{x} = [_k x_1,_k x_2,\ldots,_k x_n]$ and a set of labels for each trial $_k \lambda \in \{S_1,\ldots,S_n\}$. The goal is to build a classifier that can learn the structure of the relationships between $_k \mathbf{x}$ and $_k \lambda$. The typical procedure is shown in figure 6.5, where, for illustrative purposes, we consider $n = 2$ neurons and only two possible labels ($_k \lambda$ is a circle or a triangle, which we denote as +1 and −1, respectively, in the next paragraph). After extracting relevant features from each neuron (see figure 6.5A–B), we seek a boundary that can separate the two labels. In this case, we show a linear boundary.

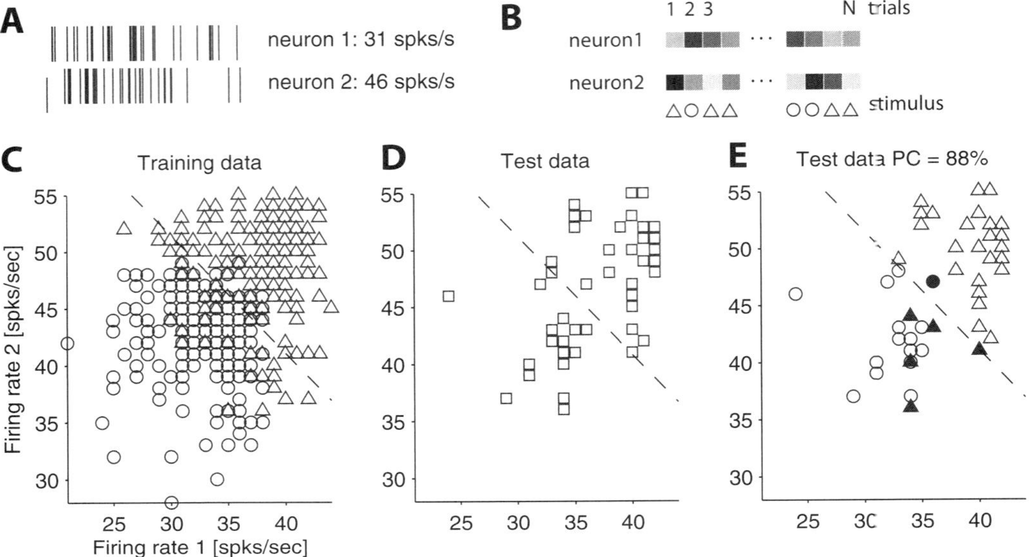

Figure 6.5
Schematic illustration of machine learning approach to decode neural ensemble data. (A) We record spike trains, and we extract relevant features for classification. In this case, we illustrate only two neurons (artificial data), and we compute the firing rate in a 1000-ms window. (B) The process is repeated for all N trials where two different stimulus conditions (here a represented by a circle and a triangle) are repeated in pseudorandom order. In each trial, we build a vector (here of size 2) that represents the ensemble response. (C) The data are separated into a training set and a test set. This plot of the training data shows the separation between the two conditions. The classifier (in this case a linear classifier) seeks to find a separation between two conditions. (D) Given test data, the classifier uses the same boundary from (C) to separate the data. Those points above the boundary are classified as triangles whereas those points below the boundary are classified as circles. (E) Classifier predictions. The erroneous labels are marked by filled symbols. In this toy example, the classifier correctly labels 88% of the points. spks/s, spikes per second; PC, percentage correct.

There is a rich literature on the mathematics of supervised learning, and multiple algorithms are available for classification.[2] A simple algorithm for classification is Fisher's linear discriminant. We seek a linear boundary characterized by weight vector α and constant b. We can describe the linear classifier as $f\left(_k\mathbf{x};\alpha,b\right)=b+\alpha_k\mathbf{x}$. In any given trial, we take the neural ensemble response, compute the dot product with the weights, and add a constant. If the resulting value is greater than 0, then we classify the point as condition $+1$, otherwise, we classify the point as condition -1. The boundary should be able to separate the means of the responses for each of the two conditions: $\mathbf{m}_{+1}=\left(1/n_{+1}\right)\sum_{k=1}^{n_{+1}}{}_k\mathbf{x}_{+1}$ and $\mathbf{m}_{-1}=\left(1/n_{-1}\right)\sum_{k=1}^{n_{-1}}{}_k\mathbf{x}_{-1}$, where n_{+1} and n_{-1} denote the number of trials with labels $+1$ and -1, respectively, and $_k\mathbf{x}_{+1}$ and $_k\mathbf{x}_{-1}$ denote the neural ensemble vectors in a trial k where the label is $+1$ or -1, respectively. In addition to separating the means, we also want to minimize the variation within each group. For this purpose, we also consider the variances for each condition:

$$\mathbf{\Gamma}_{+1} = \left\langle \left({}_k\mathbf{x}_{+1} - \mathbf{m}_{+1} \right) \left({}_k\mathbf{x}_{+1} - \mathbf{m}_{+1} \right)^T \right\rangle_k \text{ and } \mathbf{\Gamma}_{-1} = \left\langle \left({}_k\mathbf{x}_{-1} - \mathbf{m}_{-1} \right) \left({}_k\mathbf{x}_{-1} - \mathbf{m}_{-1} \right)^T \right\rangle_k \tag{6.7}$$

The Fisher linear discriminant seeks to find the value of the weights α that maximizes the following expression:

$$\hat{\mathbf{c}} = \arg \max_{\alpha} \frac{\left(\alpha^T \left(\mathbf{m}_{+1} - \mathbf{m}_{-1} \right) \right)^2}{\alpha^T . \left(\mathbf{\Gamma}_{+1} + \mathbf{\Gamma}_{-1} \right) . \alpha} \tag{6.8}$$

subject to $\|\alpha\| = 1$. It can be shown that this expression is minimized when

$$\left(\mathbf{\Gamma}_{+1} + \mathbf{\Gamma}_{-1} \right) . \alpha = \mathbf{m}_{+1} - \mathbf{m}_{-1}, \tag{6.9}$$

which can be easily solved using MATLAB. While it is always advisable to write your own code, MATLAB has an implementation of a simple linear discriminant classifier in the `classify` function. Figure 6.5C–E shows how a line can separate the data and minimize the classification error.

Support vector machines (SVMs) constitute a robust type of classifier that has proven to be successful in many neural decoding applications (as well as in many other domains). For a description of the mathematical principles behind SVM classifiers, see Vapnik (1995) and Cristianini and Shawe-Taylor (2000). There are multiple freely available implementations of SVM classifiers including one in MATLAB's `svmtrain` and related functions.

Cross-validation

A critical component of all the decoding approaches is to perform cross-validation by separating the data into a *training set* and a *test set*. Failure to do so leads to overfitting the data. Depending on the exact characteristics of the problem, dimensionality and number of trials, overfitting can lead to overoptimistic measures of performance and hence misleading conclusions.

A schematic illustration of a typical cross-validation procedure is shown in figure 6.6. Essentially, a random subset of the data is used for training (i.e., finding the classifier parameters) and the remaining data are used for evaluating the performance of the classifier. The process is also emphasized in figure 6.5: part C shows the data used to find the parameters of the Fisher linear discriminant, and the resulting boundary line was applied to novel data shown in part D to lead to the classification results shown in part E. For a longer discussion of cross-validation procedures, see Meyers and Kreiman (2011).

Real-Time Feedback from Spikes for Closed-Loop Experimentation

Our world is *continuous*. Therefore, experiments that show discrete presentations of stimuli or are broken down into blocks or trials are limited in how well they simulate real life. Wanting to better simulate real life, scientists occasionally turn to experiments which modify themselves in real time as a function of some brain activity (Fetz, 1969; Griffin et al., 2004; Cerf et al., 2010;

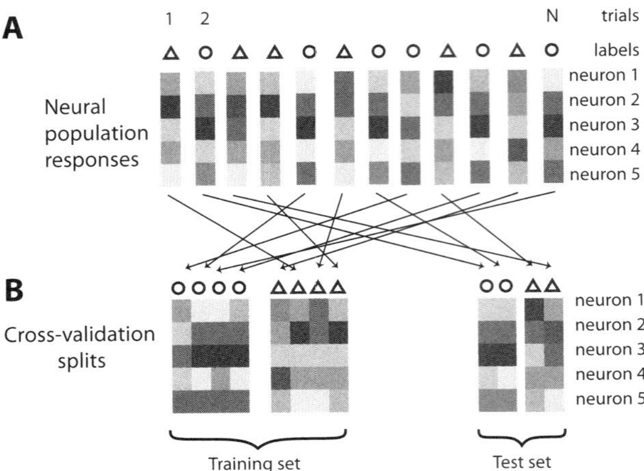

Figure 6.6
Schematic illustration of cross-validation. (A) Population of neuronal responses (here five neurons) over N pseudoran-dom labeled trials. (B) In each cross-validation run, the data are divided into a training set and a nonoverlapping test set (i.e., without replacement). In this example, 75% of the data go into the training set and 25% of the data go into the test set, but other splits are also possible. Here we ensure that the proportion of labels in the training set is equally divided between the two possible labels so that chance level in classification will be 50%. Only the training data are used to change the parameters in the classifier. Classification performance values reported should be based on the test set.

Jadhav et al., 2012; Rutishauser et al., 2013). These are experiments whose analysis is done quickly (i.e., "real time") and the outcome of such analysis influences the experiment itself. This is commonly referred to as a closed-loop (or "online") experiment. In contrast, open-loop experiments are those where the neuronal response does not influence the experiment itself. Technically, closed-loop experiments are challenging as the outcome of the data analysis is directly fed back to the subject and thus alters behavior. In contrast, in offline experiments there is no interaction between the data analysis and the experiment.

Implementing real-time experiments in humans, using spikes and single unit data, is challenging for multiple reasons—both technical and conceptual. There are a number of additional considerations relative to conducting an *offline* experiment. Below, we discuss these challenges and possible solutions. A critical parameter is the latency from a neuronal event occurring in the brain until the desired stimulus change (such as presenting a picture conditional on the firing rate of a neuron). What is considered real time? Clearly, these experiments are at best "near"-real-time experiments, as there will always be a delay. Depending on the exact type of experiment, a latency of less than 100 ms is typically acceptable as patients tend to not change their behavior because of these (Szelag et al., 2009).

Single units are very powerful and useful for real-time experiments. That is because their speed and short latencies lend themselves to altering the experiment and behavior. If we as experimenters learn of a burst of activity that started at time 0, and are able to alter the outcome of a trial in less than 100 ms following this activity, the subject is likely to interpret the entire sequence as happening in real time. More so, given the high SNR of single-unit firing, typically only a few trials are needed to decode the meaning of a certain unit's firing. That is, we can offer single trial classifiers that operate based simply on separating the *baseline* activity of the unit and *active firing*.

Challenges in Designing Real-Time Experiments

Programming Challenges Much of our programming tools were designed for offline experiments where time is effectively set by internal mechanisms that are placed in the code. That is, a typical stimulus presentation routine would present an image on the screen and then wait for 1 s before removing the image from the screen. In contrast, in online experiments we often have to hold a variable that measures the time from previous events. That is, a code block that performs the same function in a real-time experiment will often start with a line that sets the time variable to the current time, and then starts a *while* loop that asks whether one second had already passed in which the experimenter would repeatedly pull information from the brain, for example, and exit the loop to remove the image from the screen. This is mostly true for closed-loop real-time experiments where data are being analyzed continuously and processing time is sometimes a bottleneck that experimenters want to overcome by not operating sequentially but rather using the delay time (in this example 1 s) to perform some of the analyses.

Another challenge with online experimental implementation is the tight integration with the acquisition system or simulations thereof for testing and debugging. Therefore, the experimental code will have to include acquisition-system-specific calls to interact with the acquisition system to receive data. This implies that the experiment itself has to be implemented in a way that permits inserting such commands. This typically prohibits using closed-source systems such as those frequently used in visual psychophysics and a number of popular tools in experimental design, which do not support such interaction. Testing and debugging such an approach is challenging. Ideally, the same experimental code can be used for both an actual patient and a "simulated" situation where simulated data are used or previously recorded data are replayed. Finally, an additional challenge in real-time coding involves controlling the flow of experiment time in the event of unexpected circumstances.

Real-time approaches are frequently used in other fields (i.e., electrical engineering, gaming), allowing us to adapt existing and proven methods. Below, we outline two solutions that we found useful in implementing our closed-loop experiments (Cerf et al., 2010).

Event-Based Programming An efficient way to handle real-time coding is "event based programming," which uses the operating system's event queue. This allows processes and methods to "register" with an event manager, which in turn calls the registered function when a certain event has occurred ("callback"). For example, instead of having a while loop that checks at the end of each cycle whether a button was pressed, you register a function to be called when a button has been pressed with the event manager, and whenever the button is pressed, this function is called automatically. One can set functions that are based on events occurring, on the period occurrence of a particular event (say, call a function that pulls data from the brain repeatedly every millisecond or every 100 ms regardless of the linearity of the code), or events that happen a period of time following a certain event (say, run a function exactly 300 ms after a button was pressed).

Most programming languages and operating systems offer event-based processing. In the following, we use MATLAB to illustrate examples, but similar commands can be used in other environments. The main functions in MATLAB for event-based coding are: "timer" and "addlistener."

The addlistener object registers a listener with the event handler, which starts the event process. The listener uses strings to capture events (i.e., using 'ButtonPressed' as a trigger that indicates the event that marks a button press).

The "timer" function allows for setting delays or precise or periodic timing. For example, setting the 'StartDelay' attribute instructs the timer to trigger an event after a certain delay. Calling start(eventlistener) and stop(eventlistener) controls the flow of the experiment. That is, at the end of each trial, for example, we can 'stop' all listeners and then re-'start' them at the beginning of a future trial.

Finally, it is important to note the numerous variants of the timer function that really allow for optimized control of the experimental timing, with the ability to set time for a certain time, for a periodic recurring call (using the 'Period' attribute), for a specific delay, or for a delay from the occurrence of a certain future event.

Clock-Slowing Debug   Debugging real-time code is challenging. Frequently, specific events can only occur in short windows of opportunity, such as when a particular neuronal event occurs within 100 ms after the subject presses a button. When testing code—for example, by line-by-line execution—such time conditions cannot be satisfied. How can this code be debugged then? MATLAB and various other OS implementations now allow you to 'slow time.' Meaning—one can decide that every processor cycle (the metric used to count time in computers) will account for a different time unit. As in—instead of one CPU cycle accounting for 1/3500 MHz, one can decide that it will actually account for 1 s, therefore providing a slow run of everything. This allows for effective real-time debugging. Obviously the brain still works at the same speed in the real-time environment, so one needs to think of "slowing down the input to the code as well," but given that oftentimes when debugging real-time codes we are working with simulators (see the next paragraph), we should be able to either get "slower brains" by slowing the OS altogether, making our spikes simulator slow as well, or by tweaking our spike generator to simply work at a much lower sampling rate.

Simulating a Brain   Time working with the patient is the most precious of commodities in single neuron recordings, requiring careful testing before performing experiments. Below, we outline methods we found useful: (1) using prerecorded brain data, (2) using existing brain simulators, and (3) generating random spikes. The second and the third options utilize simulated data, with the difference being that one is using a simulator that was provided by a vendor or by the acquisition system (those are now widely supported by many vendors) while the other is generating simulated spikes. Simulators nowadays can go from simple (generating spikes at a Poisson distribution with a given rate for each of multiple neurons—where the parameters can often be tweaked to some extent) to much more complex simulators that actually were designed to simulate a particular brain area or neuronal behavior (i.e., a neuron which changes its firing rate based on event triggers from the environment—equivalent to its preferred stimulus being presented).[3]

Performance Issues   When discussing real-time processing, one question that arises is the notion of timing in the brain. A vast literature has explored the notion of time perception, the effects of minor delays on the perception of real time, and the psychology of "now." We won't address any of these questions in this chapter other than to state that studies (Szelag & Poppel, 2000) suggest that with small variability humans perceive events that happen with a delay of over 100 ms as certainly lagging. That is, if your keyboard would respond to your key presses with the tiny delay of 100 ms, this would surely be noticeable by you and presumably lead you to use a different keyboard. Therefore, when we come to design real-time experiments, and depending on the questions and task, we have to set some boundaries to what constitutes real time and what delays we accept in our code. Given that the short delay between the moment in which a neuron in the brain spiked to the moment it somehow changed the experiment and was perceived by the patient is in milliseconds, we need to make sure that all our analyses and according experimental modification happen within that small timescale. There are a number of performance considerations that should be taken into account in order to minimize the latency between neuronal responses and the desired effects.

First, frequently the analysis of data from an individual recording channel is identical to that from other channels. Thus, parallel processing can be used to take advantage of the recent increase in the availability of multiple CPUs/GPUs/cores. Second, there are many speed improvements that can be utilized by appropriate programming techniques. For example, in MATLAB, preallocating arrays rather than having those grow in size during runtime can affect speed greatly. Usage of correct logical operators (| for array elements versus || for scalar values) improves performance. When using compiled languages such as C or Java, the compiler itself has a set of flags that can be used to improve real-time performance at the expense of memory.

Generally, real-time programs tend to be harder to read because they occasionally improve efficiency at the expense of proper coding. However, some simple tricks can aid in improving and analyzing the codes' performance bottlenecks. CPU timing metrics like the 'tic'/'toc' commands in MATLAB indicate locations where runtime operations spend more time than others. In addition, many languages including MATLAB and C have a variety of "profiling" tools that enable the debugging of code timing following a run.

Here are a few typical bottlenecks in real-time experiments:

Networking Conducting closed-loop experiments frequently requires using multiple hardware systems (such as for acquisition, data analysis, and stimulus display) that need to exchange large quantities of data among them. This requires a fast and dedicated network among those machines. Additional parameters that affect network congestion and speed include the type and amount of data used. Acquisition systems deliver data of several types: raw continuous data, filtered continuous data, spike waveforms after a threshold was applied by the acquisition system, clustered spikes/shapes after they were thresholded and sorted, only the spike peak times, or just the number of spikes within a window (firing rate). Additionally, the amount of data is also influenced by the size of the window of time used to retrieve data, the sampling rate, and the number of channels. Some systems allow the user to control these parameters remotely while others do not. For instance, at present the Blackrock system allows only a choice between continuous, or spikes and shapes; and data are only delivered for all recorded channels together at the sampling rate at which the data were recorded (i.e., there is no possibility to subselect a range of channels). We advise using a local area network with its own routers and switches and short cabling to allow for rapid data transfer.

Computer performance Performance of the hardware/software system itself can greatly affect the speed and precision of online analyses. Open processes, small paging files, or unexpected update checks can harm our experiments. There are tools to remove all but the essential processes when crucial experiments are run, and multiple companies specialize in real-time performance tweaking. First, if possible, minimize the displays. Many acquisition applications show spikes and continuous data on the screen, which can slow down processing. Also, channels which do not yield useful data should be removed from the recording and analysis entirely to save bandwidth and computing power. Also, depending on the analysis performed, the sampling rate can be reduced. Finally, visualization can be minimized to reduce the load. One additional recommendation is to keep the configuration file used for each and every experiment together with the experiment files.

Online Data Access through Application Programming Interfaces (APIs)

The authors of this chapter have worked with two acquisition systems thus far: Neuralynx Cheetah, and Blackrock's Neuroport. The following briefly outlines their methods and what consideration one needs to take into account when handling those.

Blackrock's Neuroport—CBMex

Without going into specific details of the system (which are likely to change rapidly) we will briefly discuss the main logic of the CBMex environment and the main considerations to account for when using it.

CBMex was not written by Blackrock. This means that parts of the code were tailored to very specific needs rather than general users. Currently CBMex is supported only on PC computers with the requirement of a matching CBMex to each Firmware version of Neuroport (i.e., if the Neuroport firmware is upgraded, CBMex must be upgraded also).

CBMex provides the standard commands required from an online application—connecting to the Neuroport locally or remotely ("cbmex('open')"), syncing the clocks of the acquisition system and the experimental machine ("cbmex('time')"), which is the effective synchronization of the remote and acquisition systems, the start of a recording buffer ("cbmex('trialconfig')"), and the polling of data ("cbmex('trialdata')"). Data is polled from the last time the buffer was emptied in every poll, and the parameters of the command determine whether one wants the continuous data in full or simply the spikes structure. The window of time used is a buffer that can get full if data isn't pulled often enough. When the buffer is full, data is overwritten in a circular fashion. The documentation says that pulling every 3 s is safe, but tests we have done show that based on standard computers out there, one can in fact pull data from even much longer windows (~10 s) and still not lose any data. The connection and data transfer is done via the User Datagram Protocol (UDP), and the protocol as well as the actual CBMex C code are readily available.

Notably, the interaction between the Neuroport system and the PC (currently the only OS) computer which manages the acquisition (start, stop, file names, etc.) can be done using cbmex as well ("cbmex('fileconfig')"). CBMex allows for sending triggers from the code to the system, alerting on certain events as well, and altogether gives the basic required functionality.

One major conceptual problem in the current version of the system has to do with certain race conditions that can exist between the cbmex initialization and the acknowledgment from the Neuroport that often results in a crash or freezing of the interaction and an ultimate crash of the MEX interpreter and, subsequently, of MATLAB. This is caused if two cbmex commands are sent in sequence but arrive in conflicting order—say, cbmex('open') followed by comex('trailconfig')—which often results in a crash. The simple workaround involves imposing a short pause between the commands, which is long enough to ensure the arrival of one before the other (this is a UDP-related problem that is often occurring in protocols of that nature). The above code would then be translated to the following: cbmex('open'); pause(0.1); comex('trialconfig'); this is enough to solve this problem.

Neuralynx's Cheetah—NetCom   Neuralynx offers a .NET-based API "NetCom" to interact with the acquisition system. It can therefore interact with any language that supports .NET, including MATLAB. NetCom acts as a client to the Cheetah acquisition server. However, contrary to the Neuroport, which broadcasts its data over the network to the 'network address' (x.x.x.128), NetCom uses a single communication channel and requires registering to open the data. This is effectively similar still to Neuroport since the cbmex('open') command effectively operates as a single channel of transport. Only a single Netcom client can directly communicate with the acquisition system. However, Neuralynx provides an additional program called the "Netcom router" that permits multiple clients to connect to the acquisition system.

Table 6.1
Comparison of the features of online data access provided by the two Food and Drug Administration–approved systems at time of writing

Feature	Blackrock Neuroport	Neuralynx Cheetah
Ability to select a subset of objects rather than all existing objects	–	+
Multiple clients can connect	+	+ (With NetCom router)
Company-provided and supported code	–	+
Support	–	+
Ability to read separately video data, events, spikes data, etc. (Blackrock only events, spikes)	–	+
Documentation	Partial; not used much and not debugged much	Partial, not used much but exists; the documentation file format is Windows only
Support for other platforms	– (Only PC at present)	All .NET platforms (effectively only PC)
Minimal latency possible	1 ms	15 ms (due to 512 samples transfer block size for continuous data)
Ability to receive data at lower sampling rate or with different filters than recorded	–	+

NetCom provides commands such as "NlxConnectToServer" to connect to the acquisition system or router or "NlxGetNewCSCData" to receive new data.

Netcom's strategy is to separate the connection to the server from the data streams themselves, which allows for exchange of system commands before choosing which data to use. Netcom follows a subscription model: Each client can specify which data channel(s) to subscribe to and then subsequently poll data from only this subset of channels. In comparison to Neuroport, NetCom provides more choices to users such as choice of channels and objects to load and does not require the polling of all the objects for any given query. For example, users can first request a list of available objects using "NlxGetCheetahObjectsAndTypes()" (where objects include continuous and spike channels, Event (TTL) channels, video channels, etc.). This is then followed by a choice of which data to subscribe to by indicating to NetCom the relevant streams. Data is streamed in blocks of 512 samples, which can create some difficulties requiring the client to wait for the next buffer in order to get a full spike shape if it happened to be at the edge of a 512-sample block. Table 6.1 compares the features of online data access provided by the Neuroport and Cheetah.

Data Analyses Challenges

Real-Time Experiment Decisions   Unlike offline experiments, which can be analyzed in multiple ways after finishing the experiment, online experiments suffer (and benefit) from the fact that the analysis has to be structured before running the experiment. That is because the experiment flow is part of the analysis and is affected by it. If an experiment tests some learning effect, for instance, and makes the trials harder or easier based on the performance on previous trials, then nearly every experimental run is going to be different for each patient and each session, based on the performance in that very run. That is, the experiment will rarely be exactly the same when repeated twice. Thus the analysis has to take this difference into account and still be able to draw overall conclusions about the population. For example, if learning affects the performance of the patient in the next trial, then the experimental information that led to it must be used for the analysis.

Choices of parameters and analysis methods cannot be changed post hoc, as is usually done. Here is an example of this: In an experiment run by one of the authors (MC) the decision was made that the analysis of spikes would be done on multiunits rather than on online sorted data since the sorting time would slow the processing too much, which could affect the actual runtime decisions (the patient would become aware of the fact that there is a delay between his or her thoughts and their manifestation on the experimental computer; see Cerf et al., 2010). The decision was therefore made to use only unsorted spikes for the experiment. The analysis of existing data showed that this would introduce a certain level of noise, but we felt that the patients would be able to perform well nevertheless. Once the data were collected, we proceeded to perform detailed offline analysis including careful spike sorting to look at the single units that drove the results. While the results were a little different (2% improvement in the performance would have been achieved from single units) we cannot with confidence know the extent of the actual improvement that would have been observed in reality. We cannot say that using single units would have been simply an improvement on using the multiunits, because the additional hidden variable here is how it would affect the patient's performance. It is conceivable that running the experiment with single units would make the patient "work harder" or try different strategies to control the particular brain–machine interface that was used in the experiment, and therefore the performance would have been much improved (or vice versa: maybe the patient would have lost interest and tried less hard since things would be great early on). Simply put, there is no way to draw a conclusion on what would have happened were the experiment conditional on different data than the ones used in the actual experiment.

Choice of Data to be Analyzed Similar to the previous point, we feel the need to emphasize that, more so than in other experiments, online experiments can benefit significantly from prior knowledge on the properties of the neurons one records from.

That is, if one has the capacity (time, patient time, human hours, etc.) to arrive at the patient's room prior to the experiment, acquire data before the beginning of the experiment, and tailor the variables to the particular experiments, this always proves invaluable. In our ideal experiments we set the thresholds and sorting algorithms for the particular patient's needs before experiments start. Then the experiment is run, and the parameters are saved and fixed and used throughout.

Here's a concrete example of such: The Cheetah system, as was mentioned earlier, is able to separate the channels read during the experiments and transfer to the analysis computer only a small set of relevant channels. In the study by Cerf et al. (2010) it became clear to us that we would need to look at the raw data from the acquisition system and filter/threshold/analyze it locally ourselves rather than relying on the filtering done by the acquisition system. Transferring the raw data of a few seconds of multiple channels can be taxing on the network and could lead to an overwhelming strain on the acquisition system that could end up delaying processing or affecting the experiment. Therefore, we began every experiment by a prescreening of the units to be used in the experiment, and we decreased the dimensionality of our units to only channels that had uniquely responsive units (i.e., units responding to stimuli which were not repeated in other channels). This typically decreased the number of channels we looked at from 64 to about 5. This allowed us to focus on smaller amounts of data, and we could even focus on tweaking some of our decoding parameters to the nature of those channels (mainly, how we normalized the firing rates). This only came from knowing our data well and, particularly, knowing that some units in areas like the parahippocampal cortex have higher firing rates and will result in more demanding bandwidth needs than others; this even allowed us to separate in the code the retrieval of channels at different times within our 20-ms window to make sure we controlled the flow of information and analyses. These are the types of knowledge that are hard to change later as suggested in the previous section but could improve the analyses significantly.

Experimental choices for real-time data processing include using multiunits or single units; using raw data or only spikes, LFPs, or firing rates; using all channels or a smaller number of channels; acquiring and sampling all channels or only some (usually one would want all here at the highest sampling rate unless this really poses a strain on the acquisition system); and deciding what network architecture to use.

Bandwidth and Processing Time   In the context of real-time experiments the following variables can impact performance:

Bandwidth = <number of channels> x <window of activity> x <sampling rate>
Data arrival latency = <bandwidth> / <network congestion >
Processing time = <data arrival latency> + <filtering time> + <spike detection time> + [spike-sorting time]

The main choices we can control are the number of channels used for the analysis and the window of activity used. The bandwidth is also governed by the type and formatting of the data. We often consider three data types: the entire continuous data (LFPs), filtered or unfiltered; the thresholded data (spikes); or the firing rate. Bandwidth requirements differ significantly among these.

Storage of Experimental Data   Unlike offline experiments, where the analyses can be fully reproduced postexperiment, the variables in real-time experiments are part of the experiment. As such, it is essential in real-time experiments to maintain as many of the environmental conditions and variables as possible. In order to do so, saving all the data every few trials is recommended.

On the acquisition system side, storing data in the most raw form possible is advantageous. Some systems offer two ways of storing data to facilitate this—in a compact format for future analysis, in addition to a raw format. For example, the Cheetah system saves the filtered continuous data in individual files, one for each recording channel. However, an additional file ("NDMA file") can also be stored, which contains all recorded data in its raw form before filtering (similar to the structure used by Neuroport). This file is very large (for example, 40 GB/h for a 64-channel system) but can be used to fully replicate the experiment offline ("replay"). Such data replay is an excellent means for system development.

Experimental Challenges

Finally, we will address some of the challenges that need to be noted simply due to the running of an experiment online.

Architecture   Figure 6.7 (plate 3) illustrates a typical online experimental setup. Wires lead from the patient's brain directly to the acquisition system, from which the data are transferred over the network to an additional computer (either directly to the experimental computer or, preferably, to a dedicated computer for data analysis), and then, after a decision is made based on the previous data, ultimately fed into the experimental computer to drive the machine in the brain–machine interactive setup.

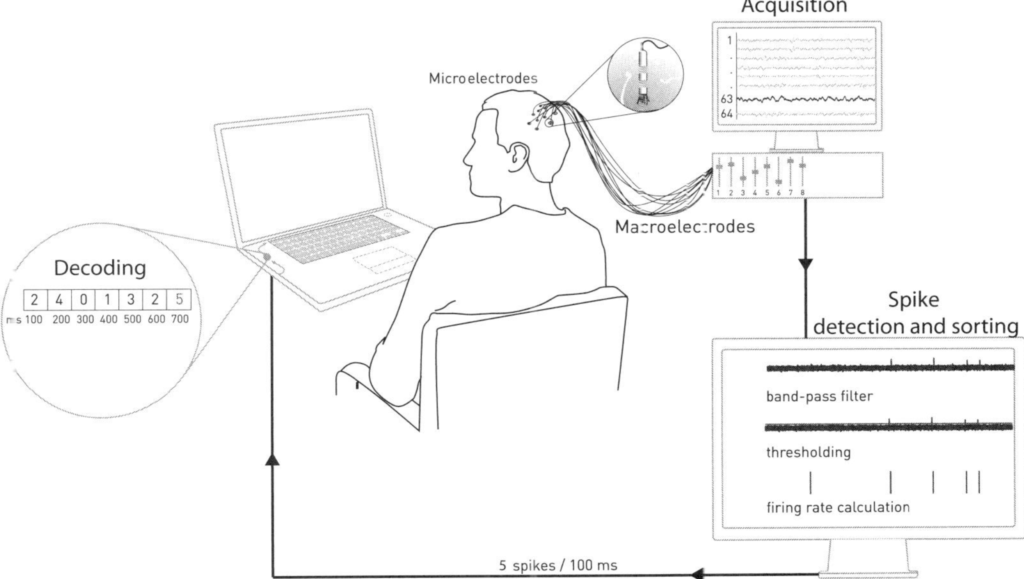

Figure 6.7 (plate 3)
Illustration of a typical online experimental setup.

Real-time experiments present a technical challenge beyond simply setting up a laptop in front of the patient. Instead, a cluster of computers is typically required. In the authors' experience, these challenges end up being difficult to handle. Problems include setting up a network locally, making sure that all computers communicate with each other, and trying to replicate the conditions as much as possible each time (use the same bandwidth, use the same cables, use the same network architecture, etc.).

One versus Multiple Trials   An ideal online experiment is one in which decisions effectively are made in runtime using the data from only one trial. That is, some occurrence in the environment is analyzed in real time and affects the experimental flow immediately. This is hard to do. Decoders often need a significant number of preceding training trials before they can arrive at the stage where a choice can be made over one trial in real time, and many codes will never reach this threshold due to a number of possible problems (decoder, data recorded, noise, etc.). This means that the experimenter will have to make choices in the experiment as to what level of decoding performance is required, how many trials are used for training, and then how many trials are used to make a decision. Each experiment has different needs, and the data drive much of these decisions, but one of the main advantages of single neuron recording is that we can aim for single trial decoding of choice and behavior, unlike with other methods.

In his classical "volition" experiment, for instance, Benjamin Libet could see the principal result (the readiness potential) only by averaging dozens of electroencephalography (EEG) trials. This is still true to date even with the improvement of EEG acquisition methods. Analysis based on LFPs (or ECoG) of single trials is possible (Chen et al., 2013; Maoz et al., 2012) but suffers from the same limitations to some degree. However, recent online experimentation in real-time single neuron recordings in humans (Cerf et al., 2010) and analysis of offline experiments *as if* they were online—to test the potential ability to have analyzed data in single trials (Quiroga et al., 2007; Fried et al., 2011)—suggest that the data have the potential of being ideal for these types of experiments.

Training Length   While the decoding literature suggests ways to estimate the optimal level of training a decoder requires in order to reach best results in the eventual testing phase, online experiments often cannot abide by these calculations. This is merely because we are also bound from above by the patient and clinical constraints. One way to decide when to end the training phase and advance to the testing phase is simply to set a fixed predetermined performance threshold and lock the parameters (or keep updating them to improve performance) once the threshold is met. This, however, could result in some runs of the experiment that never arrive at the testing phase. An alternative is, of course, to ignore the performance altogether and just set a fixed number of trials after which the decoder is locked no matter what and is used for testing regardless. The authors have used both methods in the past and currently propose a combination of the two which is not ideal in most experimental conditions but is the optimal one for real-time experiments in humans: fixing a threshold for ideal move to testing, in addition to setting a maximal number of trials at which one shifts to testing regardless of the performance.

Offline Analysis of Online Data One additional recommendation for conducting real-time experiments is to ultimately analyze the data from the online experiment offline later as if the experiment was not run in a closed loop. Such analysis done by Cerf et al. (2010) showed, in fact, that the performance could have improved only slightly from using single units (rather than multiunits), different filtering, and wider windows of brain data polling. While one cannot change any of the analysis parameters post hoc in an online experiment, one can surely use this information for improved future analyses and for future reporting to determine the ideal parameter for potential online experiments replicating the results or building upon this knowledge. The authors encourage any experimenter considering future online experiments to ultimately report both the results of the online experiment and the results of offline data analyses.

Philosophical and Ethical Considerations in Real-Time Closed-Loop Experiments

Chapter 4 discussed important ethical considerations with respect to recording data from human patients. Here we extend some of those considerations in the particular context of real-time closed-loop experiments.

While our behavior is controlled by our brain at all times, we have the subjective experience (perhaps illusory) of an "interaction" between our brain and our reflective consciousness that is very familiar to us. We operate in the world and feel we are the agents of our behavior. Online experiments profoundly change this access architecture by giving patients direct feedback on their brain activity. There are currently probably only a few dozen humans that were actually able to directly hear the activity of their own brain and manipulate their behavior accordingly. We do not know if and how this feeling and unique experience changes behavior. It may very well be that just the sheer access the individual has to his or her brain allows for changes in neuronal patterns. Simply put, it could be that a person who moves a cursor on a screen based on the activity of a neuron is not just *reading* the neuron's activity but actually modifying the activity in real time. We currently have no way to separate the two cases or test this difference, but as experimenters we should be aware of this concern and think of ways to figure out what is entailed when giving a person external access to his or her thoughts.

In our online experiments, we pride ourselves on the fact that the patients' behavior can be manipulated by the experiment itself and feel that we gain unique knowledge about the way in which the brain operates. However, one needs to remember that while this constitutes acquiring a unique and unprecedented access to the brain, this is also a very atypical situation, which is not necessarily representative of normal behavior. One should also note that, in our experience, all (100%) of our patients who participated in online experiments provided feedback indicating that the experience was fascinating and remarkable for them.

When and Why to Consider Online Experiments

A few things make online experiments advisable for scientists venturing into the world of single neuron recordings in humans:

1. Unlike many noninvasive methods, single neuron recordings can make use of real-time information. Functional magnetic resonance imaging, for instance, does not have sufficient temporal resolution for real-time experiments (with delays of seconds after neuronal activity manifested in the blood-oxygen-level-dependent signal); EEG experiments often require averaging over numerous repetitions in order to arrive at a result that is easily registered in a single trial in the direct readout of a neuron.

2. Online experiments are still new and are in many ways an uncharted territory for scientists. This is especially true for human electrophysiology ones. Thus there is ample room to develop and refine new methods.

3. Online experiments are typically more engaging for the patients. Given the unique nature of access that the patients have to their own brain, our experience suggests that they are often more likely to find these engaging and try harder and with more enthusiasm to explore the level of access they get to their brain. Real-time experiments tend to be more satisfying for the patients.

4. Online experiments allow us to address questions that are at the heart of the most exciting inquiries in neuroscience, such as the nature of free will.

5. Finally, real-time experiments give us access to internal processes in a way that is not simply the equivalent of an offline experiment, just done online, but is actually a practice that provides us with the ability to interact with neurons in a way that circumvents some of the internal network flow. By taking the output of one neuron and effectively using it as the input to a new system, we actually, in a simplified way, create connections in the brain that otherwise do not exist. For example, by taking the output of memory cells in the hippocampus and giving the patient direct access to these neurons (as done in Cerf et al., 2010), the network doesn't just get an additional source of information to interact with but effectively simulates a set of connections from the memory to the visual system, such that input from one feeds into the other. This is something that is not in the original connectivity map but is rather a set of connections that we the experimenters created effectively. Controlling and wisely choosing which two systems to connect can actually allow us to study interactions that otherwise cannot be tested.

Software

On the website (www.humansingleunit.org) we provide open source software as well as links for many of the algorithms and methodology presented in this chapter. In particular, we include:

• A comprehensive open-source closed-loop framework called *StimOMatic* (Rutishauser et al., 2013). It offers a flexible software framework that provides a plug-in architecture to facilitate development of closed-loop experiments. All components discussed in this section are implemented (stimulus display, real-time analysis, continuous and spike-data analysis), and data is processed in parallel on many CPUs/cores in parallel. New functionality can be added by implementing new plug-ins specific to the experiment. These plug-ins are small and self-contained, enabling experimenters to focus on the task at hand. In addition, we provide careful evaluation of the latencies and bottlenecks of the different system components.

• Documented object-oriented event-based code used for the experiments in Cerf et al. (2010).

• A list of links to several freely available algorithms for machine learning and decoding analyses.

• A simulator which allows for setting up of multiple neurons (up to 512 channels at near real time (see note 3).

Notes

1. In some cases, investigators assume that averaging over trials is equivalent to averaging over multiple independent neurons (Parker, A. J., and Newsome, W. T. (1998), "Sense and the single neuron: Probing the physiology of perception," *Annual Review of Neuroscience, 21,* 227–277). The extent to which this approximation holds depends on multiple assumptions and remains a topic of debate in the field.

2. The boundary in figure 6.5 was computed using MATLAB's classify function. A list of links to several freely available algorithms can be found in http://tinyurl.com/klabdecoding.

3. We provide one such simulator on the website www.humansingleunit.org which allows for setting up of multiple neurons (up to 512 channels at near real time), each with its own baseline firing rate, preferred stimulus firing rate, duration, and latency) and the ability to generate single/multiunits per channel. The control of the virtual appearance of the preferred stimulus is governed by external events that can be sent, via TCP/IP, serial port, or direct command, to the listening process, which will alter the particular preferring neurons' firing rate.

References

Anastassiou, C. A., Perin, R., Markram, H., & Koch, C. (2011). Ephaptic coupling of cortical neurons. *Nature Neuroscience, 14,* 217–223.

Bankman, I. N., Johnson, K. O., & Schneider, W. (1993). Optimal detection, classification, and superposition resolution in neural waveform recordings. *IEEE Transactions on Bio-Medical Engineering, 40,* 836–841.

Berens, P. (2009). CircStat: A MATLAB Toolbox for Circular Statistics. *Journal of Statistical Software, 31,* 1–21.

Bishop, C. M. (1995). *Neural networks for pattern recognition.* Oxford: Clarendon Press.

Buzsáki, G. (2004). Large-scale recording of neuronal ensembles. *Nature Neuroscience, 7,* 446–451.

Buzsáki, G. (2006). *Rhythms of the brain.* Oxford: Oxford University Press.

Buzsáki, G., Anastassiou, C. A., & Koch, C. (2012). The origin of extracellular fields and currents—EEG, ECoG, LFP and spikes. *Nature Reviews. Neuroscience, 13,* 407–420.

Canolty, R. T., Edwards, E., Dalal, S. S., Soltani, M., Nagarajan, S. S., Kirsch, H. E., et al. (2006). High gamma power is phase-locked to theta oscillations in human neocortex. *Science, 313,* 1626–1628.

Caplan, J. B., Madsen, J. R., Raghavachari, S., & Kahana, M. J. (2001). Distinct patterns of brain oscillations underlie two basic parameters of human maze learning. *Journal of Neurophysiology, 86,* 368–380.

Cerf, M., Thiruvengadam, N., Mormann, F., Kraskov, A., Quiroga, R. Q., Koch, C., et al. (2010). On-line, voluntary control of human temporal lobe neurons. *Nature, 467,* 1104–1108.

Chen, L., Madhavan, R., Rapoport, B., & Anderson, W. (2013). Real-time brain oscillation detection and phase-locked stimulation using autoregressive spectral estimation and time-series forward prediction. *IEEE Transactions on Bio-Medical Engineering, 60,* 753–762.

Choi, J. H., Jung, H. K., & Kim, T. (2006). A new action potential detector using the MTEO and its effects on spike sorting systems at low signal-to-noise ratios. *IEEE Transactions on Bio-Medical Engineering, 53,* 738–746.

Cristianini, N., & Shawe-Taylor, J. (2000). *An introduction to support vector machines and other kernel-based learning methods.* Cambridge: Cambridge University Press.

Csicsvari, J., Hirase, H., Czurko, A., & Buzsáki, G. (1998). Reliability and state dependence of pyramidal cell–interneuron synapses in the hippocampus: An ensemble approach in the behaving rat. *Neuron, 21,* 179–189.

Donoho, D. L., & Johnstone, I. M. (1994). Ideal spatial adaptation by wavelet shrinkage. *Biometrika, 81,* 425–455.

Fetz, E. E. (1969). Operant conditioning of cortical unit activity. *Science, 163,* 955–958.

Fisher, N. I. (1993). *Statistical analysis of circular data.* Cambridge: Cambridge University Press.

Fried, I., Mukamel, R., & Kreiman, G. (2011). Internally generated preactivation of single neurons in human medial frontal cortex predicts volition. *Neuron, 69,* 548–562.

Fries, P., Reynolds, J. H., Rorie, A. E., & Desimone, R. (2001). Modulation of oscillatory neuronal synchronization by selective visual attention. *Science, 291,* 1560–1563.

Fries, P., Roelfsema, P. R., Engel, A. K., Konig, P., & Singer, W. (1997). Synchronization of oscillatory responses in visual cortex correlates with perception in interocular rivalry. *Proceedings of the National Academy of Sciences of the United States of America, 94,* 12699–12704.

Gabbiani, F., & Koch, C. (1999). Principles of spike train analysis. In C. Koch & I. Segev (Eds.), *Methods in neuronal modeling: From synapses to networks* (pp. 313–360). Cambridge, MA: MIT Press.

Gibson, S., Judy, J. W., & Markovi, D. (2012). Spike sorting: The first step in decoding the brain. *IEEE Signal Processing Magazine, 29,* 124–143.

Gold, C., Henze, D. A., Koch, C., & Buzsáki, G. (2006). On the origin of the extracellular action potential waveform: A modeling study. *Journal of Neurophysiology, 95,* 3113–3128.

Grasse, D. W., & Moxon, K. A. (2010). Correcting the bias of spike field coherence estimators due to a finite number of spikes. *Journal of Neurophysiology, 104,* 548–558.

Griffin, A. L., Asaka, Y., Darling, R. D., & Berry, S. D. (2004). Theta-contingent trial presentation accelerates learning rate and enhances hippocampal plasticity during trace eyeblink conditioning. *Behavioral Neuroscience, 118,* 403–411.

Harris, K. D., Henze, D. A., Csicsvari, J., Hirase, H., & Buzsáki, G. (2000). Accuracy of tetrode spike separation as determined by simultaneous intracellular and extracellular measurements. *Journal of Neurophysiology, 84,* 401–414.

Henze, D. A., Borhegyi, Z., Csicsvari, J., Mamiya, A., Harris, K. D., & Buzsáki, G. (2000). Intracellular features predicted by extracellular recordings in the hippocampus in vivo. *Journal of Neurophysiology, 84,* 390–400.

Hill, D. N., Mehta, S. B., & Kleinfeld, D. (2011). Quality metrics to accompany spike sorting of extracellular signals. *Journal of Neuroscience, 31,* 8699–8705.

Hung, C. P., Kreiman, G., Poggio, T., & DiCarlo, J. J. (2005). Fast readout of object identity from macaque inferior temporal cortex. *Science, 310,* 863–866.

Jacobs, J., Kahana, M. J., Ekstrom, A. D., & Fried, I. (2007). Brain oscillations control timing of single-neuron activity in humans. *Journal of Neuroscience, 27,* 3839–3844.

Jadhav, S. P., Kemere, C., German, P. W., & Frank, L. M. (2012). Awake hippocampal sharp-wave ripples support spatial memory. *Science, 336,* 1454–1458.

Jarvis, M. R., & Mitra, P. P. (2001). Sampling properties of the spectrum and coherency of sequences of action potentials. *Neural Computation, 13,* 717–749.

Joshua, M., Elias, S., Levine, O., & Bergman, H. (2007). Quantifying the isolation quality of extracellularly recorded action potentials. *Journal of Neuroscience Methods, 163,* 267–282.

Kay, S. M. (1993). *Fundamentals of statistical signal processing.* Englewood Cliffs, NJ: PTR Prentice-Hall.

Kim, K. H., & Kim, S. J. (2003). A wavelet-based method for action potential detection from extracellular neural signal recording with low signal-to-noise ratio. *IEEE Transactions on Bio-Medical Engineering, 50,* 999–1011.

Le Van Quyen, M., Foucher, J., Lachaux, J. P., Rodriguez, E., Lutz, A., Martinerie, J., et al. (2001). Comparison of Hilbert transform and wavelet methods for the analysis of neuronal synchrony. *Journal of Neuroscience Methods, 111,* 83–98.

Lewicki, M. S. (1998). A review of methods for spike sorting: The detection and classification of neural action potentials. *Network, 9*, R53–R78.

Logothetis, N. K. (2002). The neural basis of the blood-oxygen-level-dependent functional magnetic resonance imaging signal. *Philosophical Transactions of the Royal Society of London, 357*, 1003–1037.

Manning, J. R., Jacobs, J., Fried, I., & Kahana, M. J. (2009). Broadband shifts in local field potential power spectra are correlated with single-neuron spiking in humans. *Journal of Neuroscience, 29*, 13613–13620.

Maoz, U., Ye, S., Ross, I., Mamelak, A., & Koch, C. (2012). Predicting action content on-line and in real time before action onset: An intracranial human study. *Advances in Neural Information Processing Systems, 25*, 881–889.

Meyers, E., & Kreiman, G. (2011). Tutorial on pattern classification in cell recording. In N. Kriegeskorte & G. Kreiman (Eds.), *Understanding visual population codes* (pp. 517–538). Cambridge, MA: MIT Press.

Miller, K. J., Leuthardt, E. C., Schalk, G., Rao, R. P. N., Anderson, N. R., Moran, D. W., et al. (2007). Spectral changes in cortical surface potentials during motor movement. *Journal of Neuroscience, 27*, 2424–2432.

Mitra, P. P., & Bokil, H. (2008). *Observed brain dynamics*. New York: Oxford University Press.

Najmi, A. H., & Sadowsky, J. (1997). The continuous wavelet transform and variable resolution time-frequency analysis. *Johns Hopkins APL Technical Digest, 18*, 134–140.

Nelson, M. J., & Pouget, P. (2010). Do electrode properties create a problem in interpreting local field potential recordings? *Journal of Neurophysiology, 10*, 2315–2317.

Nelson, M. J., Pouget, P., Nilsen, E. A., Patten, C. D., & Schall, J. D. (2008). Review of signal distortion through metal microelectrode recording circuits and filters. *Journal of Neuroscience Methods, 159*, 141–157.

Nenadic, Z., & Burdick, J. W. (2005). Spike detection using the continuous wavelet transform. *IEEE Transactions on Bio-Medical Engineering, 52*, 74–87.

Obeid, I. (2007). Comparison of spike detectors based on simultaneous intracellular and extracellular recordings. *2007 3rd International IEEE/EMBS Conference on Neural Engineering, Vols. 1 and 2*, 410–413.

Parker, A. J., & Newsome, W. T. (1998). Sense and the single neuron: Probing the physiology of perception. *Annual Review of Neuroscience, 21*, 227–277.

Poggio, T., & Smale, S. (2003). The mathematics of learning: Dealing with data. *Notices of the AMS, 50*, 537–544.

Pouzat, C., Mazor, O., & Laurent, G. (2002). Using noise signature to optimize spike-sorting and to assess neuronal classification quality. *Journal of Neuroscience Methods, 122*, 43–57.

Quiroga, R. Q. (2009). What is the real shape of extracellular spikes? *Journal of Neuroscience Methods, 177*, 194–198.

Quiroga, R. Q., Nadasdy, Z., & Ben-Shaul, Y. (2004). Unsupervised spike detection and sorting with wavelets and superparamagnetic clustering. *Neural Computation, 16*, 1661–1687.

Quiroga, R. Q., Reddy, L., Koch, C., & Fried, I. (2007). Decoding visual inputs from multiple neurons in the human temporal lobe. *Journal of Neurophysiology, 98*, 1997–2007.

Robinson, D. A. (1968). The electrical properties of metal microelectrodes. *Proceedings of the IEEE, 56*, 1065–1071.

Rutishauser, U., Kotowicz, A., & Laurent, G. (2013). A method for closed-loop presentation of sensory stimuli conditional on the internal brain-state of awake animals. *Journal of Neuroscience Methods, 215*, 139–155.

Rutishauser, U., Ross, I. B., Mamelak, A. N., & Schuman, E. M. (2010). Human memory strength is predicted by theta-frequency phase-locking of single neurons. *Nature, 464*, 903–907.

Rutishauser, U., Schuman, E. M., & Mamelak, A. N. (2006). Online detection and sorting of extracellularly recorded action potentials in human medial temporal lobe recordings, in vivo. *Journal of Neuroscience Methods, 154*, 204–224.

Rutishauser, U., Schuman, E. M., & Mamelak, A. N. (2008). Activity of human hippocampal and amygdala neurons during retrieval of declarative memories. *Proceedings of the National Academy of Sciences of the United States of America, 105*, 329–334.

Schmitzer-Torbert, N., Jackson, J., Henze, D., Harris, K., & Redish, A. D. (2005). Quantitative measures of cluster quality for use in extracellular recordings. *Neuroscience, 131*, 1–11.

Siegel, M., Warden, M. R., & Miller, E. K. (2009). Phase-dependent neuronal coding of objects in short-term memory. *Proceedings of the National Academy of Sciences of the United States of America, 106*, 21341–21346.

Singer, J., & Kreiman, G. (2012). Introduction to statistical learning and pattern classification. In N. Kriegeskorte & G. Kreiman (Eds.), *Visual population codes: Toward a Common Multivariate Framwork for Cell Recording and Functional Imaging* (pp. 497–516). Cambridge, MA: MIT Press.

Szelag, E., Dreszer, J., Lewandowska, M., & Szymaszek, A. (2009). Neural representation of time and timing processes. In E. Kraft, B. Gulyás, & E. Pöppel (Eds.), *Neural correlates of thinking* (pp. 187–199). New York: Springer.

Szelag, E., & Poppel, E. (2000). Temporal perception: A key to understanding language. *Behavioral and Brain Sciences*, *23*, 52.

Torrence, C., & Compo, G. P. (1998). A practical guide to wavelet analysis. *Bulletin of the American Meteorological Society*, *79*, 61–78.

Vapnik, V. N. (1995). *The nature of statistical learning theory*. New York: Springer.

Viskontas, I. V., Ekstrom, A. D., Wilson, C. L., & Fried, I. (2007). Characterizing interneuron and pyramidal cells in the human medial temporal lobe in vivo using extracellular recordings. *Hippocampus*, *17*, 49–57.

Wilson, M. A., & McNaughton, B. L. (1993). Dynamics of the hippocampal ensemble code for space. *Science*, *261*, 1055–1058.

Wiltschko, A. B., Gage, G. J., & Berke, J. D. (2008). Wavelet filtering before spike detection preserves waveform shape and enhances single-unit discrimination. *Journal of Neuroscience Methods*, *173*, 34–40.

Womelsdorf, T., Fries, P., Mitra, P. P., & Desimone, R. (2006). Gamma-band synchronization in visual cortex predicts speed of change detection. *Nature*, *439*, 733–736.

Zaghloul, K. A., Blanco, J. A., Weidemann, C. T., McGill, K., Jaggi, J. L., Baltuch, G. H., et al. (2009). Human substantia nigra neurons encode unexpected financial rewards. *Science*, *323*, 1496–1499.

Zanos, T. P., Mineault, P. J., & Pack, C. C. (2011). Removal of spurious correlations between spikes and local field potentials. *Journal of Neurophysiology*, *105*, 474–486.

II COGNITIVE NEUROSCIENCE FINDINGS AND INSIGHTS

7 Single Neuron Correlates of Declarative Memory Formation and Retrieval in the Human Medial Temporal Lobe

Ueli Rutishauser, Erin M. Schuman, and Adam N. Mamelak

A key function of the nervous system is to rapidly assess incoming sensory information with respect to previous experience. Many behaviors are conditional on both current and past experience. Such conditional branching requires memory, which allows an animal to learn. The ability to learn rapidly is crucial for survival (Sokolov, 1963). As a result, mammalian brains evolved many specialized learning systems that operate in parallel and jointly support behavior (Knowlton & Squire, 1993; Schacter & Tulving, 1994; Poldrack et al., 2001; Squire, 2004). One such system provides declarative memory, which enables the rapid transformation of experiences (episodes) into long-term memories (Tulving, 2002). These memories can later be accessed (i.e., declared) either by free recall or by familiarity. In humans, the ability to rapidly form new memories even after a single learning trial depends on the medial temporal lobe (MTL), in particular the hippocampus and its surrounding structures (Squire & Alvarez, 1995; Squire et al., 2004). The capacity to store new information after a single exposure of the declarative memory system is large. For example, studies show that human subjects can easily encode 10,000 novel stimuli after a single exposure lasting only seconds each (Standing et al., 1970; Standing, 1973). The neural mechanisms that support this capacity are poorly understood.

There are prominent animal models of MTL-dependent memory; for example spatially tuned hippocampal neurons in rodents are thought to support spatial memory. Episodic memory and the conscious experience of declaring a memory, however, are assessed by verbal report and thus are impossible to measure directly in animals. Episodic memory has been suggested to be largely unique to humans, with some features already present in primates (Hampton, 2001; but see Fortin et al., 2004). Human single unit recordings are uniquely positioned to contribute toward our understanding of the neural mechanisms of MTL-dependent memories above and beyond what can be learned from animal models. In this chapter, we review insights that have been obtained from single unit recordings in awake behaving humans engaged in declarative memory tasks, together with experimental considerations useful for the design of future memory experiments.

Experimental Design

The series of experiments summarized in this review are based on a common experimental paradigm (with exceptions noted). Here, we will outline the design criteria as well as important practical limits that are useful when designing memory experiments with subjects for intracranial recordings.

Our goal was to assess declarative memory for novel (never seen before) images. To allow meaningful simultaneous behavior and electrophysiological analysis, pictures were presented one at a time. This facilitates analysis as neuronal responses can be attributed directly to a stimulus. In contrast, presenting multiple stimuli at the same time makes meaningful analysis difficult. For example, asking patients to choose between two presented stimuli where one is familiar and one novel makes it impossible to distinguish a novelty from a familiarity response without taking into account other variables such as gaze position. We presented complex novel visual stimuli (photographs) for a short period of time (2–4 s) on the screen of a notebook computer situated at the patient's bedside (see figure 7.1). Patients were asked to carefully look at the stimuli for a later memory test. In some experiments, patients were also asked to remember other attributes of the novel stimuli such as where they were shown on the screen (spatial recall, see figure 7.1A) or what the background color of the screen was when the images were shown (associative recall). We asked a simple question about each stimulus at the end of a trial, such as whether an animal was present or not. Learning trials where this question was answered incorrectly were rare (typically <2%). We hypothesize that this control question encouraged attention to the stimuli, but we did not verify this experimentally. After viewing the to-be-learned stimuli, subjects were engaged in an attention-demanding nonmemory task to prevent active rehearsal. The delay period between encoding and testing was 10–30 minutes, accounted for by both the distraction task and rest.

Following the delay, subjects were shown another sequence of images. These were either identical to the images seen during learning (familiar images) or novel (not seen before). After each image, patients were asked to indicate whether they had seen the image before or not (see figure 7.1). In some tasks, patients had to make a forced choice between new and old (figure 7.1A) whereas in others this was combined with a 1–6 confidence rating (figure 7.1B). Also, in tasks with a recollection component, subjects were asked to indicate additional attributes about the learning experience such as where the image was on the screen when it was first seen (quadrant of the screen or left/right side of the screen). The 1–6 confidence rating is most powerful because it allows the calculation of behavioral receiver operating characteristics (ROCs) that permit the careful quantification of the patient's behavior. We found that patients generally have a good sense of the quality of their memories and they like to indicate whether they are simply guessing or are certain of their memory.

A critical question is what number and difficulty of images can be presented to patients. Keeping in mind that many epilepsy patients have memory problems due to their seizures (Hermann et al., 1997; Viskontas et al., 2000; Helmstaedter, 2002), a wide spectrum of memory

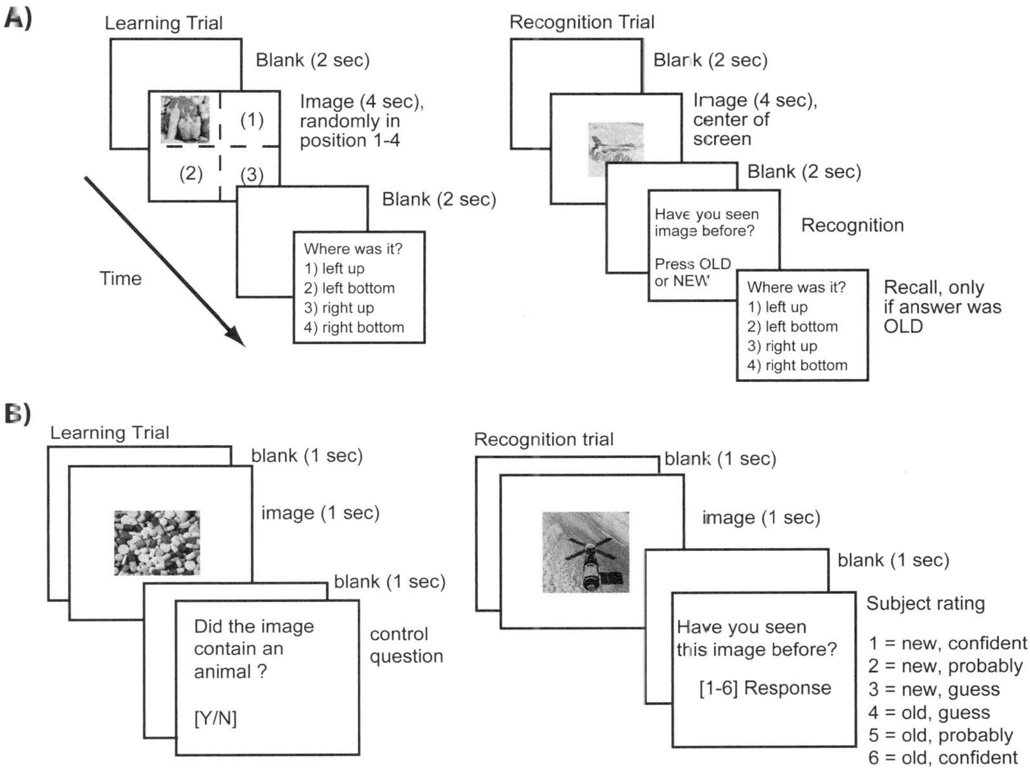

Figure 7.1
Structure of the task. (A) Version with no recall component. (B) Version with spatial recall component. Images shown to patients were in color but here are shown in black and white for illustration only.

abilities is typically seen in the patient population. For example, in a sample of 9 patients (see table S1 in Rutishauser et al., 2010) we found that the average scores on the memory components of the Wechsler Memory Scale—Revised (WMS–R) neuropsychological test were significantly below what would be expected in a normal control group. However, we also note that there was substantial ability with some patients performing much above what would be expected. In order to relate the neurophysiological response to behavior, we require both significantly above-chance performance as well as a sufficient number of error trials. Thus, the stimulus set has to facilitate good-enough but not perfect performance. We found that this is not feasible with a fixed stimulus set. Rather, we assessed the memory abilities of each patient based on the patient's neuropsychological evaluation (WMS–R scale) as well as a short practice version of the test. Based on this assessment, we then administered a stimulus set of appropriate difficulty. The difficulty of the stimulus sets is modified by changing the within-category differences between images. That is, for the example of cars, an easy stimulus set will have several images of cars which are obviously different from each other where a difficult set will have cars which are harder to distinguish (i.e., several red compact cars). We presented the following numbers of stimuli per learning block: 12 with spatial recall with 4 possible choices (25% chance level), 50 with a recall component with 2 possible choices (50% chance level), and 100 without a recall component (familiarity only). During the retrieval block, we presented 24 images (12 new, 12 old) with recall with 4 choices, 50 (25 new, 25 old) with recall with 2 choices, and 100 (50 new, 50 old) for recognition only. Stimuli were presented either for 2 or 4 s each (depending on the experiment). We found that generally, 4 s is the upper limit for stimulus presentation as many patients will find it too slow. To minimize possible motor artifacts, we implemented the task such that explicit question screens were displayed after each image (such as "Did you see this image before?"). If answer keys were pressed before this question appeared, the response was not accepted.

To motivate patients to perform the memory tasks, we made sure that tasks were not too hard (which is frustrating) as well as providing real-time feedback. At the end of every block, a display immediately showed performance metrics such as percentage correct. We found that this greatly encourages participation and patients will work hard to improve their scores. For example, we observed patients who were only reluctantly performing the short practice test but who, after seeing that feedback was displayed at the end (and noted by the experimenter), would perform better and pay more attention in subsequent tasks.

Due to the unique clinical population and small sample sizes, the relevance for results derived from this population for the normal population is not immediately clear. To facilitate comparisons by others, we consider it useful to present the neuropsychological evaluation of the cognitive abilities of the experimental subjects based on standardized tests. The scores on well-known neuropsychological tests allow clinical readers to assess how the behavioral and neuronal observations relate to other populations with different cognitive and memory abilities. We generally observe that our experimental subjects perform below average but well within the normal range on such tests such as the WMS–R or Wechsler Adult Intelligence Scale–III (Wechsler, 1997a, 1997b). Both tests are routinely performed as part of the clinical workup of patients, and we find it useful to collect and report scores.

Novelty Responses and Single Trial Learning in the Human MTL

The assessment of stimulus novelty is a prerequisite for many kinds of learning (Kohonen & Lehtio, 1981; Stark & Squire, 2000; Welzl et al., 2001; Li et al., 2003; Davis et al., 2004; Yamaguchi et al., 2004). Physiological variables such as frontal scalp event-related potentials and skin conductance differ shortly after onset of novel compared with familiar stimuli. This difference is abolished in MTL lesion patients (Knight, 1996). Also, behavioral experiments with hippocampal lesion subjects show that one of their primary deficits is the absence of a boost in memory performance normally triggered by salient stimulus attributes. In one study, MTL lesion subjects showed reduced (but above chance) recognition memory performance but, strikingly, had a complete lack of performance improvement for items shown together with task-irrelevant novel attributes. In contrast, normal control subjects had a substantial gain in memory performance for these items (Kishiyama et al., 2004). This property of memory was originally described by von Restorff (1933). Together, this indicates that the "von Restorff effect" (Wallace, 1965; Einsbour & George, 1974; Hunt, 1995; Parker et al., 1998) is driven by neuronal mechanisms that reside in the MTL.

An intriguing feature of a novelty response is its plasticity—a stimulus is novel only once, so if it is seen again, the response should either be diminished or completely abolished because the stimulus is now familiar. Novelty responses are thus a sensitive measure to quantify learning and plasticity.

Neurons that respond differently when subjects are presented with an identical stimulus repeatedly are good candidates to explore the neural mechanisms of rapid learning. Questions of interest for neuronal responses that express a rapid change between the first and second time a stimulus is presented are as follows: (1) direction of change (is there an increase or decrease in firing rate?), (2) tuning of the response (is it stimulus-specific or abstract?), and (3) is the magnitude of the change fixed, or does it relate to the memory/learning experience in a specific way?

We identified a subpopulation of single neurons (see chapter 6 for how neurons were isolated) in the hippocampus and the amygdala that showed striking differences in their spiking response when comparing novel with familiar stimuli (Rutishauser et al., 2006; Rutishauser et al., 2008). There were two subgroups of cells, one which increased firing (relative to baseline) for novel stimuli only and a second group which increased their firing rate only when familiar stimuli were presented. In total, approximately 20% of recorded cells showed this behavior (40 out of 244 from 6 subjects in one study). Figure 7.2 (plate 4) shows an example. The difference could be seen as soon as 10 minutes after presenting the novel stimulus (the shortest period tested) and can be seen at least 24 hours later (the longest period tested). The tuning of the cells was not stimulus specific as novelty and familiarity sensitive neurons responded to stimuli of different categories (such as animals, houses, trees, landscapes, toys, fruits). That is, the same cell would indicate the novelty of a stimulus regardless of which category the stimulus was from. This is in contrast to the subpopulation of cells which show highly selective responses to particular familiar stimuli such as a particular person or place (see chapter 8 in this volume). These cells are visually selective (to visual categories, individuals, or specific objects) but nevertheless also show changes in firing rate as a function of repeated presentation even when subjects are only passively viewing stimuli without an explicit memory task and shorter time intervals between repetitions (Pedreira et al., 2010). Interestingly, similar novelty/familiarity sensitive responses in the monkey MTL that have been described were also not stimulus selective (Rolls et al., 1982; Riches et al., 1991; Fahy et al., 1993; Rolls et al., 1993; Xiang & Brown, 1998). Viskontas et al. performed a similar task to ours and found novelty and familiarity sensitive cells in the parahippocampal gyrus, entorhinal cortex, hippocampus, and amygdala (Viskontas et al., 2006). These cells were similarly abstract and untuned. Interestingly, Viskontas et al. (2006) also found

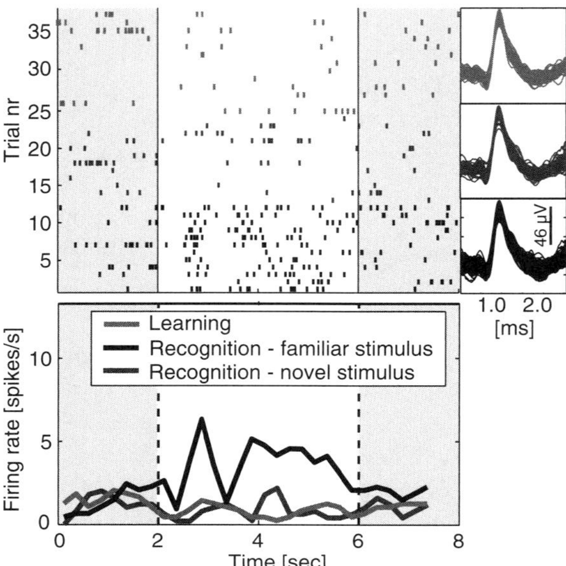

Figure 7.2 (plate 4)
Example of a neuron in the hippocampus that increases its firing rate to familiar but not to novel stimuli (a "familiarity detector"). The same images were presented twice, initially during learning (green trials) and later again during retrieval (red trials). Also during retrieval other, novel stimuli (blue trials) were shown. Trials were randomly intermixed but are shown sorted for display purposes only. The waveforms associated with the trials of the three different categories are shown on the right, confirming that the same neuron was present in both parts of the experiment. The average-firing-rate histogram at the bottom summarizes the firing-rate difference between novel (green) and familiar (red) trials. Notice that the visual input for both trials was the same. The only difference was that stimuli had been seen before (red trials). This cell is thus a memory cell. nr, number. Modified from Rutishauser et al. (2006).

a subset of cells which signaled novelty selectively only for a particular category (scenes or faces). Together with our results, it seems that there are at least two distinct classes of novelty-sensitive neurons in the human MTL: abstract and visually tuned. The first class of untuned general novelty detectors could serve to signal the significance of stimuli during the acquisition of new memories (Lisman & Otmakhova, 2001). It has been suggested that such neurons trigger dopaminergic release through projections to the ventral tegmental area (Lisman & Grace, 2005), which in turn strengthens plasticity in the hippocampus.

MTL Responses during Behavioral Recognition Memory Errors

Errors in recognition memory lead to either forgetting a stimulus (mistakenly identifying it as new) or falsely recognizing a stimulus as familiar (a false memory). What is the source of such errors? In the case of forgetting, was there no memory established in the MTL, that is, did the stimulus truly appear to be novel from the point of view of MTL activity, or did the patient s mply press the wrong button? Alternatively, the MTL memory signal could have been present but was not taken into consideration by circuits elsewhere that lead to the motor decision. Similar questions arise for false memories (Schacter & Dodson, 2001). Was there truly a response in the memory systems of the brain that indicated that the stimulus was familiar, or was the source of this error rather in decision making or motor output?

We investigated these questions by single trial analysis of the neuronal responses we had previously identified based on the behaviorally correct trials. The error trials were statistically independent as they were not previously considered for selection of the memory neurons. We proceeded to train a population decoder that had access to all simultaneously recorded novelty and familiarity sensitive neurons in a particular session. The decoding algorithm takes into account the spike counts of all neurons and decides whether the currently viewed stimulus is new or old. Neurons coded this information very reliably: The decoder could tell (for correct trials) whether the stimulus was new or old on a single trial basis with an accuracy of 75% based on spike counts of a single previously identified novelty/familiarity neuron. Performance increased to 93% if six neurons were considered (see figure 4 in Rutishauser et al., 2006, for details). We proceeded to use the same decoder to analyze the error trials and found that for 75% of error trials, the decoder predicted what the correct response would have been (but which was not given by the subject). Having access to the spiking activity of MTL neurons that are sensitive to novelty thus allows a simple decoder (that counts the number of spikes) to outperform the patient engaged in the task. The neurons thus have better memory than the patient demonstrated behaviorally, showing that information about the forgotten memories remained available. Together with evidence that these neurons do not represent motor actions or decisions (discussed below), we hypothesize that the apparent "failure" to recognize the stimuli is due to decision making about the memories rather than simply pressing the wrong button. Regarding a stimulus as novel if only weak evidence is available is optimal in some situations, and this decision bias is likely a function of the different types of errors are rewarded/punished. This suggests that our decoder outperformed the patient because it had a different decision threshold than the patient. This hypothesis remains to be explored.

MTL Responses during Free Recall of Spatial Location

Episodic memories allow us to remember not only whether we have seen something before but also where and when. Thus, an episodic memory combines multiple pieces of information into a unitary memory. When recalling an experience, we are vividly aware that we have personally experienced this event and can recall many attributes associated with it (Tulving, 2002). However, for some memories, recollection fails us, and all that remains is a strong feeling of familiarity. This is often summarized as the "butcher on the bus" phenomenon (Mandler, 1980). In this case, recognition memory correctly labeled the person as familiar, but no associated attributes could be retrieved. We asked patients to remember for a sequence of stimuli both which pictures they had seen but also where they were presented on the screen. Stimuli appeared randomly in one of four quadrants on the screen. Later, during retrieval, stimuli were shown at the center of the screen and patients were asked to indicate the location the stimulus was shown in when it was first presented if they indicated that a stimulus was familiar. This experiment served as the behavioral paradigm to investigate the neural footprints of recognition with and without associated spatial recollection.

We compared trials where both recognition and recollections succeeded with trials where only recognition succeeded. We refer to the first and second category as R+ and R–, respectively (see figure 7.1B). The same criteria to select a population of novelty and familiarity signaling neurons as described in the previous section was used in this analysis (114 out of the 412 well-separated units from a total of 17 sessions from 8 patients distinguished novel from familiar stimuli; see Rutishauser et al., 2008). The response was significantly different between R+ and R– trials for novelty as well as familiarity signaling neurons. The differences, however, were of different polarity for the two classes of neurons. Neurons that increased their firing rate for familiar stimuli had a higher firing rate for R+ trials compared to R– trials. In contrast, neurons that increased their firing rate for novel stimuli had a lower firing rate for R+ trials compared to R– trials (see figure 7.3, plate 5). This indicates that both classes of neurons fire proportionally to the strength of the memory but with opposite slopes. This leads to the following strength-of-memory-gradient hypothesis: The stronger the memory, the stronger (weaker) the firing response for familiarity (novelty) neurons is, respectively (see figure 7.3A, plate 5). This gradient would start with errors (stimuli forgotten, i.e., a very weak memory), continue with stimuli that were only recognized but had no recollection (R–), and ends with stimuli that were both recognized and recollected (R+). We used a single trial measure of the response of each neuron to quantify this gradient as described below.

We defined a response index $R_{I,j} = (N(i) - mean(NEW))/(mean(baseline))\ C_j\ 100\%$ where $N(i)$ is the number of spikes after stimulus onset in trial i, j is the neuron index, and C_j is equal to –1 for neurons that increase their rate for novel stimuli and equal to 1 for neurons that increase their rate for familiar stimuli. The response index quantifies how different the response to a familiar image was relative to baseline and the response to novel stimuli. It is, by definition, negative for neurons that increase their firing rate in response to novel stimuli. To allow pooling of both classes, the response index was multiplied by the constant C_j as indicated.

Some patients with poor memory could not recollect the spatial location above chance level whereas others performed significantly above chance ($77 \pm 6\%$ vs. $35 \pm 4\%$; chance level of 25%). Both groups had good recognition memory (see figure 1 in Rutishauser et al., 2008, for details). Comparing R+ and R– trials in both groups separately, we found that R+ trials had a significantly stronger response index compared to R– trials (67% vs. 45%; see figure 7.3B, C, plate 5, for statistics). The 22% difference in firing rate was due to successful spatial recollection. Patients with at-chance recollection performance still had R+ as well as R– trials either due to some remaining memory or due to chance. In contrast to the above-chance patients, however, there was no difference between R+ and R– trials in this patient group (response index of 45% vs. 46%; see figure 7.3D, plate 5). The at-chance recollection patient group's ability to

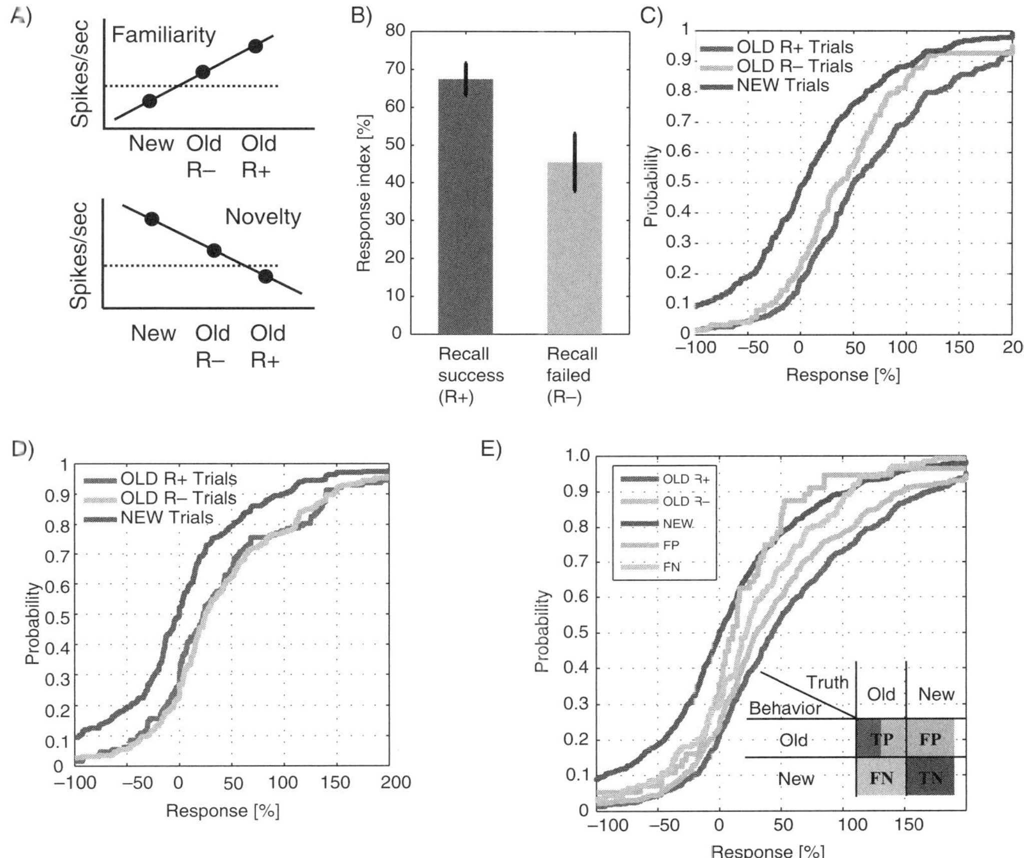

Figure 7.3 (plate 5)
(A) Continuous strength of memory representation hypothesis illustrated for neurons that increase their firing rate in response to the presentation of familiar (top) and novel (bottom) stimuli. (B) Average response index for trials with successful (R+) and failed (R−) recall. The difference was significantly different (p < 0.01). Error bars are ±SE over n = 386 and 123 trials, respectively. (C) The same data as shown in (B) but replotted as a cumulative distribution. Notice the two shifts to the right of the entire distribution: the first due to familiarity (to R−) and the other by recollection (to R+). (D) For patients who could not recollect, there was no significant difference between R+ and R− trials (p = 0.53, Kolmogorov–Smirnov test). (E) Activity during errors reflects true memory rather than behavior. Response indices were significantly different between forgotten stimuli and novel stimuli (p < 0.001) as well as between falsely remembered stimuli and novel stimuli (p < 0.01) (both Kolmogorov–Smirnov test). Modified from Rutishauser et al. (2008). TP, true positive; FP, false positive; FN, false negative; TN, true negative.

recognize stimuli was statistically not different from the above-chance recollection group. Thus, the inability to recollect in the at-chance patients was likely not due to a general capacity reduction in memory but rather a selective deficit to recollect. We thus conclude that (1) novelty-sensitive MTL neurons code for both familiarity- as well as recollection-based memories and (2) this difference was abolished in patients which were not able to recollect, indicating that their MTL was not able to support spatial recall.

Trials where patients made errors also followed the same gradient (see figure 7.3E, plate 5) for both false negatives (FNs; forgotten) and false positives (FPs; falsely remembering). In both cases, the response index was significantly larger compared to the response to novel stimuli (14% and 19%, respectively). At the same time, the response was significantly weaker in both cases relative to familiar stimuli that were correctly recognized (see figure 7.3E, plate 5). This pattern of activity during behavioral errors is consistent with the idea that the neurons represent the strength of a memory continuously.

An additional conclusion that can be drawn from the error trial analysis is a distinction between motor responses and memory signals. There are separate buttons on the keyboard that subjects press to indicate whether they think a stimulus is new or old. Subjects often use two different fingers or hands for these buttons. It is thus conceivable that, rather than a memory signal, the signals reported here are motor-related signals tuned to a particular motor action. However, the error trials argue against this interpretation: The button pressed for R+, R−, and FP signals was the same ("Old" button) as was the button pressed for true negative (TN) and FN trials ("New" button). The neuronal responses clearly covary with the memory type rather than with the motor output (see figure 7.3E, plate 5). Such motor-output-related neurons (such as button presses) have been documented in the human MTL (Halgren et al., 1978; Heit et al., 1988, 1990), and this distinction is thus important. Similarly, the same distinction also argues against these signals' representing a decision about the memory as the decision was equivalent to the motor output.

Novelty and Familiarity Responses in Epileptic Areas

Was the neuronal response reported here influenced by changes induced by epilepsy? All subjects for this study have been diagnosed with epilepsy and had electrodes implanted bilaterally for seizure localization. Thus, typically there are electrodes in areas which are later deemed nonepileptic (if the seizure onset is well localized). This situation allows us to directly compare memory-related responses in tissue that was later declared epileptic with tissue that was nonepileptic. We found a comparable (but stronger) response in this "healthy" neuron population (see figure 7.4D). Similarly, we find that neurons from the "to be resected" tissue still exhibited a response to old stimuli (see figure 7.4E). This response was, however, weaker, and there was no significant difference between recollected and not-recollected stimuli. The absence of a significant difference in response to R+ and R− trials in epileptic tissue indicates that MTL areas which are in the seizure origin are damaged and have only reduced capacity to support memory

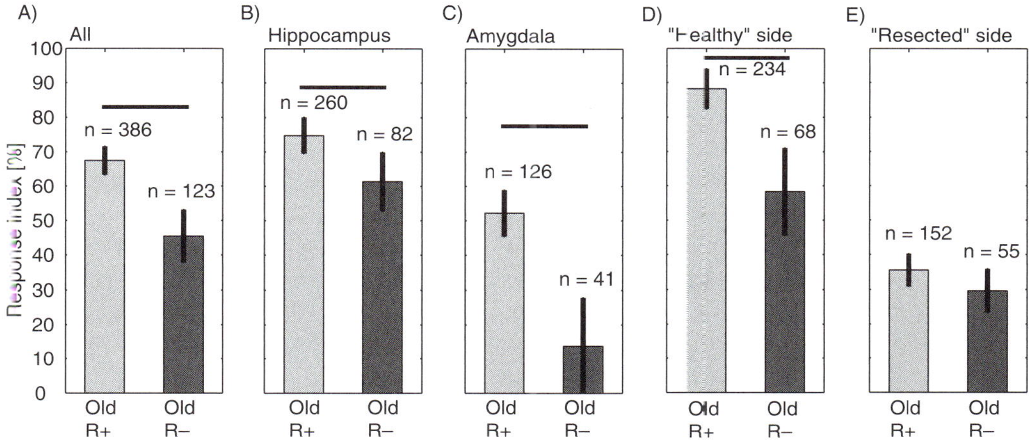

Figure 7.4
Novelty and familiarity response differences in epileptic areas. In this figure, only neurons from sessions with a 30-minute delay and successful recollection (30-minute R+) are included. (A) All trials from all areas for comparison. (B) Only trials from hippocampal neurons. (C) Only trials from amygdala neurons. (D) Only trials from the "healthy" hemisphere, that is, areas which were outside of the area of seizure onset. (E) Only trials from neurons in the eventually resected hemisphere. In (A–D), the response to R+ compared to R− trials is significantly different ($P < 0.05$, two-tailed Kolmogorov–Smirnov test; cf. figure 7.3B, plate 5). The response in (E) is not significantly different. Modified from Rutishauser et al. (2008).

function. This finding also indicates that it might be possible to use indices of novelty responses for diagnostic purposes, for example, for seizure localization or determination of whether the nonepileptic areas are capable of supporting memory as preparation for surgery. A neuron could show a reduced response without being part of the epileptogenic area itself because it receives less input from other neurons that are in the epileptogenic area. However, the measure reported above compared only hemispheres rather than more fine-grained anatomical areas. At present it is unknown whether a reduced response implies that the neuron is inside the epileptogenic zone or whether neighboring neurons can also exhibit this deficit. Our results indicate, however, that such transfer across hemispheres is minimal. Whether this is representative of normal brain organization or the result of reorganization in response to chronic epilepsy remains unknown.

Prediction of Memory Formation by Theta Phase Locking during Encoding

Are responses to novel stimuli indicative of whether a memory for a stimulus will later be available? If so, that would indicate that novelty responses are directly involved in the process of learning. Many factors modulate the probability that a memory for a stimulus will be formed. Examples include attention, arousal, motivation, and saliency of the stimulus (Paller & Wagner, 2002). Structurally, the modification of synaptic circuits by plasticity mechanisms is thought to underlie memory formation (Martin et al., 2000). How synaptic plasticity leads to memories, however, remains poorly understood. A possibly link between circuit function and learning is suggested by findings based on the theta rhythm, a prominent oscillation in the 3–8 Hz frequency range that has been linked to memory formation in several species. For example, animals that have more prelearning baseline theta activity will learn a task faster compared to animals who have less baseline theta activity (Berry and Thompson, 1978; Seager et al., 2002). The conditional presentation of stimuli selectively when theta activity is high increases the learning rate (Seager et al., 2002; Griffin et al., 2004). Extracellular stimulation applied to induce plasticity has a differential effect conditional on whether it is applied at the peak or trough of ongoing theta oscillations in rats; LTP is induced only when the stimulation is applied when the theta oscillation is at a particular phase (Pavlides et al., 1988; Holscher et al., 1997; Hyman et al., 2003; McCartney et al., 2004). These findings from the animal literature indicate that theta oscillations and the timing of neuronal activity relative to the ongoing theta oscillations have a strong influence on plasticity as well as learning, suggesting the possibility that the two are functionally linked by theta oscillations. Whether neuronal activity triggered by visual stimulation (rather than electrical) would similarly depend on theta oscillations to be transformed into a long-term memory, however, is unclear. Based on the animal experimental data discussed above, we hypothesized that the phase locking of individual neurons relative to ongoing theta oscillations would predict whether a memory was formed for a novel stimulus.

The human brain exhibits prominent theta oscillations (Huh et al., 1990; Kahana et al., 1999; Cantero et al., 2003), and some neurons in the MTL fire preferentially at a particular phase of ongoing theta oscillations (Jacobs et al., 2007; Rutishauser et al., 2010). The power of theta oscillations measured on the scalp or on the cortical surface with intracranial EEG itself can be predictive of whether a memory is formed or not (Klimesch et al., 1996; Sederberg et al., 2003). In a set of 305 neurons recorded in the amygdala (n = 185) and hippocampus (n = 120) of 9 patients we found that 21% were significantly phase locked to the ongoing theta oscillation (see chapter 6 for methods to estimate phase locking). These theta phase-locked cells thus had an increased probability of firing spikes at a particular phase of the ongoing theta oscillation (see figure 7.5 for an example). We quantified the strength of phase locking in particular trials based on the spike-triggered average (STA) and the spike–field coherence (SFC; see chapter 6 for details). Individual neurons showed striking differences in their STA (figure 7.6B) as well as SFC (figure 7.6A) in the theta band between trials that were later remembered compared to trials that were later forgotten. Notice that the SFC quantifies the strength of phase locking regardless of the phase that the locking occurs to. Thus, what is relevant for whether a memory will be formed or not is whether the preferred phase of a particular neuron was followed faithfully and not the absolute phase. While the SFC of the neurons was predictive of memory formation, we found no significant difference between the firing rates. While a substantial proportion of neurons increased their firing rate in response to the to-be-learned stimuli (21%), the extent of that increase was not indicative of whether a memory was formed or not. What was predictive, in contrast, was whether the spikes that were fired were phase locked to ongoing theta or not. Note that in a different task, where patients remembered word pairs (associative memory) by

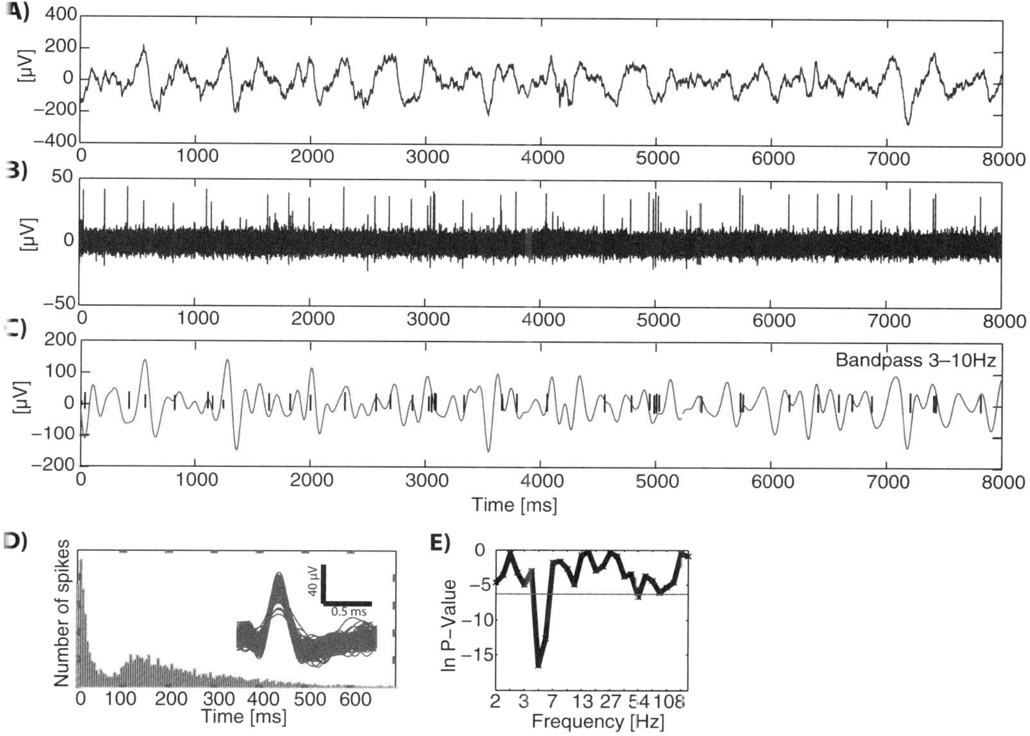

Figure 7.5
Theta activity in the human medial temporal lobe. (A–C) Example recording from a microwire, shown filtered with a 2-Hz high-pass filter (A), in the spike band with a 300-Hz high-pass filter (B), and filtered in the theta range, 3–10 Hz. (C). (D) Waveforms and interstimulus interval (ISI) of a single neuron isolated on the same channel shown in (A). Notice the bimodal ISI, indicating modulation by a slow rhythm. (E) Phase locking of the neuron shown in (D) evaluated by a Raleigh test as a function of frequency. The horizontal line indicates the threshold for significance at p = 0.0018 (Bonferroni corrected p = 0.05). The neuron was maximally phase locked at 4.8 Hz. Ln, logarithm. Adapted from Rutishauser et al. (2010).

repeatedly seeing the same pair of words, neurons were found in which the firing rate at encoding predicted later free recall success (Cameron et al., 2001). There are many differences in the two tasks, including the use of familiar stimuli (well-known words) and no recognition component (there was only free recall) that could account for the differences. Future experiments are needed to tease apart the differences between free recall alone or together with recognition, a potentially rich set of experiments left to be done.

Patients also indicated what the confidence of their memories was during a later memory test. Based on the reported confidence, we classified the learning trials as either "later forgotten," "later weak memory," or "later strong memory." Interestingly, the SFC at the time of learning was already indicative of whether a memory was later strong or weak. Pairwise comparisons

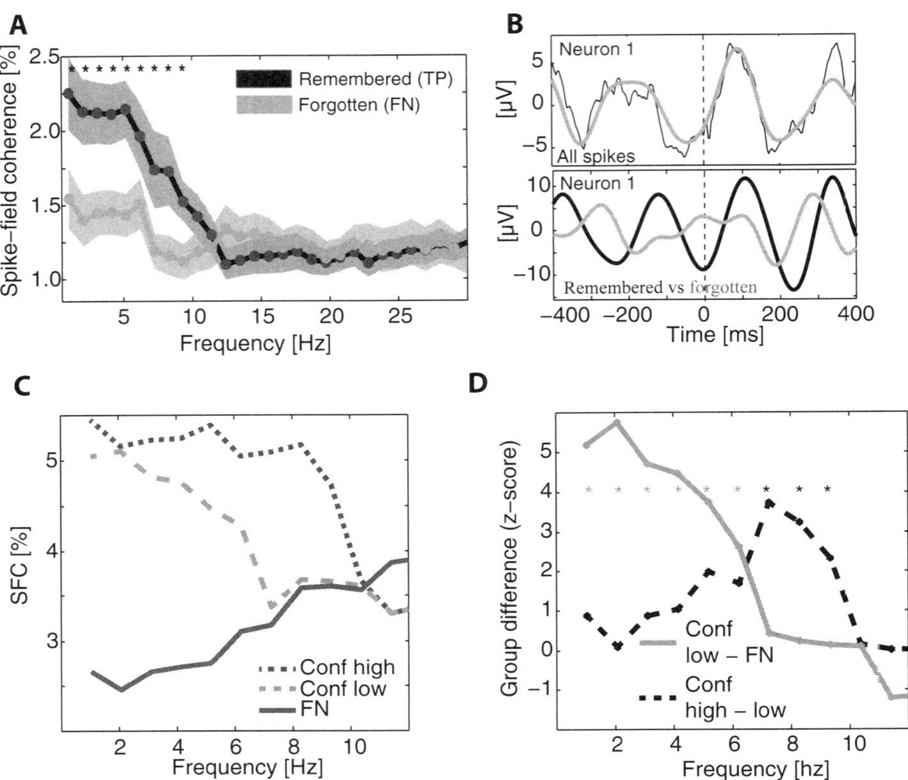

Figure 7.6
Spike–field coherence (SFC) and spike-triggered average (STA) differences between later remembered and later forgotten learning trials. (A) The SFC as a function of frequency was significantly different in the range of 2–10 Hz (q < 0.05, false-discovery rate corrected for 100 multiple comparisons). (B) Example STA from a single neuron in the amygdala. Top shows the raw trace (black) and 3–8 Hz filtered trace (gray), computed using all n = 578 spikes. Bottom shows the STA computed from all spikes fired during later remembered trials (black) and later forgotten trials (gray). (C) SFC as a function of frequency and retrieval confidence (Conf). (D) Pairwise difference between the SFC of different retrieval confidence (low, high) and forgotten stimuli. Differences are plotted in units of z scores relative to the null distribution as estimated by bootstrap statistics. Adapted from Rutishauser et al. (2010).

between weak and forgotten items showed that significant SFC in the <6 Hz range was predictive. A further comparison between later weak and strong memories indicated that a further SFC difference in the 6–9 Hz range was predictive of strong memories (see figure 7.6C, D). This suggests that there are possibly two encoding processes at work, one which operates in the <6 Hz range and one which operates in the 6–9 Hz range. Rats have two distinct theta rhythms of slightly different frequencies: type I and type II or atropine sensitive and insensitive theta, respectively (Buzsáki, 2002). Whether different classes of theta exist in humans is unknown, but the differential phase locking discussed above suggests that this might be the case. While we did not ask patients to recollect additional attributes about the stimuli, an intriguing possibility is that the strong memories were associated with recollection whereas the weak memories did not lead to storage of additional information that could later be recollected. This speculation remains to be verified experimentally.

The difference in SFC between later remembered and forgotten trials was approximately 50% (see figure 7.6A). How big of a difference in spike-timing accuracy does this imply? Using simulations, we estimated that a reduction in SFC of 50% results from adding spike-timing noise with a standard deviation of 20 ms (normally distributed; see figure 7.7). Thus, the difference between a memory and no memory was correlated with a spike-timing accuracy difference of approximately 20 ms, which is comparatively little relative to the cycle length of a theta oscillation which is 200–300 ms. This suggests that a currently unknown mechanism exists that can modulate spike timing accuracy at this degree of accuracy. Future experiments are needed to

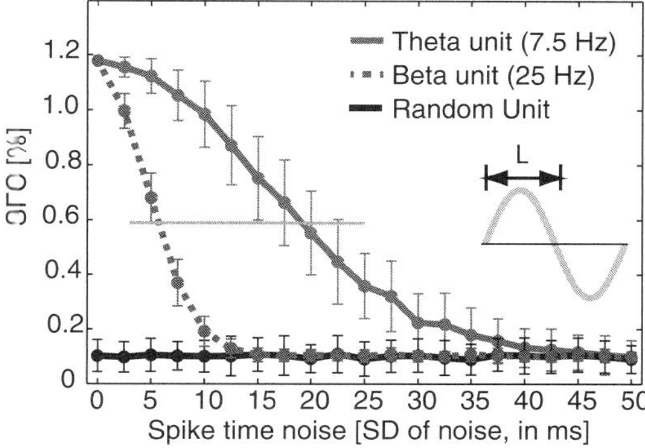

Figure 7.7
Spike–field coherence (SFC) of simulated neurons that are phase locked to different rhythms either perfectly (spike time noise = 0) or imperfectly (spike time noise > 0). Note how the SFC decays as spike timing accuracy decreases. The SFC is reduced by 50% (horizontal line) for spike-timing jitter equal to 30% of the half-cycle length (the time it takes the oscillation from peak to trough, as illustrated by L in the inset) of the underlying oscillation (gray: 7.5 Hz, half-cycle length of 66.7 ms; dashed gray: half-cycle length of 20 ms). Thus, for the theta oscillation (gray) the 50% point is reached for jitter of SD = 20 ms and for the beta unit (dashed gray) for SD = 6 ms. Phase locking becomes indistinguishable from no phase locking if the noise exceeds ~75% of the half-cycle length (gray: 44 ms; dashed gray: 13 ms). Each data point represents the SFC calculated for 1000 spikes. Error bars shown are ±SD for 100 simulations. Adapted from Rutishauser et al. (2010).

identify what specifically can modulate this accuracy and whether it is related to cognitive mechanisms such as attention, motivation, or arousal.

Open Questions and Conclusions

Many human single neuron studies, including those presented here, have reported little functional difference between neurons in the amygdala and hippocampus (figure 7.4B, C shows an example). The proportion of neurons that increased their firing rate in response to novel stimuli was larger in the hippocampus compared to the amygdala, but this difference was not prominent (Rutishauser et al., 2008). This is also true of neurons selective for visual categories, individuals or objects, which have been observed in a number of areas of the MTL including the hippocampus, amygdala, and entorhinal cortex (Kreiman et al., 2000). While some categories like spatial layouts or faces were more likely to be represented in some structures, all could generally be found in all areas. Similarly, there is an increased proportion of neurons selective for animals in the right amygdala, compared to the left amygdala, and bilateral hippocampus and entorhinal cortex (Mormann et al., 2011). One study did identify striking latency differences between the parahippocampal cortex and other MTL areas (Mormann et al., 2008), but the latencies in amygdala and hippocampal neurons were very similar (see also chapter 9). Another study found that almost 50% of amygdala neurons are sensitive to faces and a subset of those respond only to whole faces rather than parts in the amygdala (Rutishauser et al., 2011b) (see chapter 14). Such detailed study of face responses has not been performed so far for other parts of the human MTL, but an intriguing possibility is that the amygdala is specifically responsive to faces to a larger degree than other MTL structures. An earlier study on emotional responses found facial-emotion-related responses also in the hippocampus and entorhinal cortex (Fried et al., 1997), indicating that these are not exclusive to the amygdala. It remains an open question as to why the responses studies so far are so similar in different parts of the MTL. The higher likelihood of finding some types of neurons (such as face- or animal-responsive neurons) in structures such as the amygdala indicates that a promising approach to studying specialization is to compare incidence rates of certain neurons rather than finding neurons exclusively present in one area but not the other. It is also likely that a reason for the failure to strongly disambiguate the structures is that the tasks used so far are fairly general and have thus not been powerful enough. Clearly all parts of the MTL participate in some form in memory formation, and widespread neuronal responses related to visual stimuli relevant for learning have also been reported in rat and primate animal recordings (Suzuki & Eichenbaum, 2000). It is thus likely that in any given learning task, all areas contribute in some form. Carefully designed behavioral tasks will be needed to delineate the distinct contributions each area makes to this process.

It is also an open question as to whether the cells discussed in this chapter are structurally different. This pertains both to cells which are visually tuned/untuned as well as novelty- and familiarity-selective cells. The firing rates of the neurons described were generally low to very low, indicating that most likely the large majority were pyramidal cells. However, it remains to be investigated whether there is something special about these cells or whether any pyramidal cell in the areas under investigation can potentially become novelty or familiarity signaling or whether any cell can become tuned to a visual category. Some cell types can be distinguished by their extracellular waveform, and cells can be labeled as putatively inhibitory and excitatory (pyramidal) based on their waveform (Mitchell et al., 2007; Viskontas et al., 2007). Using this method, Ison et al. (2011) found that both putative inhibitory (narrow waveform) as well as excitatory (wide waveform) cells can be visually selective but that putative excitatory cells are more selective. From a theoretical point of view this is compatible with cooperative–competitive winner-take-all models, which form the basis of much theoretical work on the emergence of functional tuning of neurons embedded in cortical microcircuits (Douglas et al., 1995; Douglas & Martin, 2004; Rutishauser & Douglas, 2009; Rutishauser et al., 2011a). These theories predict that the tuning in any dimension, be it novelty/familiarity or visual category, will be narrower for excitatory neurons. In contrast, the inhibitory neurons receive input from and project to different excitatory neurons with different tuning and are thus less tuned. This model has received experimental support in V1, where orientation-tuned cells have confirmed this prediction both functionally and in terms of anatomical connectivity (Bock et al., 2011).

Memory-related tasks have been a key focus for intracranial recordings, starting with the earliest published recordings (Halgren et al., 1978; Halgren & Wilson, 1985; Heit et al., 1988, 1990). Verbal report, free recall, and judgment of confidence are crucial to memories and are largely unique to humans. Our mechanistic understanding of how learning produces memories and how we retrieve them is poor, particularly considering how much is known about synaptic mechanisms of plasticity that are thought to underlie memory, but this link has been largely unexplored (Martin et al., 2000). Work on the basic physiology of memory with human patients can make fundamental contributions toward understanding at the level of neuronal circuits. This will facilitate development of better learning methods as well as treatments for the learning problems that so often accompany diseases of the brain, including epilepsy. We would like to conclude this chapter with heartfelt thanks to all the patients with epilepsy who have volunteered their time and effort in a time of great distress to contribute to our studies, without which none of this work could have been done.

References

Berry, S. D., & Thompson, R. F. (1978). Prediction of learning rate from the hippocampal electroencephalogram. *Science, 200*, 1298–1300.

Bock, D. D., Lee, W. C., Kerlin, A. M., Andermann, M. L., Hood, G., Wetzel, A. W., et al. (2011). Network anatomy and in vivo physiology of visual cortical neurons. *Nature, 471*, 177–182.

Buzsáki, G. (2002). Theta oscillations in the hippocampus. *Neuron, 33*, 325–340.

Cameron, K. A., Yashar, S., Wilson, C. L., & Fried, I. (2001). Human hippocampal neurons predict how well word pairs will be remembered. *Neuron, 30*, 289–298.

Cantero, J. L., Atienza, M., Stickgold, R., Kahana, M. J., Madsen, J. R., & Kocsis, B. (2003). Sleep-dependent theta oscillations in the human hippocampus and neocortex. *Journal of Neuroscience, 23*, 10897–10903.

Davis, C. D., Jones, F. L., & Derrick, B. E. (2004). Novel environments enhance the induction and maintenance of long-term potentiation in the dentate gyrus. *Journal of Neuroscience, 24,* 6497–6506.

Douglas, R. J., Koch, C., Mahowald, M., Martin, K. A., & Suarez, H. H. (1995). Recurrent excitation in neocortical circuits. *Science, 269,* 981–985.

Douglas, R. J., & Martin, K. A. (2004). Neuronal circuits of the neocortex. *Annual Review of Neuroscience, 27,* 419–451.

Fahy, F. L., Riches, I. P., & Brown, M. W. (1993). Neuronal activity related to visual recognition memory: Long-term memory and the encoding of recency and familiarity information in the primate anterior and medial inferior temporal and rhinal cortex. *Experimental Brain Research, 96,* 457–472.

Fortin, N. J., Wright, S. P., & Eichenbaum, H. (2004). Recollection-like memory retrieval in rats is dependent on the hippocampus. *Nature, 431,* 188–191.

Fried, I., MacDonald, K. A., & Wilson, C. L. (1997). Single neuron activity in human hippocampus and amygdala during recognition of faces and objects. *Neuron, 18,* 753–765.

Griffin, A. L., Asaka, Y., Darling, R. D., & Berry, S. D. (2004). Theta-contingent trial presentation accelerates learning rate and enhances hippocampal plasticity during trace eyeblink conditioning. *Behavioral Neuroscience, 118,* 403–411.

Halgren, E., Babb, T. L., & Crandall, P. H. (1978). Activity of human hippocampal formation and amygdala neurons during memory testing. *Electroencephalography and Clinical Neurophysiology, 45,* 585–601.

Halgren, E., & Wilson, C. L. (1985). Recall deficits produced by afterdischarges in the human hippocampal formation and amygdala. *Electroencephalography and Clinical Neurophysiology, 61,* 375–380.

Hampton, R. R. (2001). Rhesus monkeys know when they remember. *Proceedings of the National Academy of Sciences of the United States of America, 98,* 5359–5362.

Heit, G., Smith, M. E., & Halgren, E. (1988). Neural encoding of individual words and faces by the human hippocampus and amygdala. *Nature, 333,* 773–775.

Heit, G., Smith, M. E., & Halgren, E. (1990). Neuronal activity in the human medial temporal lobe during recognition memory. *Brain, 113,* 1093–1112.

Helmstaedter, C. (2002). Effects of chronic epilepsy on declarative memory systems. *Progress in Brain Research, 135,* 439–453.

Hermann, B. P., Seidenberg, M., Schoenfeld, J., & Davies, K. (1997). Neuropsychological characteristics of the syndrome of mesial temporal lobe epilepsy. *Archives of Neurology, 54,* 369–376.

Holscher, C., Anwyl, R., & Rowan, M. J. (1997). Stimulation on the positive phase of hippocampal theta rhythm induces long-term potentiation that can be depotentiated by stimulation on the negative phase in area CA1 in vivo. *Journal of Neuroscience, 17,* 6470–6477.

Huh, K., Meador, K. J., Lee, G. P., Loring, D. W., Murro, A. M., King, D. W., et al. (1990). Human hippocampal EEG: Effects of behavioral activation. *Neurology, 40,* 1177–1181.

Hunt, R. (1995). The subtlety of distinctiveness: What von Restorff really did. *Psychonomic Bulletin & Review, 2,* 105–112.

Hyman, J. M., Wyble, B. P., Goyal, V., Rossi, C. A., & Hasselmo, M. E. (2003). Stimulation in hippocampal region CA1 in behaving rats yields long-term potentiation when delivered to the peak of theta and long-term depression when delivered to the trough. *Journal of Neuroscience, 23,* 11725–11731.

Ison, M. J., Mormann, F., Cerf, M., Koch, C., Fried, I., & Quiroga, R. Q. (2011). Selectivity of pyramidal cells and interneurons in the human medial temporal lobe. *Journal of Neurophysiology, 106,* 1713–1721.

Jacobs, J., Kahana, M. J., Ekstrom, A. D., & Fried, I. (2007). Brain oscillations control timing of single-neuron activity in humans. *Journal of Neuroscience, 27,* 3839–3844.

Kahana, M. J., Sekuler, R., Caplan, J. B., Kirschen, M., & Madsen, J. R. (1999). Human theta oscillations exhibit task dependence during virtual maze navigation. *Nature, 399,* 781–784.

Kinsbour, M., & George, J. (1974). Mechanism of word-frequency effect on recognition memory. *Journal of Verbal Learning and Verbal Behavior, 13,* 63–69.

Kishiyama, M. M., Yonelinas, A. P., & Lazzara, M. M. (2004). The von Restorff effect in amnesia: The contribution of the hippocampal system to novelty-related memory enhancements. *Journal of Cognitive Neuroscience, 16,* 15–23.

Klimesch, W., Doppelmayr, M., Russegger, H., & Pachinger, T. (1996). Theta band power in the human scalp EEG and the encoding of new information. *Neuroreport, 7,* 1235–1240.

Knight, R. (1996). Contribution of human hippocampal region to novelty detection. *Nature, 383*, 256–259.

Knowlton, B. J., & Squire, L. R. (1993). The learning of categories: Parallel brain systems for item memory and category knowledge. *Science, 262*, 1747–1749.

Kohonen, T., & Lehtio, P. (1981). Storage and processing of information in distributed associative memory systems. In G. Hinton & J. Anderson (Eds.), *Parallel models of associative memory* (pp. 105–143). Hillsdale, NJ: Lawrence Erlbaum Associates.

Kreiman, G., Koch, C., & Fried, I. (2000). Category-specific visual responses of single neurons in the human medial temporal lobe. *Nature Neuroscience, 3*, 946–953.

Li, S., Cullen, W. K., Anwyl, R., & Rowan, M. J. (2003). Dopamine-dependent facilitation of LTP induction in hippocampal CA1 by exposure to spatial novelty. *Nature Neuroscience, 6*, 526–531.

Lisman, J. E., & Grace, A. A. (2005). The hippocampal–VTA loop: Controlling the entry of information into long-term memory. *Neuron, 46*, 703–713.

Lisman, J. E., & Otmakhova, N. A. (2001). Storage, recall, and novelty detection of sequences by the hippocampus: Elaborating on the SOCRATIC model to account for normal and aberrant effects of dopamine. *Hippocampus, 11*, 551–568.

Mandler, G. (1980). Recognizing—The judgment of previous occurrence. *Psychological Review, 87*, 252–271.

Martin, S. J., Grimwood, P. D., & Morris, R. G. (2000). Synaptic plasticity and memory: An evaluation of the hypothesis. *Annual Review of Neuroscience, 23*, 649–711.

McCartney, H., Johnson, A. D., Weil, Z. M., & Givens, B. (2004). Theta reset produces optimal conditions for long-term potentiation. *Hippocampus, 14*, 684–687.

Mitchell, J. F., Sundberg, K. A., & Reynolds, J. H. (2007). Differential attention-dependent response modulation across cell classes in macaque visual area V4. *Neuron, 55*, 131–141.

Mormann, F., Dubois, J., Kornblith, S., Milosavljevic, M., Cerf, M., Ison, M., et al. (2011). A category-specific response to animals in the right human amygdala. *Nature Neuroscience, 14*, 1247–1249.

Mormann, F., Kornblith, S., Quiroga, R. Q., Kraskov, A., Cerf, M., Fried, I., et al. (2008). Latency and selectivity of single neurons indicate hierarchical processing in the human medial temporal lobe. *Journal of Neuroscience, 28*, 8865–8872.

Paller, K. A., & Wagner, A. D. (2002). Observing the transformation of experience into memory. *Trends in Cognitive Sciences, 6*, 93–102.

Parker, A., Wilding, E., & Akerman, C. (1998). The von Restorff effect in visual object recognition memory in humans and monkeys: The role of frontal/perirhinal interaction. *Journal of Cognitive Neuroscience, 10*, 691–703.

Pavlides, C., Greenstein, Y. J., Grudman, M., & Winson, J. (1988). Long-term potentiation in the dentate gyrus is induced preferentially on the positive phase of theta-rhythm. *Brain Research, 439*, 383–387.

Pedreira, C., Mormann, F., Kraskov, A., Cerf, M., Fried, I., Koch, C., et al. (2010). Responses of human medial temporal lobe neurons are modulated by stimulus repetition. *Journal of Neurophysiology, 103*, 97–107.

Poldrack, R. A., Clark, J., Pare-Blagoev, E. J., Shohamy, D., Creso Moyano, J., Myers, C., et al. (2001). Interactive memory systems in the human brain. *Nature, 414*, 546–550.

Riches, I. P., Wilson, F. A., & Brown, M. W. (1991). The effects of visual stimulation and memory on neurons of the hippocampal formation and the neighboring parahippocampal gyrus and inferior temporal cortex of the primate. *Journal of Neuroscience, 11*, 1763–1779.

Rolls, E. T., Cahusac, P. M., Feigenbaum, J. D., & Miyashita, Y. (1993). Responses of single neurons in the hippocampus of the macaque related to recognition memory. *Experimental Brain Research, 93*, 299–306.

Rolls, E. T., Perrett, D. I., Caan, A. W., & Wilson, F. A. (1982). Neuronal responses related to visual recognition. *Brain, 105*, 611–646.

Rutishauser, U., & Douglas, R. J. (2009). State-dependent computation using coupled recurrent networks. *Neural Computation, 21*, 478–509.

Rutishauser, U., Douglas, R. J., & Slotine, J. J. (2011a). Collective stability of networks of winner-take-all circuits. *Neural Computation, 23*, 735–773.

Rutishauser, U., Mamelak, A. N., & Schuman, E. M. (2006). Single-trial learning of novel stimuli by individual neurons of the human hippocampus–amygdala complex. *Neuron, 49*, 805–813.

Rutishauser, U., Ross, I. B., Mamelak, A. N., & Schuman, E. M. (2010). Human memory strength is predicted by theta-frequency phase-locking of single neurons. *Nature, 464*, 903–907.

Rutishauser, U., Schuman, E. M., & Mamelak, A. N. (2008). Activity of human hippocampal and amygdala neurons during retrieval of declarative memories. *Proceedings of the National Academy of Sciences of the United States of America, 105*, 329–334.

Rutishauser, U., Tudusciuc, O., Neumann, D., Mamelak, A. N., Heller, A. C., Ross, I. B., et al. (2011b). Single-unit responses selective for whole faces in the human amygdala. *Current Biology, 21*, 1654–1660.

Schacter, D. L., & Dodson, C. S. (2001). Misattribution, false recognition and the sins of memory. *Philosophical Transactions of the Royal Society of London. B, 356*, 1385–1393.

Schacter, D. L., & Tulving, E. (1994). *Memory systems.* Cambridge, MA: MIT Press.

Seager, M. A., Johnson, L. D., Chabot, E. S., Asaka, Y., & Berry, S. D. (2002). Oscillatory brain states and learning: Impact of hippocampal theta-contingent training. *Proceedings of the National Academy of Sciences of the United States of America, 99*, 1616–1620.

Sederberg, P. B., Kahana, M. J., Howard, M. W., Donner, E. J., & Madsen, J. R. (2003). Theta and gamma oscillations during encoding predict subsequent recall. *Journal of Neuroscience, 23*, 10809–10814.

Sokolov, E. N. (1963). Higher nervous functions: The orienting reflex. *Annual Review of Physiology, 25*, 545–580.

Squire, L. R. (2004). Memory systems of the brain: A brief history and current perspective. *Neurobiology of Learning and Memory, 82*, 171–177.

Squire, L. R., & Alvarez, P. (1995). Retrograde amnesia and memory consolidation: A neurobiological perspective. *Current Opinion in Neurobiology, 5*, 169–177.

Squire, L. R., Stark, C. E., & Clark, R. E. (2004). The medial temporal lobe. *Annual Review of Neuroscience, 27*, 279–306.

Standing, L. (1973). Learning 10,000 pictures. *Quarterly Journal of Experimental Psychology, 25*, 207–222.

Standing, L., Conezio, J., & Haber, R. N. (1970). Perception and memory for pictures: Single-trial learning of 2500 visual stimuli. *Psychonomic Science, 19*, 73–74.

Stark, C. E., & Squire, L. R. (2000). Functional magnetic resonance imaging (fMRI) activity in the hippocampal region during recognition memory. *Journal of Neuroscience, 20*, 7776–7781.

Suzuki, W. A., & Eichenbaum, H. (2000). The neurophysiology of memory. *Annals of the New York Academy of Sciences, 911*, 175–191.

Tulving, E. (2002). Episodic memory: From mind to brain. *Annual Review of Psychology, 53*, 1–25.

Viskontas, I. V., Ekstrom, A. D., Wilson, C. L., & Fried, I. (2007). Characterizing interneuron and pyramidal cells in the human medial temporal lobe in vivo using extracellular recordings. *Hippocampus, 17*, 49–57.

Viskontas, I. V., Knowlton, B. J., Steinmetz, P. N., & Fried, I. (2006). Differences in mnemonic processing by neurons in the human hippocampus and parahippocampal regions. *Journal of Cognitive Neuroscience, 18*, 1654–1662.

Viskontas, I. V., McAndrews, M. P., & Moscovitch, M. (2000). Remote episodic memory deficits in patients with unilateral temporal lobe epilepsy and excisions. *Journal of Neuroscience, 20*, 5853–5857.

von Restorff, H. (1933). Uber die Wirkung von Bereichsbildungen im Spurenfeld. *Psychologische Forschung, 18*, 299–342.

Wallace, W. P. (1965). Review of the historical, empirical, and theoretical status of the Von Restorff phenomenon. *Psychological Bulletin, 63*, 410–424.

Wechsler, D. (1997a). *Wechsler Adult Intelligence Scale* (3rd ed.). San Antonio, TX: The Psychological Corporation.

Wechsler, D. (1997b). *Wechsler Memory Scale* (3rd ed.). San Antonio, TX: The Psychological Corporation.

Welzl, H., D'Adamo, P., & Lipp, H. P. (2001). Conditioned taste aversion as a learning and memory paradigm. *Behavioural Brain Research, 125*, 205–213.

Xiang, J. Z., & Brown, M. W. (1998). Differential neuronal encoding of novelty, familiarity and recency in regions of the anterior temporal lobe. *Neuropharmacology, 37*, 657–676.

Yamaguchi, S., Hale, L. A., D'Esposito, M., & Knight, R. T. (2004). Rapid prefrontal–hippocampal habituation to novel events. *Journal of Neuroscience, 24*, 5356–5363.

8 Visual Cognitive Adventures of Single Neurons in the Human Medial Temporal Lobe

Florian Mormann, Matias J. Ison, Rodrigo Quian Quiroga, Christof Koch, Itzhak Fried, and Gabriel Kreiman

Overview

Here we summarize some of the work examining visual cognition through recordings of single neurons in the human medial temporal lobe (MTL). These recordings were performed in patients with pharmacologically intractable epilepsy as part of the procedure to determine the seizure focus for surgical resection (as described in chapter 3). These surgical procedures provide a rare opportunity to listen to the activity of neural circuits in the human brain while subjects are awake and perform different types of cognitive tasks (Ojemann, 1997; Crick et al., 2004; Engel et al., 2005; Kreiman, 2007; Mukamel & Fried, 2012; Suthana et al., 2012). Throughout this chapter, we emphasize some of the opportunities, challenges and solutions, and progress and open questions, as well as some of the scientific and clinical implications of the work.

Medial Temporal Lobe Connectivity

We start by providing a brief overview of the neuroanatomical connections of the MTL. Most of our understanding about detailed connectivity is based on tracing studies in rodents or nonhuman primates. There has been increased interest recently in the use of noninvasive methods based on diffusion tensor imaging and related techniques to study connectivity patterns in the human brain (e.g., Marcus et al., 2011), but the most detailed studies of connectivity patterns have been conducted in monkeys and rodents.

The human MTL comprises three anatomical structures: amygdala, hippocampus, and parahippocampal gyrus. The parahippocampal gyrus features three functionally and histologically different types of cortex; these are—from posterior to anterior—parahippocampal, perirhinal, and entorhinal cortex. Studies on cytoarchitectonic organization and connectivity in nonhuman primates and rodents have established the entorhinal cortex as the main input region for the hippocampus. Neuronal signals predominantly propagate from superficial layers of the entorhinal cortex through specific hippocampal subregions (dentate gyrus $\rightarrow$ CA3 $\rightarrow$ CA1 $\rightarrow$ subiculum) back to deep entorhinal layers (Squire et al., 2004; Amaral & Lavenex, 2007). The entorhinal cortex receives its major inputs from the perirhinal and parahippocampal cortices (Suzuki & Amaral, 1994), which in turn have reciprocal connections to various multimodal

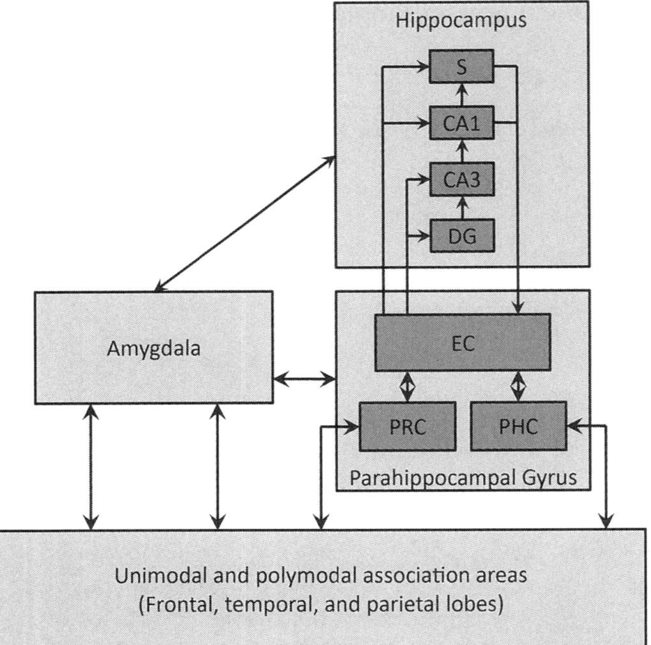

Figure 8.1
Schematic diagram of medial temporal lobe structures and their connectivity. S, subicular complex; DG, dentate gyrus; CA1, CA3, the CA fields of the hippocampus; EC, entorhinal cortex; PRC, perirhinal cortex; PHC, parahippocampal cortex. Adapted from Squire et al. (2004) and Suzuki (1996).

association areas in the frontal, temporal, and parietal lobe (see figure 8.1). The amygdala takes a special role within the MTL in that it has connections to virtually every other brain region and represents one of the major hubs in brain connectivity (Young et al., 1994).

The hippocampus, amygdala, entorhinal cortex, and parahippocampal cortex are frequent targets for implanted electrodes in epilepsy patients (see chapter 3). The perirhinal cortex is less often targeted by clinical procedures due to higher interindividual variability of its exact localization (Insausti et al., 1998).

The Medial Temporal Lobe and Memory

The functional role of neurons in the human MTL is still largely unknown. Insight about human MTL function has been derived mostly from lesion studies. The best known case is H.M., an epilepsy patient who underwent neurosurgical bilateral MTL resection in 1953 and subsequently suffered from a predominantly anterograde amnesia in the sense that he could never again encode explicit memories while still being able to recall memories from before his surgery (Scoville & Milner, 1957; Squire, 2009). From this and related cases, neuroscientists have learned that the

MTL plays a crucial role in memory formation. Rutishauser and colleagues review the role of single neurons in the human brain during several tasks that involve memory formation in chapter 7. Given the prominent role of the hippocampus and surrounding structures in declarative memory, the memory for facts and events that can later be consciously and explicitly recalled, it is important to ponder upon the relationship between the bewildering complexity of responses that we describe below and the consolidation of memories. It is tempting to speculate that some of the responses to complex stimuli described in this chapter constitute a backbone to form novel associations, to detect novelty, and ultimately to form long-term memories. We discuss this point further at the end of the chapter.

One Hat Does Not Fit All

An important cautionary note is pertinent here. While we seek to find unifying principles for MTL function, we should not ignore the fact that the MTL constitutes a large span of the brain, with multiple distinct structures, each with its own inputs and outputs. The history of biology teaches us once and again about the marvelous specificity encountered at all levels, from the molecular to the cellular to the neural circuit level. Thus, we need to sharpen our tools and our description to separately describe the activity of neurons in different structures. Documenting such specificity is not always easy given the difficulties inherent to precisely localizing electrodes and in achieving large sample numbers in our studies. We discuss below some important examples of specificity and their implications for elucidating the function and interactions among MTL structures.

Methodological Considerations

Patients with epilepsy who do not respond to medication go through a battery of tests to evaluate their suitability for resective surgery (Engel, 1996; Ojemann, 1997). As discussed in more detail in chapters 3 and 4, it is very important to describe the procedures and tests at full length to the patient (and to the families when appropriate) so that they can provide their informed consent about the surgical protocol, electrode implantation, and cognitive testing. The patients independently consent to the clinical procedures and any research-related procedures including cognitive testing. A description of the surgical procedure is provided in chapter 3. In addition to other institutional requirements, any person interested in working with patients must have a thorough understanding of the ethical considerations raised in chapter 4.

The quality of signals recorded from microwire electrodes depends on multiple factors. Neurophysiology in a hospital setting is not an easy endeavor, and there are abundant sources of electrical noise (many of which can appear to be deceptively similar to real spikes). Some of the critical factors include the electrode design and impedance and the cables and hardware used to amplify, filter, and condition the signals, or the presence of other electrical equipment in the room. These topics are discussed in more detail in chapter 5. Several aspects of the algorithms and software developed to process neurophysiological signals are discussed in chapter 6.

After electrodes have been implanted and all the hardware has been set up, several important considerations can help smooth the procedures for cognitive evaluation and testing. Of utmost importance is to understand that the goal of the whole procedure is to address the patients' seizure events. Thus, all clinical considerations must take priority over the research efforts. In addition to guaranteeing the patients' safety and conducting any necessary clinical procedures and evaluations, it is critical to understand the patients' needs, unique situations, and comforts. The patient is staying in a hospital, for several days, away from home, surrounded by a plethora of strange people that come in and out, with electrodes and cables implanted in his or her brain, with severely limited mobility, sometimes sleep deprived, sometimes in pain, sometimes mentally or emotionally affected by the withdrawal of antiepileptic medication, and with all sorts of equipment in his/her milieu. While we do not want to undermine the talent of the whole team of neurosurgeons, neurologists, nurses, technicians, engineers, and investigators, we doff our hats to the patients. They are the heroes of our scientific and clinical efforts. In spite of all the difficulties, they put forward their best faces and efforts to help advance medicine and push the frontiers of science.

It is essential for investigators to be aware of the environment and respect the patients' need for privacy, rest, and comfort. Investigators must also clearly convey the procedures to the patients. Whenever appropriate, these procedures also need to be communicated to the family. Investigators must constantly check for patient fatigue and ease during the entire testing procedure. Most patients are willing to perform several paradigms per day with ample breaks in between. For this purpose it has proven important to restrict the duration of each individual paradigm to a maximum of 30 to 45 minutes in order to prevent fatigue, which could compromise continued participation. Due to a number of factors, not all patients will be able to complete all types of tests. It should be clearly understood that participation in experiments occurs on a strictly voluntary basis and that patients are free to pause, halt, or quit experiments at any time. On some occasions, this may entail adaptability and the design of adequate tests to meet the requirements of the ongoing conditions. When patients see that effort is taken to meet their individual needs, it is our experience that they are usually more than willing to participate in cognitive experiments, as they rightly feel that they participate in advancing science and potentially helping future patients.

Almost all experiments discussed here have involved presenting visual stimuli for a certain amount of time (e.g., among many others, Heit et al., 1988; Fried et al., 1997; Kreiman et al., 2000a; Quian Quiroga et al., 2005; Rutishauser et al., 2006; Mormann et al., 2008; Ison et al., 2011; Mormann et al., 2011; Rutishauser et al., 2011). Stimuli are typically presented on a laptop computer situated in front of the patient at a distance of approximately 50 to 80 cm. The receptive field sizes of neurons in the human MTL are not exactly known. However, extrapolating from intracranial field potential studies in humans (Yoshor et al., 2007; Agam et al., 2010) and from single cell studies in monkey inferior temporal cortex (ITC; Logothetis & Sheinberg, 1996; Tanaka, 1996), it is likely that many receptive fields span many degrees of visual angle and include the fovea. Largely, stimuli subtending ~2–10 degrees of visual angle have been presented at the fovea. Little do we know about the preferences of neurons in the human MTL (with some notable exceptions highlighted below). Because the type of stimuli often used in "low-level" vision experiments (e.g., Gabor patches, curvatures, etc.) are not directly meaningful to the subjects, typically, investigators have used a large battery of "natural" stimuli including animals, human faces (famous or not, depicting specific emotions or not), scenes, household objects, and abstract patterns. Stimuli have been presented in color or gray scale. Stimuli are presented for as short as 33 ms and as long as several seconds. Unless strongly required by the characteristics of the task, it is advisable to use event-related designs with a randomized order of presentation to avoid confounding factors. We discuss the response properties elicited by visual stimuli in further detail below.

In some cases, investigators have followed a two-step procedure whereby they first present a battery of pictures and then select the ones that elicit stronger responses for further investigation. In these cases, neural responses are evaluated immediately after the session or even in real time. A smaller subset of stimuli that elicit selective responses in one or more of the neurons can then be used for subsequent experimental paradigms. In some cases, the pictures presented in these sessions can be chosen based on the likings and interests of the patients. Asking patients, for example, about their favorite TV shows, their hobbies, sports, political interests, places they have visited, and so on is helpful in selecting stimuli that are meaningful for the patients. Experience has shown that the more familiar and relevant a stimulus is for a patient, the higher the likelihood of finding a MTL neuron responding to this stimulus. This anecdotal notion was confirmed in a quantitative study in which pictures of people that were personally known to the patient such as family members, friends, or medical personnel were more likely to elicit neuronal responses than famous people not known personally, which in turn were more likely to elicit responses than people unknown to the patient (Viskontas et al., 2007). In several recordings, selective responses were even observed to pictures of the experimenters themselves who had been unknown to the patients until a day or two before the recordings (e.g., Viskontas et al., 2007; Quian Quiroga et al., 2009; Cerf et al., 2010). These reports are evidence for the strong degree of plasticity in the MTL and will be an important subject for future research. When neurons recorded from particular microwires change their response behavior between recording sessions, it is difficult to decide whether this is due to micromovements of the wires (i.e., recording from different neurons) or is due to altered response tuning within the same neuron (i.e., plasticity). On the other hand, anecdotal evidence shows that at least in some cases selective response behavior in a recording channel (i.e., a "similar-looking" unit responding to the same single person in a set of over 100 stimuli) remains constant over five days or more.

Fewer experiments have investigated the responses to auditory stimuli. Chords, speech, and complex pieces of music have been presented to exceptionally rare cases of patients with microwires implanted in auditory cortex (Mukamel et al., 2005; Nir et al., 2007; Bitterman et al., 2008). Most of these studies examining responses to nonverbal acoustic stimuli focused on neurons outside the MTL and are therefore beyond the scope of this chapter. In patients with MTL microwires, use of auditory stimuli has included tones and words spoken by a computer voice, for example, names of persons, objects, and so on (see "A Plethora of Visually Evoked Responses" below).

A large variety of tasks have been implemented to study how the MTL orchestrates different aspects of cognition. As a general rule, even when the main interest centers around the physiological responses to the stimuli, it is advisable to have subjects perform a simple (unrelated) task as opposed to passive viewing conditions under which subjects might doze off or not pay attention to the stimuli. For example, after image offset, subjects may have to answer a simple question such as whether the picture contained certain features or belonged to a particular stimulus category. Such a simple task requires them to attend to the pictures. Every stimulus presentation is preceded by a fixation cross for a few hundred milliseconds to assess baseline firing activity.

Most epilepsy patients are capable of participating in cognitive paradigms. Depending on the location of the epileptic focus, patients may exhibit localization-related neuropsychological deficits such as mnemonic deficits in MTL epilepsy or executive function deficits in frontal lobe epilepsy. In dubious cases it may be advisable to perform a cognitive screening prior to implanting electrodes in order to assess patients' capability to participate in the planned research. Moreover, a battery of neuropsychological test results is usually available to the attending physician to help decide whether or not a patient may be suited for a particular research paradigm.

When taking advantage of the unique access to studying the brain activity of patients undergoing epilepsy monitoring, one should always realize that these patients usually have a functionally and structurally altered brain, which is what causes them to suffer from epileptic seizures in the first place. These alterations, however, are focally confined, which is why neurosurgical removal of the seizure-generating focal region ideally leads to freedom from seizures. It is therefore reasonable to assume that the majority of the brain behaves like any normal human brain. Nevertheless, researchers are well-advised to always ensure that their reported findings remain valid after excluding data obtained from these focal regions, particularly if a hemispheric asymmetry of findings is observed (Mormann et al., 2011).

A Plethora of Visually Evoked Responses

Background: Inferior Temporal Cortex

With the exception of primary visual cortex (area V1), the stimulus features that lead to high firing rates in neurons along the ventral visual stream remain poorly understood (Logothetis & Sheinberg, 1996; Tanaka, 1996; Riesenhuber & Poggio, 2000; Connor et al., 2007; Serre et al., 2007). This has not precluded investigators from correlating neuronal responses with the presentation of simple and complex shapes including faces, fractals, paper clips, and many more. The ventral visual stream is organized in a coarsely hierarchical fashion. At the highest echelon of the visual hierarchy, ITC, neurons typically show strong responses to complex shapes (Rolls, 1991; Logothetis & Sheinberg, 1996; Tanaka, 1996). The neurophysiological responses in monkey ITC have latencies of approximately 100 ms (Eskandar et al., 1992; Keysers et al.,

2001; Hung et al., 2005). Interestingly, ITC neurons show a significant degree of robustness to transformations in properties of the image including scale, position, some degree of rotation, color, and other cues (Rolls, 1991; Logothetis et al., 1994; Ito et al., 1995; Logothetis et al., 1995; Logothetis & Sheinberg, 1996; Grill-Spector et al., 1998). As succinctly outlined above (see "Medial Temporal Lobe Connectivity"), it is likely that ITC signals constitute the main input to the MTL structures discussed in this chapter (including direct and indirect input depending on the particular substructure within the MTL).

While the ITC in nonhuman primates has been investigated rather extensively, electrophysiological studies in this brain region have been rather rare in humans. In fact, it is not even clear what the human homologue of monkey ITC actually is. Evidence from intracranial field potentials (but not single units) recorded from several areas along the ventral occipital and temporal lobe show strong selectivity to complex shapes, tolerance to object transformations, and short latencies, properties that are coarsely comparable to the results reported in the macaque monkey ITC (Allison et al., 1999; McCarthy et al., 1999; Liu et al., 2009; Agam et al., 2010). Likewise, the hierarchical clustering of different stimuli, based on similar neuronal response patterns, is preserved across species—that is, the same stimuli tend to group together both in humans and in monkeys (Kriegeskorte et al., 2008). These functional homologies can in turn be used to map anatomical regions subserving the same specific functions between human and nonhuman primates (e.g., Tsao et al., 2008a). Yet, these functional similarities should not be interpreted to imply direct homology at the anatomical, cellular, or connectivity levels between human and monkey structures, species whose last common ancestor lived about 30 million years ago. More work is needed to establish homologies between macaque studies and human studies to better understand the inputs to the MTL.

Responses to Complex Shapes

A variety of responses to visual shapes have been demonstrated in the human MTL (Heit et al., 1988; Kreiman et al., 2000a; Mormann et al., 2008). Given the strong inputs from ITC, some MTL neurons may also inherit some of the dependence on low-level image properties such as contrast (Steinmetz et al., 2011).

Action potentials recorded from extracellular microwires in the human MTL exhibit regional differences in amplitude (see figure 8.2A). Given the dependence on action potential amplitude on the distance between the microwire and neuronal soma (see chapter 6), the origin of this difference might be related to the local density of different areas in the MTL. If the distance between the tip of the electrode and a given cell was larger in amygdala than in hippocampus, then the average amplitude would be expected to be larger in the hippocampus (since the amplitude of an extracellularly recorded spike decreases as a function of distance; see chapter 6). Interestingly, in mice the neuronal density in the hippocampus (CA1, CA2/3) has been found to be twice as large as the neuronal density in amygdala, strongly supporting this finding (von Bohlen und Halbach and Unsicker, 2002). In the human brain, a study with patients with either hippocampal sclerosis or lesional temporal lobe epilepsy found evidence that the density in the hippocampal region is lower than the density in the entorhinal cortex (Dawodu & Thom, 2005). This is also consistent with the larger amplitude exhibited by entorhinal cortex neurons. The difference with respect to hippocampal cells was marginally significant (Wilcoxon rank-sum test, p = 0.04).

The spontaneous firing rates in the human MTL are typically quite low (mean of 0.6 spikes/s, SD = 0.9 spikes/s for putative pyramidal cells; Ison et al., 2011; figure 8.2C, D). The baseline firing rates before stimulus presentation show differences across different parts of the MTL (e.g., Mormann et al., 2008, figure 8.2C) and are consistent with those reported in the rodent and macaque literature. Intriguingly, some of the MTL neurons exhibit a remarkably low firing rate (with less than one spike per minute) but show transient increased firing in response to a preferred stimulus (e.g., Quian Quiroga et al., 2005; figure 8.2D). This sparse response suggests that some of the single electrode studies where investigators try to move an electrode to isolate neurons with high sustained firing rate could be missing some of the extremely "shy" neurons (see the discussion in Shoham et al., 2006). In contrast, microelectrodes used for single unit recordings in humans cannot be moved and thus may provide a more "unbiased" sample of units including those with low firing rates.

The MTL contains various neuron classes (Lein et al., 2004), including a majority of (excitatory) principal cells and several types of (inhibitory) interneurons. Investigators have used electrophysiological features like the baseline firing activity and spike duration to separate putative pyramidal cells from putative interneurons (e.g., Ison et al., 2011, figure 8.2B). In general, putative interneurons defined in this fashion show more graded responses to multiple stimuli, which has been interpreted to reflect a plausible mechanism to generate sparse representations by combining cell populations with different selectivity (Ison et al., 2011).

Invariance Properties

Critical to visual recognition as well as to the formation of episodic memories is the need to build representations that discount many of the basic metric properties of the stimuli as they are usually irrelevant to the task at hand (Squire & Zola-Morgan, 1991; Wagner et al. 1999; Dudai,

Figure 8.2
Properties of medial temporal lobe (MTL) neurons. (A) Mean waveforms for visually selective putative pyramidal cells (N = 155) recorded from the hippocampus (H; N = 49), entorhinal cortex (EC; N = 32), parahippocampal cortex (PHC; N = 22), and amygdala (A; N = 52). Bands show the mean ±SEM. Entorhinal neurons show the highest amplitudes, followed by hippocampal neurons. Waveforms recorded from the amygdala and parahippocampal cortex exhibit smaller amplitudes. Cell identity was determined with k-means clustering based on firing rate and spike width (Ison et al., 2011). (B) Normalized waveforms of putative pyramidal cells and interneurons. (C) Mean baseline firing rate (FR) for MTL putative pyramidal cells (N = 155). Error bars denote SEM. (D) Mean peak FR over all stimuli. (E) Response latency for MTL putative pyramidal cells responding to one or more stimuli. Latencies were defined as the first time point where the instantaneous firing rate (IFR) crossed a threshold given by 3 SD above the mean baseline IFR. (F) Selectivity of putative pyramidal cells. Stimulus selectivity was assessed using the selectivity index introduced by Quian Quiroga et al. (2005). This measure of selectivity is close to 0 for uniformly distributed random firings and approaches 1 the more selective the neuron is. (G) Spike detection threshold estimated from the median absolute amplitudes of the high-pass filtered signal for MTL responsive units (single and multiunits, N = 579). (H) Mean spike amplitude for MTL responsive units.

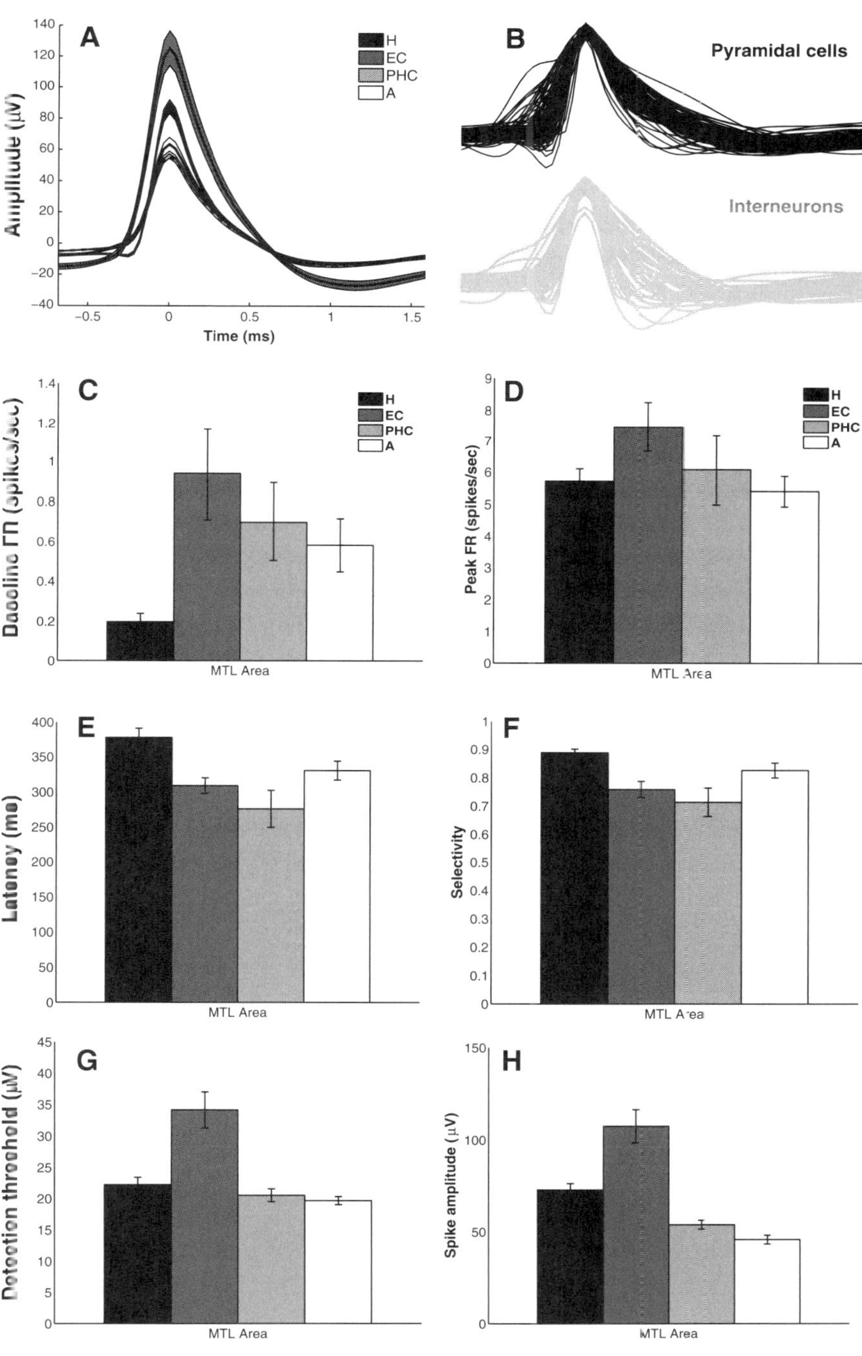

2012). Indeed, neurons in the human MTL show a remarkable degree of tolerance to many image transformations. We discuss a few striking examples here.

Some neurons respond to multiple different images that are quite distinct at the pixel level but which belong to the same category. The word "category" signifies many different things. In a remarkable study describing the activity of neurons in the macaque ITC and prefrontal cortex, Miller's group parametrically varied the shape of various synthetic images of cats and dogs and trained monkeys to separate these into two visual categories (Freedman et al., 2001; Meyers et al., 2008). What's particularly interesting about this study is that the distance (at least defined in pixel space) between two stimuli belonging to the same "category" was the same as the distance between two stimuli belonging to different categories (see also Sigala & Logothetis, 2002), thus showing the task-dependent categorical nature of the responses.

In human cognitive studies, the word "category" is often taken to imply "semantic" categories such as humans, animals, objects, and so on. In one of the first studies describing visual responses in the human MTL, the authors described single neuron responses to subsets of the presented stimuli that constituted semantic categories such as animals, household objects, cars, spatial scenes, and famous and unknown faces (Kreiman et al., 2000a). Extending these findings, a recent study described a specific response preference for the category of animals in neurons from the right, but not left, human amygdala (Mormann et al., 2011). An example of such a response to "animals" is shown in figure 8.3 (plate 6). This categorical selectivity was confirmed in a separate functional magnetic resonance imaging (fMRI) experiment in healthy subjects and appeared to be independent of emotional dimensions such as valence or arousal. A plausible evolutionary explanation is that the phylogenetic importance of animals, which could represent either predators or prey, has resulted in neural adaptations for the dedicated processing of these biologically salient stimuli in a phylogenetically old structure like the amygdala.

Two cautionary notes should be mentioned here. First, as emphasized above, only a limited number of stimuli were presented and we cannot be sure that the neuron would respond to all possible images of animals. Second, it is possible that stimuli belonging to the same category share basic metric properties that distinguish them from stimuli in other categories.

Another seminal study described neurons representing a specific semantic concept in the human MTL (Quian Quiroga et al., 2005). An example of this type of responses is shown in figure 8.4 (plate 7). These cells may respond to very different pictures of a given person that the patient is familiar with. The notion that simple metric or pixel properties can describe those responses is even harder to reconcile with the activity of some neurons that also responded to the written or spoken name (Quian Quiroga et al., 2005; Quian Quiroga et al., 2009). These responses to visual stimuli, text, and voices do not directly bring us any closer to understanding the origin of this type of sparse, selective, and invariant neuronal preference. Under several strong assumptions, investigators were able to estimate that a single person is likely to be represented by up to a few million neurons (but possibly a much smaller number) and that, conversely, a given neuron can represent tens to hundreds of different people (Waydo et al., 2006). Indeed, this type of encoding of semantic information in such an explicit and specific way is more in

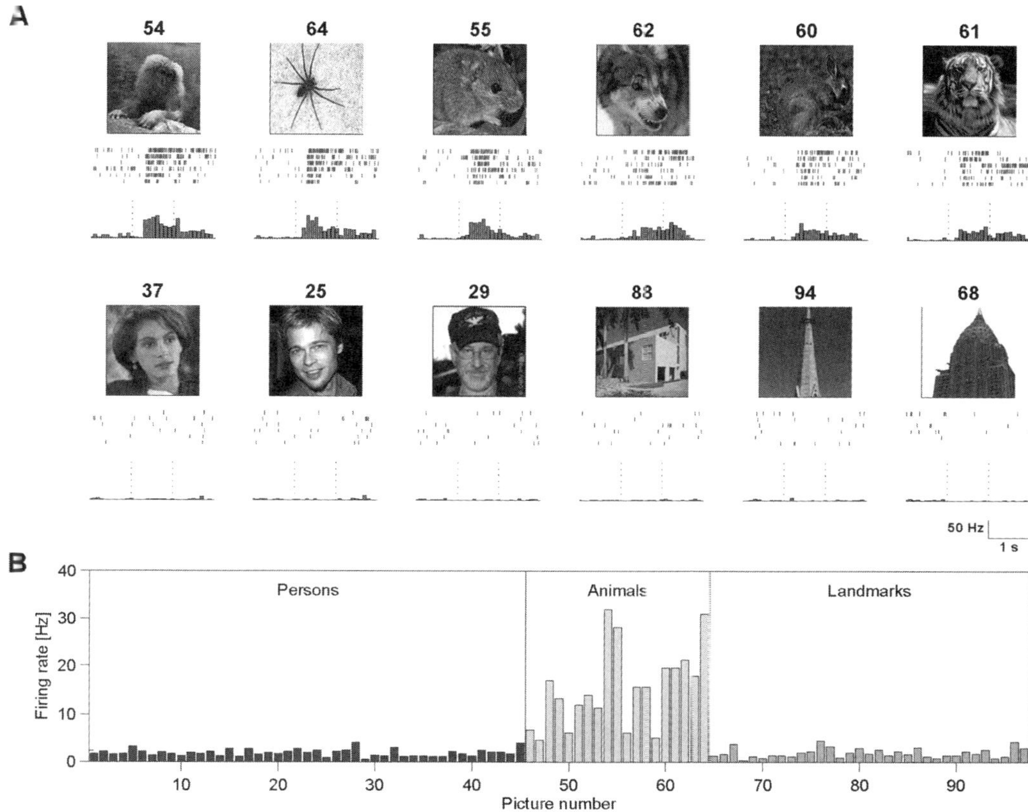

Figure 8.3 (plate 6)

A single unit in the amygdala activated by multiple different pictures containing animals. (A) Responses of a neuron in the right amygdala to pictures from different stimulus categories, presented in randomized order. Here we show the responses to 12 of the 97 pictures shown in this experiment. For each picture, the corresponding raster plots (trial order is from top to bottom) and peristimulus time histograms are given. Vertical dashed lines indicate image onset and offset (1 s apart). (B) The mean response firing rates of this neuron between image onset and offset across six presentations for all individual pictures. Pictures of persons, animals, and landmarks are denoted by brown, yellow, and cyan bars, respectively. Modified from Mormann et al. (2011).

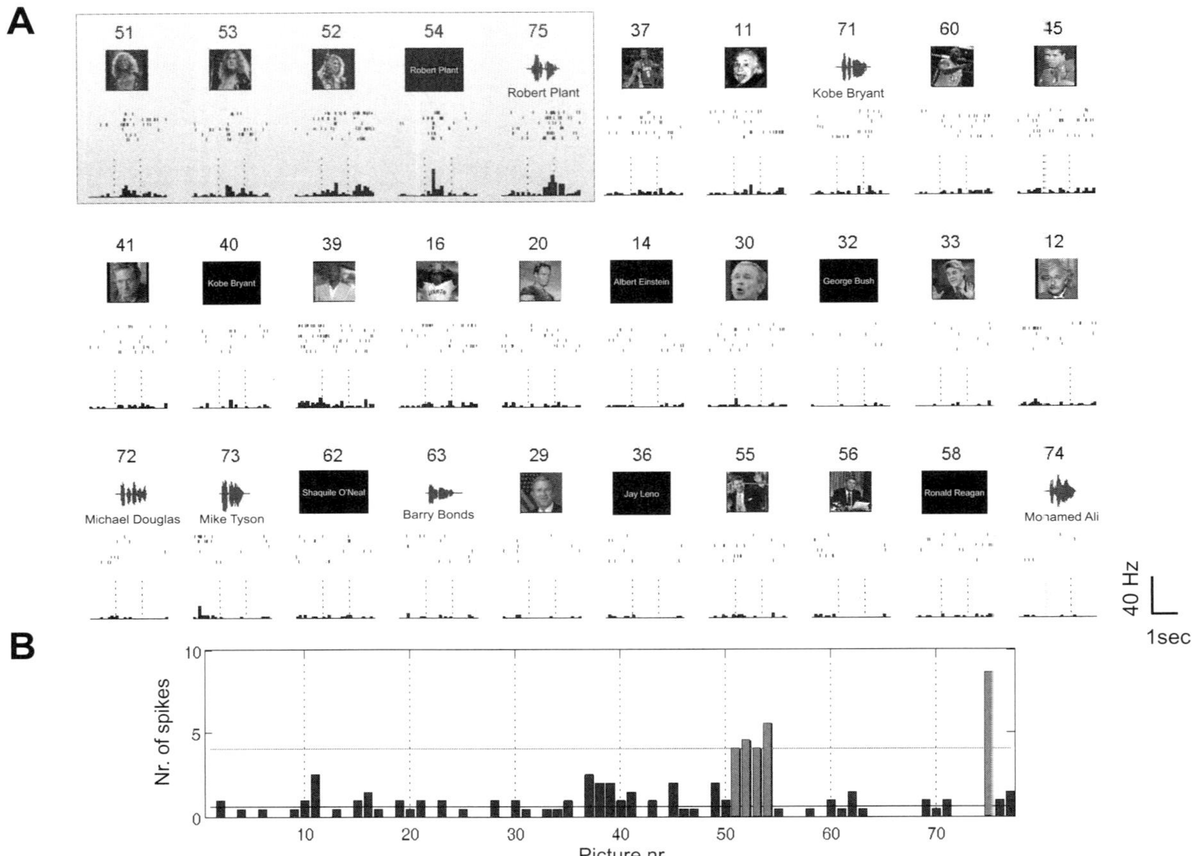

Figure 8.4 (plate 7)
A single unit in the entorhinal cortex responding to multiple different stimuli representing the same individual person. (A) A neuron in the entorhinal cortex that responded selectively to pictures of singer Robert Plant (stimulus 51, 53, and 52), as well as to his written (stimulus 54) and spoken (stimulus 75) name. There were no significant responses to any other picture, sound, or text presentations. For space reasons, only the largest 30 (out of 78) responses are displayed. In each case the raster plots for the six trials, peristimulus time histograms, and the corresponding pictures are shown. The vertical dotted lines mark picture onset and offset (1 s apart). (B) Median number (Nr) of spikes (across trials) for all stimuli. Presentations of Robert Plant are marked with red bars. Stimulus numbers correspond to the ones shown above each picture in (A). The gray horizontal line shows 5 SD above the baseline threshold used for defining significant responses. Modified from Quian Quiroga et al. (2009).

agreement with a sparse coding theory than a distributed code (Barlow, 1972; Quian Quiroga et al., 2008; Quian Quiroga & Kreiman, 2010). The bewildering nature of this type of responses is illustrated by the variety of names that have been used to describe these neurons including "grandmother cells," "concept cells," "Jennifer Aniston cells" (referring to the best known such example), "visually selective cells," "sparse but distributed cells," "semantic neurons," and many more (Quian Quiroga et al., 2005; Quian Quiroga et al., 2008; Quian Quiroga & Kreiman, 2010).

Ultimately, one of the major challenges in this type of recording is that we only have a limited amount of recording time (30 to 45 minutes for a typical experiment). Under such conditions, it is very difficult to assess with any degree of certainty what the neurons truly respond to given the lack of a systematic and theoretical understanding of the origins of such responses. In early visual areas including the retina and primary visual cortex, investigators often display tuning curves where the responses of a neuron show a smooth variation according to specific stimulus parameters such as its position or its orientation. A problem in determining the precise tuning curves of semantic neurons in the MTL is to find a suitable parameterization of the stimulus space—that is, among all the possible stimuli that could in theory be presented. In other words, it is difficult to determine the semantic distance between two stimuli, or categories of stimuli, particularly for complex shapes that are not designed to quantitatively evaluate differences and distances (Freedman et al., 2001; Serre et al., 2007; Pinto et al., 2008; Huth et al., 2012). The problems of determining the preferences of neurons outside of the sensory periphery are common to neurophysiological recordings across many other brain areas and species. Yet, they are compounded in human studies by the short duration of any one experiment in any one patient.

Response Latencies

An intriguing aspect of the MTL responses is the rather long response latency compared to response times in the monkey and cognitive processing times in humans. Neurons in entorhinal cortex, hippocampus, and amygdala typically show latencies in the range of ~300 to ~500 ms after stimulus onset (Kreiman et al., 2000a; Mormann et al., 2008; Ison et al., 2011). Responses in parahippocampal cortex have an onset latency between ~200 and ~400 ms and are, on average, ~120 ms earlier than in the other three regions. This is considerably longer than the latencies in the macaque ITC. One possibility is that latencies in the human brain are longer (perhaps due to a larger brain). However, intracranial field potential recordings in the human ITC yield response latencies of ~100 to 150 ms (McCarthy et al., 1999; Liu et al., 2009), which are only slightly longer than the single unit or local field potential (LFP) latencies in the macaque ITC. The nature of the intracranial field potential signals is poorly understood, but direct comparisons of spikes and LFPs reveal similar latencies in ITC (Kreiman et al., 2006; Nielsen et al., 2006) as well as other areas (Katzner et al., 2009; Rasch et al., 2009; Burns et al., 2010). Another possibility is that human ITC does not project directly to the human MTL, or at least that there are many more synapses in between. Little can we currently say about this possibility. Yet, it is intriguing to note that recordings in macaque MTL (entorhinal cortex, hippocampus, amygdala) also show significantly longer latencies (150–250 ms) than in ITC (70–150 ms), although not as long as the ones in humans (Miyashita et al., 1989; Rolls et al., 1989; Suzuki et al., 1997; Wirth et al., 2003; Yanike et al., 2004) (see also table 1 in Mormann et al., 2008). The human parahippocampal cortex shows shorter latencies and may therefore be involved in object recognition or at least receive direct input from areas involved in object recognition, while the substantially longer latencies in entorhinal cortex, hippocampus, and amygdala may reflect the need for recurrent processing and further computations (see "Discussion" below).

Some recent studies have proposed a hierarchy of activation latencies within the MTL (Mormann et al., 2008; Ison et al., 2011). It should be noted that the relative latencies across these different steps are much more widely spaced apart than the relative latencies described in the macaque visual system hierarchy (e.g., Schmolesky et al., 1998). On the other hand, a direct correlation between response latency and stimulus selectivity as an indicator for hierarchical processing was found not only between but also within human MTL regions, at least for the parahippocampal cortex, entorhinal cortex, and hippocampus (Mormann et al., 2008). Hierarchical processing along the visual cortex has been widely debated, and there is no shortage of feedback loops, horizontal connections, bypass routes, and other complications. Yet, the coarse hierarchy has led to initial steps en route toward building systematic computational models of visual recognition (Fukushima, 1980; Rolls, 1991; Riesenhuber & Poggio, 1999; Serre et al., 2007). One can only hope that, to the extent that there is some hierarchy in the MTL, this may also help point to initial computational rules and models.

Regional Specificity

Lumping together all neurons within the MTL is a major oversimplification. As we collect more data and probe the system in more sophisticated ways, it is likely that we will encounter more and more differentiation. Plenty of evidence from the rodent and macaque neurophysiology literature suggests that the parahippocampal gyrus, hippocampus, entorhinal cortex, and amygdala have their own properties and caprices. Each of these areas can further be subdivided into specific computational domains including different layers in the entorhinal cortex, different areas in the hippocampus (e.g., CA3, CA1, dentate gyrus), and different nuclei in the amygdala (Lavenex & Amaral, 2000). Traditionally, there have been two challenges in further elucidating the different roles of neurons in these different structures. The first one simply concerns the low number of neurons in each area. The more we split, the fewer neurons we end up with in each area for rigorous statistical analysis. The second one concerns the difficulty in precise localization of the microelectrodes within each area (Ekstrom et al., 2008).

In spite of these difficulties, many studies have now documented important differences across these different areas. As mentioned in the previous section, parahippocampal neurons respond significantly earlier than those in other studied MTL regions (Mormann et al., 2008). Parahippocampal neurons also differ from entorhinal, hippocampal, and amygdala cells in that they tend to respond to a larger percentage of the presented stimuli, thus exhibiting lower response selectivity, whereas selectivity is highest for principal cells in hippocampus (Mormann et al., 2008;

Ison et al., 2011). Finally, the parahippocampal cortex is the only one among the four MTL regions studied whose neurons are not modulated by repeated stimulus presentation in the sense of repetition suppression (Pedreira et al., 2010). With respect to sensory modalities, the parahippocampal cortex again takes a special role by exhibiting no auditory responses and accordingly ro semantic invariance to auditory stimuli (Quian Quiroga et al., 2009).

Regional preferences for certain classes of stimuli, possibly indicative of domain-specific processing, are found in several MTL regions. For instance, neurons in the right, but not left, amygdala show an increased response to the category of animals (Mormann et al., 2011). Unpublished anecdotal evidence furthermore suggests that neurons in the parahippocampal cortex preferentially respond to spatial scenes (perhaps reflecting the neuronal basis of the notion proposed by blood flow studies in the so-called parahippocampal place area; Epstein et al., 1999) whereas entorhinal neurons tend to show a slight category preference for objects, in agreement with functional imaging studies in humans (Litman et al., 2009).

Given that the MTL does not seem to be directly involved in visual recognition (see below for details), it seems intriguing that neurons would show this type of preference, lateralization, or other type of specificity to anthropomorphically defined "categories" and that the stimuli in those studies shared specific features that correlated with the interests of the neurons recorded from. For example, responses in the amygdala to animals could perhaps relate to the evolutionary relevance of these stimuli rather than any representation of the shape of animals. We hope that future studies will reveal further insights about the interpretation of this type of complex response.

Lesion Studies

As noted above (see "The Medial Temporal Lobe and Memory"; see also chapters 3 and 7 and Scoville and Milner, 1957; Squire et al., 2004), bilateral removal of the hippocampus and surrounding structures leads to deficits in memory consolidation. In spite of significant efforts in the field, no impairments have been observed in visual recognition or visual perception (Squire et al., 2004; Shrager et al., 2006). In the few cases where visual deficits are described, they typically can be ascribed to a learning component in the task or extension of the lesion to perirhinal and ITC. It seems likely that the type of responses described above may therefore not be concerned with the process of recognition per se (see further discussion of this point below).

In patients with congenital bilateral lesions of the amygdala (Siebert et al., 2003), recognition of emotional facial expressions is impaired due to altered processing of facial features and a relative neglect of the eye region (Adolphs et al., 2005). The possibility of studying acute impairment of amygdala function may arise in patients with acute limbic encephalitis who often present with bilateral MRI lesions in the amygdala and anterior hippocampus during early stages of the disease (Bien & Elger, 2007). While it is difficult to dissociate amygdala and hippocampal involvement in these lesions, they still offer a unique possibility to study amygdala lesions in the absence of long-term adaptation processes that can be expected for congenital lesions.

Visual Imagery

Given the striking responses to a wide variety of stimulus transformations described above, several investigators wondered whether it would be possible to completely remove the stimulus and evaluate whether neuronal responses in the MTL could be modulated by asking subjects to volitionally create mental imagery (Kreiman et al., 2000b; Gelbard-Sagiv et al., 2008; Cerf et al., 2010).

These mental imagery exercises constitute good examples of tasks that are difficult to implement in animal models. Still, several investigators have conjectured that animals create mental images when they need to remember information for several seconds. For example, in delayed-match-to-sample tasks, correct performance relies on subjects' ability to retain information during the delay period. It is tempting to speculate that subjects (humans and nonhumans) are mentally rehearsing/holding pertinent information during the delay. Indeed, several studies have demonstrated that neurons can show selective modulation during such conditions in the macaque ITC, MTL, and prefrontal cortex (Miyashita & Chang, 1988; Miyashita, 1993; Higuchi & Miyashita, 1996; Naya et al., 1996; Rao et al., 1997; Suzuki et al., 1997; Rainer et al., 1999; Naya & Suzuki, 2011).

While recording the activity from a neuron that showed enhanced responses to a particular image A compared to another image B, investigators asked subjects to mentally imagine A or B in the absence of any visual stimulus upon listening to a tone that was associated with A or another tone that indicated B. They observed that MTL neurons modulated their spiking activity in the absence of a visual stimulus and that this modulation showed the same preferences as the responses elicited by a visual input (Kreiman et al., 2000b). In another study, subjects were shown brief video clips and were later asked to freely recall aspects of those video clips. Again, MTL neurons showed activation patterns in the absence of a visual stimulus and in the absence of any instructing tone. Those activation events were correlated with the contents of their recall imagery as described by the subjects and were also consistent with the neuronal responses elicited by visual presentation (Gelbard-Sagiv et al., 2008). In a more recent instantiation and extension of these findings, investigators showed that they can decode these volitionally initiated changes in firing rate in real time (Cerf et al., 2010). Furthermore, the patients were able to willfully increase or decrease the firing rate of these neurons in a highly selective manner. That is, the firing neurons of at least some of the MTL neurons are at least under partial voluntary control.

Perhaps not surprisingly, the responses elicited in the absence of a visual stimulus were typically smaller in magnitude and showed larger trial-to-trial variability than those evoked by visual stimulation. However, these differences may be partly attributable to the precision by which we can control the onset of visual stimuli and the relative imprecision by which we can characterize the onset of recall. Even in the presence of equally strong responses, variability in the onset times could lead to the appearance of weaker and delayed responses.

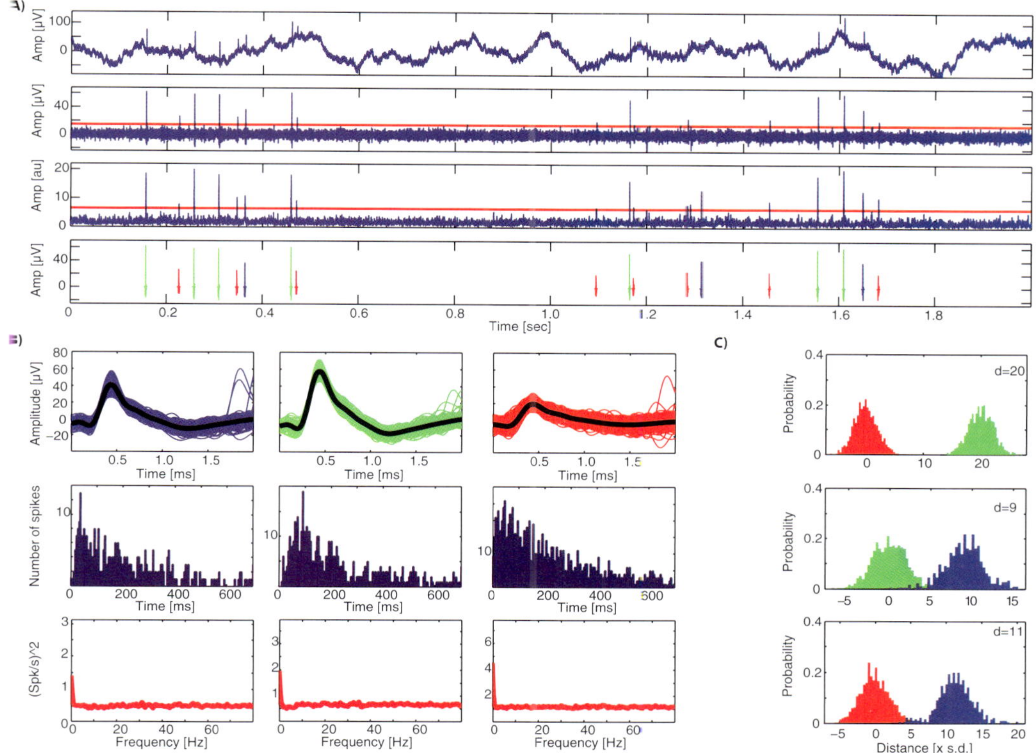

Plate 1 (figure 6.1)

Spike detection and sorting example. Shown are recordings obtained from a single wire implanted in the right anterior cingulate cortex of a human epileptic patient. The detection and sorting shown was done automatically using OSort. (A) From top to bottom: raw data (2 Hz high pass), band-pass filtered (300 Hz–3 kHz; red line is 4 times the estimated standard deviation; see text), energy-signal used for spike detection (line shows $5 \times SD$ as used for spike detection in this example), and detected and sorted spikes (color indicates cluster identity, as computed by OSort). (B) Metrics of three of the identified clusters: raw waveforms (top), interspike interval (ISI; middle), and power spectrum of the spike train (bottom). All clusters were well separated, with the percentage of ISIs < 3 ms equal to 0%, 0%, and 0.19% respectively. (C) Pairwise projection test for all possible combinations shows that the clusters are well separated. The distance d is indicated (see text). Amp, amplitude; Nr, number; Spk/s, spikes per second.

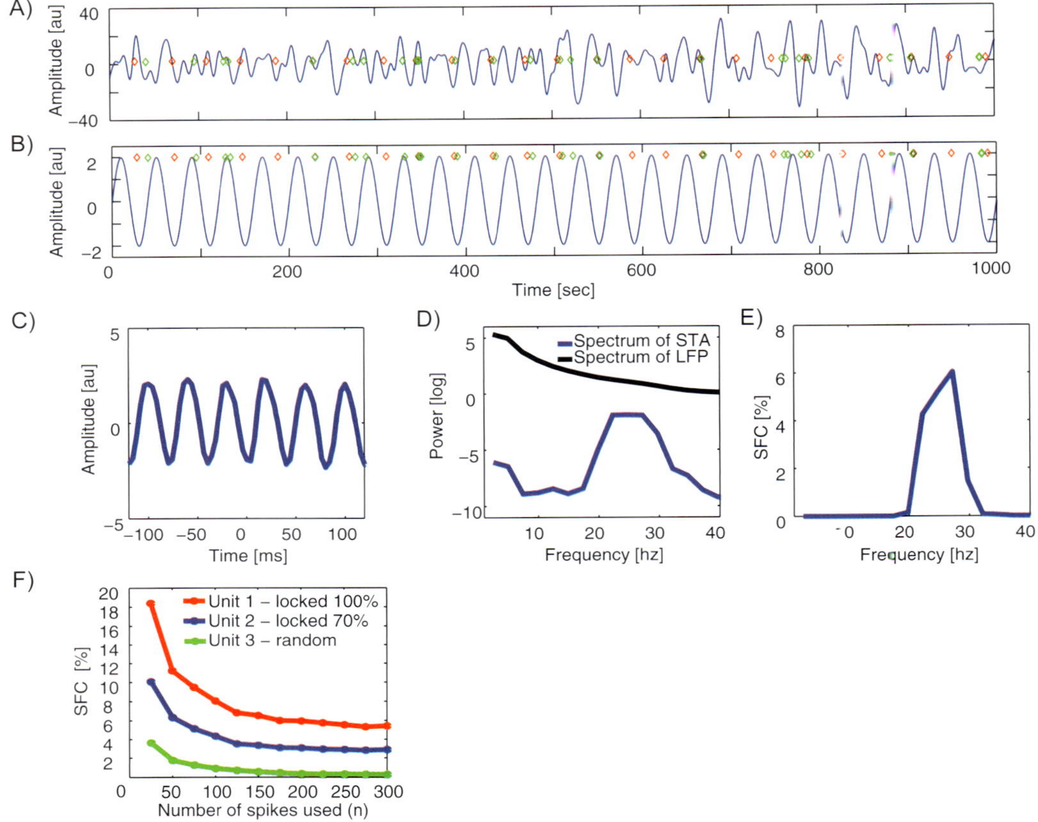

Plate 2 (figure 6.4)
The steps of calculating the spike–field coherence (SFC), illustrated using simulated data. (A) Simulated extracellular. Shown is a band-pass-filtered version of the simulated raw signal (10–300 Hz). (B) One part of (A) is a 25-Hz oscillation of amplitude 2. The red and green dots indicate the spikes of two simulated neurons. (C) The spike-triggered average (STA) for the red simulated neuron shown in (B). (D) The power spectrum of the STA (blue) shown in (C). (E) The SFC as a function of frequency. (F) The mean peak height of the SFC at 25 Hz as shown in (E) as a function of the number of spikes used for its calculation. Adapted from Rutishauser et al. (2010). LFP, local field potential; au, arbitrary units.

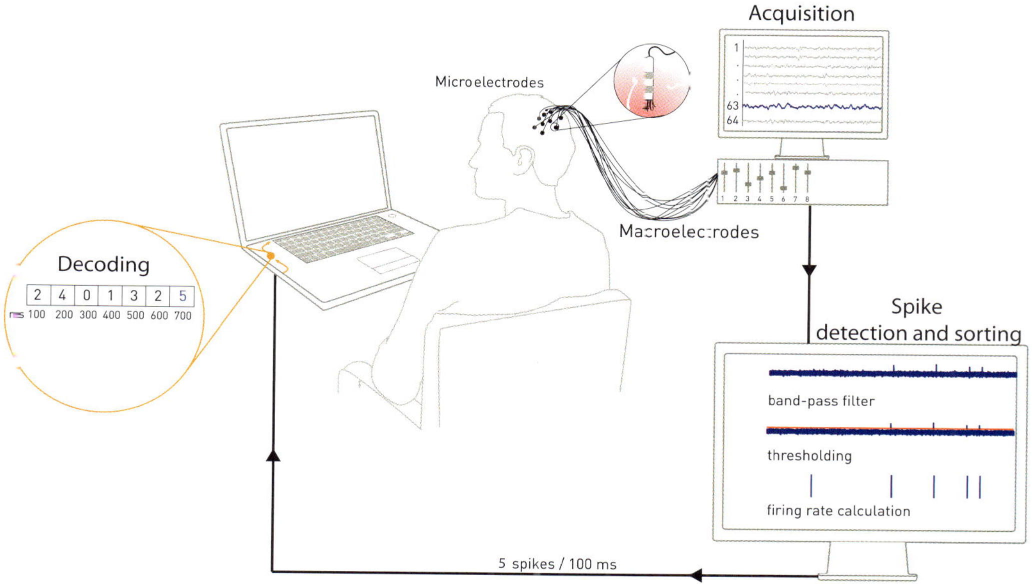

Plate 3 (figure 6.7)
Illustration of a typical online experimental setup.

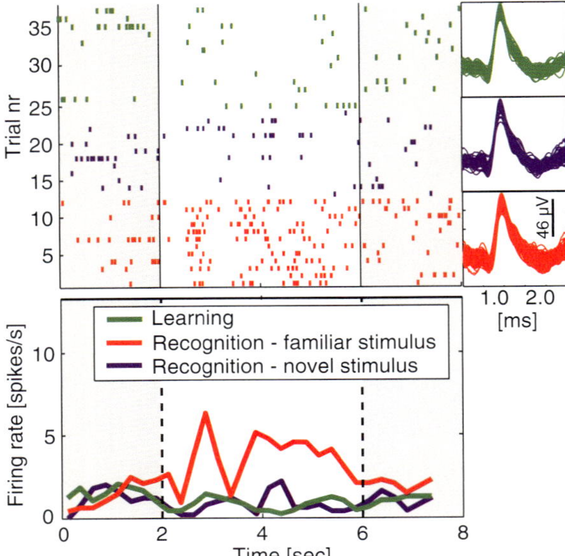

Plate 4 (figure 7.2)
Example of a neuron in the hippocampus that increases its firing rate to familiar but not to novel stimuli (a "familiarity detector"). The same images were presented twice, initially during learning (green trials) and later again during retrieval (red trials). Also during retrieval other, novel stimuli (blue trials) were shown. Trials were randomly intermixed but are shown sorted for display purposes only. The waveforms associated with the trials of the three different categories are shown on the right, confirming that the same neuron was present in both parts of the experiment. The average-firing-rate histogram at the bottom summarizes the firing-rate difference between novel (green) and familiar (red) trials. Notice that the visual input for both trials was the same. The only difference was that stimuli had been seen before (red trials). This cell is thus a memory cell. nr, number. Modified from Rutishauser et al. (2006).

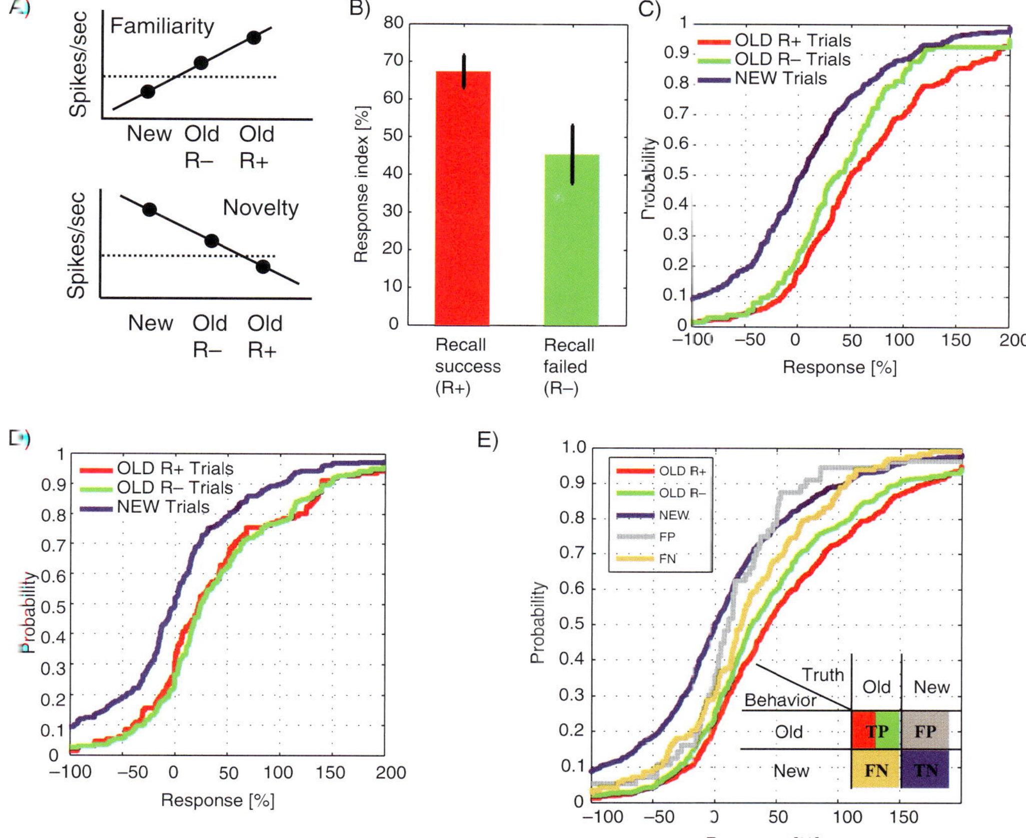

Plate 5 (figure 7.3)

(A) Continuous strength of memory representation hypothesis illustrated for neurons that increase their firing rate in response to the presentation of familiar (top) and novel (bottom) stimuli. (B) Average response index for trials with successful (R+) and failed (R−) recall. The difference was significantly different (p < 0.01). Error bars are ±SE over n = 386 and 123 trials, respectively. (C) The same data as shown in (B) but replotted as a cumulative distribution. Notice the two shifts to the right of the entire distribution: the first due to familiarity (to R−) and the other by recollection (to R+). (D) For patients who could not recollect, there was no significant difference between R+ and R− trials (p = 0.53, Kolmogorov–Smirnov test). (E) Activity during errors reflects true memory rather than behavior. Response indices were significantly different between forgotten stimuli and novel stimuli (p < 0.001) as well as between falsely remembered stimuli and novel stimuli (p < 0.01) (both Kolmogorov–Smirnov test). Modified from Rutishauser et al. (2008). TP, true positive; FP, false positive; FN, false negative; TN, true negative.

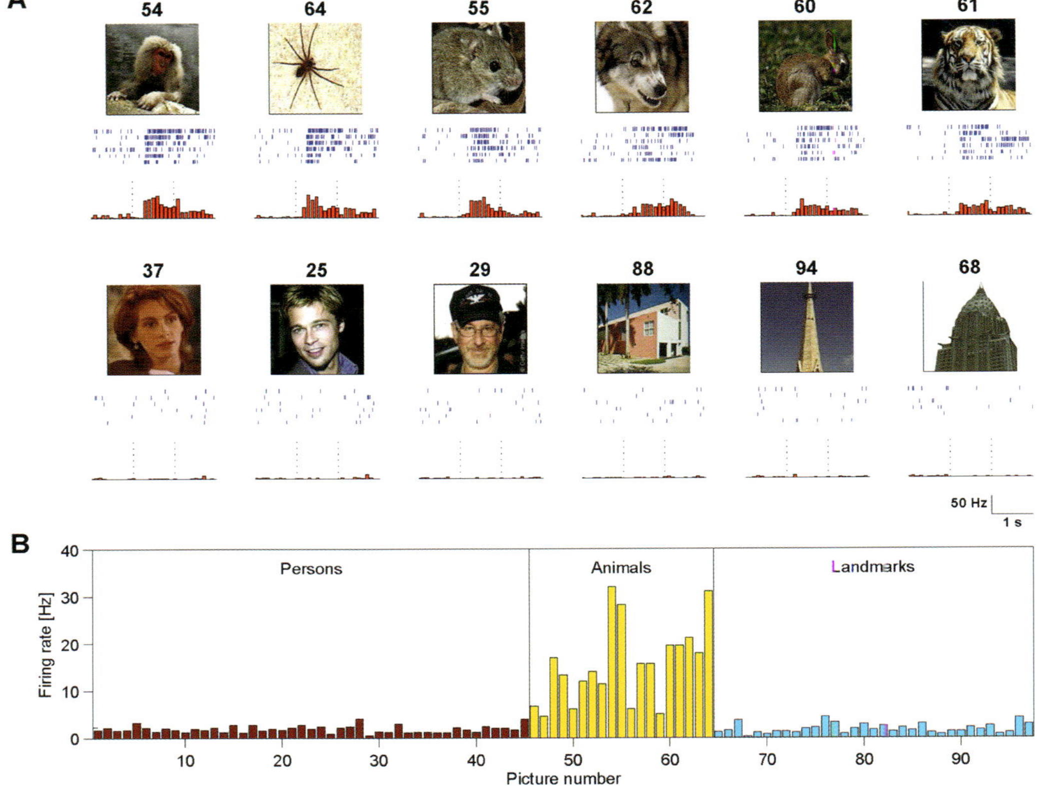

Plate 6 (figure 8.3)
A single unit in the amygdala activated by multiple different pictures containing animals. (A) Responses of a neuron in the right amygdala to pictures from different stimulus categories, presented in randomized order. Here we show the responses to 12 of the 97 pictures shown in this experiment. For each picture, the corresponding raster plots (trial order is from top to bottom) and peristimulus time histograms are given. Vertical dashed lines indicate image onset and offset (1 s apart). (B) The mean response firing rates of this neuron between image onset and offset across six presentations for all individual pictures. Pictures of persons, animals, and landmarks are denoted by brown, yellow, and cyan bars, respectively. Modified from Mormann et al. (2011).

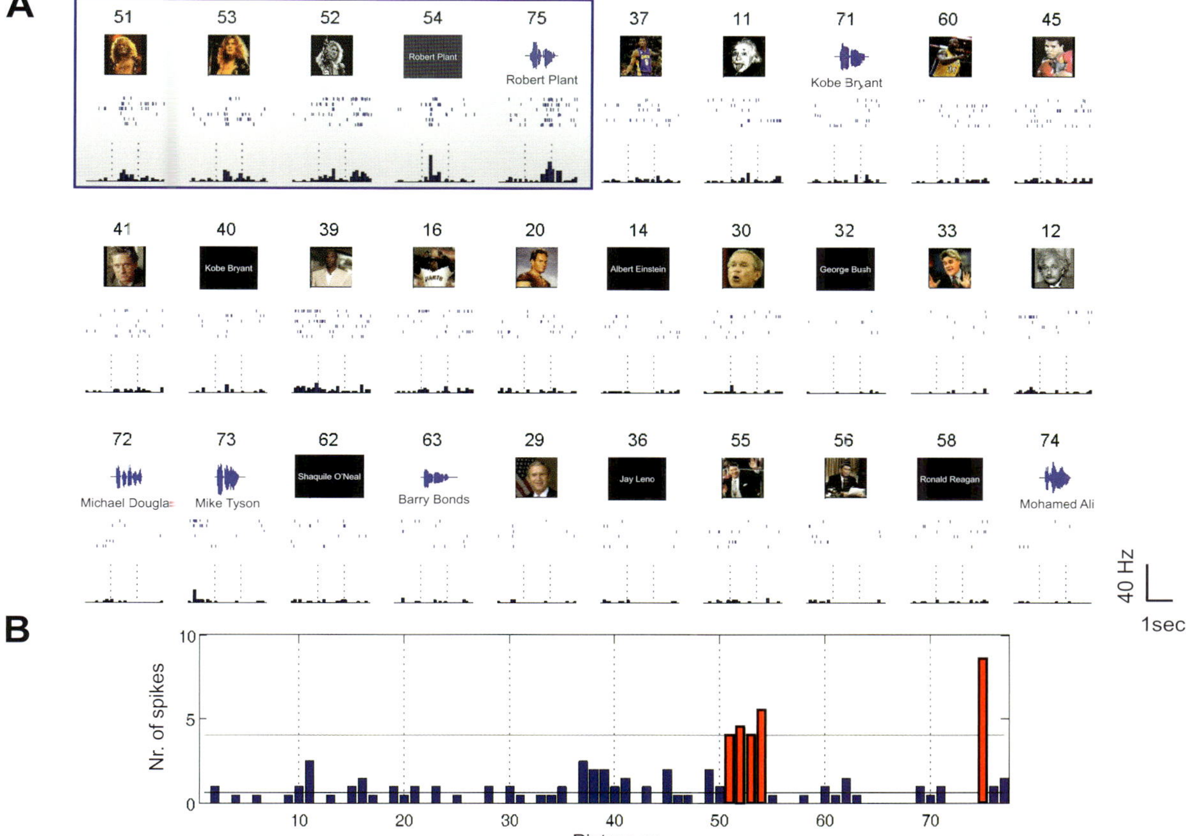

Plate 7 (figure 8.4)

A single unit in the entorhinal cortex responding to multiple different stimuli representing the same individual person. (A) A neuron in the entorhinal cortex that responded selectively to pictures of singer Robert Plant (stimulus 51, 53, and 52), as well as to his written (stimulus 54) and spoken (stimulus 75) name. There were no significant responses to any other picture, sound, or text presentations. For space reasons, only the largest 30 (out of 78) responses are displayed. In each case the raster plots for the six trials, peristimulus time histograms, and the corresponding pictures are shown. The vertical dotted lines mark picture onset and offset (1 s apart). (B) Median number (Nr) of spikes (across trials) for all stimuli. Presentations of Robert Plant are marked with red bars. Stimulus numbers correspond to the ones shown above each picture in (A). The gray horizontal line shows 5 SD above the baseline threshold used for defining significant responses. Modified from Quian Quiroga et al. (2009).

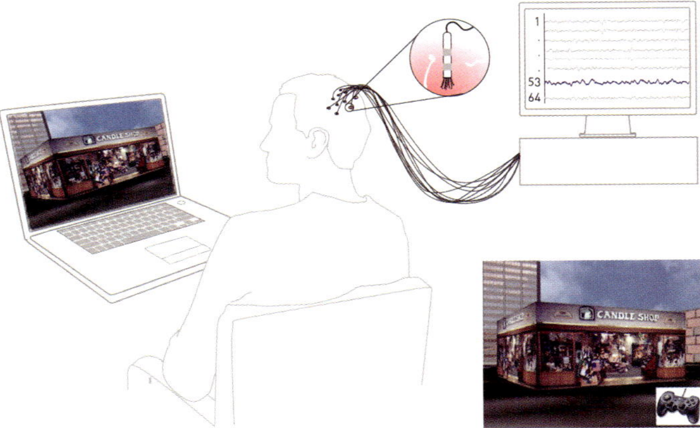

Plate 8 (figure 9.2)

Example experimental setup for intracranial recordings during spatial navigation. Participants drive through a virtual environment, picking up randomly located passengers and delivering them to stores. Insets: Snapshot of an example target store within the virtual driving game. Participants use the joystick shown to navigate to stores such as the Candle Shop that is depicted.

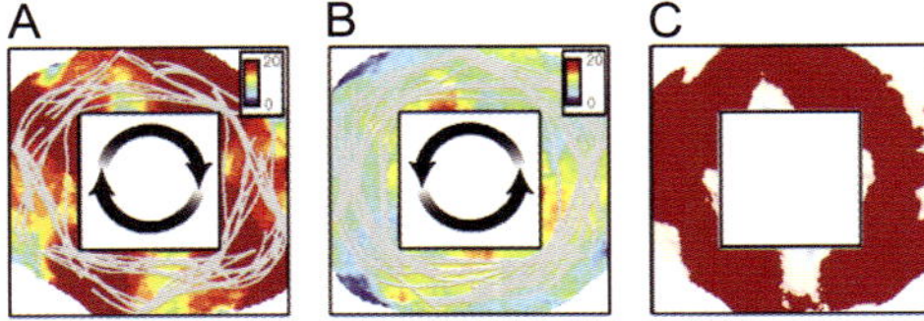

Plate 9 (figure 9.4)

An example entorhinal neuron that increases in firing rate while the participant is driving in a clockwise direction (A) compared to a (B) counterclockwise direction. (C) Shown is the difference in average firing rate of this entorhinal cell across all clockwise compared to counterclockwise driving trials. Hotter colors indicate higher firing rates. Adapted with permission from Jacobs et al. (2010).

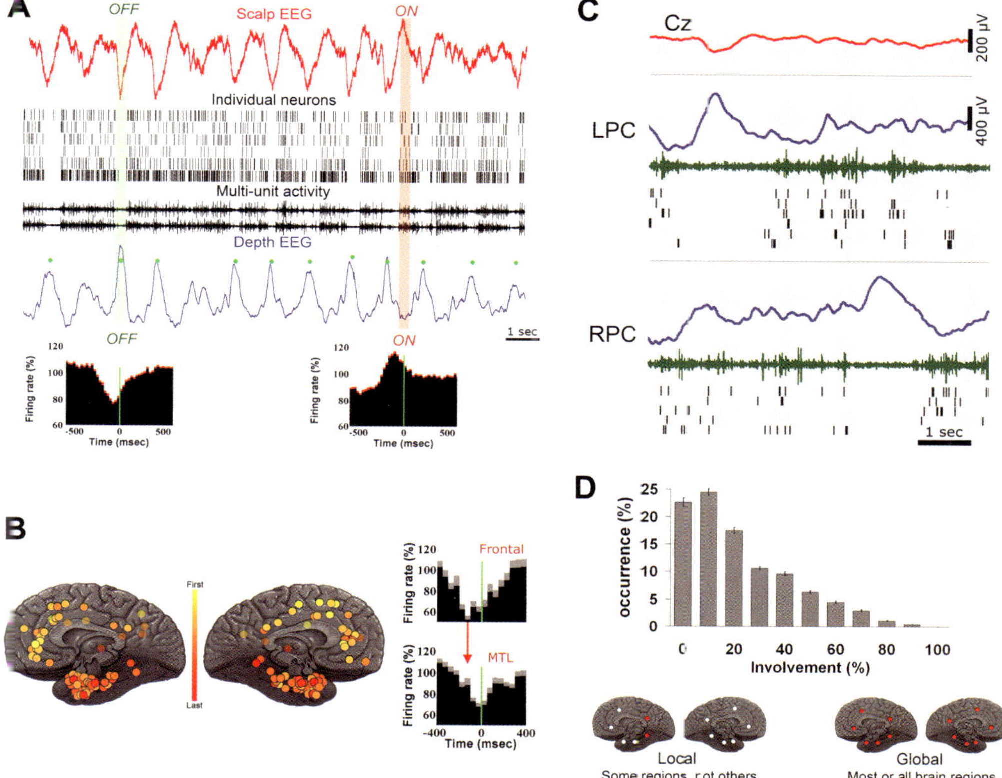

Plate 10 (figure 10.3)

Regional slow waves in human sleep. (A) Neuronal activity underlying slow waves in human sleep. Electrical brain activity across 15 s of deep non–rapid eye movement (NREM) sleep. Top (red), scalp electroencephalography (EEG); bottom (blue), intracranial depth EEG in entorhinal cortex. Green dots, individual slow waves that are automatically detected and separated from pathological events. Black, multiunit activity (MUA) and action potentials of six neurons. Vertical green bar, OFF periods of inactivity. Vertical orange bar, ON periods of neuronal silence. Bottom insets, an analysis across 600 units confirms that neurons increase and decrease their activity in concert with local electrical fields. (B) Slow waves have a tendency to propagate along typical paths. Left, each circle denotes a depth electrode, and its color marks the typical slow-wave timing at that location. Right, average unit activity in frontal cortex (top, n = 76) and medial temporal lobe (MTL) (bottom, n = 155), triggered by the same scalp slow waves. Note that, on average, slow waves and underlying neuronal activity occur earliest in the frontal lobe, about 200 ms later in the temporal lobe, and finally in the hippocampus. (C) An example of local sleep slow waves occurring at different times in left posterior cingulate cortex (LPC) and right posterior cingulate cortex (RPC). Rows (top to bottom) depict activity in scalp EEG (Cz, red), left and right posterior cingulate. Blue, depth EEG; green, MUA; black lines, single unit spikes. White shadings mark local OFF periods. (D) The vast majority of slow waves occur locally. Distribution of slow-wave involvement (percentage of monitored brain structures expressing each wave) shows that global slow waves are quite rare. Modified from Nir et al. (2011).

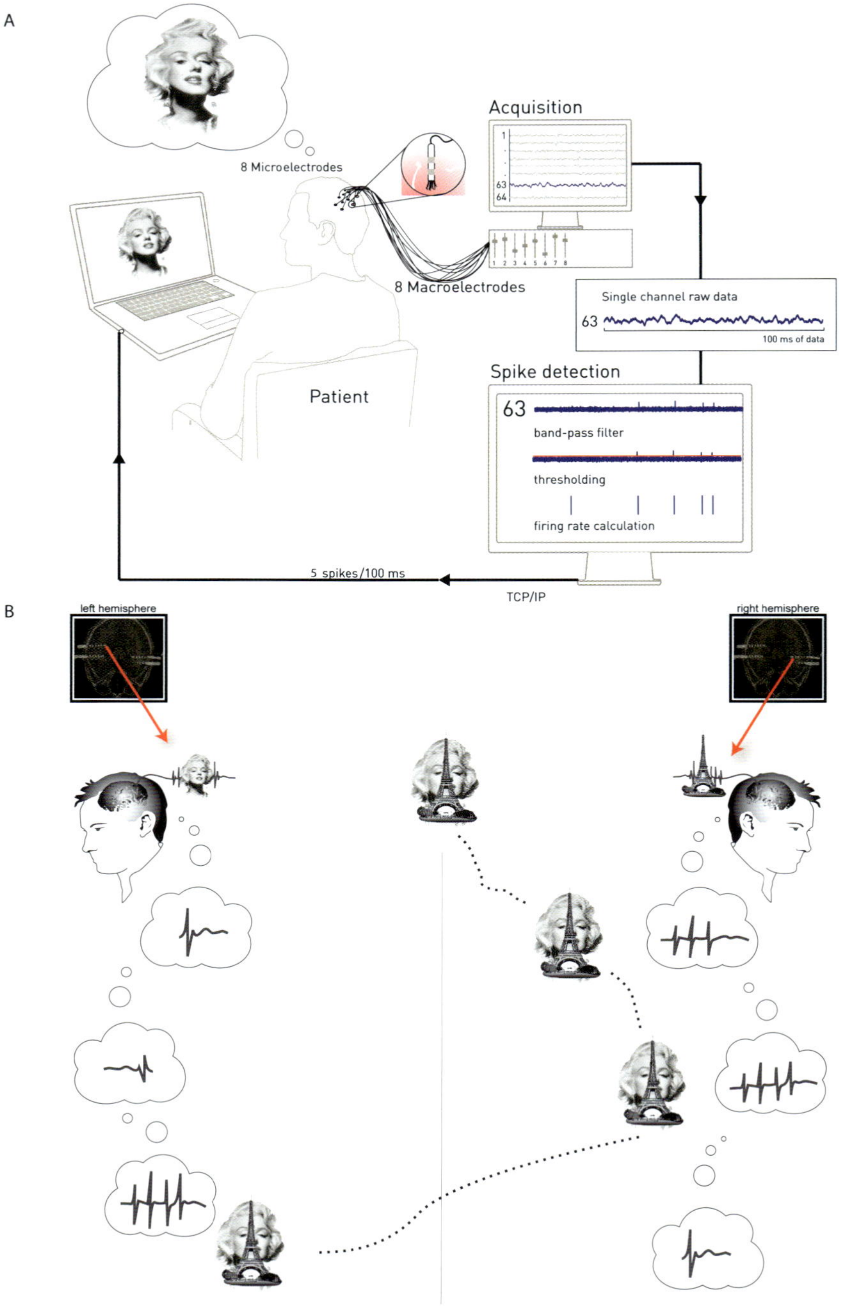

Plate 11 (figure 11.4)

(a) Recording from intracranial electrodes, neurons are identified that respond to a specific concept: in this instance, a cell responsive to the image of Marilyn Monroe. This cell increases its firing rate to the image or thought of Monroe. (b) This cell is then pitted against a different cell which was found to represent the Eiffel Tower. The two images are superimposed, and the subject is asked to bring the image of Monroe to maximum visibility. The visibility of the image is controlled by real-time decoding of the activity of each neuron relative to the other neuron and its own baseline. In this example, we show a case where the subject initially begins to fail the experiment—the firing of the Eiffel Tower neuron increases and the visibility of the tower increases, creating negative feedback. However, the subject is able to exert control and, by concentrating on the internal thought of Monroe, is able to override this sensory input and increase the firing rate of the Monroe neuron and decrease that of the Eiffel Tower neuron, bringing the image of Monroe to visibility. The scans show the location of the respective electrodes within the brain.

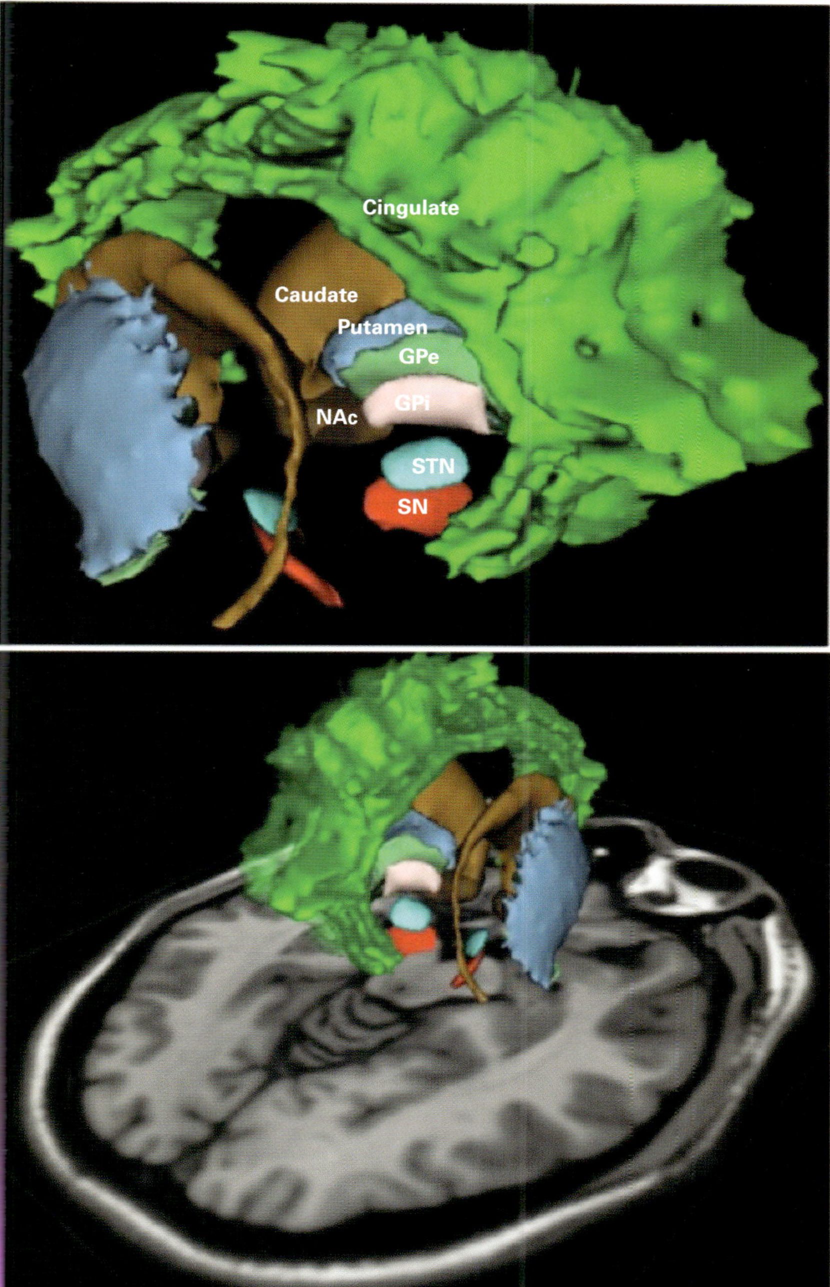

Plate 12 (figure 12.1)

Anatomical relationship of corticostriatal structures implicated in reward processing, reinforcement learning, and decision making. The cingulate cortex is thought to play an important role in cognitive functions, including reward anticipation, error detection, and decision making. The nucleus accumbens (NAc) is frequently studied for its role in addiction, but it also participates in learning and motivation through its connections with dopaminergic neurons, limbic areas, and prefrontal cortex. The substantia nigra (SN) sends dopaminergic signals to the striatum and cortex and encodes prediction error signals that drive learning. STN, subthalamic nucleus; GPe, globus pallidus externa; GPi, globus pallidus interna. Images provided by Kirk Finnis, Ph.D., and Medtronic, Inc.

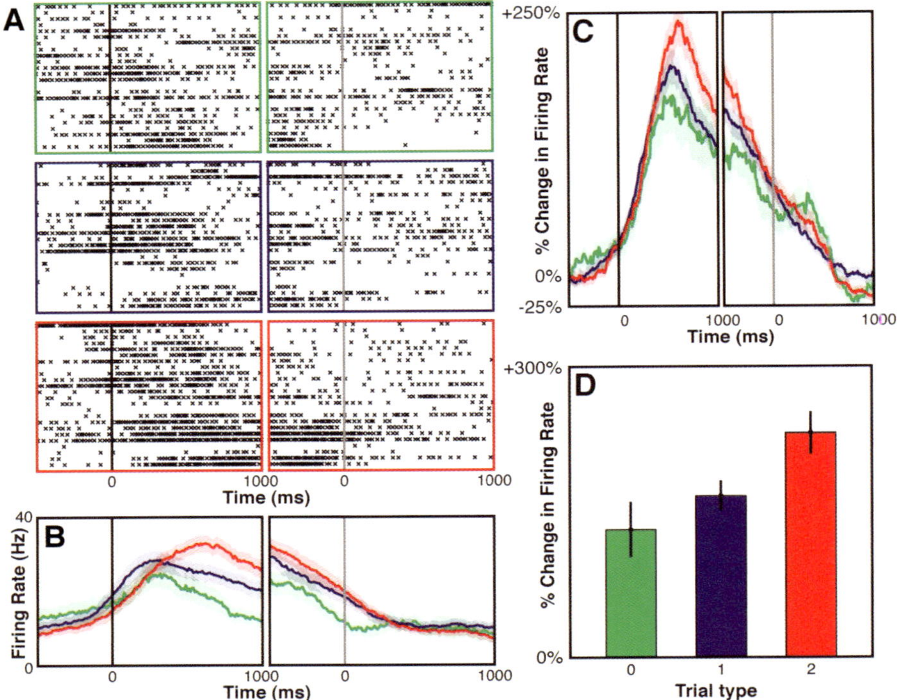

Plate 13 (figure 12.2)
Individual and population neuronal responses. (A) Example neuron showing modulation of firing based on cue-related interference. Rasters for Type 0 (green), 1 (blue), and 2 (red) trials are shown aligned to the cue (black line) and choice (gray line). (B) Average firing rates of the same neuron, demonstrating increasing firing with increasing interference. Error bars (SEM) are depicted with shading. (C) Average firing of all cue-related neurons. (D) Same as in (C), but showing activity averaged within a 200-ms-wide window centered 500 ms after the cue. Neuronal firing increased with cognitive interference (p = 0.02, analysis of variance), correlating with reaction time. From Sheth et al. (2012).

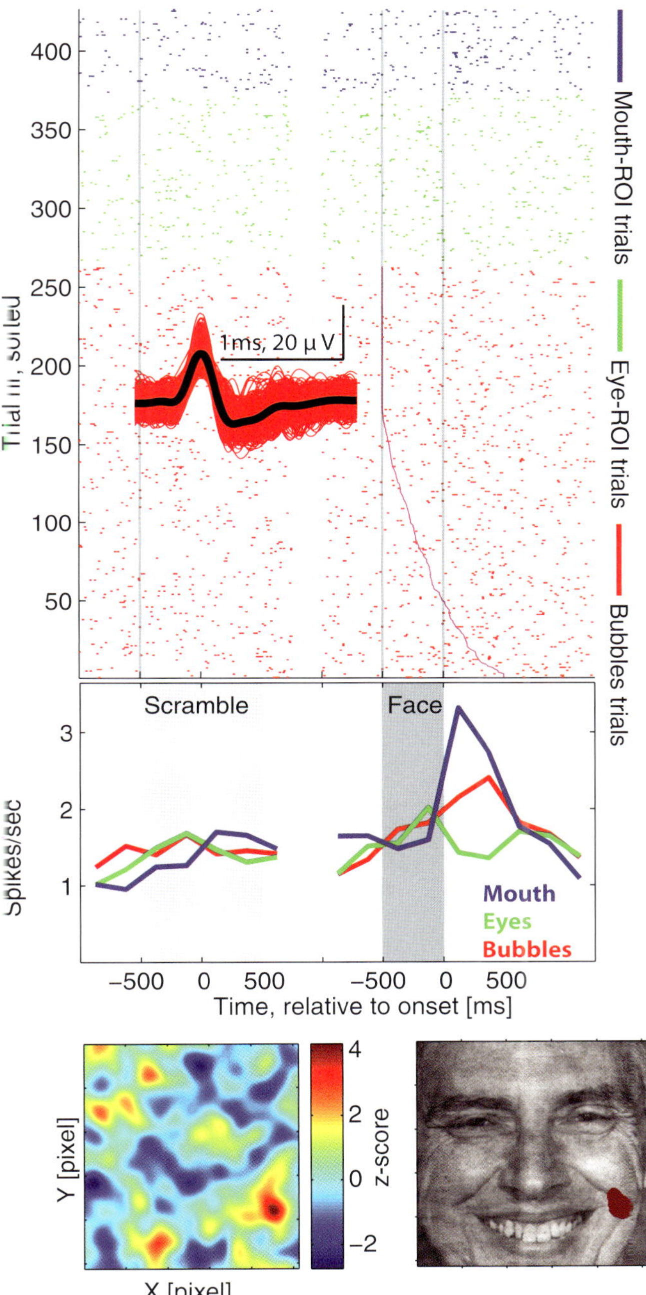

Plate 14 (figure 13.3)

Single neuron example of an amygdala neuron with a part-selective response. Shown are the raster (top) with the wave-forms superimposed, the poststimulus time histogram (middle), and the neuronal classification image (bottom). The significant portion of the classification image is superimposed on a face. The neuronal classification image is located close to the mouth, and the neuron also responds to faces where the mouth region of interest (ROI) is revealed. Nr, number.

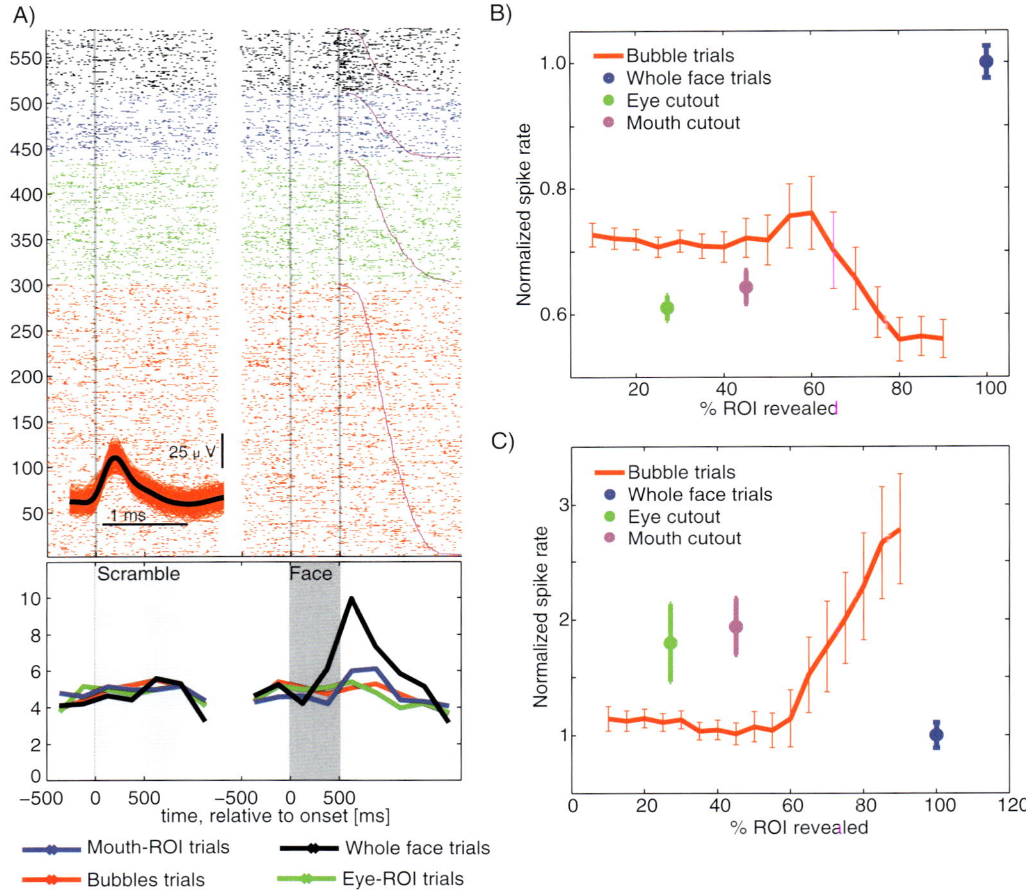

Plate 15 (figure 13.4)

Whole face selectivity of amygdala neurons. (A) Single neuron example of a whole-face selective neuron. Trials are shown ordered according to category (color code) and reaction time (magenta line). The stimulus (face) was on the screen for 500 ms. The superimposed waveforms show the waveform of each spike displayed in the raster; y-axis is in units of trial number (top) and spikes/second (bottom). (B, C) Quantification of nonlinear response to whole faces (WFs) and parts for neurons that increase their firing rate to WFs (B) and those that decrease (C). ROI, region of interest. Modified from Rutishauser et al. (2011b).

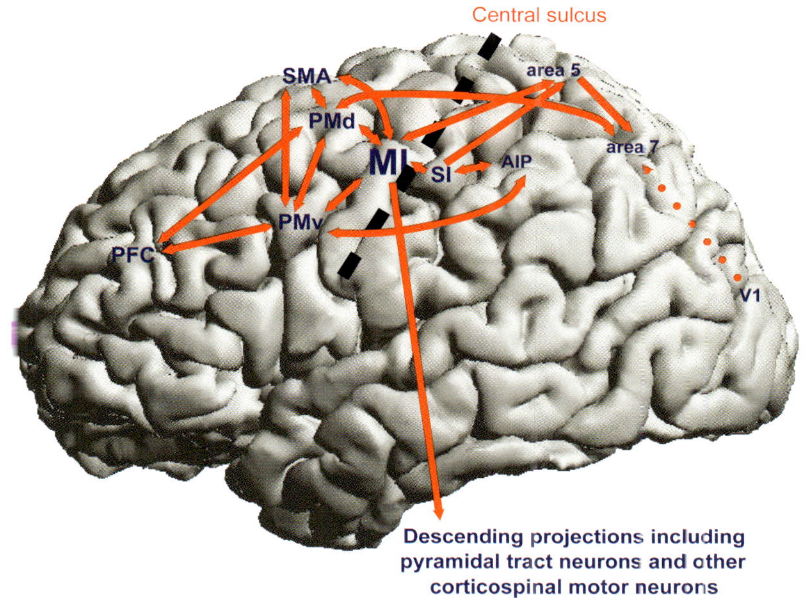

Plate 16 (figure 17.2)
A simplified schematic of cortical regions and connected pathways involved in control of reaching and grasping actions; see text. SMA, supplementary motor area; PMd, dorsal premotor cortex; PMv, ventral premotor cortex; MI, primary motor cortex; SI, primary somatosensory cortex; AIP, anterior intraparietal region; PFC, prefrontal cortex; V1 primary visual cortex. Adapted from figures in Martin (2003), and Scott (2004), and from connectivity results in monkeys from Vogt and Pandya (1978), Matelli et al. (1986), Pandya and Yeterian (1996), and Dum and Strick (2005).

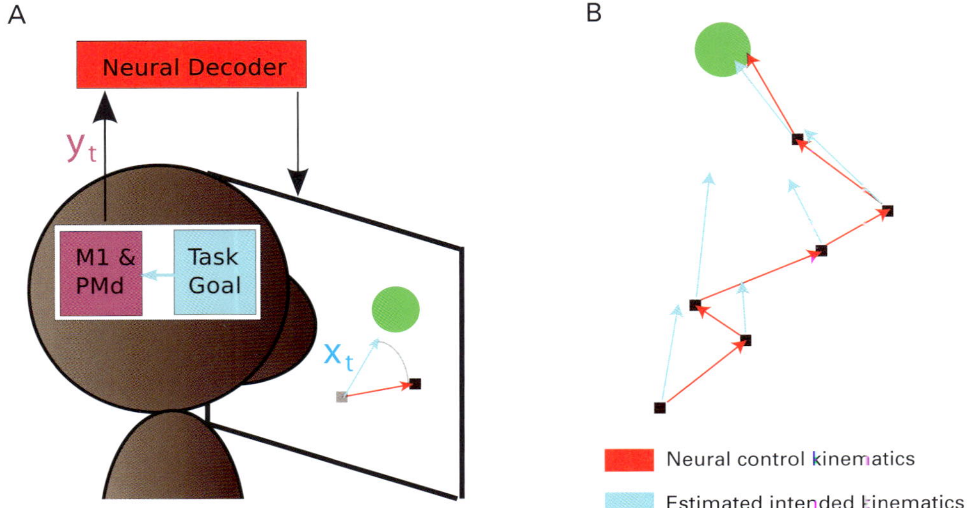

Plate 17 (figure 17.3)
Generating an "intention-based" kinematic training set. (A) The user is engaged in online control with a neural cursor. During each moment in the session, the neural decoder drives the cursor with a velocity shown as a red vector. Gilja et al. assumed that the monkey intended the cursor to generate a velocity towards the target in that moment, so following data collection the researchers rotate this vector to generate an estimate of intended velocity, shown as a blue vector. Note that this blue vector was not rendered on the screen as part of the experiment but is drawn there just to aid in explanation. This new set of kinematics is the training set used to train the control algorithm. M1, primary motor cortex; PMd, dorsal premotor cortex. (B) An example of this transformation applied to successive cursor updates. Figure reproduced and legend modified from Gilja et al. (2012) with permission from *Nature Neuroscience*.

Consciousness

Several investigators have tried to elucidate the neuronal correlates of consciousness (e.g., among many others, Crick, 1994; Crick & Koch, 1995; Rees et al., 2002; Koch, 2005). Arguably one of the greatest challenges for scientific reductionism, the quest involves trying to explain the subjective world of feelings, perceptions, and other sensations purely based on the activity of neurons and their interactions. In other words, while many centers, departments, and textbooks still discuss the mind and the brain as two separate entities (a remnant of Descartes' dualism), the hope is to unify these two descriptions into a single one where the mind can be explained by biological circuitry. Given the difficulties inherent in examining subjective perception in animal models, the welcome opportunity to investigate the human brain from the inside holds the potential to offer important insights in this field.

An important experimental paradigm in the field has been the study of binocular rivalry (Leopold & Logothetis, 1999; Blake & Logothetis, 2002). When two different stimuli, A and B, are shown to the two eyes, perception typically alternates in a seemingly random fashion between the two percepts. Several investigators have been intrigued by the notion of finding where these subjective percepts take place in the brain given that the stimulus is essentially constant in the sensory periphery. A series of elegant experiments by Logothetis's group has shown that there is a progression in the proportion of neurons that correlate with subjective perception from almost none in early areas such as lateral geniculate nucleus and V1 (Lehky & Maunsell, 1996; Leopold & Logothetis, 1996) all the way to almost all in ITC (Sheinberg & Logothetis, 1997). Consistent with these findings, most of the visually selective neurons in the human MTL also modulated their responses with the alternating percept (Kreiman et al., 2002; see figure 8.5). Similar conclusions were reached in a study where stimuli were flashed for brief periods of time and then instantly masked to avoid retinal afterimages; MTL neurons fired if and only if subjects were aware of the stimulus (Quiroga et al., 2008).

A particularly intriguing aspect of consciousness is the notion of free will. At any given time, we have a strong feeling that we are the owners of our actions, that we can choose to go right or left as we please. The feeling of "authorship" or "will" can be considered to be another example of a conscious sensation, and it is therefore legitimate to ask where/when/how this content of consciousness is encoded. Libet introduced a simple experimental paradigm to examine volition that has led to endless discussions in the literature (Libet, 1985). In brief, subjects were instructed to tap their index finger at will while monitoring the internal time when they felt the "urge" to execute the movement. By relating this internal decision time with averaged scalp signals from electroencephalographic recordings, Libet argued that major changes in brain oscillations take place hundreds of milliseconds, and in some cases seconds, before subjects became aware of their own intentions to move.

Following Libet's observation, Fried et al. (1991) reported that stimulation in the supplementary motor area elicited in subjects a sensation of urge to perform particular motor acts. More

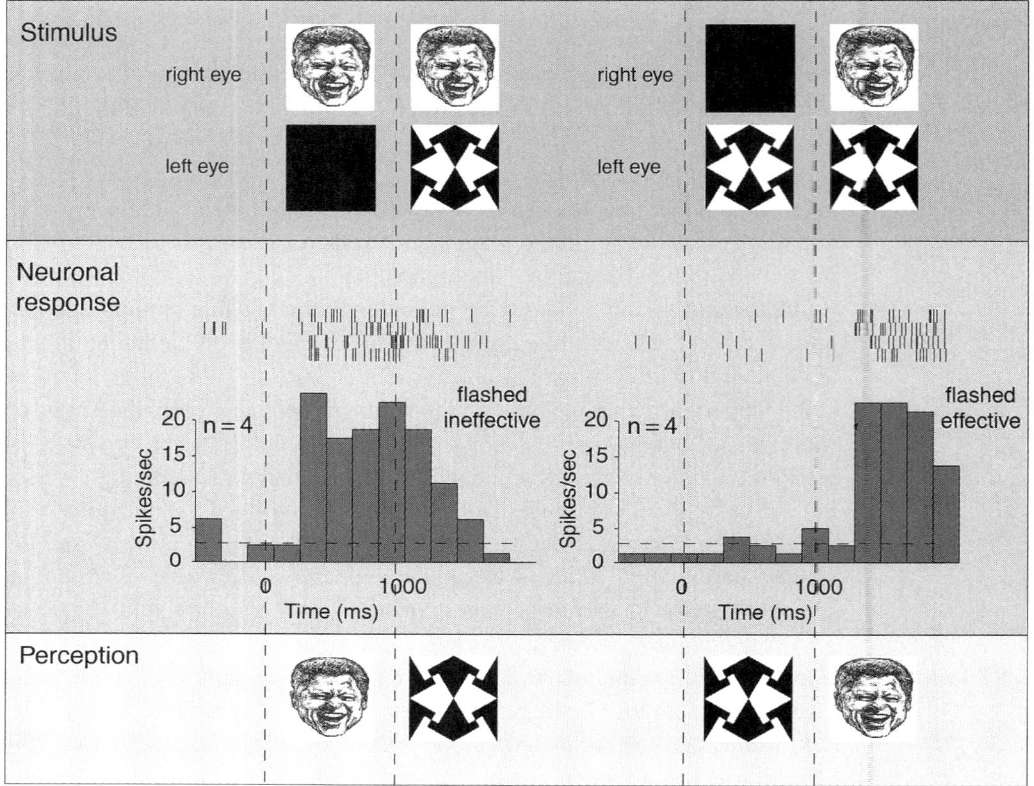

Figure 8.5

Right amygdala neuron that follows the subjective percept. (A) Stimulus presentation: An image is presented monocularly for 1000 ms. The same image then is flashed onto the same ipsilateral eye while a different image is flashed to the contralateral eye for 500 ms. (A1) A picture of Clinton is presented monocularly, and a black-and-white pattern is later flashed onto the other eye. (A2) The pattern is presented monocularly, and Clinton is flashed to the contralateral eye (right). Images were chosen randomly with the constraint that the two pictures could not belong to the same category. (C) Subjective percept. (C1) The picture of Clinton is suppressed during the binocular period by the pattern. (C2) The picture of Clinton perceptually suppresses the pattern. (B) Neural response. This neuron responded selectively to Clinton among 49 different stimuli. (B1) Responses when Clinton was shown monocularly and a different ineffective stimulus was flashed. Although there seems to be elevated firing activity during the binocular period, it is not statistically significant after taking into account the response latency. (B2) Responses when an ineffective stimulus was shown monocularly and the picture of Clinton was flashed and perceptually suppressed the ineffective stimulus. Bin size, 200 ms. The dashed lines denote the onset of the monocular image and the flash. The horizontal dashed line shows the baseline activity of the neuron (2.8 Hz). The number of repetitions of each stimulus is shown above the histogram. Modified from (Kreiman et al., 2002).

recently, Fried and colleagues showed that neurons in the medial frontal cortex (particularly those in the supplementary motor area, presupplementary motor areas, and dorsal/rostral aspects of the anterior cingulate) showed strong changes that predicted when subjects would decide to execute a movement in this paradigm (in single trials; Fried et al., 2011). In contrast, MTL neurons showed only marginal above-chance performance in this task (tentatively interpreted by the authors as a consequence of the memory component of the task).

The relationship between consciousness and MTL activity is called into question by neurological and lesion studies. While bilateral removal of the hippocampus can lead to severe impairment in memory consolidation, no changes have been documented in the moment-to-moment awareness or consciousness of these subjects (Scoville & Milner, 1957; Milner et al., 1968; Squire et al., 2004). Bilateral removal of the amygdala can lead to significant impairment in interpreting emotions, but it also does not seem to directly affect conscious processing (Adolphs et al., 2005). In other words, neurological evidence (as well as lesion studies in animal models) seems to argue that MTL is not necessary for consciousness.

In fact, some investigators have argued that causality works in the opposite direction here. Perhaps activity in the hippocampus and entorhinal cortex is not necessary for conscious perception but consciousness is necessary for activation of these areas. That is, we propose that information reaching the hippocampus and its input structures has already been processed in such a way that unconscious activity is filtered out. At the very least, it seems likely that conscious awareness may enhance the probability of recalling information and provide flexibility to what, when, and how information is consolidated into memories (Lisman & Sternberg, 2013).

Certain aspects of content do seem to require an intact amygdala (also see chapter 13). In particular, as noted above, recognition of fearful facial expressions is impaired by lesions to the amygdala (Adolphs et al., 1994). Other evidence also seems to imply that unconscious activity may reach the amygdala. For example, in a functional neuroimaging study, investigators found amygdala activation (i.e., blood flow changes) even when the faces denoting emotional expressions were perceptually suppressed (Morris et al., 1998; Whalen et al., 1998). Amygdala blood-oxygen-level-dependent activation by emotional facial expressions was also reported in a subject with cortical blindness (Pegna et al., 2005). These findings have led researchers to hypothesize the existence of a subcortical pathway to the amygdala that bypasses visual cortex and provides rapid activation of the amygdala by fear-related stimuli prior to conscious perception to prime early visual processing of these stimuli (Phelps & LeDoux, 2005; Ohman et al., 2007). However, there is no clear evidence for such a pathway in humans and the notion of rapid subcortical signals to the amygdala seems controversial (Pessoa & Adolphs, 2010; Cauchoix & Crouzet, 2013). The neuronal responses to animals in the right human amygdala (Mormann et al., 2011) were independent of emotional stimulus content, and the visual response latencies in the human amygdala are similar to those found in other regions in the temporal lobe (Mormann et al., 2008; Liu et al., 2009). These responses therefore do not support the notion of a distinct fast subcortical route and are more likely to be generated along the cortical object recognition pathway.

Discussion: Trying to Put All the Pieces Together Again

Sparse Coding

For the single unit recordings currently possible in humans, it is important to realize that the brain's networks of neurons are vastly undersampled. For example, recording from 64 microwires, one can obtain a total yield of tens of single and multiunits from various regions, which represents just a tiny sample of the several billion neurons the human brain consists of (Herculano-Houzel, 2009). Hence, it is legitimate to ask, "What can we actually expect to learn despite this vast spatial undersampling?"

This, in turn, depends on how our environment is represented by the neuronal activity in our brains. The question of how the brain encodes semantic information such as the identity of objects or persons has been a key focus in neuroscience for decades. Two opposing theories have been formulated. One theory assumes representation by a complex network in which a large number of neurons encode a certain concept in their entirety (distributed coding). Another theory describes a sparse coding in which just a few neurons encode an explicit representation of a specific concept. Taken to the extreme, this type of coding could be implemented by a single cell representing a single semantic concept, for example, that of one's grandmother, which is why this idea has been termed "grandmother cell" (Gross, 2002). A grandmother cell would thus be activated whenever one sees their grandmother, hears about her, talks to her on the phone, thinks of her, and so on.

Under multiple assumptions, it has been argued that distributed coding is far more cost-efficient than sparse coding. For example, a network of 8 neurons, half of which are activated by any given stimulus, can encode and distinguish between 70 different stimuli, whereas a sparse network of 8 neurons can only encode 8 stimuli in an independent on/off manner. For an efficient distributed code, one would also expect each neuron in the distributed network to respond to a substantial fraction of all presented stimuli. In this way, the selectivity of observed responses can provide an indication about which of the two types of coding is present.

Semantic neurons—that is, neurons which encode the semantic contents of a stimulus in an invariant manner—have been described in the human MTL, but not in other regions such as orbitofrontal cortex or anterior cingulate from which human single units have also been recorded. Given that a distributed code appears to be more cost-efficient than a sparse representation of abstract semantic concepts, an obvious question is why the brain would allow itself the "luxury" of an explicit, sparse, and invariant code for semantic information. This apparent luxury might be better understood once we realize how episodic memory formation, one of the main functions of the human MTL, might be connected to semantic representations.

You Shall Remember This

The process of encoding episodic memories consists of associating pieces of semantic information (what happened where with whom involved and so on) in a defined temporal order. Lesion studies in humans have shown that structures in the MTL are essential for the encoding of episodic memories (Squire et al., 2004; Squire et al., 2007; Squire, 2009; Milner et al., 1968, Scoville & Millner, 1957). Representations of semantic information at the single unit level are frequently found in these very structures and thus might provide a unique opportunity to investigate how our brain links pieces of semantic information together into episodic memories (Quiroga, 2012). To learn novel information in high-dimensional space, it pays to have sparse representations, since they are easy to manipulate. It is also easy to learn a new high-level concept (i.e., a new movie star) without having to change all the low-level representations (i.e., to move all of the boundaries between the categories). In rodent navigation one can distinguish path integration (dead reckoning navigation) along a linear track from landmark navigation (map navigation) in a two-dimensional environment. The majority of place cells found on a linear track are unidirectional, that is, they respond only during runs in one of the two directions and in this sense are context dependent. Place cells in a two-dimensional environment, on the other hand, are omnidirectional, that is, independent of the direction from which the animal approaches the place field and thus also independent of the temporal context. These two phenomena bear a certain analogy to episodic and semantic memory formation, respectively. Episodic memories are always context dependent whereas semantic memories are context invariant and can emerge via generalization of recurring context-dependent experiences (Buzsáki, 2005).

Based on this analogy, one may hypothesize that the basic neural mechanisms during the exploration of a new environment in rodents could also play a role for encoding memories. A phenomenon of particular interest in this context is the reactivation of sequences of single neuron activity observed in rodents. Here sequences of place cell activity (see chapter 9) that were activated during the exploration of an environment are "replayed" during a subsequent resting period and particularly during sleep (Skaggs & McNaughton, 1996; Louie & Wilson, 2001). These replayed sequences exhibit a high temporal compression and can occur in forward and reverse order (Nadasdy et al., 1999; Foster & Wilson, 2006; Diba & Buzsáki, 2007).

In rodents the replay of these sequences typically occurs during brief periods (ca. 100 ms) of high-frequency activity termed sharp wave/ripples in the LFP and is assumed to serve for consolidation of spatial memories by transferring the sequences into cortex for permanent storage (O'Neill et al., 2010). By analogy one may hypothesize that consolidation of episodic memories in humans is likewise achieved by reactivation of a previously experienced sequence of semantic neurons in the homologous brain structures. These sequences might serve to link information of an experienced episode such as who, what, where, and so forth and to bring them into a defined temporal order (Howard & Kahana, 1999).

Apparent Lack of Topography

Essentially all of sensory neocortex in macaque monkeys is characterized by elegant topography (less clearly so in the mouse neocortex). Nearby neurons share similar properties such as visual field proximity or correlated wavelength sensitivity in primary visual cortex blobs. Particularly striking demonstrations of this topography abound in the visual system where nearby neurons display related receptive field structures and hence form a retinotopic map of visual space (Hubel & Wiesel, 1962; Wandell, 1995). Similar claims have been made in higher cortical areas (e.g., Mountcastle, 1957; Tanaka, 1996; Kreiman et al., 2006; Tsao et al., 2008b). Several suggestions have been advanced to describe the advantage of topographical arrangements of neurons including the minimization of wire lengths.

Many investigators have tried to find evidence for topographic mapping in the hippocampus and surrounding structures with no success. Two nearby neurons can show completely different place fields in rodents (O'Keefe & Nadel, 1978) and may respond to distinct stimuli in humans. For instance, a neuron in the right human amygdala may respond to various animals while another neuron recorded on the same microwire may respond to a single individual person (Mormann et al., 2011).

While the semantically invariant neurons found in the human MTL are in many ways reminiscent of rodent place cells, they do not appear to be confined to the hippocampus but instead are found also in the entorhinal cortex, amygdala, and—with a lesser degree of invariance—in the parahippocampal cortex. The function of certain cell types in the MTL of rodents shows a clear topography and has been well characterized, for example, grid cells in entorhinal cortex (Hafting et al., 2005; Killian et al., 2012), head direction cells in the subiculum (Taube et al., 1990), and hippocampal place cells (O'Keefe & Dostrovsky, 1971). However, it remains unclear to what extent functional analogies between these spatial responses and the semantic responses observed in human studies exist.

Amygdala and Emotions

The amygdala has been implicated in the processing of biologically relevant information, including both aversive (e.g., threat- or fear-related) and appetitive (e.g., reward-related) stimuli (Baxter & Murray, 2002; Phelps & LeDoux, 2005; Adolphs, 2008; Pessoa, 2008; see chapter 13). Earlier notions that the amygdala might be specialized to elicit or mediate fear responses (LeDoux, 1996) have been supplemented by more abstract accounts whereby the amygdala processes ambiguity or unpredictability in the environment (Herry et al., 2007) and mediates an organism's vigilance and arousal (Davis & Whalen, 2001). Electrophysiological recordings in monkeys have found amygdala neurons that respond selectively to complex stimuli. These include single neurons that respond to faces (Leonard et al., 1985), as well as cells responding selectively to the identity or expression of monkey faces or to human faces or objects (Gothard et al., 2007). Furthermore, amygdala neurons in the macaque were shown to encode the reward value of stimuli (Nishijo et al., 1988; Paton et al., 2006) during reinforcement learning in a highly adaptive manner that is modulated by attention and arousal (Belova et al., 2007). In humans, neuroimaging studies of the amygdala argue for a broad role in processing stimuli that are strongly rewarding or punishing (Sander et al., 2003; Ohman et al., 2007). This plethora of diverse findings has left it unclear what stimulus categories the amygdala might encode, limiting current theories of amygdala function.

In human fMRI memory studies, increased blood flow in the amygdala during encoding was found to correlate with improved memory formation (e.g., Canli et al., 2000). In addition, the phase locking of human amygdala neurons to ongoing theta oscillations was found to predict memory formation (Rutishauser et al., 2011, see also chapter 7). It is thus conceivable that the amygdala plays a key role in filtering out biologically relevant information for memory encoding and consolidation. One should be aware, though, that the amygdala consists of many nuclei that have different connections and subserve diverse functions. With our electrophysiological studies in humans, we are currently only tapping the surface of this complex functionality. More work will be needed in our single unit studies to understand the interactions between emotions, memory formation, and the responses to complex stimuli described here.

Open Questions

The proverbial parable of blind people examining an elephant and drawing different conclusions depending on their vantage point can hardly be more appropriate here. Take the recordings from the hippocampus from different investigators. Some will claim, "These neurons represent space!" (e.g., see chapter 9). Others will adamantly point to how well these neurons can help predict what people will remember (e.g., see chapter 7). Yet others will point to the strong degree of tolerance of some of these neurons in response to different images of the same familiar person (e.g., this chapter). This is, of course, only an oversimplification, and there could be even more descriptions if we included other manuscripts, other species, and other experimental paradigms. It is tempting to speculate that many of these observations could be described within a common theoretical framework. Perhaps this is only part of our will to push Occam's principle, but it seems worth trying.

While we hope that future studies will help unify and consolidate our understanding of different responses elicited by distinct tasks and paradigms, we should not lose track of the important anatomical and functional demarcations among different MTL structures. Several of these differences have been pointed out above (see "One hat does not fit all" and "Regional Specificity"). Sometimes when the differences are not black and white, the small numbers of neurons sampled may preclude our drawing sharp differences across areas. Pooling data across studies (whenever possible) may help ameliorate such practical difficulties. Meta-analyses have been quite frequent and useful in other domains, and we are just beginning to accumulate sufficient data in the field where such comparisons may yield interesting insights.

We suggested above that there might be a distinction between the hippocampus and the amygdala in terms of the extent to which unconscious information reaches those areas. If indeed unconscious information can reach the amygdala but not the hippocampus and surrounding structures (a highly tentative notion that will require further investigation), this may provide interesting opportunities to examine differential inputs to the two structures and how those inputs correlate with the contents of consciousness.

Neurons in the MTL take visual information from the highest echelons of visual cortex and can perform seemingly magic transformations. Hippocampal neurons show a remarkable degree of abstraction in responding to pictures but also to text, with visual input or during visual imagery, detecting associations between stimuli, how novel they are and whether they should be remembered or not. In this sense, the rhinal–hippocampus axis seems to be a fabulous arena to examine the transformation of sensory information into semantic and cognitive information. A large volume of studies has focused on purely sensory aspects (e.g., what is the receptive field size of neurons in ITC and how do these neurons respond with more than one stimulus within their receptive field?). Another growing volume of studies, partly summarized in this chapter, has focused on purely cognitive aspects (e.g., can neurons respond to abstract properties of the stimuli or even in the absence of a stimulus?). The necessary and sufficient transformation that can link these two worlds, the strictly sensory one and the inner cognitive domain, seems to be a critical theme for further investigation.

Clinical Implications

For many years, clinicians performed drastic cuts along the corpus callosum in their efforts to treat seizures. Clinical practice was influenced by the elegant studies of Roger Sperry, inspired by the work describing the organization of the visual and language systems (Sperry, 1982). One can only hope that the growing body of neurophysiological recordings will help inspire and inform further improving ways to treat epilepsy. This may take the form of further quantifying and documenting the effects of removing the epileptogenic areas in order to improve surgical approaches. Most neurophysiologists are surprised by the notion that unilateral removal of relatively large spans of MTL cortex does not seem to lead to very obvious cognitive deficiencies although some loss of verbal (predominant in left MTL) or nonverbal (predominant in right MTL) memory function can be detected in almost all cases when sensitive neuropsychological tests are used (Gleissner et al., 2002). This striking observation may well speak to the enormous amount of redundancy in the network, or it may be ascribed to plasticity and adaptation postresection (although even acute studies do not seem to reveal major deficits; see however Bartsch et al., 2010). Alternatively, it may very well be that we have not yet examined the right type of behavior. In parallel to the work of Sperry mentioned above, many behavioral studies had failed to find any cognitive deficit after callosotomies. Neurophysiological recordings may help provide clues about the functions of neurons in different parts of the MTL and thus inspire direct behavioral evaluation. In addition, microelectrode recordings from seizure-generating brain regions will provide fuel for studies investigating the mechanisms involved in seizure generation, propagation, and termination, both at the level of LFPs (Stead et al., 2010; Jacobs et al., 2012) and unit activity (chapter 18).

Learning and memory challenges are common in many epileptic patients. Additionally, the hippocampus is one of the key areas that show significant deterioration in patients with Alzheimer's disease. Neurophysiological recordings in the MTL can lead to a better understanding of how neural circuits orchestrate learning and memory formation and may thus provide novel ideas to help alleviate cognitive deficits in patients suffering from epilepsy, dementia, and other cognitive challenges that depend on the MTL.

Acknowledgments

While many of the authors of this chapter have now traveled the world, the work described here centers around research that was performed at the California Institute of Technology and the University of California, Los Angeles, between 1998 and 2010 under the tutelage of CK and IF.

References

Adolphs, R. (2008). Fear, faces, and the human amydgala. *Current Opinion in Neurobiology, 18*, 166–172.

Adolphs, R., Gosselin, F., Buchanan, T., Tranel, D., Schyns, P., & Damasio, A. R. (2005). A mechanism for impaired fear recognition after amygdala damage. *Nature, 433*, 68–72.

Adolphs, R., Tranel, D., Damasio, H., & Damasio, A. (1994). Impaired recognition of emotion in facial expressions following bilateral damage to the amygdala. *Nature, 372*, 669–672.

Agam, Y., Liu, H., Pappanastassiou, A., Buia, C., Golby, A. J., Madsen, J. R., et al. (2010). Robust selectivity to two-object images in human visual cortex. *Current Biology, 20*, 872–879.

Allison, T., Puce, A., Spencer, D., & McCarthy, G. (1999). Electrophysiological studies of human face perception: I. Potentials generated in occipitotemporal cortex by face and non-face stimuli. *Cerebral Cortex, 9*, 415–430.

Amaral, D. G., & Lavenex, P. (2007). Hippocampal neuroanatomy. In P. Andersen, R. G. M. Morris, D. G. Amaral, T. Bliss, & J. O'Keefe (Eds.), *The hippocampus book*. New York: Oxford University Press.

Barlow, H. (1972). Single units and sensation: A neuron doctrine for perception. *Perception, 1*, 371–394.

Bartsch, T., Schonfeld, R., Muller, F. J., Alfke, K., Leplow, B., Aldenhoff, J., et al. (2010). Focal lesions of human hippocampal CA1 neurons in transient global amnesia impair place memory. *Science, 328*, 1412–1415.

Baxter, M. G., & Murray, E. A. (2002). The amygdala and reward. *Nature Reviews. Neuroscience, 3*, 563–573.

Belova, M. A., Paton, J. J., Morrison, S. E., & Salzman, C. D. (2007). Expectation modulates neural responses to pleasant and aversive stimuli in primate amygdala. *Neuron, 55*, 970–984.

Bien, C. G., & Elger, C. E. (2007). Limbic encephalitis: A cause of temporal lobe epilepsy with onset in adult life. *Epilepsy & Behavior, 10*, 529–538.

Bitterman, Y., Mukamel, R., Malach, R., Fried, I., & Nelken, I. (2008). Ultra-fine frequency tuning revealed in single neurons of human auditory cortex. *Nature, 451*, 197–201.

Blake, R., & Logothetis, N. (2002). Visual competition. *Nature Reviews. Neuroscience, 3*, 13–21.

Burns, S. P., Xing, D., & Shapley, R. M. (2010). Comparisons of the dynamics of local field potential and multiunit activity signals in macaque visual cortex. *Journal of Neuroscience, 30*, 13739–13749.

Buzsáki, G. (2005). Theta rhythm of navigation: Link between path integration and landmark navigation, episodic and semantic memory. *Hippocampus, 15*, 827–840.

Canli, T., Zhao, Z., Brewer, J., Gabrieli, J. D., & Cahill, L. (2000). Event-related activation in the human amygdala associates with later memory for individual emotional experience. *Journal of Neuroscience, 20*, RC99.

Cauchoix, M., & Crouzet, S. M. (2013). How plausible is a subcortical account of rapid visual recognition? *Frontiers in Human Neuroscience, 7*, 39.

Cerf, M., Thiruvengadam, N., Mormann, F., Kraskov, A., Quiroga, R. Q., Koch, C., et al. (2010). On-line, voluntary control of human temporal lobe neurons. *Nature, 467*, 1104–1108.

Connor, C. E., Brincat, S. L., & Pasupathy, A. (2007). Transformation of shape information in the ventral pathway. *Current Opinion in Neurobiology, 17*, 140–147.

Crick, F. (1994). *The astonishing hypothesis*. New York: Simon & Schuster.

Crick, F., & Koch, C. (1995). Are we aware of neural activity in primary visual cortex? *Nature, 375*, 121–123.

Crick, F., Koch, C., Kreiman, G., & Fried, I. (2004). Consciousness and neurosurgery. *Neurosurgery, 55*, 273–282.

Davis, M., & Whalen, P. J. (2001). The amygdala: Vigilance and emotion. *Molecular Psychiatry, 6*, 13–34.

Dawodu, S., & Thom, M. (2005). Quantitative neuropathology of the entorhinal cortex region in patients with hippocampal sclerosis and temporal lobe epilepsy. *Epilepsia, 46*, 23–30.

Diba, K., & Buzsáki, G. (2007). Forward and reverse hippocampal place-cell sequences during ripples. *Nature Neuroscience, 10*, 1241–1242.

Dudai, Y. (2012). The restless engram: Consolidations never end. *Annual Review of Neuroscience, 35*, 227–247.

Ekstrom, A., Suthana, N., Behnke, E., Salamon, N., Bookheimer, S., & Fried, I. (2008). High-resolution depth electrode localization and imaging in patients with pharmacologically intractable epilepsy. *Journal of Neurosurgery, 108*, 812–815.

Engel, A. K., Moll, C. K., Fried, I., & Ojemann, G. A. (2005). Invasive recordings from the human brain: Clinical insights and beyond. *Nature Reviews. Neuroscience, 6*, 35–47.

Engel, J. (1996). Surgery for seizures. *New England Journal of Medicine, 334*, 647–652.

Epstein, R., Harris, A., Stanley, D., & Kanwisher, N. (1999). The parahippocampal place area: Recognition, navigation, or encoding? *Neuron, 23*, 115–125.

Eskandar, E. N., Richmond, B. J., & Optican, L. M. (1992). Role of inferior temporal neurons in visual memory: I. Temporal encoding of information about visual images, recalled images, and behavioral context. *Journal of Neurophysiology, 68*, 1277–1295.

Foster, D. J., & Wilson, M. A. (2006). Reverse replay of behavioural sequences in hippocampal place cells during the awake state. *Nature, 440*, 680–683.

Freedman, D., Riesenhuber, M., Poggio, T., & Miller, E. (2001). Categorical representation of visual stimuli in the primate prefrontal cortex. *Science, 291*, 312–316.

Fried, I., Katz, A., McCarthy, G., Sass, K. J., Williamson, P., Spencer, S. S., et al. (1991). Functional organization of human supplementary motor cortex studied by electrical stimulation. *Journal of Neuroscience, 11*, 3656–3666.

Fried, I., MacDonald, K. A., & Wilson, C. (1997). Single neuron activity in human hippocampus and amygdala during recognition of faces and objects. *Neuron, 18*, 753–765.

Fried, I., Mukamel, R., & Kreiman, G. (2011). Internally generated preactivation of single neurons in the human brain predicts volition. *Neuron, 69*, 548–562.

Fukushima, K. (1980). Neocognitron: A self organizing neural network model for a mechanism of pattern recognition unaffected by shift in position. *Biological Cybernetics, 36*, 193–202.

Gelbard-Sagiv, H., Mukamel, R., Harel, M., Malach, R., & Fried, I. (2008). Internally generated reactivation of single neurons in human hippocampus during free recall. *Science, 322*, 96–101.

Gleissner, U., Helmstaedter, C., Schramm, J., & Elger, C. E. (2002). Memory outcome after selective amygdalohippocampectomy: A study in 140 patients with temporal lobe epilepsy. *Epilepsia, 43*, 87–95.

Gothard, K. M., Battaglia, F. P., Erickson, C. A., Spitler, K. M., & Amaral, D. G. (2007). Neural responses to facial expression and face identity in the monkey amygdala. *Journal of Neurophysiology, 97*, 1671–1683.

Grill-Spector, K., Kushnir, T., Edelman, S., Itzchak, Y., & Malach, R. (1998). Cue-invariant activation in object-related areas of the human occipital lobe. *Neuron, 21*, 191–202.

Gross, C. (2002). Genealogy of the "grandmother cell." *Neuroscientist, 8*, 512–518.

Hafting, T., Fyhn, M., Molden, S., Moser, M. B., & Moser, E. I. (2005). Microstructure of a spatial map in the entorhinal cortex. *Nature, 436*, 801–806.

Heit, G., Smith, M. E., & Halgren, E. (1988). Neural encoding of individual words and faces by the human hippocampus and amygdala. *Nature, 333*, 773–775.

Herculano-Houzel, S. (2009). The human brain in numbers: A linearly scaled-up primate brain. *Frontiers in Human Neuroscience, 3*, 31.

Herry, C., Bach, D. R., Esposito, F., Di Salle, F., Perrig, W. J., Scheffler, K., et al. (2007). Processing of temporal unpredictability in human and animal amygdala. *Journal of Neuroscience, 27*, 5958–5966.

Higuchi, S., & Miyashita, Y. (1996). Formation of mnemonic neuronal responses to visual paired associates in inferotemporal cortex is impaired by perirhinal and entorhinal lesions. *Proceedings of the National Academy of Sciences of the United States of America, 93*, 739–743.

Howard, M. W., & Kahana, M. J. (1999). Contextual variability and serial position effects in free recall. *Journal of Experimental Psychology. Learning, Memory, and Cognition, 25*, 923–941.

Hubel, D. H., & Wiesel, T. N. (1962). Receptive fields, binocular interaction and functional architecture in the cat's visual cortex. *Journal of Physiology, 160*, 106–154.

Hung, C., Kreiman, G., Poggio, T., & DiCarlo, J. (2005). Fast read-out of object identity from macaque inferior temporal cortex. *Science, 310*, 863–866.

Huth, A. G., Nishimoto, S., Vu, A. T., & Gallant, J. L. (2012). A continuous semantic space describes the representation of thousands of object and action categories across the human brain. *Neuron, 76*, 1210–1224.

Insausti, R., Juottonen, K., Soininen, H., Insausti, A., Partanen, K., Vainio, P., et al. (1998). MR volumetric analysis of the human entorhinal, perirhinal and temporopolar cortices. *AJNR. American Journal of Neuroradiology, 19*, 659–671.

Ison, M. J., Mormann, F., Cerf, M., Koch, C., Fried, I., & Quiroga, R. Q. (2011) Selectivity of pyramidal cells and interneurons in the human medial temporal lobe. *Journal of Neurophysiology, 106*, 1713–1721.

Ito, M., Tamura, H., Fujita, I., & Tanaka, K. (1995). Size and position invariance of neuronal responses in monkey inferotemporal cortex. *Journal of Neurophysiology, 73*, 218–226.

Jacobs, J., Staba, R., Asano, E., Otsubo, H., Wu, J. Y., Zijlmans, M., et al. (2012). High-frequency oscillations (HFOs) in clinical epilepsy. *Progress in Neurobiology, 98*, 302–315.

Katzner, S., Nauhaus, I., Benucci, A., Bonin, V., Ringach, D. L., & Carandini, M. (2009). Local origin of field potentials in visual cortex. *Neuron, 61*, 35–41.

Keysers, C., Xiao, D. K., Foldiak, P., & Perret, D. I. (2001). The speed of sight. *Journal of Cognitive Neuroscience, 13*, 90–101.

Killian, N. J., Jutras, M. J., & Buffalo, E. A. (2012). A map of visual space in the primate entorhinal cortex. *Nature, 491*, 761–764.

Koch, C. (2005). *The quest for consciousness* (1st ed.). Los Angeles: Roberts & Company Publishers.

Kreiman, G. (2007). Single neuron approaches to human vision and memories. *Current Opinion in Neurobiology, 17*, 471–475.

Kreiman, G., Fried, I., & Koch, C. (2002). Single neuron correlates of subjective vision in the human medial temporal lobe. *Proceedings of the National Academy of Sciences of the United States of America, 99*, 8378–8383.

Kreiman, G., Hung, C., Quian Quiroga, R., Kraskov, A., Poggio, T., & DiCarlo, J. (2006). Object selectivity of local field potentials and spikes in the inferior temporal cortex of macaque monkeys. *Neuron, 49*, 433–445.

Kreiman, G., Koch, C., & Fried, I. (2000a). Category-specific visual responses of single neurons in the human medial temporal lobe. *Nature Neuroscience, 3*, 946–953.

Kreiman, G., Koch, C., & Fried, I. (2000b). Imagery neurons in the human brain. *Nature, 408*, 357–361.

Kriegeskorte, N., Mur, M., Ruff, D. A., Kiani, R., Bodurka, J., Esteky, H., et al. (2008). Matching categorical object representations in inferior temporal cortex of man and monkey. *Neuron, 60*, 1126–1141.

Lavenex, P., & Amaral, D. G. (2000). Hippocampal–neocortical interaction: A hierarchy of associativity. *Hippocampus, 10*, 420–430.

LeDoux, J. (1996). *The emotional brain*. New York: Simon & Schuster.

Lehky, S. R., & Maunsell, J. H. R. (1996). No binocular rivalry in the LGN of alert monkeys. *Vision Research, 36*, 1225–1234.

Lein, E. S., Zhao, X., & Gage, F. H. (2004). Defining a molecular atlas of the hippocampus using DNA microarrays and high-throughput in situ hybridization. *Journal of Neuroscience, 24*, 3879–3889.

Leonard, C. M., Rolls, E. T., Wilson, F. A. W., & Baylis, G. C. (1985). Neurons in the amygdala of the monkey with responses selective for faces. *Behavioural Brain Research, 15*, 159–176.

Leopold, D. A., & Logothetis, N. K. (1996). Activity changes in early visual cortex reflect monkeys' percepts during binocular rivalry. *Nature, 379*, 549–553.

Leopold, D. A., & Logothetis, N. K. (1999). Multistable phenomena: Changing views in perception. *Trends in Cognitive Sciences, 3*, 254–264.

Libet, B. (1985). Unconscious cerebral initiative and the role of conscious will in voluntary action. *Behavioral and Brain Sciences, 8,* 529–566.

Lisman, J., & Sternberg, E. J. (2013). Habit and nonhabit systems for unconscious and conscious behavior: Implications for multitasking. *Journal of Cognitive Neuroscience, 25,* 273–283.

Litman, L., Awipi, T., & Davachi, L. (2009). Category-specificity in the human medial temporal lobe cortex. *Hippocampus, 19,* 308–319.

Liu, H., Agam, Y., Madsen, J. R., & Kreiman, G. (2009). Timing, timing, timing: Fast decoding of object information from intracranial field potentials in human visual cortex. *Neuron, 62,* 281–290.

Logothetis, N. K., Pauls, J., Bulthoff, H. H., & Poggio, T. (1994). View-dependent object recognition by monkeys. *Current Biology, 4,* 401–414.

Logothetis, N. K., Pauls, J., & Poggio, T. (1995). Shape representation in the inferior temporal cortex of monkeys. *Current Biology, 5,* 552–563.

Logothetis, N. K., & Sheinberg, D. L. (1996). Visual object recognition. *Annual Review of Neuroscience, 19,* 577–621.

Louie, K., & Wilson, M. (2001). Temporally structured replay of awake hippocampal ensemble activity during rapid eye movement sleep. *Neuron, 29,* 145–156.

Marcus, D. S., Harwell, J., Olsen, T., & Van Essen, D. C. (2011). The human connectome project informatics platform. *Frontiers in Neuroscience, 5.*

McCarthy, G., Puce, A., Belger, A., & Allison, T. (1999). Electrophysiological studies of human face perception: II. Response properties of face-specific potentials generated in occipitotemporal cortex. *Cerebral Cortex, 9,* 431–444.

Meyers, E., Freedman, D., Kreiman, G., Miller, E., & Poggio, T. (2008). Dynamic population coding of category information in ITC and PFC. *Journal of Neurophysiology, 100,* 1407–1419.

Milner, B., Corkin, S., & Teuber, H. (1968). Further analysis of the hippocampal amnesic syndrome: 14-year follow-up study of H. M. *Neurpsychologia, 6,* 215–234.

Miyashita, Y. (1993). Inferior temporal cortex: Where visual perception meets memory. *Annual Review of Neuroscience, 16,* 245–263.

Miyashita, Y., & Chang, H. S. (1988). Neuronal correlate of pictorial short-term memory in the primate temporal cortex. *Nature, 331,* 68–71.

Miyashita, Y., Rolls, E. T., Cahusac, P. M. B., Niki, H., & Feigenbaum, J. D. (1989). Activity of hippocampal formation neurons in the monkey related to a conditional spatial response task. *Journal of Neurophysiology, 61,* 669–678.

Mormann, F., Dubois, J., Kornblith, S., Milosavljevic, M., Cerf, M., Ison, M., et al. (2011). A category-specific response to animals in the right human amygdala. *Nature Neuroscience, 14,* 1247–1249.

Mormann, F., Kornblith, S., Quiroga, R. Q., Kraskov, A., Cerf, M., Fried, I., et al. (2008). Latency and selectivity of single neurons indicate hierarchical processing in the human medial temporal lobe. *Journal of Neuroscience, 28,* 8865–8872.

Morris, J. S., Ohman, A., & Dolan, R. J. (1998). Conscious and unconscious emotional learning in the human amygdala. *Nature, 393,* 467–470.

Mountcastle, V. B. (1957). Modality and topographic properties of single neurons of cat's somatic sensory cortex. *Journal of Neurophysiology, 20,* 408–434.

Mukamel, R., & Fried, I. (2012). Human intracranial recordings and cognitive neuroscience. *Annual Review of Psychology, 63,* 511–537.

Mukamel, R., Gelbard, H., Arieli, A., Hasson, U., Fried, I., & Malach, R. (2005). Coupling between neuronal firing, field potentials, and fMRI in human auditory cortex. *Science, 309,* 951–954.

Nadasdy, Z., Hirase, H., Czurko, A., Csicsvari, J., & Buzsaki, G. (1999). Replay and time compression of recurring spike sequences in the hippocampus. *Journal of Neuroscience, 19,* 9497–9507.

Naya, Y., Sakai, K., & Miyashita, Y. (1996). Activity of primate inferotemporal neurons related to a sought target in a paired-association task. *Proceedings of the National Academy of Sciences of the United States of America, 93,* 2664–2669.

Naya, Y., & Suzuki, W. (2011). Integrating what and when across the primate medial temporal lobe. *Science, 333,* 4.

Naya, Y., Sakai, K., & Miyashita, Y. (1996). Activity of primate inferotemporal neurons related to a sought target in a paired-association task. *Proceedings of the National Academy of Sciences of the United States of America, 93,* 2664–2669.

Nielsen, K., Logothetis, N., & Rainer, G. (2006). Dissociation between LFP and spiking activity in macaque inferior temporal cortex reveals diagnostic parts-based encoding of complex objects. *Journal of Neuroscience, 26,* 9639–9645.

Nir, Y., Fisch, L., Mukamel, R., Gelbard-Sagiv, H., Arieli, A., Fried, I., et al (2007). Coupling between neuronal firing rate, gamma LFP, and BOLD fMRI is related to inter-neuronal correlations. *Current Biology, 17,* 1275–1285.

Nishijo, H., Ono, T., & Nishino, H. (1988). Single neuron responses in amygdala of alert monkeys during complex sensory stimulation with affective significance. *Journal of Neuroscience, 8,* 3570–3583.

O'Keefe, J., & Dostrovsky, J. (1971). The hippocampus as a spatial map: Preliminary evidence from unit activity in the freely-moving rat. *Brain Research, 34,* 171–175.

O'Keefe, J., & Nadel, L. (1978). *The hippocampus as a cognitive map.* London: Clarendon Press.

Ohman, A., Carlsson, E., Lundgvist, D., & Ingvar, M. (2007). On the unconscious subcortical origin of human fear. *Physiology & Behavior, 92,* 180–185.

Ojemann, G. A. (1997). Treatment of temporal lobe epilepsy. *Annual Review of Medicine, 48,* 317–328.

O'Neill, J., Pleydell-Bouverie, B., Dupret, D., & Csicsvari, J. (2010). Play it again: Reactivation of waking experience and memory. *Trends in Neurosciences, 33,* 220–229.

Paton, J. J., Belova, M. A., Morrison, S. E., & Salzman, C. D. (2006). The primate amygdala represents the positive and negative value of visual stimuli during learning. *Nature, 439,* 865–870

Pedreira, C., Mormann, F., Kraskov, A., Cerf, M., Fried, I., Koch, C., et al. (2010). Responses of human medial temporal lobe neurons are modulated by stimulus repetition. *Journal of Neurophysiology, 103,* 97–107.

Pegna, A. J., Khateb, A., Lazeyras, F., & Seghier, M. L. (2005). Discriminating emotional faces without primary visual cortices involves the right amygdala. *Nature Neuroscience, 8,* 24–25.

Pessoa, L. (2008). On the relationship between emotion and cognition. *Nature Reviews. Neuroscience, 9,* 148–158.

Pessoa, L., & Adolphs, R. (2010). Emotion processing and the amygdala: From a "low road" to "many roads" of evaluating biological significance. *Nature Reviews. Neuroscience, 11,* 773–783.

Phelps, E. A., & LeDoux, J. (2005). Contributions of the amygdala to emotion processing: From animal models to human behavior. *Neuron, 48,* 175–187.

Pinto, N., Cox, D. D., & DiCarlo, J. J. (2008). Why is real-world visual object recognition hard? *PLoS Computational Biology, 4,* e27.

Quian Quiroga, R., Kraskov, A., Koch, C., & Fried, I. (2009). Explicit encoding of multimodal percepts by single neurons in the human brain. *Current Biology, 19,* 1308–1313.

Quian Quiroga, R., & Kreiman, G. (2010). Measuring sparseness in the brain. *Psychological Review, 117,* 291–297.

Quian Quiroga, R., Kreiman, G., Koch, C., & Fried, I. (2008). Sparse but not "Grandmother-cell" coding in the medial temporal lobe. *Trends in Cognitive Sciences, 12,* 87–91.

Quian Quiroga, R., Reddy, L., Kreiman, G., Koch, C., & Fried, I. (2005). Invariant visual representation by single neurons in the human brain. *Nature, 435,* 1102–1107.

Quiroga, R. (2012). Concept cells: The building blocks of declarative memory functions. *Nature Reviews. Neuroscience, 13,* 587–597.

Quiroga, R. Q., Mukamel, R., Isham, E. A., Malach, R., & Fried, I. (2008). Human single-neuron responses at the threshold of conscious recognition. *Proceedings of the National Academy of Sciences of the United States of America, 105,* 3599–3604.

Rainer, G., Rao, S., & Miller, E. (1999). Prospective coding for objects in primate prefrontal cortex. *Journal of Neuroscience, 19,* 5493–5505.

Rao, S., Rainer, G., & Miller, E. (1997). Integration of what and where in the primate prefrontal cortex. *Science, 276,* 821–824.

Rasch, M., Logothetis, N. K., & Kreiman, G. (2009). From neurons to circuits: Linear estimation of local field potentials. *Journal of Neuroscience, 29,* 13785–13796.

Rees, G., Kreiman, G., & Koch, C. (2002). Neural correlates of consciousness in humans. *Nature Reviews. Neuroscience*, *3*, 261–270.

Riesenhuber, M., & Poggio, T. (1999). Hierarchical models of object recognition in cortex. *Nature Neuroscience*, *2*, 1019–1025.

Riesenhuber, M., & Poggio, T. (2000). Models of object recognition. *Nature Neuroscience*, *3*(Suppl), 1199–1204.

Rolls, E. (1991). Neural organization of higher visual functions. *Current Opinion in Neurobiology*, *1*, 274–278.

Rolls, E. T., Miyashita, Y., Cahusac, P. M. B., Kesner, R. P., Niki, H., Feigenbaum, J. D., et al. (1989). Hippocampal neurons in the monkey with activity related to the place in which a stimulus is shown. *Journal of Neuroscience*, *9*, 1835–1845.

Rutishauser, U., Mamelak, A. N., & Schuman, E. M. (2006). Single-trial learning of novel stimuli by individual neurons of the human hippocampus–amygdala complex. *Neuron*, *49*, 805–813.

Rutishauser, U., Ross, I. B., Mamelak, A. N., Schuman, E. M. (2011). Human memory strength is predicted by theta-frequency phase-locking of single neurons. *Nature*, *464*, 903–907.

Rutishauser, U., Tudusciuc, O., Neumann, D., Mamelak, A., Heller, C., Ross, I., et al. (2011). Single-unit responses selective for whole faces in the human amygdala. *Current Biology*, *21*, 1654–1660.

Sander, D., Gradman, J., & Zalla, T. (2003). The human amygdala: An evolved system for relevance detection. *Reviews in the Neurosciences*, *14*, 303–316.

Schmolesky, M., Wang, Y., Hanes, D., Thompson, K., Leutgeb, S., Schall, J., et al. (1998). Signal timing across the macaque visual system. *Journal of Neurophysiology*, *79*, 3272–3278.

Scoville, W. B., & Milner, B. (1957). Loss of recent memory after bilateral hippocampal lesions. *Journal of Neurology, Neurosurgery, and Psychiatry*, *20*, 11–21.

Serre, T., Kreiman, G., Kouh, M., Cadieu, C., Knoblich, U., & Poggio, T. (2007). A quantitative theory of immediate visual recognition. *Progress in Brain Research*, *165C*, 33–56.

Sheinberg, D. L., & Logothetis, N. K. (1997). The role of temporal areas in perceptual organization. *Proceedings of the National Academy of Sciences of the United States of America*, *94*, 3408–3413.

Shoham, S., O'Connor, D. H., & Segev, R. (2006). How silent is the brain: Is there a "dark matter" problem in neuroscience? *Journal of Comparative Physiology. A, Neuroethology, Sensory, Neural, and Behavioral Physiology*, *192*, 777–784.

Shrager, Y., Gold, J. J., Hopkins, R. O., & Squire, L. R. (2006). Intact visual perception in memory-impaired patients with medial temporal lobe lesions. *Journal of Neuroscience*, *26*, 2235–2240.

Siebert, M., Markowitsch, H. J., & Bartel, P. (2003). Amydgala, affect and cognition: Evidence from 10 patients with Urbach-Wiethe disease. *Brain*, *126*, 2627–2637.

Sigala, N., & Logothetis, N. (2002). Visual categorization shapes feature selectivity in the primate temporal cortex. *Nature*, *415*, 318–320.

Skaggs, W. E., & McNaughton, B. L. (1996). Replay of neuronal firing sequences in rat hippocampus during sleep following spatial experience. *Science*, *271*, 1870–1873.

Sperry, R. (1982). Some effects of disconnecting the cerebral hemispheres. *Science*, *217*, 1223–1226.

Squire, L., & Zola-Morgan, S. (1991). The medial temporal lobe memory system. *Science*, *253*, 1380–1386.

Squire, L. R. (2009). The legacy of patient H.M. for neuroscience. *Neuron*, *61*, 6–9.

Squire, L. R., Stark, C. E., & Clark, R. E. (2004). The medial temporal lobe. *Annual Review of Neuroscience*, *27*, 279–306.

Squire, L. R., Wixted, J. T., & Clark, R. E. (2007). Recognition memory and the medial temporal lobe: A new perspective. *Nature Reviews. Neuroscience*, *8*, 872–883.

Stead, M., Bower, M., Brinkmann, B. H., Lee, K., Marsh, W. R., Meyer, F. B., et al. (2010). Microseizures and the spatiotemporal scales of human partial epilepsy. *Brain*, *133*, 2789–2797.

Steinmetz, P. N., Cabrales, E., Wilson, M. S., Baker, C. P., Thorp, C. K., Smith, K. A., et al. (2011). Neurons in the human hippocampus and amygdala respond to both low- and high-level image properties. *Journal of Neurophysiology*, *105*, 2874–2884.

Suthana, N., Haneef, Z., Stern, J., Mukamel, R., Behnke, E., Knowlton, B., et al. (2012). Memory enhancement and deep-brain stimulation of the entorhinal area. *New England Journal of Medicine, 366*, 502–510.

Suzuki, W., Miller, E., & Desimone, R. (1997). Object and place memory in the macaque entorhinal cortex. *Journal of Neurophysiology, 78*, 1062–1081.

Suzuki, W. A. (1996). Neuroanatomy of the monkey entorhinal, perirhinal and parahippocampal cortices: Organization of cortical inputs and interconnections with amygdala and striatum. *Seminars in Neuroscience, 8*, 3–12.

Suzuki, W. A., & Amaral, D. G. (1994). Perrirhinal and parahippocampal cortices of the macaque monkey: Cortical afferents. *Journal of Comparative Neurology, 350*, 497–533.

Tanaka, K. (1996). Inferotemporal cortex and object vision. *Annual Review of Neuroscience, 19*, 109–139.

Taube, J. S., Muller, R. U., & Ranck, J. B. J. (1990). Head-direction cells recorded from the postsubiculum in freely moving rats: I. Description and quantitative analysis. *Journal of Neuroscience, 10*, 420–435.

Tsao, D. Y., Moeller, S., & Freiwald, W. A. (2008a). Comparing face patch systems in macaques and humans. *Proceedings of the National Academy of Sciences of the United States of America, 105*, 19514–19519.

Tsao, D. Y., Schweers, N., Moeller, S., & Freiwald, W. A. (2008b). Patches of face-selective cortex in the macaque frontal lobe. *Nature Neuroscience, 11*, 877–879.

Viskontas, I. V., Ekstrom, A. D., Wilson, C. L., & Fried, I. (2007). Characterizing interneuron and pyramidal cells in the human medial temporal lobe in vivo using extracellular recordings. *Hippocampus, 17*, 49–57.

von Bohlen und Halbach, O., & Unsicker, K. (2002). Morphological alterations in the amygdala and hippocampus of mice during ageing. *European Journal of Neuroscience, 16*, 2434–2440.

Wagner, A. D., Koutstaal, W., & Schacter, D. L. (1999). When encoding yields remembering: Insights from event-related neuroimaging. *Philosophical Transactions of the Royal Society of London. Series B, Biological Sciences, 354*, 1307–1324.

Wandell, B. A. (1995). *Foundations of vision*. Sunderland, MA: Sinauer Associates.

Waydo, S., Kraskov, A., Quian Quiroga, R., Fried, I., & Koch, C. (2006). Sparse representation in the human medial temporal lobe. *Journal of Neuroscience, 26*, 10232–10234.

Whalen, P. J., Rauch, S. L., Etcoff, N. L., McInerney, S. C., Lee, M. B., & Jenike, M. A. (1998). Masked presentations of emotional facial expressions modulate amygdala activity without explicit knowledge. *Journal of Neuroscience, 18*, 411–418.

Wirth, S., Yanike, M., Frank, L. M., Smith, A. C., Brown, E. N., & Suzuki, W. A. (2003). Single neurons in the monkey hippocampus and learning of new associations. *Science, 300*, 1578–1581.

Yanike, M., Wirth, S., & Suzuki, W. A. (2004). Representation of well-learned information in the monkey hippocampus. *Neuron, 42*, 477–487.

Yoshor, D., Bosking, W. H., Ghose, G. M., & Maunsell, J. H. (2007). Receptive fields in human visual cortex mapped with surface electrodes. *Cerebral Cortex, 17*, 2293–2302.

Young, M. P., Scannell, J. W., Burns, G. A., & Blackmore, C. (1994). Analysis of connectivity: Neural systems in the cerebral cortex. *Reviews in the Neurosciences, 5*, 227–250.

Navigating Our Environment: Insights from Single Neuron Recordings in the Human Brain

Nanthia Suthana and Itzhak Fried

Successfully navigating an environment is a task critical for everyday function and is thought to involve the formation of a mental representation or "cognitive map" of space (Tolman, 1948). Evidence suggests that the hippocampus and associated structures within the medial temporal lobe (MTL) may support the formation of a cognitive map of space; patients with bilateral MTL damage cannot learn new information associated with everyday experienced events (e.g., they are unable to learn how to navigate a new city; Goodrich-Hunsaker et al., 2010; King et al., 2004; Spiers et al., 2001; Maguire et al., 1996). Furthermore, processing the spatial information or context associated with novel experiences may assist in the binding of items within an event into a stable memory representation, or episodic memory (Eichenbaum, 2004; Burgess et al., 2002; Holscher et al., 2003). Further support for the importance of MTL structures in spatial navigation is seen in the spatial disorientation and losing one's way in familiar surroundings due to MTL damage sustained from neurological disorders such as Alzheimer's disease (Mapstone et al., 2003; Monacelli et al., 2003; Kalova et al., 2005).

Processing of Spatial Information within the MTL

The human MTL consists of the hippocampus and adjacent parahippocampal, entorhinal, and perirhinal cortices. The hippocampus can be further separated into distinct subregions including CA fields 1–4, dentate gyrus, and subiculum. Anatomical studies in nonhuman primates have characterized MTL circuitry, which is largely conserved across mammalian species (Manns & Eichenbaum, 2006). For a simplified diagram of MTL circuitry see figure 9.1. The "parahippocampal region" has sometimes been used to describe an inclusive region combining entorhinal, perirhinal, and the parahippocampal cortices (Witter et al., 2000). Sensory information from widespread cortical areas enters the hippocampus through initial synaptic connections via the perirhinal and parahippocampal cortices (Amaral & Witter, 1995; Duvernoy, 1998; Suzuki & Amaral, 1994). While dorsal visual areas dominate projections to the perirhinal cortex, caudal visual areas in addition to superior temporal, posterior parietal, and retrosplenial areas innervate the parahippocampal cortex (Van Hoesen, 1982; Murray et al., 2000). The efferent connectivity of the perirhinal cortex and parahippocampal cortex suggest that the former may be involved in

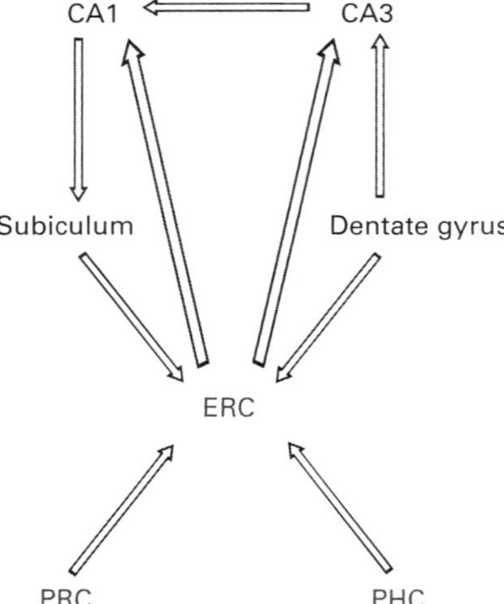

Figure 9.1
Shown is a simplified diagram of medial temporal lobe (MTL) circuitry. The MTL consists of hippocampal subregions (CA1, CA3, dentate gyrus, and subiculum) shown in black and adjacent cortical regions (entorhinal cortex [ERC], perirhinal cortex [PRC], parahippocampal cortex [PHC]) shown in gray.

processing the nonspatial identity of stimuli while the latter may be more involved in processing of spatial information (Suzuki & Amaral, 1994). Parallel pathways from perirhinal cortex and parahippocampal cortex then converge onto neurons within the entorhinal cortex (ERC). From the ERC, both types of information then synapse onto hippocampal subregions dentate gyrus (via the perforant path), CA3, and CA1 (via the direct path; Amaral & Insausti, 1990). Object relevant-information (via perirhinal cortex) versus spatially relevant information (via parahippocampal cortex) ends up in the anterior and posterior portions of the hippocampus, respectively (Suzuki & Amaral, 1994).

Using Virtual Reality to Study Human Brain Activity during Spatial Navigation

Studies investigating brain activity during human spatial navigation have largely relied on neuroimaging methodologies that use noninvasive measurement of brain activity such as electroencephalography (EEG) or functional magnetic resonance imaging (fMRI). Many of these studies have shown that the hippocampus and parahippocampal cortex increase in activation during navigation of a virtual spatial environment compared to rest or nonspatial tasks (Suthana et al., 2009, 2011; Aguirre et al., 1998; Maguire et al., 1998; Weniger et al., 2010). However, brain activity measured in these studies is indirect, rather reflecting the electrical activity from a population of neurons (e.g., EEG activity) or the oxygenation of blood (i.e., blood-oxygenated-level-dependent [BOLD] activity) delivered to a population of neurons within an area of usually 2–3 cubic millimeters, and on occasion as small as 1 cubic millimeter (Ekstrom, 2010). Intracranial recording of single neurons and local field potential (LFP) activity provides a more direct method of studying neuronal activity during cognition. Given the invasive nature of intracranial recordings, however, recordings can only be obtained from patients for clinical reasons, such as evaluation for epileptic seizures. Studies with epilepsy patients over the past decade have revealed fascinating characteristics of single neuron and LFP activity within the human MTL during spatial navigation of a virtual environment. One challenge in studying spatial navigation in patients who are implanted with intracranial electrodes is that they are connected to large recording devices and real-world navigation is not feasible, so, although participants in intracranial studies are confined to hospital beds, computer technology has allowed for the development of virtual reality environments that can simulate the real world. In fact, rodent studies have recently begun using virtual reality environments as well to study rodent spatial navigation where neurons exhibit similar firing patterns compared to navigation of a real-world environment (Harvey et al., 2009; Chen et al., 2013). Apart from the obvious lack of vestibular and proprioceptive components during virtual navigation, neuronal mechanisms between virtual and real-world navigation appear to be similar, if not identical, although some differences have been investigated (Chen et al., 2013). Intracranial studies over the last several years have been able to record single neuron and LFP activity in the human MTL during spatial navigation of virtual environments (Kahana et al., 1999; Caplan et al., 2003). A common virtual navigation game used in human intracranial research studies is *Yellow Cab* (Caplan et al., 2003), in which participants play the role of a taxicab driver, picking up passengers randomly placed throughout an environment and delivering them to specific stores (e.g., the Candle Shop; see figure 9.2, plate 8). Using a joystick

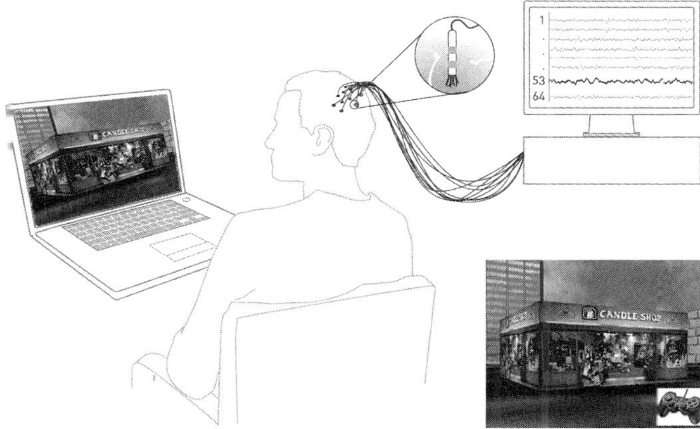

Figure 9.2 (plate 8)
Example experimental setup for intracranial recordings during spatial navigation. Participants drive through a virtual environment, picking up randomly located passengers and delivering them to stores. Insets: Snapshot of an example target store within the virtual driving game. Participants use the joystick shown to navigate to stores such as the Candle Shop that is depicted.

to control driving, participants can navigate a virtual city viewed on a laptop screen while brain activity is simultaneously recorded (see figure 9.2, plate 8). In the *Yellow Cab* game participants are also awarded points ("virtual payment") for each delivery and are thus motivated to learn the spatial layout of the environment; using the shortest route to a target goal destination results in less time per delivery and thus more points are earned. In this way, spatial navigation processes including learning and memory can be studied in the hospital room setting while brain activity is continuously recorded.

Local Field Potential Activity during Spatial Navigation

Intracranial recording of LFP activity during spatial navigation using *Yellow Cab* has shown that theta (4–8 Hz) oscillations increase in amplitude (power) during active virtual navigation compared to periods of inactivity or stillness (Caplan et al., 2003). Increased power in the theta band associated with virtual movement occurs within the hippocampus (Ekstrom et al., 2007) but also occurs within other brain regions such as the parahippocampal region and frontal cortex with an overall larger increase within areas of the right compared to the left hemisphere (Caplan et al., 2003). Hemispheric asymmetry of theta activity is consistent with another intracranial LFP study, also using *Yellow Cab*, which found an increase in the amplitude of gamma oscillatory activity within right-hemisphere regions during virtual navigation compared to stillness (Jacobs et al., 2010). Since theta power increases do not seem to be limited to the hippocampus, and neither do they differentiate between goal-directed navigation (e.g., "Deliver passenger to Candle Shop") versus seeking-related navigation (e.g., "Find a passenger"), it has been suggested that theta oscillations may play a role in attention and sensorimotor integration across distant brain areas. Recent results using a virtual version of the Morris water maze show that theta amplitude recorded from the posterior portion of the hippocampus is correlated with navigational performance (Cornwell et al., 2008). In the Morris water maze participants navigate a joystick around a circular pool of water to find a hidden submerged platform. During navigation of this task, hippocampal and parahippocampal amplitude of theta activity is similarly increased during goal-directed navigation compared to aimless movements (Cornwell et al., 2008). Overall, results show that virtual reality used alongside intracranial recordings offers a unique and informative method of studying spatial navigation in humans where theta oscillations within the MTL may play an important role. Various spatial navigation studies report changes in the amplitude of theta oscillations while recent results from memory studies suggest the phase of theta oscillations may also play an important role (Rutishauser et al., 2011; Jacobs et al., 2007; Canolty and Knight, 2010; Suthana et al., 2012). Specifically, the phase of theta may be tightly coupled with single neuron spiking activity and the amplitude of gamma oscillations, which may play an important role in successful memory formation (Rutishauser et al., 2011; Canolty and Knight, 2010). Thus, future studies exploring specific patterns related to the phase of the ongoing theta oscillation may provide additional information regarding the relationship between theta activity, spatial navigation, and related learning and memory processes.

Single Neuron Activity during Spatial Navigation

Single neuron firing rate (spiking) activity can be recorded in addition to LFP activity from intracranial microwire electrodes, thereby offering a unique window into how single cells can support behaviors such as spatial navigation. On occasion studies can combine recording of single neuron and LFP activity with fMRI within the same individuals to see how different measurements of brain activity may be related and contribute to navigational abilities. In 2007, Ekstrom and colleagues simultaneously recorded LFPs and single neuron activity within patients during spatial navigation and, more recently, compared these electrophysiological signals to fMRI BOLD activity in the same participants (Ekstrom et al., 2009). Results showed a divergence between single neuron spiking and LFP activity while LFP and BOLD activity were strongly correlated within the hippocampus during spatial navigation. These results suggest that single neurons within the MTL perhaps can contain unique information aside from local field activity. In fact, this is the case, where single neurons within the hippocampus and parahippocampal region can code for very specific types of spatial information during navigation. In 2003, Ekstrom and colleagues reported results from single neuron recordings from patients playing the *Yellow Cab* driving game (Ekstrom et al., 2003). Results showed that a proportion of neurons within the hippocampus increased in firing rate when the patient was in a specific location within the city. These *place-responsive* or "place" cells increased in firing rate when the participant was in a particular virtual place independent of view or trajectory (e.g., facing or driving east) and remapped depending on the participant's goal (e.g., delivering a passenger to the Candle Shop vs. to the House of Pizza). Although place-responsive cells made up only 24% of total hippocampal cells recorded, this percentage was significantly higher than found in other brain areas such as the frontal cortex, parahippocampal region, or the amygdala, suggesting location selectivity may be a specific function of hippocampal neurons (see figure 9.3A). These findings are consistent with place-responsive neurons recorded from the rodent brain (Chen et al., 2013; for further discussion of cross-species comparisons, see the "Future Directions" section below). *View-selective* neurons, those responsive to a particular view, were also found, albeit in a significantly higher proportion within the parahippocampal region compared to the hippocampus (see figure 9.3B). These results suggest important functional differences may exist between the human hippocampus and parahippocampal region during spatial navigation.

In another study, using a similar navigational task, participants drove either in a clockwise or counterclockwise direction around a virtual city. Of over 1000 neurons recorded from across several brain regions, Jacobs et al. (2010) found an interesting dissociation between hippocampal and entorhinal neuron activity. While hippocampal neurons were *place selective*, ERC neurons preferred the direction the participant was driving. In other words, ERC neurons were *path selective*, increasing in firing rate during either clockwise or counterwise driving directions (figure 9.4, plate 9; Jacobs et al., 2010). Although *path-selective* cells were found in brain regions other than the ERC, such as the hippocampus, parahippocampal cortex, or frontal cortex, the proportion of ERC cells that showed path selectivity was significantly higher than other

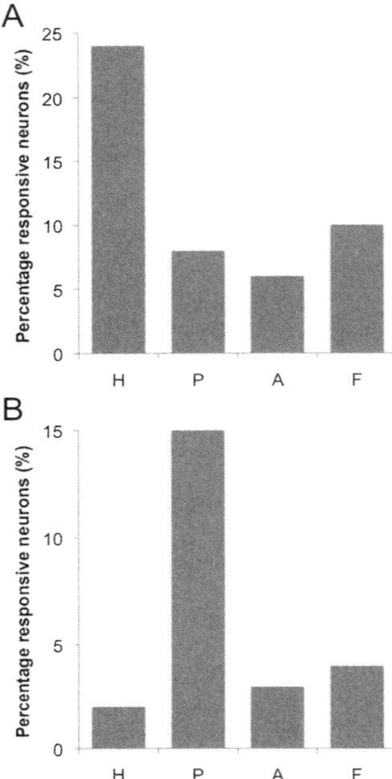

Figure 9.3
Shown is the percentage of total responsive neurons within the hippocampus (H), parahippocampal region (P), amygdala (A), and frontal lobe (F). (A) A significantly higher percentage of hippocampal neurons were place selective as compared to other brain areas shown. (B) A significantly higher percentage of parahippocampal (entorhinal and perirhinal) neurons were view selective as compared to other brain areas shown. Adapted from Ekstrom et al. (2003).

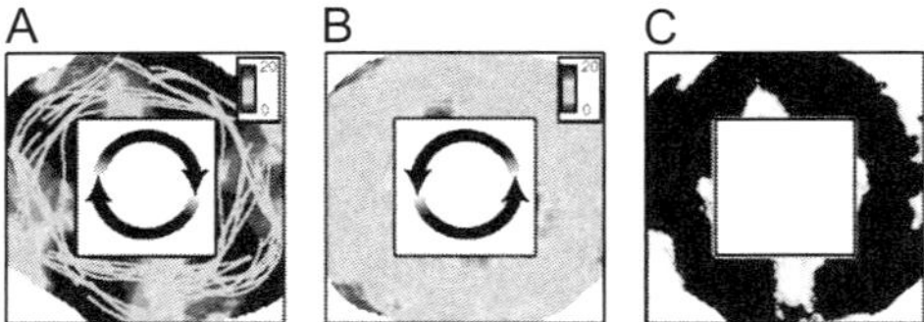

Figure 9.4 (plate 9)
An example entorhinal neuron that increases in firing rate while the participant is driving in a clockwise direction (A) compared to a (B) counterclockwise direction. (C) Shown is the difference in average firing rate of this entorhinal cell across all clockwise compared to counterclockwise driving trials. Hotter colors indicate higher firing rates. Adapted with permission from Jacobs et al. (2010).

brain areas, suggesting a unique function for the ERC in processing directional information. Another interesting finding showed that while hippocampal place cells increased in activity to a single unique location within the environment, ERC path cells exhibited increased activity throughout the entire lap of the environment, suggesting these cells may be processing aspects of the context (i.e., direction) of a continuous navigational event. Recently, in an open environment navigational task, a small but significant percentage (14%) of human ERC cells were found to exhibit sixfold (60°) symmetry in their firing rate arranged in a triangular grid across a virtual environment (Jacobs et al., 2013). These "grid-like" cells have previously been found in the rodent entorhinal cortex (Hafting et al., 2005) illustrating homologous spatial coding mechanisms across species.

Future Directions

Research studying brain activity during spatial navigation uses virtual navigation for its obvious ease of use in a hospital or laboratory room setting. However, is neuronal activity the same during real versus virtual navigation? Real-world navigation relies on visual as well as vestibular cues whereas virtual navigation relies heavily on visual cues alone. Vestibular cues obviously play an important role during spatial navigation since humans with bilateral vestibular damage exhibit severe spatial memory deficits (Hüfner et al., 2007; Brandt et al., 2005) and lesions of bilateral afferents in rodents disrupt hippocampal place cell firing fields (Stackman et al., 2002; Russell et al., 2003). However, recent electrophysiological data in rodents suggest there are striking similarities in the neuronal mechanisms underlying visual versus movement driven navigation (Chen et al., 2013). Human navigation is largely dependent on visual cues, and it has been shown that humans can learn spatial layouts from visual cues alone and can transfer spatial information learned in a virtual setting to a matched real-world environment (Redlick et al., 2001; Warren et al., 2001; Witmer et al., 1996). Nonetheless, with advancements in intracranial recording technology, wireless electrodes capable of recording brain activity during real-world exploration will provide an exciting novel method for studying single neuron and LFP activity during spatial navigation.

Intracranial recording of hippocampal neurons has only recently begun to address the differences of neurons in separate hippocampal subfields (i.e., CA3, CA1, dentate gyrus, or subiculum). Rodent studies suggest neurons within different hippocampal subregions exhibit divergent properties during spatial navigation (Lee et al., 2004; Leutgeb et al., 2004; Taube et al., 1990). Future intracranial studies will be able to better address these subregional differences in humans, using higher Tesla MRIs and improved resolution methods to localize electrodes within the human hippocampus (Ekstrom et al., 2008).

There is an abundant literature of results from in vivo electrophysiological recordings in rodents during active exploration of an environment. For example, single neuron activity in rodents correlates strongly to a given phase of ongoing theta (4–8 Hz) LFP oscillations during rodent navigation (O'Keefe and Recce, 1993) and similar increases in the amplitude of theta oscillations, suggesting convergence of mechanisms across species (Buzsáki, 2005; O'Keefe and Recce, 1993). Human hippocampal place cells exhibit properties similar to those of rodent hippocampal place cells, which were first discovered in 1971 (O'Keefe & Dostrovsky, 1971; Wilson & McNaughton, 1993). Other types of cells have been found in the rodent that have yet to be characterized in humans—for example, subicular "head-direction cells," which code for the animal's head direction (Taube et al., 1990). Whether results from rodent studies can be directly mapped onto humans is unclear since subtle anatomical differences in MTL circuitry between species do exist (Suzuki & Amaral, 1994; Saleem & Tanaka, 1996). Furthermore, how exactly these types of findings may relate to episodic memory is still unclear since it is impossible to directly probe a rodent's recollection of a previously navigated environment. Another difference between rodent and human findings is that the exact frequency band of theta oscillations in humans may appear to also include part of the delta band (e.g., 3 Hz), which includes frequencies between 1 and 4 Hz (Jacobs et al., 2010; Rutishauser et al., 2011; Cornwell et al., 2008; Ekstrom et al., 2005; Jacobs et al., 2007).

Conclusions

Research opportunities with patients implanted with intracranial recording electrodes for evaluation of epilepsy have provided invaluable information regarding how the brain supports our ability to negotiate space. Development of computer-simulated virtual environments has provided an easier way of studying mental processes that underlie real-world navigation in a clinically restricted research setting. Intracranial recording results suggest the theta band and upper delta bands play an important role during navigation and that single neurons within the MTL carry spatially relevant information indicative of the possible existence of a cognitive map of space. Many unanswered questions remain including how these findings relate to learning and memory functions of the medial temporal regions and whether they are truly reflective of neural processes engaged during real-world active navigation. Future improvements in electrode development, wireless technology, and detailed analyses of spiking–LFP relationships yield exciting areas for future exploration into brain mechanisms underlying our ability to navigate the world around us.

References

Aguirre, G. K., Zarahn, E., & D'Esposito, M. (1998). The variability of human, BOLD hemodynamic responses. *NeuroImage*, *8*, 360–369.

Amaral, D. G., & Insausti, R. (1990). The hippocampal formation. In J. K. Mai & G. Paxinos (Eds.), *The human nervous system* (pp. 711–756). San Diego, CA: Academic Press.

Amaral, D. G., & Witter, M. P. (1995). *Hippocampal formation*. New York: Academic Press.

Brandt, T., Schautzer, F., Hamilton, D., Brüning, R., Markowitsch, H. J., Kalla, R., et al. (2005). Vestibular loss causes hippocampal atrophy and impaired spatial memory in humans. *Brain*, *128*, 2732–2741.

Burgess, N., Maguire, E. A., & O'Keefe, J. (2002). The human hippocampus and spatial and episodic memory. *Neuron*, *35*, 625–641.

Buzsáki, G. (2005). Theta rhythm of navigation: Link between path integration and landmark navigation, episodic and semantic memory. *Hippocampus, 15*, 827–840.

Canolty, R. T., & Knight, R. T. (2010). The functional role of cross-frequency coupling. *Trends in Cognitive Sciences, 14*, 506–515.

Caplan, J. B., Madsen, J. R., Schulze-Bonhage, A., Aschenbrenner-Scheibe, R., Newman, E. L., & Kahana, M. J. (2003). Human theta oscillations related to sensorimotor integration and spatial learning. *Journal of Neuroscience, 23*, 4726–4736.

Chen, G., King, J. A., Burgess, N., & O'Keefe, J. (2013). How vision and movement combine in the hippocampal place code. *Proceedings of the National Academy of Sciences USA, 110*, 378–83.

Cornwell, B. R., Johnson, L. L., Holroyd, T., Carver, F. W., & Grillon, C. (2008). Human hippocampal and parahippocampal theta during goal-directed spatial navigation predicts performance on a virtual Morris water maze. *Journal of Neuroscience, 28*, 5983–5990.

Duvernoy, H. M. (1998). *The human hippocampus: Functional anatomy, vascularization, and serial sections with MRI.* Berlin: Springer.

Eichenbaum, H. (2004). Hippocampus: Cognitive processes and neural representations that underlie declarative memory. *Neuron, 44*, 109–120.

Ekstrom, A. (2010). How and when the fMRI BOLD signal relates to underlying neural activity: The danger in dissociation. *Brain Research. Brain Research Reviews, 62*, 233–244.

Ekstrom, A., Suthana, N., Behnke, E., Salamon, E., Bookheimer, S., & Fried, I. (2008). High-resolution depth electrode localization and imaging in patients with pharmacologically intractable epilepsy. *Journal of Neurosurgery, 108*, 812–815.

Ekstrom, A., Suthana, N., Millett, D., Fried, I., & Bookheimer, S. (2009). Correlation between BOLD fMRI and theta-band local field potentials in the human hippocampal area. *Journal of Neurophysiology, 101*, 2668–2678.

Ekstrom, A., Viskontas, I., Kahana, M., Jacobs, J., Upchurch, K., Bookheimer, S., et al. (2007). Contrasting roles of neural firing rate and local field potentials in human memory. *Hippocampus, 17*, 606–617.

Ekstrom, A. D., Caplan, J. B., Ho, E., Shattuck, K., Fried, I., & Kahana, M. J. (2005). Human hippocampal theta activity during virtual navigation. *Hippocampus, 15*, 881–889.

Ekstrom, A. D., Kahana, M. J., Caplan, J. B., Fields, T. A., Isham, E. A., Newman, E. L., et al. (2003). Cellular networks underlying human spatial navigation. *Nature, 425*, 184–188.

Goodrich-Hunsaker, N. J., Livingstone, S. A., Skelton, R. W., & Hopkins, R. O. (2010). Spatial deficits in a virtual water maze in amnesic participants with hippocampal damage. *Hippocampus, 20*, 481–491.

Hafting, T., Fyhn, M., Molden, S., Moser, M. B., & Moser, E. I. (2005). Microstructure of a spatial map in the entorhinal cortex. *Nature, 436*, 801–806.

Harvey, C. D., Collman, F., Dombeck, D. A., & Tank, D. W. (2009). Intracellular dynamics of hippocampal place cells during virtual navigation. *Nature, 461*, 941–946.

Holscher, C., Rolls, E. T., & Xiang, J. (2003). Perirhinal cortex neuronal activity related to long-term familiarity memory in the macaque. *European Journal of Neuroscience, 18*, 2037–2046.

Hüfner, K., Hamilton, D., Kalla, R., Stephan, T., Glasauer, S., Ma, J., et al. (2007). Spatial memory and hippocampal volume in humans with unilateral vestibular deafferentation. *Hippocampus, 17*, 471–485.

Jacobs, J., Kahana, M. J., Ekstrom, A. D., & Fried, I. (2007). Brain oscillations control timing of single neuron activity in humans. *Journal of Neuroscience, 27*, 3839–3844.

Jacobs, J., Kahana, M. J., Ekstrom, A. D., Mollison, M. V., & Fried, I. (2010). A sense of direction in human entorhinal cortex. *Proceedings of the National Academy of Sciences of the United States of America, 107*, 6487–6492.

Jacobs, J., Weidemann, C. T., Miller, J. F., Solway, A., Burke, J. F., Wei, X. X., Suthana, N., Sperling, M. R., Sharan, A. D., Fried, I., & Kahana, M. J. (2013). Direct recordings of grid-like neuronal activity in human spatial navigation. *Nature Neuroscience, 16*, 1188–1190.

Kahana, M. J., Sekuler, R., Caplan, J. B., Kirschen, M., & Madsen, J. R. (1999) Human theta oscillations exhibit task dependence during virtual maze navigation. *Nature, 399*, 781–784.

Kalova, E., Vlcek, K., Jarolimova, E., & Bures, J. (2005). Allothetic orientation and sequential ordering of places is impaired in early stages of Alzheimer's disease: Corresponding results in real space tests and computer tests. *Behavioural Brain Research, 159*, 175–186.

King, J. A., Trinkler, I., Hartley, T., Vargha-Khadem, F., & Burgess, N. (2004). The hippocampal role in spatial memory and the familiarity–recollection distinction: A case study. *Neuropsychology, 18*, 405–417.

Lee, I., Yoganarasimha, D., Rao, G., & Knierim, J. J. (2004). Comparison of population coherence of place cells in hippocampal subfields CA1 and CA3. *Nature, 430*, 456–459.

Leutgeb, S., Leutgeb, J. K., Treves, A., Moser, M. B., & Moser, E. I. (2004). Distinct ensemble codes in hippocampal areas CA3 and CA1. *Science, 305*, 1295–1298.

Maguire, E. A., Burgess, N., Donnett, J. G., Frackowiak, R. S., Frith, C. D., & O'Keefe, J. (1998). Knowing where and getting there: A human navigation network. *Science, 280*, 921–924.

Maguire, E. A., Burke, T., Phillips, J., & Staunton, H. (1996). Topographical disorientation following unilateral temporal lobe lesions in humans. *Neuropsychologia, 34*, 993–1001.

Manns, J. R., & Eichenbaum, H. (2006). Evolution of declarative memory. *Hippocampus, 16*, 795–808.

Mapstone, M., Steffenella, T. M., & Duffy, C. J. (2003). A visuospatial variant of mild cognitive impairment: Getting lost between aging and AD. *Neurology, 60*, 802–808.

Monacelli, A. M., Cushman, L. A., Kavcic, V., & Duffy, C. J. (2003). Spatial disorientation in Alzheimer's disease: The remembrance of things passed. *Neurology, 61*, 1491–1497.

Murray, E. A., Bussey, T. J., Hampton, R. R, & Saksida, L. M. (2000). The parahippocampal region and object identification [review]. *Annals of the New York Academy of Science, 911*, 166–174.

O'Keefe, J., & Dostrovsky, J. (1971). The hippocampus as a spatial map: Preliminary evidence from unit activity in the freely-moving rat. *Brain Research, 34*, 171–175.

O'Keefe, J., & Recce, M. L. (1993). Phase relationship between hippocampal place units and the EEG theta rhythm. *Hippocampus, 3*, 317–330.

Redlick, F. P., Jenkin, M., & Harris, L. R. (2001). Humans can use optic flow to estimate distance of travel. *Vision Research, 41*, 213–219.

Russell, N. A., Horii, A., Smith, P. F., Darlington, C. L., & Bilkey, D. K. (2003). Long-term effects of permanent vestibular lesions on hippocampal spatial firing. *Journal of Neuroscience, 23*, 6490–6498.

Rutishauser, U., Tudusciuc, O., Neumann, D., Mamelak, A. N., Heller, A. C., Ross, I. B., et al. (2011). Single-unit responses selective for whole faces in the human amygdala. *Current Biology, 21*, 1654–1660.

Saleem, K. S., & Tanaka, K. (1996). Divergent projections from the anterior inferotemporal area TE to the perirhinal and entorhinal cortices in the macaque monkey. *Journal of Neuroscience, 16*, 4757–4775.

Spiers, H. J., Burgess, N., Maguire, E. A., Baxendale, S. A., Hartley, T., Thompson, P. J., et al. (2001). Unilateral temporal lobectomy patients show lateralized topographical and episodic memory deficits in a virtual town. *Brain, 124*, 2476–2489.

Stackman, R. W., Clark, A. S., & Taube, J. S. (2002). Hippocampal spatial representations require vestibular input. *Hippocampus, 12*, 291–303.

Suthana, N., Ekstrom, A., Moshirvaziri, S., Knowlton, B., & Bookheimer, S. (2011). Dissociations within human hippocampal subregions during encoding and retrieval of spatial information. *Hippocampus, 21*, 694–701.

Suthana, N., Haneef, Z., Stern, J., Mukamel, R., Behnke, E., Knowlton, B., et al. (2012). Memory enhancement and deep-brain stimulation of the entorhinal area. *New England Journal of Medicine, 366*, 502–510

Suthana, N. A., Ekstrom, A. D., Moshirvaziri, S., Knowlton, B., & Bookheimer, S. Y. (2009). Human hippocampal CA1 involvement during allocentric encoding of spatial information. *Journal of Neuroscience, 29*, 10512–10519.

Suzuki, W. A., & Amaral, G. (1994). Perirhinal and parahippocampal cortices of the macaque monkey: Cortical afferents. *Journal of Comparative Neurology, 350*, 497–533.

Taube, J. S., Muller, R. U., & Ranck, J. B., Jr. (1990). Head-direction cells recorded from the postsubiculum in freely moving rats: II. Effects of environmental manipulations. *Journal of Neuroscience, 10*, 436–447.

Tolman, E. C. (1948). Cognitive maps in rats and men. *Psychological Review, 55*, 189–208.

Van Hoesen, G. (1982). The parahippocampal gyrus: New observations regarding its cortical connections in the monkey. *Trends in Neurosciences, 5*, 345–350.

Warren, W. H., Jr., Kay, B. A., Zosh, W. D., Duchon, A. P., & Sahuc, S. (2001). Optic flow is used to control human walking. *Nature Neuroscience, 4*, 213–216.

Weniger, G., Kay, B. A., Zosh, W. D., Duchon, A. P., & Sahuc, S. (2010). The human parahippocampal cortex subserves egocentric spatial learning during navigation in a virtual maze. *Neurobiology of Learning and Memory*, *93*, 46–55.

Wilson, M. A., & McNaughton, B. L. (1993). Dynamics of the hippocampal ensemble code for space. *Science*, *261*, 1055–1058.

Witmer, B. G., Bailey, J. H., Kner, B. W. & Parsons, K. C. (1996). Virtual spaces and real world places: Transfer of route knowledge. *International Journal of Human–Computer Studies*, *45*, 413–428.

Witter, M. P., Wouterlood, F. G., Naber, P. A., & Van Haeften, P. (2000). Anatomical organization of the parahippocampal–hippocampal network. *Annals of the New York Academy of Sciences*, *911*, 1–24.

10 Microelectrode Studies of Human Sleep

Yuval Nir, Michel Le Van Quyen, Giulio Tononi, and Richard J. Staba

Sleep is a pervasive, universal, and fundamental behavior and is present in every animal species where it has been studied (Cirelli & Tononi, 2008). All available evidence indicates that sleep is restorative for brain function and has a vital role for supporting cognition (Hobson, 2005; Banks & Dinges, 2007). For example, memory consolidation may be optimally performed offline when we are disconnected from the environment (Stickgold & Walker, 2007; Diekelmann & Born, 2010). While significant progress has been made in understanding cellular (Gilestro et al., 2009; Liu et al., 2010; Bushey et al., 2011) and behavioral effects of sleep (Banks & Dinges, 2007; Stickgold & Walker, 2007; Killgore, 2010), the contribution of specific activity patterns in sleep to cognitive restoration remains unclear (Wilson & McNaughton, 1994; Nadasdy et al., 1999; Tononi & Cirelli, 2006; Stickgold & Walker, 2007; Diekelmann & Born, 2010). In addition, sleep brings about dramatic changes in consciousness—we remain largely disconnected from external sensory stimuli; at times our perceptual awareness reduces to impoverished fragments that cannot be recalled and reported while at other times intrinsic activity gives rise to rich dream representations (Hobson & Pace-Schott, 2002; Nir & Tononi, 2010). Despite sleep's being a basic behavior occupying a third of our life and affecting cognition, mood, and health, the mechanisms underlying the interplay between sleep and cognition remain unclear.

At present, our understanding of sleep and the manner by which it affects cognition and perception reflects a massive gap between human behavioral studies and electrophysiology investigations, typically in rodents. While such a gap is inherent to many fields of cognitive neuroscience, it is especially evident in sleep research, where technical difficulties have precluded extensive use of functional imaging studies in humans (Nofzinger, 2005; Dang-Vu et al., 2010). In addition, although sleep was an important research focus of early single unit primate studies (Evarts, 1964; Steriade & Deschenes, 1973), such studies have become increasingly rare. Thus, patient studies offer a unique opportunity to bridge the gap between human behavior and rodent electrophysiology. In fact, the value of such sleep recordings goes beyond understanding sleep itself. Spontaneous brain activity in sleep offers a unique window into the activity of functional networks that transcends the ability of patients and healthy individuals to follow elaborate tasks during daytime. It is also possible to record continuously over many hours in which noise and movement play a minimal role and address many questions in systems neuroscience such as the relation between various brain oscillations and neuronal activity (Buzsáki, 2006).

Direct brain recordings during sleep in patients with epilepsy, and single unit studies in particular, constitute an invaluable opportunity for elucidating the relation between sleep and cognition. Such studies permit the investigation of simultaneously recorded neuronal activity from multiple brain areas bilaterally and provide sampling of activity across cortical and subcortical structures that is rarely achieved in animal studies. By recording simultaneously from multiple brain regions, patient studies have the potential to reveal regional diversity in the properties of sleep oscillations such as their spatial topography, spectral characteristics, precise timing and propagation, and phase coherence. Naturally, in patients with suspected temporal lobe epilepsy, recording sites are dictated by clinical considerations and usually encompass mostly medial limbic structures. Fortunately, these brain regions play a pivotal role in supporting many activity patterns in sleep such as slow waves (Murphy et al., 2009; Nir et al., 2011) and sleep spindles (Andrillon et al., 2011) and could provide important information about hotly debated issues such as corticohippocampal "dialogue" in sleep (Buzsáki, 1998). From a practical standpoint, recordings carried out during sleep are well tolerated by patients and do not interfere with visits and other activities as can be the case with daytime recordings.

Another important and intriguing aspect is the close link between sleep and epilepsy (Dinner & Lüders, 2001). Any investigator who embarks on patient sleep recordings should be well aware of this intimate and complex relationship. On one hand, sleep data are an exceptional opportunity for understanding epileptogenesis, and in some presurgical cases changes in the rate of interictal epileptiform discharges (IEDs) from waking to sleep can provide important diagnostic information on the location of the seizure onset zone (Lieb et al., 1980; Sammaritano et al., 1991). However, special care must be exerted when using patient sleep data for making inferences about normal sleep physiology. These considerations will be discussed in the "Epilepsy and Sleep" section.

Overall, single unit patient sleep studies have already given rise to important discoveries about sleep neurophysiology, and important experience has been gained in terms of how to approach sleep data and minimize technical challenges and confounds of epilepsy. However, there is much to be done yet, and exciting future directions await this field.

Patient Sleep Studies: Technical Considerations

Patient sleep studies are best conducted as continuous full-night recordings that combine microelectrodes and wideband electroencephalography (EEG) intracerebral data with standard polysomnography (PSG; see figure 10.1). While data acquisition using intracerebral depth electrodes is described in detail in chapters 3–6, PSG refers to a set of noninvasive measures that are used for standardized scoring of sleep–wake stages. Such scoring is indispensable for comparing sleep recordings across nights, across patients, and in relation to benchmark findings in healthy populations. Ideally, PSG should include electrooculogram (EOG), electromyogram (EMG), electrocardiogram, video monitoring, and scalp EEG from multiple derivations in consistent locations. Whenever possible, PSG data should be perfectly synced with intracerebral depth EEG/microwire recordings. It is highly advantageous to be able to use PSG data not only for sleep staging in 30-s intervals but to relate more precisely intracerebral activities to noninvasive measures such as EEG graphoelements, eye movements, and behavior (see figure 10.1).

Detailed established guidelines for PSG setup and analysis are widely available (Iber et al., 2007). In brief, signals should be acquired at a sampling rate of 500 Hz or higher and referenced to electrodes with minimal brain activity contribution (mastoids or electrodes pasted on earlobes). EEG should be obtained from multiple derivations since different elements (e.g., slow waves, sleep spindles, and alpha activity) are best detected in different locations (frontal, centroparietal, and occipital, respectively). Two EOG electrodes should be pasted, below the left and above the right canthi, and referenced to contralateral reference electrodes. EOG signals are important to delineate sleep onset (often associated with slow eye movements), periods of rapid eye movement (REM) sleep, and brief awakenings that may be associated with blinks and rapid/ irregular eye movements. At least two EMG electrodes are placed to record chin muscle activity and are typically referenced to each other. EMG helps detect brief arousals, and it is indispensable in separating periods of REM sleep from those of wakefulness with closed eyes. It is recommended to calibrate and verify EOG and EMG signals by asking the patient to move his or her eyes in all directions, blink, and clench teeth before beginning data acquisition.

Basic research involving presurgical patients with epilepsy conducted on a hospital ward constitutes a challenging and potentially inhospitable environment for consolidated sleep that could influence when and how sleep recordings are carried out. During the clinical evaluation, microelectrode single unit studies performed shortly after surgical depth electrode implantation could benefit from greater signal-to-noise ratio and lower occurrence of seizures that can disrupt sleep (Mendez & Radtke, 2001; Foldvary, 2002). By contrast, a longer postoperative recovery interval often corresponds with greater patient comfort and consolidated sleep due to habituation to the hospital room that might be advantageous to the sleep study. Investigators must consider the anticipated duration of the clinical evaluation that can last between one and two weeks, which typically provides sufficient opportunity to schedule sleep recording(s). In studies that seek to generalize sleep findings to healthy populations, it is important to consider tapering of antiepileptic drugs (AEDs) and sleep deprivation that are often prescribed clinically in depth patients who have low seizure frequency. Both are intended to increase the occurrence of seizures and other epileptiform discharges that may interrupt habitual sleep patterns (Peled & Lavie, 1986). Accordingly, optimal timing for sleep studies in our experience at UCLA is 48–72 hours after electrode implantation and prior to tapering of AEDs or sleep deprivation. In addition, it is important to avoid sleep studies in close proximity (<12 hours) to daytime seizures that can affect sleep architecture (Foldvary, 2002), as well as carefully document medication, seizures, and sleep history (e.g., naps). Up to one third of patients with refractory seizures have sleep-related breathing disorders, for example, obstructive sleep apnea syndrome (OSAS), that are associated with midsleep arousals and sleep fragmentation that should also be taken into consideration (Malow et al., 2000). Finally, ambient noise, unexpected visitors, and routine clinical procedures (e.g., nursing staff taking vital signs) may lead to poor sleep hygiene. Therefore, a successful sleep study often hinges on managing these details and coordination among clinical and research personnel.

Routine data preprocessing includes filtering (EEG, above 0.3 Hz; EOG, 0.3–35 Hz; EMG, 10–100 Hz) and sleep scoring upon visualization of EEG, EOG, EMG, and video (Iber et al., 2007). Ideally, spike sorting should be done on entire full-night recordings as a whole, so that the activity of the same neuron or small neuronal populations can be compared and analyzed throughout sleep and possibly also in surrounding intervals of wakefulness. In practice, more than half a million putative action potentials could be detected in individual channels in such recordings, and existing clustering algorithms often use template matching to simplify computational load (Quian Quiroga et al., 2004). At any rate, the stability of unit recordings throughout long sleep recordings should not be taken for granted; rather, it is necessary to carefully inspect the consistency of waveforms and interspike interval distributions (see, e.g., figure S3B in Nir et al., 2011).

Epilepsy and Sleep

Effects of Epilepsy, Medication, and Comorbidity on Sleep

Patients with epilepsy, compared to healthy controls, are two times more likely to complain of sleep disturbances, chiefly excessive daytime sleepiness (EDS) and insomnia that can have a negative influence on quality of life measures (de Weerd et al., 2004; Piperidou et al., 2008). Furthermore, with respect to type of epilepsy, patients with nocturnal frontal lobe epilepsy report greater tiredness after awakening and more frequent spontaneous midsleep awakenings (Vignatelli et al., 2006). Refractory temporal lobe epilepsy is associated with similar sleep disturbances and in some cases abnormal sleep efficiency and microarchitecture (Crespel et al., 2000). Other studies show that sleep disruptions are more pronounced on nights with seizures compared

Figure 10.1
Experimental setup for patient sleep studies. (A) Setup for polysomnography (PSG) includes electrooculogram (EOG), electromyogram (EMG), scalp electroencephalogram (EEG) from multiple derivations, and video monitoring. (B) Diagram of flexible probes used for concomitant recording of depth EEG (platinum contacts) and local field potential (LFP)/unit activity (microwires) synced with PSG data. (C) Overview of typical depth electrode locations in encompassing multiple brain regions seen from medial view. Abbreviations: OF, orbitofrontal cortex; AC, anterior cingulate; SM, supplementary motor; PC/P, posterior cingulate/parietal cortex; PH, parahippocampal gyrus; HC, hippocampus; E, entorhinal cortex; Am, amygdala. (D) Representative examples of 30-s PSG data used for sleep scoring in wake stages, N2, N3, and rapid eye movement (REM) sleep. (E) Representative PSG-based sleep; sleep efficiency corresponds to total sleep time per time in bed. Sleep latency is to stage 2. WASO refers to waking after sleep onset; SWS, slow-wave sleep (N2 + N3); NREM, non–rapid eye movement. (F) Representative hypnogram (time course of sleep–wake stages across sleep in one individual). (G) Representative average power spectra of scalp EEG computed separately in stages N2, N3, and REM sleep; note high power in slow-wave (<4 Hz) and spindle (10–16 Hz) range in NREM sleep. (H) example of data acquired during 6 s of NREM sleep. Rows (top to bottom) show activity in scalp EEG (Cz), depth EEG (entorhinal cortex), multiunit activity (MUA) in one microwire, and spiking activity in six isolated single units (black bars). Modified from Nir et al. (2011).

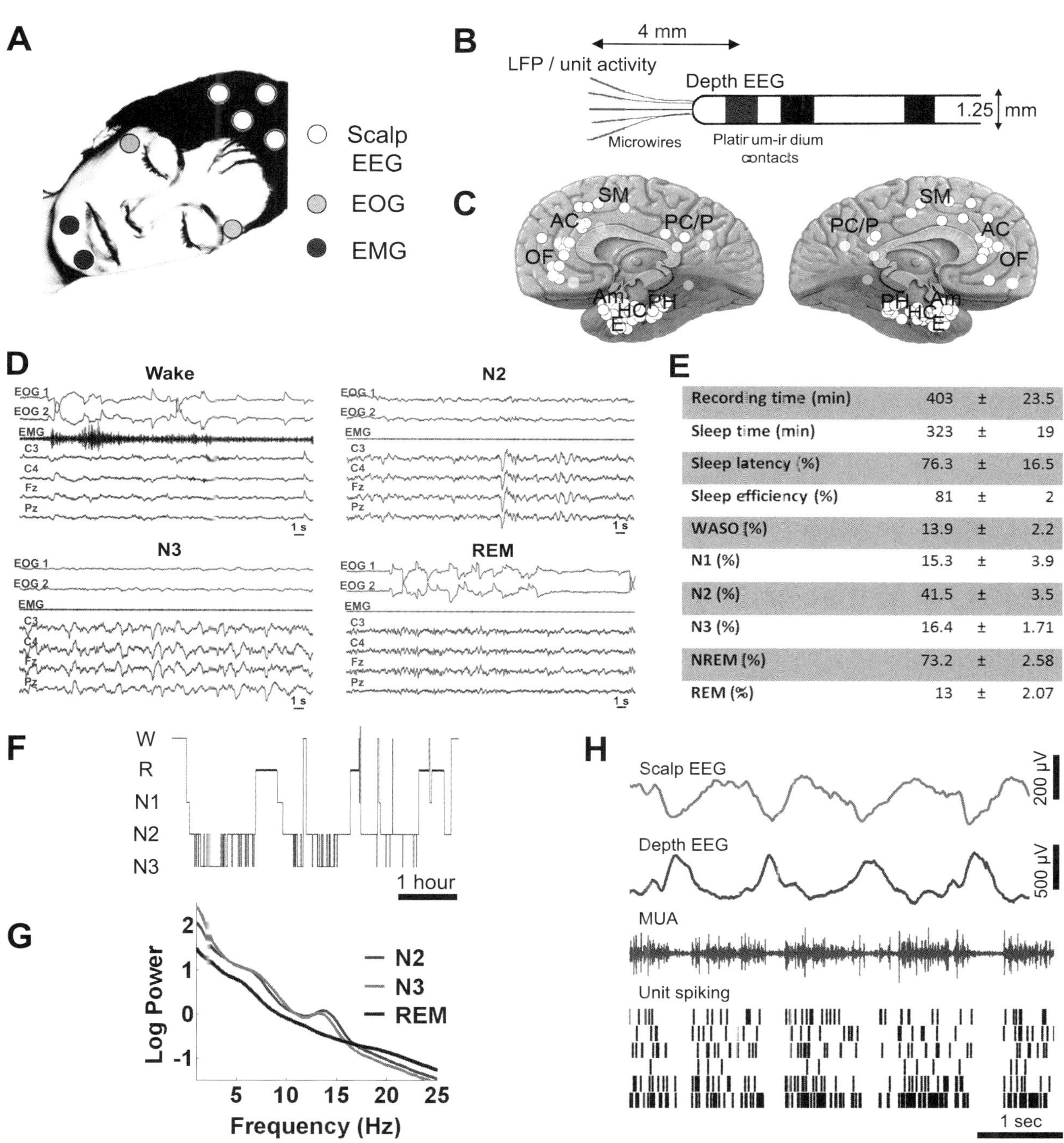

to seizure-free nights (Foldvary-Schaefer & Grigg-Damberger, 2009; Carrion et al., 2010), but importantly, EDS decreases and quality of sleep improves in patients with good postsurgical seizure outcome (Carrion et al., 2010). Results from these studies suggest seizures can disrupt sleep and highlights the importance of seizure control in restoration of normal sleep.

Other factors affecting sleep include AEDs and comorbidities. First-generation AEDs such as barbiturates and benzodiazepines facilitate sleep onset but can reduce the amount of REM sleep, increase daytime somnolence, and lower sleep quality compared to second-generation AEDs (Bazil, 2003). Compared to healthy controls, patients with poorly controlled seizures have a higher prevalence of depression and anxiety (Stefanello et al., 2010), which can have detrimental effects on sleep (van Mill et al., 2010). OSAS and corresponding complaint of EDS is also associated with epilepsy. OSAS associated with epilepsy could be due to AED-related weight gain, prescribed barbiturates or benzodiazepines that reduces upper airway rigidity, or in some cases the use of vagal nerve stimulation that can effect respiration and exacerbate OSAS. The link between EDS and OSAS does not appear to be due to sleep apnea or sleep disruption (Roure et al., 2008) although treatment of OSAS reduces the rate of overnight respiratory disturbances and increases sleep stability as well as improving EDS (Sforza & Krieger, 1992; Conradt et al., 1998). Patients with epilepsy receiving therapy for OSAS had a reduction of IEDs (Oliveira et al., 2000), while other studies noted that some patients had a 50% or greater reduction in seizures frequency (Vaughn et al., 1996; Malow et al., 1997; Malow et al., 2003).

Effects of Sleep on Epileptogenicity
Some seizures and IEDs occur more frequently during sleep and particularly during episodes of non-REM (NREM) compared to REM sleep. For example, seizures associated with autosomal-dominant nocturnal frontal lobe epilepsy occur only during sleep whereas seizures associated with benign epilepsy with centrotemporal spikes occur up to 80% of the time during sleep. In addition, epileptic encephalopathy with continuous spike-and-wake during sleep is defined by the occurrence of diffuse electrical status epilepticus that occupies up to 85% of NREM sleep. In general, focal seizures arising from frontal lobe occur more frequently during sleep than temporal lobe seizures (Bazil & Walczak, 1997; Crespel et al., 2000) while the latter type of seizures are more likely to secondarily generalize during NREM sleep compared to REM sleep (Herman et al., 2001).

In most types of epilepsy IEDs occur more frequently during NREM sleep than REM sleep. It is hypothesized that during NREM sleep, synchrony of neuronal discharges within thalamo-cortical networks increases cortical excitability and facilitates the spread of focal IEDs to remote brain areas (Steriade et al., 1994). By contrast, transition to REM sleep or arousal from NREM sleep is associated with a reduction in thalamocortical synchronization and spatial restriction of IEDs. Consistent with the presumed sleep-related changes in cortical excitability, studies of human hippocampal single neurons found interictal firing rates and propensity for bursting are highest during NREM sleep and lowest during REM sleep (Staba et al., 2002c). In relation to epileptogenicity, single neurons in the medial temporal lobe (MTL) ipsilateral to the seizure onset have higher firing and burst rates as well as greater synchrony of discharges than neurons in contralateral MTL sites (Staba et al., 2002b). In this same study the greatest differences in firing properties and synchrony with respect to sites of seizure onset were associated with NREM and REM sleep. This latter finding could reflect the relatively greater autonomy of primary epileptogenic brain areas compared to remote sites that do not support seizure genesis (Gentilono et al., 1975; Lieb et al., 1980).

Interictal bursts of pathological high-frequency oscillations (HFOs; 200–500 Hz) are also found in human MTL ipsilateral to seizure onset (see figure 10.2A, B; Bragin et al., 1999a; Bragin et al., 1999b; Staba et al., 2002a) and in neocortical sites capable of generating seizures although often of lower spectral frequency (Worrell et al., 2004). Single neuron discharges increase significantly during pathological HFOs, and there is evidence to suggest pathological HFOs reflect spontaneous bursts of abnormally synchronous unit discharges that appear as local population spikes (figure 10.2C, D; Bragin et al., 2002; Bragin et al., 2011). Similar to the sleep-related changes in MTL single neuron discharges described in the preceding paragraph,

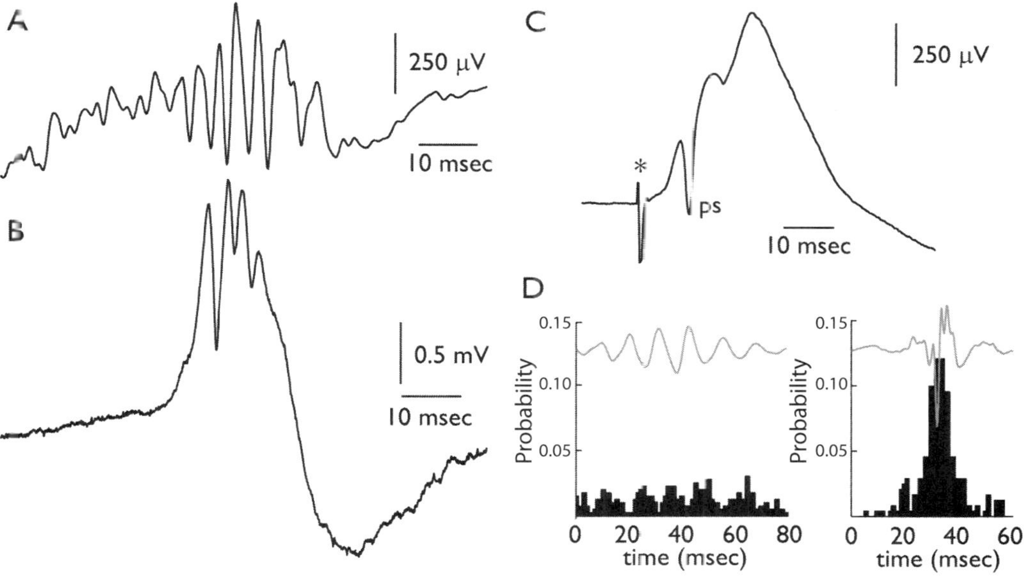

Figure 10.2
High-frequency oscillations (HFOs) in epileptic hippocampus. (A) Pathological HFO (~333 Hz) recorded from a microelectrode positioned in hippocampus ipsilateral to seizure onset of a patient with medial temporal lobe epilepsy and hippocampal sclerosis. (B) Spontaneous interictal electroencephalography spike with superimposed pathological HFO (~300) recorded from microelectrode in entorhinal cortex. (C) Average (n = 12) evoked response recorded from microelectrode positioned in hippocampus during electrical stimulation (indicated by asterisk) from adjacent clinical electrode in entorhinal cortex. Note population spike (ps) superimposed on large-amplitude postsynaptic potential. (D) Peri-event histogram of multiunit discharge during ripple-frequency HFO (left) and pathological HFO (right) in entorhinal cortex. Averaged HFOs (n = 10) recorded from same distal microwire and neuronal discharges recorded from adjacent microwires spaced 0.5 mm apart in entorhinal cortex. Modified from Staba (2012).

pathological HFOs occur more frequently during NREM sleep and rates remain elevated during REM sleep compared to waking, which contrasts with the occurrence of presumably normal ripple-frequency HFOs (80–160 Hz) in human MTL and ripples in normal nonprimate hippocampus associated with REM sleep episodes (Buzsáki et al., 1992; Staba et al., 2004). Overall, NREM sleep facilitates the occurrence of epileptiform discharges whereas mechanisms governing REM sleep that typically restrict neuronal synchrony are less effective inside primary epileptogenic brain areas.

Neurophysiology of Human Sleep

Patient studies of human sleep, and microelectrode recordings in particular, have significantly furthered our understanding of sleep electrophysiology by investigating different sleep oscillations such as slow waves, sleep spindles, gamma and ripple oscillations, and ultraslow neuronal fluctuations. The manner in which such sleep oscillations may contribute to memory consolidation is an area gaining increasing attention.

Sleep Slow Waves

The most prominent electrophysiological events in sleep are slow waves and related K-complexes—isolated high-amplitude waves that are triggered by external or internal stimuli (Colrain, 2005). Animal studies have established that such waves reflect a bistability of thalamocortical neurons undergoing a slow oscillation (<1 Hz) between active ("UP") and inactive ("DOWN") states, and that these waves group and modulate other neuronal oscillations (Steriade et al., 1993; Contreras & Steriade, 1995; Destexhe et al., 2007; Crunelli & Hughes, 2010). Recently, microelectrode studies confirmed that in humans slow waves are similarly associated with underlying neuronal bistability (see figure 10.3A, plate 10) oscillating between active and inactive states (Cash et al., 2009; Csercsa et al., 2010; Le Van Quyen et al., 2010; Nir et al., 2011), as was found in natural sleep of rodents (Vyazovskiy et al., 2009b) and cats (Chauvette et al., 2011).

Importantly, such studies also revealed exciting findings that were not expected based on the noninvasive studies and animal literature. While slow oscillations are remarkably synchronous when examined in brain slices (Sanchez-Vives & McCormick, 2000) and in animals under anesthesia (Chauvette et al., 2011), human studies focusing on natural sleep and recording in many regions in parallel revealed that most sleep slow waves and the underlying active and inactive neuronal states occur locally (see figure 10.3C, D, plate 10) where some regions can be active while others are silent (Nir et al., 2011). Furthermore, in some cases wake-like and sleep-like activity patterns can coexist for longer durations in different cortical areas, and such activities may underlie NREM parasomnias such as sleepwalking (Terzaghi et al., 2009; Nobili et al., 2011).

It was also found that slow waves have a tendency to propagate along typical paths (see figure 10.3B, plate 10), from medial prefrontal cortex to the medial temporal lobe (MTL) through the cingulate gyrus and neighboring structures (Nir et al., 2011), which constitute an anatomical

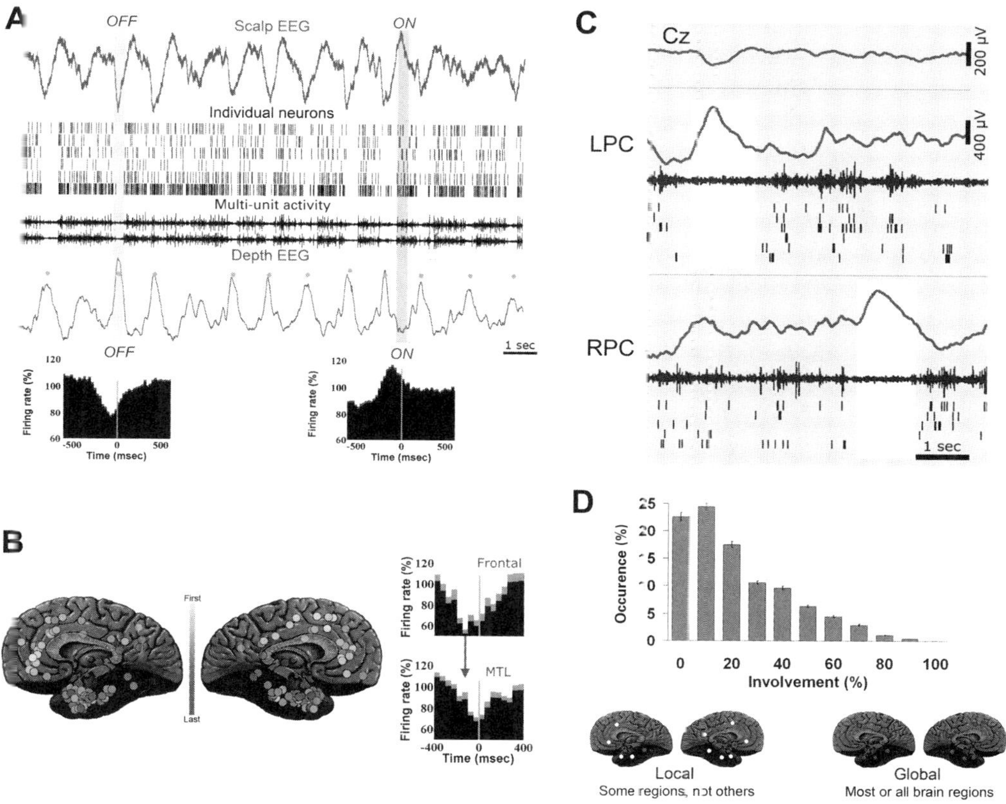

Figure 10.3 (plate 10)

Regional slow waves in human sleep. (A) Neuronal activity underlying slow waves in human sleep. Electrical brain activity across 15 s of deep non–rapid eye movement (NREM) sleep. Top (red), scalp electroencephalography (EEG); bottom (blue), intracranial depth EEG in entorhinal cortex. Green dots, individual slow waves that are automatically detected and separated from pathological events. Black, multiunit activity (MUA) and action potentials of six neurons. Vertical green bar, OFF periods of inactivity. Vertical orange bar, ON periods of neuronal silence. Bottom insets, an analysis across 600 units confirms that neurons increase and decrease their activity in concert with local electrical fields. (B) Slow waves have a tendency to propagate along typical paths. Left, each circle denotes a depth electrode, and its color marks the typical slow-wave timing at that location. Right, average unit activity in frontal cortex (top, n = 76) and medial temporal lobe (MTL) (bottom, n = 155), triggered by the same scalp slow waves. Note that, on average, slow waves and underlying neuronal activity occur earliest in the frontal lobe, about 200 ms later in the temporal lobe, and finally in the hippocampus. (C) An example of local sleep slow waves occurring at different times in left posterior cingulate cortex (LPC) and right posterior cingulate cortex (RPC). Rows (top to bottom) depict activity in scalp EEG (Cz, red), left and right posterior cingulate. Blue, depth EEG; green, MUA; black lines, single unit spikes. White shadings mark local OFF periods. (D) The vast majority of slow waves occur locally. Distribution of slow-wave involvement (percentage of monitored brain structures expressing each wave) shows that global slow waves are quite rare. Modified from Nir et al. (2011).

backbone of anatomical fibers (Hagmann et al., 2008). Such propagation was previously suggested by high-density EEG and animal studies (Massimini et al., 2005; Volgushev et al., 2006; Murphy et al., 2009; Vyazovskiy et al., 2009a; Riedner et al., 2011). Slow waves also exhibit complex propagation patterns at a local scale (Hangya et al., 2011).

In addition, important insights were gained about slow-wave propagation *within* the MTL, where noninvasive imaging is limited. By and large, cortical slow waves precede hippocampal waves, revealing a sequential propagation from the parahippocampal gyrus, through entorhinal cortex, to hippocampus (Nir et al., 2011), in line with previous animal studies (Sirota et al., 2003; Isomura et al., 2006; Hahn et al., 2007; Ji & Wilson, 2007) and with a recent study of human depth EEG (Wagner et al., 2010). As for the direction of corticohippocampal dialogue in sleep (Buzsáki, 1998), it was found that at times of hippocampal ripples (associated with the "replay" of activity in cell assemblies during sleep in rodents; Diekelmann & Born, 2010), local effects of hippocampal output can be observed within the MTL in terms of increased unit activity (Nir et al., 2011) as well as gamma bursts in parahippocampal gyrus (Le Van Quyen et al., 2010). However, ripples were not associated with detectable effects in the medial prefrontal cortex (Nir et al., 2011), a primary projection zone of hippocampal output in primates. On the whole, during NREM sleep neural activity propagates predominantly from the neocortex to the hippocampus. Future studies are needed to determine whether within this robust corticohippocampal broadcast there may be islands of hippocampocortical transmission that may be functionally relevant for memory consolidation.

Sleep Spindles

Sleep spindles are the other hallmark oscillation of NREM sleep; they are waxing-and-waning 10–16 Hz oscillations lasting 0.5–2 s and are believed to mediate many sleep-related functions (De Gennaro & Ferrara, 2003). Recent intracerebral human studies (Andrillon et al., 2011; Peter-Derex et al., 2012) revealed that spindle frequency is topographically organized with a sharp transition between fast (13–15 Hz) centroparietal spindles and slow (9–12 Hz) frontal spindles occurring 200 ms later on average (see figure 10.4). As was the case for slow waves, most spindles occur locally, thereby showing that constrained intracerebral communication is an important feature of sleep. It was also found that spindle frequency changes along with sleep depth, reflecting the level of thalamocortical hyperpolarization at any given time and that robust firing rate modulations were surprisingly weak during sleep spindles (Andrillon et al., 2011). On the whole, patient studies revealed changes in spindle occurrence, frequency, and timing between regions and across sleep (Andrillon et al., 2011; Peter-Derex et al., 2012). Some of this heterogeneity (e.g., slow frontal vs. fast centroparietal spindles) was observed also with noninvasive scalp measurements (Anderer et al., 2001; De Gennaro & Ferrara, 2003; Schabus et al., 2007; Ferrarelli et al., 2010) whereas several other novel aspects such as timing differences between brain regions, frequency changes across sleep, and the lack of robust firing rate modulations (Andrillon et al., 2011) were previously unknown.

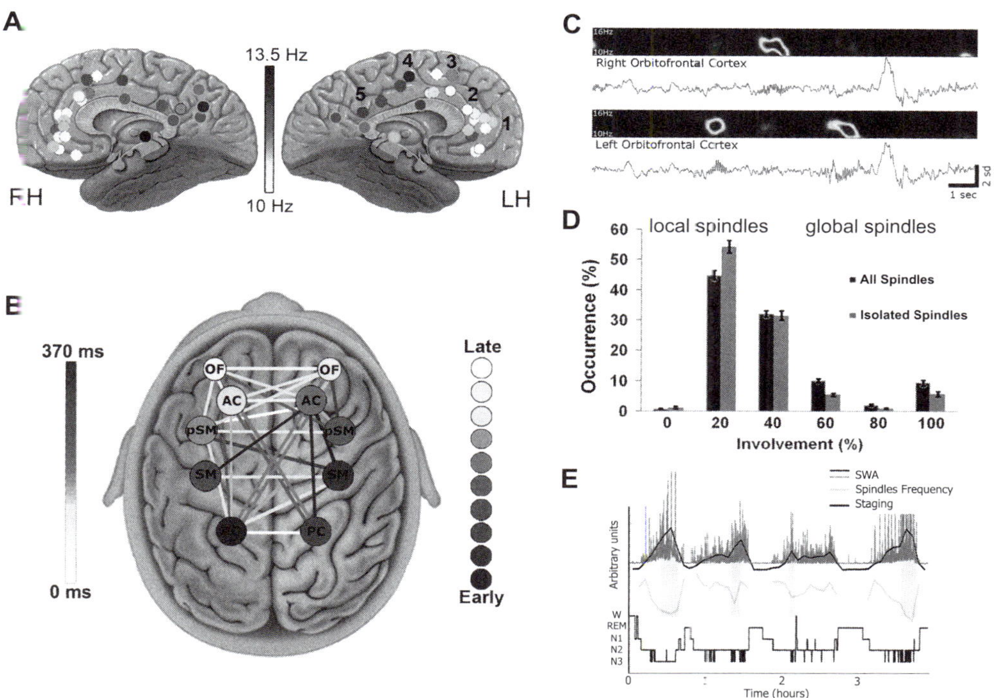

Figure 10.4

Human sleep spindles. (A) Average frequency of spindles across the medial brain; note the contrast between slow (9–12 Hz) frontal spindles and fast (13–16 Hz) centroparietal spindles. RH, right hemisphere; LH, left hemisphere. (B) Fast centroparietal spindles precede slow frontal spindles by 200 ms on average: a graph showing a quantitative analysis of the order in which spindles are detected across multiple regions (node color) and the mean temporal delays between each pair of regions (edge color). (C) Example of a local sleep spindle as seen in depth EEG across bilateral orbitofrontal cortex along with corresponding spectrograms in the spindle frequency range (9–16 Hz) during 15 s of slow-wave sleep. (D) Most sleep spindles are local. Distribution of involvement (percentage of monitored brain structures expressing each spindle) for all spindles (dark bars) and for isolated spindles (lighter bars). Note that 73% of spindles are observed in 50% of electrodes, indicating that most spindles are local. (E) Spindle frequency reflects sleep depth. Representative time course of slow-wave activity (SWA) and spindle frequency dynamics throughout sleep in the anterior cingulate of one individual. Note that spindle frequency is lowest in deep sleep when SWA is highest and increases toward transitions to rapid eye movement (REM) sleep. Modified from Andrillon et al. (2011).

Gamma and Ripple Oscillations during Slow-Wave Sleep

Gamma oscillations (40–120 Hz) are usually associated with waking functions such as sensory binding (Singer & Gray, 1995), attention (Fries et al., 2001), or encoding/retrieval of memory traces (Montgomery & Buzsáki, 2007) and have been shown to be closely related to correlated neuronal activity in humans during wakefulness (Nir et al., 2007). These oscillations are also present during slow-wave sleep, as shown by extensive evidence from in vivo (Steriade et al., 1996; Grenier et al., 2001; Isomura et al., 2006; Mena-Segovia et al., 2008) and in vitro (Dickson et al., 2003; Compte et al., 2008) recordings of the rodent and feline cortex. Such experiments demonstrated that gamma oscillations occur preferentially during the active (UP) component of the slow wave—characterized by rhythmic cycles of synaptically mediated depolarization—and disappear during the hyperpolarized (DOWN) phase. Recent microelectrode studies in the human cortex during sleep have confirmed that gamma oscillations are reliably associated with EEG slow waves and with a marked increase in local cellular discharges (Cash et al., 2009; Le Van Quyen et al., 2010; figure 10.5A). These gamma oscillations frequently appeared at about the same time in different cortical areas, including homotopic regions, forming large spatial patterns (see figure 10.5B). Similar sleep gamma oscillations were also recently reported using intracranial macroelectrodes (Valderrama et al., 2012), suggesting a strong local synchronization of the cellular activities. Indeed, coincident firings with millisecond precision between cells within the same cortical area were shown to be strongly enhanced during gamma oscillations (Le Van Quyen et al., 2010). Cortical gamma patterns in sleep have been suggested to briefly restore "microwake" activity (Destexhe et al., 2007; Haider & McCormick, 2009) and may be important for consolidation of memory traces acquired during previous wakefulness. Along this line, coupling between parahippocampal gamma oscillations and hippocampal ripple/sharp-wave complexes has been reported in humans (Le Van Quyen et al., 2010; figure 10.5C). Ripple oscillations (80–160 Hz) are known to coincide with reactivation of hippocampal activity patterns (Wilson & McNaughton, 1994) which could reflect information flow from the hippocampus to the cortex. In the human hippocampus and entorhinal cortex, ripples are similar to those described in nonprimate CA1 and CA3 in terms of duration and spectral frequency, bilateral occurrence in hippocampal areas, highest probability of occurrence during NREM sleep, and

Figure 10.5
Gamma oscillations during slow-wave sleep. (A) (i) Display of a single gamma episode (black arrow) appearing simultaneously, in either the raw signals or those filtered between 40 and 120 Hz, in the right and left posterior parahippocampal gyri (PHG) during slow-wave sleep. Note that gamma activities were temporally correlated with positive peaks (i.e., up deviations) of slow waves in scalp electroencephalography (EEG). LFP, local field potential. (ii) Corresponding wavelet transforms of two homotopic sites revealing nearly simultaneous gamma oscillations with distinct narrow band frequencies around 40 Hz (white arrows). (B) Examples of gamma events simultaneously recorded with 30 microelectrodes in the right and left parahippocampal gyri (PHG, ant: anterior part and post: posterior part). Note the complex spatiotemporal distribution of these activities, often involving both homotopic sides, the strong variability of involved electrodes and variable location of the starting site (triangles). (C) In individual events i and ii, hippocampal ripple/sharp-wave complexes (R) and parahippocampal gamma oscillations (gamma) were not coincident within a time window of 100 ms. In event iii, parahippocampal gamma oscillations immediately followed hippocampal ripple/sharp-wave complexes, suggesting a temporal coupling between both oscillations. MUA, multiunit activity; Hipp, hippocampal. Modified from Le Van Quyen et al. (2010).

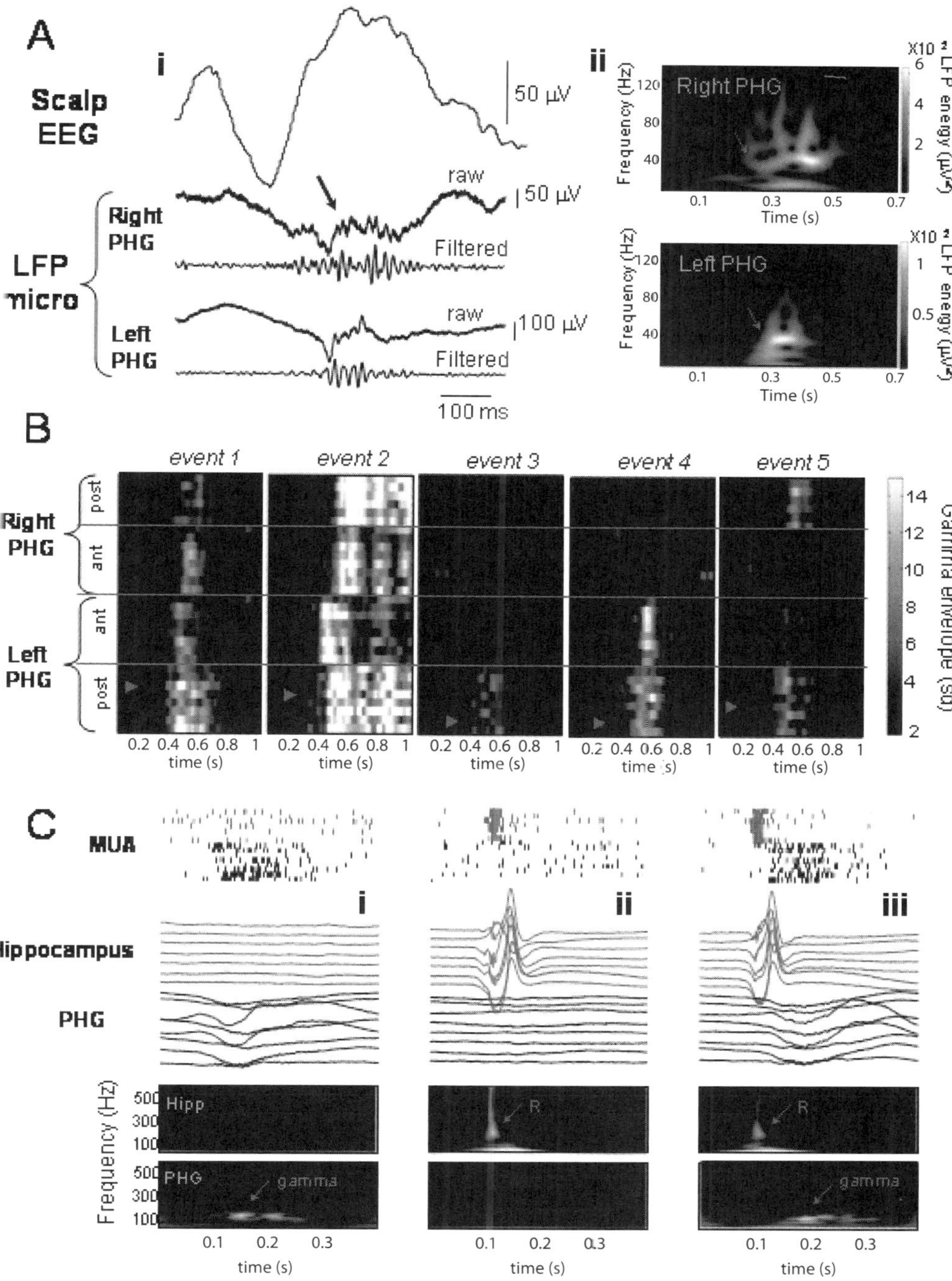
A
i
Scalp EEG
50 µV
LFP micro
Right PHG
raw
50 µV
Filtered
Left PHG
raw
100 µV
Filtered
100 ms
ii
X10²
6
4
2
LFP energy (µV²)
Right PHG
Frequency (Hz)
120
80
40
0.1 0.3 0.5 0.7
Time (s)
X10²
1
0.5
LFP energy (µV²)
Left PHG
Frequency (Hz)
120
80
40
0.1 0.3 0.5 0.7
Time (s)
B
event 1
event 2
event 3
event 4
event 5
Right PHG
post
ant
Left PHG
ant
post
0.2 0.4 0.6 0.8 1
time (s)
0.2 0.4 0.6 0.8 1
time (s)
0.2 0.4 0.6 0.8 1
time (s)
0.2 0.4 0.6 0.8 1
time (s)
0.2 0.4 0.6 0.8 1
time (s)
14
12
10
8
6
4
2
Gamma envelope (sd)
C
MUA
Hippocampus
PHG
i
ii
iii
Frequency (Hz)
500
300
100
Hipp
R
R
500
300
100
PHG
gamma
gamma
0.1 0.2 0.3
time (s)
0.1 0.2 0.3
time (s)
0.1 0.2 0.3
time (s)

minimal occurrence during REM sleep (Buzsáki et al., 1992; Staba et al., 2004; Le Van Quyen et al., 2008; figure 10.2). Thus, human microelectrode studies show that high-frequency gamma and ripple oscillations robustly occur and modulate single neuron firing during sleep. Moreover, slower fluctuations such as slow waves group and modulate faster "nested" oscillations such as cortical spindles, gamma events, and hippocampal ripples (Clemens et al., 2007; Le Van Quyen et al., 2010; Andrillon et al., 2011; Nir et al., 2011), confirming findings in animal studies (Sirota et al., 2003; Battaglia et al., 2004; Steriade, 2006). Future human microelectrode studies at the large-scale level are necessary to further characterize the full details of these local and long-range coupling of oscillations across vast cortical territories.

Ultraslow Resting-State Fluctuations in Sleep and Wakefulness

Although perception and action occur on the subsecond timescale, it has long been recognized that cortex also shows fluctuations in electrical activity with slower dynamics. Recently, resting-state ultraslow fluctuations (<0.1 Hz, at the timescale of tens of seconds) in blood oxygen level–dependent functional magnetic resonance imaging signals have gained attention as a powerful tool to study functional brain networks in health and disease (Fox & Raichle, 2007). However, the extent to which such waves reflect neuronal activity or may stem from nonneuronal sources (e.g., cardiac, respiratory) remained unclear until very recently. Recent intracerebral recordings (He et al., 2008; Nir et al., 2008) established that spontaneous ultraslow neuronal activity can indeed be detected in direct cortical recordings and that it exhibits significant correlations between nodes within the same functional system (see figure 10.6). An important open question in this field is whether such resting-state waves may reflect cognitive processes such as mind wandering, shifts in attention or mental imagery, or whether they may be more closely related to basic maintenance of synaptic contacts (Balduzzi et al., 2008). Interestingly, ultraslow waves in humans were found to persist and grow stronger in sleep. They are also present in anesthesia (Vincent et al., 2007) and to some extent in vegetative patients (Ovadia-Caro et al., 2012), thus arguing against involvement of such waves in conscious processes.

Sleep and Memory Consolidation

A promising research direction is to clarify the role sleep may have in offline consolidation of memories (Diekelmann & Born, 2010). An important postulated mechanism for such consolidation involves hippocampal sharp-wave ripples accompanying reactivation of neuronal ensembles that were active during preceding wake experience (Wilson & McNaughton, 1994; Nadasdy et al., 1999; Girardeau et al., 2009). Importantly, such high-frequency MTL events can be routinely recorded in patient sleep studies (Staba et al., 2002a). Indeed, some studies are beginning to link sleep activities with those recorded during cognitive tasks in wakefulness. Intracranial EEG studies have found that successful learning is correlated with NREM sleep oscillations including MTL ripples (Axmacher et al., 2008a; Axmacher et al., 2008b) and slow waves (Bodizs et al., 2002; Moroni et al., 2008). Future work focusing on single unit activities should further clarify the contribution of sleep to memory. Ideally, one would try to relate activity that is

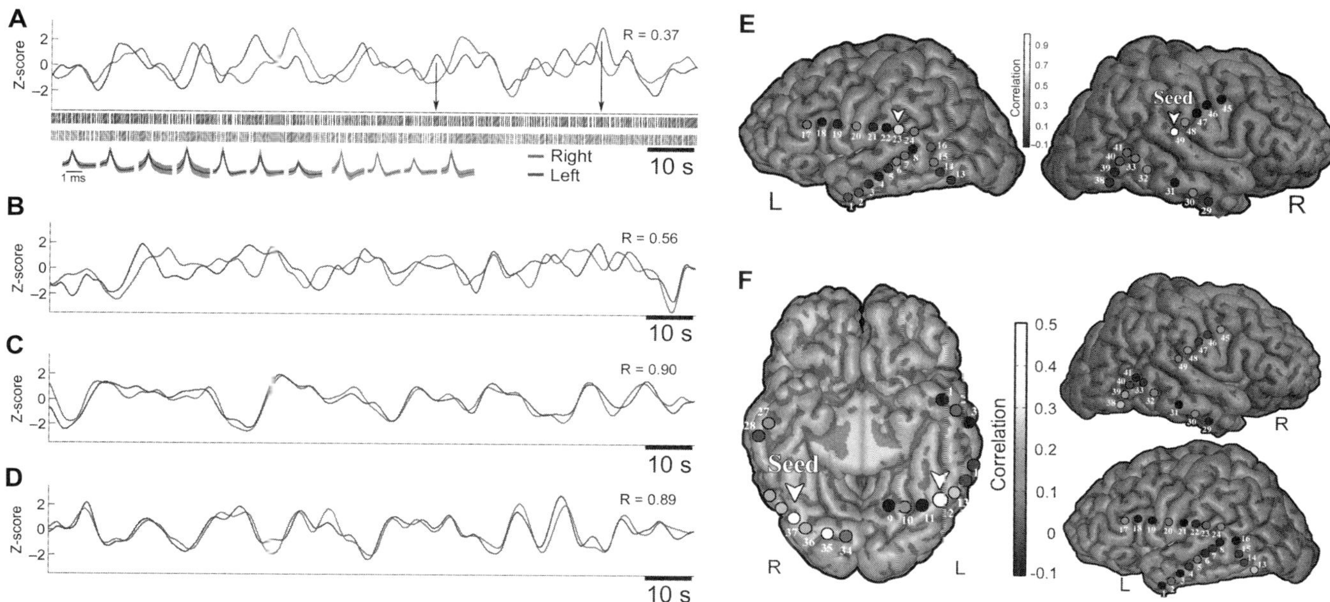

Figure 10.6

Ultraslow (<0.1 Hz) fluctuations of neuronal activity in rest and sleep. (A–D) Examples of correlated ultraslow (<0.1 Hz) fluctuations in neuronal activity between left and right auditory cortices across hemispheres. (A) Firing-rate modulations during rest in wakefulness. Vertical lines show actual spike times. Black arrows indicate the relation between time courses of slow firing-rate modulations and neuronal discharges. Waveforms of neuronal action potentials are shown below. (B) Local field potential (LFP) gamma-power modulations during rest in wakefulness. (C) LFP gamma-power modulations during rapid eye movement (REM) sleep. (D) LFP gamma-power modulations during non-REM stage N2 sleep. Note that ultraslow fluctuations are robustly correlated across hemispheres, and correlations are markedly enhanced during sleep. (E) Correlations between ultraslow fluctuations in electrocorticography (ECoG) gamma power of an auditory-related electrode ("seed," arrow) and all other electrodes reveal a highly selective spatial topography of ultraslow fluctuations. Results are shown on a cortical reconstruction seen from a lateral view. Note that a strong correlation in the contralateral hemisphere is found over the homotopic auditory cortex. L, left; R, right. (F) Correlations between ultraslow fluctuations in ECoG gamma power of a visual, face-selective electrode ("seed," arrow) and all other electrodes. Results shown using a ventral view (left) and lateral view (right). The strongest correlation is found in a contralateral face-selective region while minimal correlations are observed between visual and auditory electrodes, indicating the functional selectivity of spontaneous correlations. Modified from Nir et al. (2008).

selectively involved in the representation and learning of specific contents to reactivation in subsequent sleep and to verbal memory recall in the morning while avoiding measurements within the seizure onset zone (where ripples and associated neuronal bursts in sleep are often pathological).

Emerging New Theme: Regional Diversity of Sleep Oscillations

By analyzing simultaneous activity across multiple brain regions, microelectrode studies of human sleep demonstrated that slow waves and sleep spindles (Nir et al., 2011) as well as gamma bursts (Le Van Quyen et al., 2010) are mostly local, thereby showing that initial indications of local sleep oscillations in animals (Sirota & Buzsáki, 2005; Mohajerani et al., 2010; Vyazovskiy et al., 2011) or in patients with sleep disorders (Terzaghi et al., 2009) constitute the rule rather than the exception. Thus, an important new theme that emerged from single unit human studies is that sleep oscillations are much more heterogeneous than initially assumed (Magnin et al., 2010; Andrillon et al., 2011; Hangya et al., 2011; Nir et al., 2011). Regional diversity in occurrence, spectral, and temporal aspects of sleep oscillations was hardly observable with noninvasive human imaging, when recording from a limited number of brain regions in rodents, or when using anesthesia as a model for sleep.

Future Directions

Microelectrode studies of human sleep have provided important novel insights into human sleep neurophysiology, but so far such studies have been conducted by a limited number of research centers, and further research is bound to provide new insights on sleep function and its role in health and disease.

An intriguing future direction is the relation between dreaming and underlying brain activity. Although dreams have fascinated us since the dawn of time, their rigorous, scientific study is only a recent development (Nir & Tononi, 2010). As with other investigations of conscious processes, this research ultimately relies on reports from human subjects, and patient studies offer the unique opportunity to relate such reports to detailed measurements of brain activity. Understanding the relation between brain activity in REM sleep and wakefulness to eye movements also remains unexplored. For example, does such activity in REM sleep follow a feedforward propagation as it does upon visual stimulation in wakefulness (Mormann et al., 2008)? Or perhaps dreaming is more related to top-down processes as may be the case in mental imagery where mnemonic activity precedes that of visual representations (Takeuchi et al., 2011)?

Intracranial depth electrodes can also be adapted with microdialysis probes (Fried et al., 1999) to investigate the neurochemical modulation of sleep–wake states (Jones, 2005). Given that rodents and other mammals switch between vigilance states much more rapidly, microdialysis samples typically collected 10 minutes apart in animal studies inevitably integrate over different sleep states. By contrast, in humans it is possible to collect samples that occur exclusively in specific sleep stages, including REM sleep. Indeed, human microdialysis studies have begun to examine such changes—for example, by measuring levels of ventricular serotonin (Zeitzer et al., 2002) and extracellular adenosine (Zeitzer et al., 2006) as well as hypocretin and melanin-concentrating hormone levels across the sleep–wake cycle (Blouin et al., 2013). In this latter study, neurochemical changes in the amygdala were correlated not only with vigilance and arousal but also with emotional and social behaviors. Indeed, elucidating the relation between neuromodulation and cognitive aspects of daytime behaviors is a fascinating avenue for future research. From a technical standpoint, future advances including zero-flow microdialysis (Cavus et al., 2005) and methods with superior temporal resolution (e.g., microdialysis with faster recovery time or electrochemical sensing probes) could help bridge the gap between neuromodulatory changes (currently studied with a temporal resolution of minutes) and millisecond-precision electrophysiology. Bridging this gap could help in understanding, for example, whether the local occurrence of sleep oscillations may be accompanied by local neuromodulatory changes (for instance, via local modulation of presynaptic release; Laplante et al., 2005).

Another line of future research is to better understand responses to external sensory stimuli during sleep, and in what ways signal propagation along ascending sensory pathways differs between states of vigilance. Increasing evidence suggests that cortical responses are largely preserved across sleep in primary sensory regions (Issa & Wang, 2008; Nir et al., 2012), calling into question the traditional proposal that the thalamus does not relay peripheral signals effectively to the cortex in sleep (McCormick & Bal, 1994; Steriade, 2003). However, it remains unclear to what extent such stimuli effectively drive the activity of high-order regions and what may be the fate in sleep of late sustained components (e.g., P300 in response to deviant stimuli) that are better linked to conscious perception (Dehaene & Changeux, 2011).

Finally, other intriguing directions include further explorations of the relation between sleep and memory consolidation (see above), investigations of sleep deprivation and the cognitive effects of sleepiness (Banks & Dinges, 2007), understanding dynamics of brain activity during state transitions such as those accompanying the descent to sleep (Bodizs et al., 2005; Magnin et al., 2010), and the phenomenon of sleep inertia (Marzano et al., 2011).

References

Anderer, P., Klosch, G., Gruber, G., Trenker, E., Pascual-Marqui, R. D., Zeitlhofer, J., et al. (2001). Low-resolution brain electromagnetic tomography revealed simultaneously active frontal and parietal sleep spindle sources in the human cortex. *Neuroscience, 103*, 581–592.

Andrillon, T., Nir, Y., Staba, R. J., Ferrarelli, F., Cirelli, C., Tononi, G., et al. (2011). Sleep spindles in humans: Insights from intracranial EEG and unit recordings. *Journal of Neuroscience, 31*, 17821–17834.

Axmacher, N., Elger, C. E., & Fell, J. (2008a). Ripples in the medial temporal lobe are relevant for human memory consolidation. *Brain, 131*, 1806–1817.

Axmacher, N., Haupt, S., Fernandez, G., Elger, C. E., & Fell, J. (2008b). The role of sleep in declarative memory consolidation—Direct evidence by intracranial EEG. *Cerebral Cortex, 18*, 500–507.

Balduzzi, D., Riedner, B. A., & Tononi, G. (2008). A BOLD window into brain waves. *Proceedings of the National Academy of Sciences of the United States of America, 105*, 15641–15642.

Banks, S., & Dinges, D. F. (2007). Behavioral and physiological consequences of sleep restriction. *Journal of Clinical Sleep Medicine, 3*, 519–528.

Battaglia, F. P., Sutherland, G. R., & McNaughton, B. L. (2004). Hippocampal sharp wave bursts coincide with neocortical "up-state" transitions. *Learning & Memory (Cold Spring Harbor, N.Y.)*, *11*, 697–704.

Bazil, C. W. (2003). Effects of antiepileptic drugs on sleep structure: Are all drugs equal? *CNS Drugs*, *17*, 719–728.

Bazil, C. W., & Walczak, T. S. (1997). Effects of sleep and sleep stage on epileptic and nonepileptic seizures. *Epilepsia*, *38*, 56–62.

Blouin, A. M., Fried, I., Wilson, C. L., Staba, R. J., Behnke, E. J., Lam, H. A., et al. (2013). Human hypocretin and melanin-concentrating hormone levels are linked to emotion and social interaction. *Nature Communications*, *4*, 1547.

Bodizs, R., Bekesy, M., Szucs, A., Barsi, P., & Halasz, P. (2002). Sleep-dependent hippocampal slow activity correlates with waking memory performance in humans. *Neurobiology of Learning and Memory*, *78*, 441–457.

Bodizs, R., Sverteczki, M., Lazar, A. S., & Halasz, P. (2005). Human parahippocampal activity: Non-REM and REM elements in wake–sleep transition. *Brain Research Bulletin*, *65*, 169–176.

Bragin, A., Benassi, S. K., Kheiri, F., & Engel, J., Jr. (2011). Further evidence that pathologic high-frequency oscillations are bursts of population spikes derived from recordings of identified cells in dentate gyrus. *Epilepsia*, *52*, 45–52.

Bragin, A., Engel, J., Jr., Wilson, C. L., Fried, I., & Buzsáki, G. (1999a). High-frequency oscillations in human brain. *Hippocampus*, *9*, 137–142.

Bragin, A., Engel, J., Jr., Wilson, C. L., Fried, I., & Mathern, G. W. (1999b). Hippocampal and entorhinal cortex high-frequency oscillations (100–500 Hz) in human epileptic brain and in kainic acid–treated rats with chronic seizures. *Epilepsia*, *40*, 127–137.

Bragin, A., Wilson, C. L., Staba, R. J., Reddick, M., Fried, I., & Engel, J., Jr. (2002). Interictal high-frequency oscillations (80–500 Hz) in the human epileptic brain: Entorhinal cortex. *Annals of Neurology*, *52*, 407–415.

Bushey, D., Tononi, G., & Cirelli, C. (2011). Sleep and synaptic homeostasis: Structural evidence in *Drosophila*. *Science*, *332*, 1576–1581.

Buzsáki, G. (1998). Memory consolidation during sleep: A neurophysiological perspective. *Journal of Sleep Research*, *7*(Suppl 1), 17–23.

Buzsáki, G. (2006). *Rhythms of the brain*. New York: Oxford University Press.

Buzsáki, G., Horvath, Z., Urioste, R., Hetke, J., & Wise, K. (1992). High-frequency network oscillation in the hippocampus. *Science*, *256*, 1025–1027.

Carrion, M. J., Nunes, M. L., Martinez, J. V., Portuguez, M. W., & da Costa, J. C. (2010). Evaluation of sleep quality in patients with refractory seizures who undergo epilepsy surgery. *Epilepsy & Behavior*, *17*, 120–123.

Cash, S. S., Halgren, E., Dehghani, N., Rossetti, A. O., Thesen, T., Wang, C., et al. (2009). The human K-complex represents an isolated cortical down-state. *Science*, *324*, 1084–1087.

Cavus, I., Kasoff, W. S., Cassaday, M. P., Jacob, R., Gueorguieva, R., Sherwin, R. S., et al. (2005). Extracellular metabolites in the cortex and hippocampus of epileptic patients. *Annals of Neurology*, *57*, 226–235.

Chauvette, S., Crochet, S., Volgushev, M., & Timofeev, I. (2011). Properties of slow oscillation during slow-wave sleep and anesthesia in cats. *Journal of Neuroscience*, *31*, 14998–15008.

Cirelli, C., & Tononi, G. (2008). Is sleep essential? *PLoS Biology*, *6*, e216.

Clemens, Z., Molle, M., Eross, L., Barsi, P., Halasz, P., & Born, J. (2007). Temporal coupling of parahippocampal ripples, sleep spindles and slow oscillations in humans. *Brain*, *130*, 2868–2878.

Colrain, I. M. (2005). The K-complex: A 7-decade history. *Sleep*, *28*, 255–273.

Compte, A., Reig, R., Descalzo, V. F., Harvey, M. A., Puccini, G. D., & Sanchez-Vives, M. V. (2008). Spontaneous high-frequency (10–80 Hz) oscillations during up states in the cerebral cortex in vitro. *Journal of Neuroscience*, *28*, 13828–13844.

Conradt, R., Hochban, W., Heitmann, J., Brandenburg, U., Cassel, W., Penzel, T., et al. (1998). Sleep fragmentation and daytime vigilance in patients with OSA treated by surgical maxillomandibular advancement compared to CPAP therapy. *Journal of Sleep Research*, *7*, 217–223.

Contreras, D., & Steriade, M. (1995). Cellular basis of EEG slow rhythms: A study of dynamic corticothalamic relationships. *Journal of Neuroscience*, *15*, 604–622.

Crespel, A., Coubes, P., & Baldy-Moulinier, M. (2000). Sleep influence on seizures and epilepsy effects on sleep in partial frontal and temporal lobe epilepsies. *Clinical Neurophysiology*, *111*(Suppl 2), S54–S59.

Crunelli, V., & Hughes, S. W. (2010). The slow (<1 Hz) rhythm of non-REM sleep: A dialogue between three cardinal oscillators. *Nature Neuroscience*, *13*, 9–17.

Csercsa, R., Dombovári, B., Fabó, D., Wittner, L., Erőss, L., Entz, L., et al. (2010). Laminar analysis of slow wave activity in humans. *Brain*, *133*, 2814–2829.

Dang-Vu, T. T., Schabus, M., Desseilles, M., Sterpenich, V., Bonjean, M., & Maquet, P. (2010). Functional neuroimaging insights into the physiology of human sleep. *Sleep*, *33*, 1589–1603.

De Gennaro, L., & Ferrara, M. (2003). Sleep spindles: An overview. *Sleep Medicine Reviews*, *7*, 423–440.

Dehaene, S., & Changeux, J. P. (2011). Experimental and theoretical approaches to conscious processing. *Neuron*, *70*, 200–227.

Destexhe, A., Hughes, S. W., Rudolph, M., & Crunelli, V. (2007). Are corticothalamic "up" states fragments of wakefulness? *Trends in Neurosciences*, *30*, 334–342.

de Weerd, A., de Haas, S., Otte, A., Trenite, D. K., van Erp, G., Cohen, A., et al. (2004). Subjective sleep disturbance in patients with partial epilepsy: A questionnaire-based study on prevalence and impact on quality of life. *Epilepsia*, *45*, 1397–1404.

Dickson, C. T., Biella, G., & de Curtis, M. (2003). Slow periodic events and their transition to gamma oscillations in the entorhinal cortex of the isolated guinea pig brain. *Journal of Neurophysiology*, *90*, 39–46.

Diekelmann, S., & Born, J. (2010). The memory function of sleep. *Nature Reviews. Neuroscience*, *11*, 114–126.

Dinner, D. S., & Lüders, H. O. (Eds.). (2001). *Epilepsy and sleep: Physiological and clinical relationships*. San Diego, CA: Academic Press.

Evarts, E. V. (1964). Temporal patterns of discharge of pyramidal tract neurons during sleep and waking in the monkey. *Journal of Neurophysiology*, *27*, 152–171.

Ferrarelli, F., Peterson, M. J., Sarasso, S., Riedner, B. A., Murphy, M. J., Benca, R. M., et al. (2010). Thalamic dysfunction in schizophrenia suggested by whole-night deficits in slow and fast spindles. *American Journal of Psychiatry*, *167*, 1339–1348.

Foldvary, N. (2002). Sleep and epilepsy. *Current Treatment Options in Neurology*, *4*, 129–135.

Foldvary-Schaefer, N., & Grigg-Damberger, M. (2009). Sleep and epilepsy. *Seminars in Neurology*, *29*, 419–428.

Fox, M. D., & Raichle, M. E. (2007). Spontaneous fluctuations in brain activity observed with functional magnetic resonance imaging. *Nature Reviews. Neuroscience*, *8*, 700–711.

Fried, I., Wilson, C., Maidment, N., Engel, J., Behnke, E., Fields, T., et al. (1999). Cerebral microdialysis combined with single neuron and electroencephalographic recording in neurosurgical patients. *Journal of Neurosurgery*, *91*, 697–705.

Fries, P., Reynolds, J. H., Rorie, A. E., & Desimone, R. (2001). Modulation of oscillatory neuronal synchronization by selective visual attention. *Science*, *291*, 1560–1563.

Gentilomo, A., Colicchio, G., Pola, P., Rossi, G. F., & Scerrati, M. (1975). Brain depth recording of interictal epileptic potentials during sleep in man. In P. Levin & W. P. Koella (Eds.), *Sleep* (pp. 444–446). Basel: Karger.

Gilestro, G. F., Tononi, G., & Cirelli, C. (2009). Widespread changes in synaptic markers as a function of sleep and wakefulness in *Drosophila*. *Science*, *324*, 109–112.

Girardeau, G., Benchenane, K., Wiener, S. I., Buzsáki, G., & Zugaro, M. B. (2009). Selective suppression of hippocampal ripples impairs spatial memory. *Nature Neuroscience*, *12*, 1222–1223.

Grenier, F., Timofeev, I., & Steriade, M. (2001). Focal synchronization of ripples (80–200 Hz) in neocortex and their neuronal correlates. *Journal of Neurophysiology*, *86*, 1884–1898.

Hagmann, P., Cammoun, L., Gigandet, X., Meuli, R., Honey, C. J., Wedeen, V. J., et al. (2008). Mapping the structural core of human cerebral cortex. *PLoS Biology*, *6*, e159.

Hahn, T. T., Sakmann, B., & Mehta, M. R. (2007). Differential responses of hippocampal subfields to cortical up–down states. *Proceedings of the National Academy of Sciences of the United States of America*, *104*, 5169–5174.

Haider, B., & McCormick, D. A. (2009). Rapid neocortical dynamics: Cellular and network mechanisms. *Neuron*, *62*, 171–189.

Hangya, B., Tihanyi, B. T., Entz, L., Fabo, D., Eross, L., Wittner, L., et al. (2011). Complex propagation patterns characterize human cortical activity during slow-wave sleep. *Journal of Neuroscience*, *31*, 8770–8779.

He, B. J., Snyder, A. Z., Zempel, J. M., Smyth, M. D., & Raichle, M. E. (2008). Electrophysiological correlates of the brain's intrinsic large-scale functional architecture. *Proceedings of the National Academy of Sciences of the United States of America, 105*, 16039–16044.

Herman, S. T., Walczak, T. S., & Bazil, C. W. (2001). Distribution of partial seizures during the sleep–wake cycle: Differences by seizure onset site. *Neurology, 56*, 1453–1459.

Hobson, J. A. (2005). Sleep is of the brain, by the brain and for the brain. *Nature, 437*, 1254–1256

Hobson, J. A., & Pace-Schott, E. F. (2002). The cognitive neuroscience of sleep: Neuronal systems, consciousness and learning. *Nature Reviews. Neuroscience, 3*, 679–693.

Iber, C., Ancoli-Israel, S., Chesson, A. L., & Quan, S. F. (2007). *The AASM manual for the scoring of sleep and associated events: Rules, terminology and technical specifications.* Westchester, IL: American Association of Sleep Medicine.

Isomura, Y., Sirota, A., Ozen, S., Montgomery, S., Mizuseki, K., Henze, D. A., et al. (2006). Integration and segregation of activity in entorhinal–hippocampal subregions by neocortical slow oscillations. *Neuron, 52*, 871–882.

Issa, E. B., & Wang, X. (2008). Sensory responses during sleep in primate primary and secondary auditory cortex. *Journal of Neuroscience, 28*, 14467–14480.

Ji, D., & Wilson, M. A. (2007). Coordinated memory replay in the visual cortex and hippocampus during sleep. *Nature Neuroscience, 10*, 100–107.

Jones, B. E. (2005). Basic mechanisms of sleep–wake states. In M. H. Kryger, T. Roth, & W. C. Dement (Eds.), *Principles and practice of sleep medicine* (4th ed., pp. 136–153). Philadelphia: Elsevier.

Killgore, W. D. (2010). Effects of sleep deprivation on cognition. *Progress in Brain Research. 185*, 105–129.

Laplante, F., Morin, Y., Quirion, R., & Vaucher, E. (2005). Acetylcholine release is elicited in the visual cortex, but not in the prefrontal cortex, by patterned visual stimulation: A dual in vivo microdialysis study with functional correlates in the rat brain. *Neuroscience, 132*, 501–510.

Le Van Quyen, M., Bragin, A., Staba, R., Crepon, B., Wilson, C. L., & Engel, J., Jr. (2008). Cell type–specific firing during ripple oscillations in the hippocampal formation of humans. *Journal of Neuroscience, 28*, 6104–6110.

Le Van Quyen, M., Staba, R., Bragin, A., Dickson, C., Valderrama, M., Fried, I., et al. (2010). Large-scale microelectrode recordings of high-frequency gamma oscillations in human cortex during sleep. *Journal of Neuroscience, 30*, 7770–7782.

Lieb, J. P., Joseph, J. P., Engel, J., Jr., Walker, J., & Crandall, P. H. (1980). Sleep state and seizure foci related to depth spike activity in patients with temporal lobe epilepsy. *Electroencephalography and Clinical Neurophysiology, 49*, 538–557.

Liu, Z. W., Faraguna, U., Cirelli, C., Tononi, G., & Gao, X. B. (2010). Direct evidence for wake-related increases and sleep-related decreases in synaptic strength in rodent cortex. *Journal of Neuroscience, 30*, 8671–8675.

Magnin, M., Rey, M., Bastuji, H., Guillemant, P., Mauguiere, F., & Garcia-Larrea, L. (2010). Thalamic deactivation at sleep onset precedes that of the cerebral cortex in humans. *Proceedings of the National Academy of Sciences of the United States of America, 107*, 3829–3833.

Malow, B. A., Fromes, G. A., & Aldrich, M. S. (1997). Usefulness of polysomnography in epilepsy patients. *Neurology, 48*, 1389–1394.

Malow, B. A., Levy, K., Maturen, K., & Bowes, R. (2000). Obstructive sleep apnea is common in medically refractory epilepsy patients. *Neurology, 55*, 1002–1007.

Malow, B. A., Weatherwax, K. J., Chervin, R. D., Hoban, T. F., Marzec, M. L., Martin, C., et al. (2003). Identification and treatment of obstructive sleep apnea in adults and children with epilepsy: A prospective pilot study. *Sleep Medicine, 4*, 509–515.

Marzano, C., Ferrara, M., Moroni, F., & De Gennaro, L. (2011). Electroencephalographic sleep inertia of the awakening brain. *Neuroscience, 176*, 308–317.

Massimini, M., Ferrarelli, F., Huber, R., Esser, S. K., Singh, H., & Tononi, G. (2005). Breakdown of cortical effective connectivity during sleep. *Science, 309*, 2228–2232.

McCormick, D. A., & Bal, T. (1994). Sensory gating mechanisms of the thalamus. *Current Opinion in Neurobiology, 4*, 550–556.

Mena-Segovia, J., Sims, H. M., Magill, P. J., & Bolam, J. P. (2008). Cholinergic brainstem neurons modulate cortical gamma activity during slow oscillations. *Journal of Physiology, 586*, 2947–2960.

Mendez, M., & Radtke, R. A. (2001). Interactions between sleep and epilepsy. *Journal of Clinical Neurophysiology, 18*, 106–127.

Mohajerani, M. H., McVea, D. A., Fingas, M., & Murphy, T. H. (2010). Mirrored bilateral slow-wave cortical activity within local circuits revealed by fast bihemispheric voltage-sensitive dye imaging in anesthetized and awake mice. *Journal of Neuroscience, 30*, 3745–3751.

Montgomery, S. M., & Buzsáki, G. (2007). Gamma oscillations dynamically couple hippocampal CA3 and CA1 regions during memory task performance. *Proceedings of the National Academy of Sciences of the United States of America, 104*, 14495–14500.

Mormann, F., Kornblith, S., Quiroga, R. Q., Kraskov, A., Cerf, M., Fried, I., et al. (2008). Latency and selectivity of single neurons indicate hierarchical processing in the human medial temporal lobe. *Journal of Neuroscience, 28*, 8865–8872.

Moroni, F., Nobili, L., Curcio, G., De Carli, F., Tempesta, D., Marzano, C., et al. (2008). Procedural learning and sleep hippocampal low frequencies in humans. *NeuroImage, 42*, 911–918.

Murphy, M., Riedner, B. A., Huber, R., Massimini, M., Ferrarelli, F., & Tononi, G. (2009). Source modeling sleep slow waves. *Proceedings of the National Academy of Sciences of the United States of America, 106*, 1608–1613.

Nadasdy, Z., Hirase, H., Czurko, A., Csicsvari, J., & Buzsáki, G. (1999). Replay and time compression of recurring spike sequences in the hippocampus. *Journal of Neuroscience, 19*, 9497–9507.

Nir, Y., Fisch, L., Mukamel, R., Gelbard-Sagiv, H., Arieli, A., Fried, I., et al. (2007). Coupling between neuronal firing rate, gamma LFP, and BOLD fMRI is related to interneuronal correlations. *Current Biology, 17*, 1275–1285.

Nir, Y., Mukamel, R., Dinstein, I., Privman, E., Harel, M., Fisch, L., et al. (2008). Interhemispheric correlations of slow spontaneous neuronal fluctuations revealed in human sensory cortex. *Nature Neuroscience, 11*, 1100–1108.

Nir, Y., Staba, R. J., Andrillon, T., Vyazovskiy, V. V., Cirelli, C., Fried, I., et al. (2011). Regional slow waves and spindles in human sleep. *Neuron, 70*, 153–169.

Nir, Y., & Tononi, G. (2010). Dreaming and the brain: From phenomenology to neurophysiology. *Trends in Cognitive Sciences, 14*, 88–100.

Nir, Y., Vyazovskiy, V., Cirelli, C., Banks, M., & Tononi, G. (2012). Auditory responses and stimulus-specific adaptation are largely preserved across NREM and REM sleep in rat primary auditory cortex. *Annual meeting of the Society for Neuroscience.* New Orleans.

Nobili, L., Ferrara, M., Moroni, F., De Gennaro, L., Russo, G. L., Campus, C., et al. (2011). Dissociated wake-like and sleep-like electro-cortical activity during sleep. *NeuroImage, 58*, 612–619.

Nofzinger, E. A. (2005). Neuroimaging and sleep medicine. *Sleep Medicine Reviews, 9*, 157–172.

Oliveira, A. J., Zamagni, M., Dolso, P., Bassetti, M. A., & Gigli, G. L. (2000). Respiratory disorders during sleep in patients with epilepsy: Effect of ventilatory therapy on EEG interictal epileptiform discharges. *Clinical Neurophysiology, 111*(Suppl 2), S141–S145.

Ovadia-Caro, S., Nir, Y., Soddu, A., Ramot, M., Hesselmann, G., Vanhaudenhuyse, A., et al. (2012). Reduction in interhemispheric connectivity in disorders of consciousness. *PLoS ONE, 7*, e37238.

Peled, R., & Lavie, P. (1986). Paroxysmal awakenings from sleep associated with excessive daytime somnolence: A form of nocturnal epilepsy. *Neurology, 36*, 95–98.

Peter-Derex, L., Comte, J. C., Mauguiere, F., & Salin, P. A. (2012). Density and frequency caudo–rostral gradients of sleep spindles recorded in the human cortex. *Sleep, 35*, 69–79.

Piperidou, C., Karlovasitou, A., Triantafyllou, N., Terzoudi, A., Constantinidis, T., Vadikolias, K., et al. (2008). Influence of sleep disturbance on quality of life of patients with epilepsy. *Seizure, 17*, 588–594.

Quian Quiroga, R., Nadasdy, Z., & Ben-Shaul, Y. (2004). Unsupervised spike sorting with wavelets and superparamagnetic clustering. *Neural Computation, 16*, 1661–1687.

Riedner, B. A., Hulse, B. K., Murphy, M. J., Ferrarelli, F., & Tononi, G. (2011). Temporal dynamics of cortical sources underlying spontaneous and peripherally evoked slow waves. *Progress in Brain Research, 193*, 201–218.

Roure, N., Gomez, S., Mediano, O., Duran, J., Pena Mde, L., Capote, F., et al. (2008). Daytime sleepiness and polysomnography in obstructive sleep apnea patients. *Sleep Medicine, 9*, 727–731.

Sammaritano, M., Gigli, G. L., & Gotman, J. (1991). Interictal spiking during wakefulness and sleep and the localization of foci in temporal lobe epilepsy. *Neurology, 41*, 290–297.

Sanchez-Vives, M. V., & McCormick, D. A. (2000). Cellular and network mechanisms of rhythmic recurrent activity in neocortex. *Nature Neuroscience, 3*, 1027–1034.

Schabus, M., Dang-Vu, T. T., Albouy, G., Balteau, E., Boly, M., Carrier, J., et al. (2007). Hemodynamic cerebral correlates of sleep spindles during human non–rapid eye movement sleep. *Proceedings of the National Academy of Sciences of the United States of America, 104*, 13164–13169.

Sforza, E., & Krieger, J. (1992). Daytime sleepiness after long-term continuous positive airway pressure (CPAP) treatment in obstructive sleep apnea syndrome. *Journal of the Neurological Sciences, 110*, 21–26.

Singer, W., & Gray, C. M. (1995). Visual feature integration and the temporal correlation hypothesis. *Annual Review of Neuroscience, 18*, 555–586.

Sirota, A., & Buzsáki, G. (2005). Interaction between neocortical and hippocampal networks via slow oscillations. *Thalamus & Related Systems, 3*, 245–259.

Sirota, A., Csicsvari, J., Buhl, D., & Buzsáki, G. (2003). Communication between neocortex and hippocampus during sleep in rodents. *Proceedings of the National Academy of Sciences of the United States of America, 100*, 2065–2069.

Staba, R. J. (2012). Normal and pathologic high-frequency oscillations. In J. L. Noebels, M. Avoli, M. A. Rogawski, R. W. Olsen, & A. V. Delgado-Escueta (Eds.), *Jasper's basic mechanisms of the epilepsies* (pp. 202–212). Oxford: Oxford University Press.

Staba, R. J., Wilson, C. L., Bragin, A., Fried, I., & Engel, J., Jr. (2002a). Quantitative analysis of high-frequency oscillations (80–500 Hz) recorded in human epileptic hippocampus and entorhinal cortex. *Journal of Neurophysiology, 88*, 1743–1752.

Staba, R. J., Wilson, C. L., Bragin, A., Fried, I., & Engel, J., Jr. (2002b). Sleep states differentiate single neuron activity recorded from human epileptic hippocampus, entorhinal cortex, and subiculum. *Journal of Neuroscience, 22*, 5694–5704.

Staba, R. J., Wilson, C. L., Bragin, A., Jhung, D., Fried, I., & Engel, J., Jr. (2004). High-frequency oscillations recorded in human medial temporal lobe during sleep. *Annals of Neurology, 56*, 108–115.

Staba, R. J., Wilson, C. L., Fried, I., & Engel, J., Jr. (2002c). Single neuron burst firing in the human hippocampus during sleep. *Hippocampus, 12*, 724–734.

Stefanello, S., Marin-Leon, L., Fernandes, P. T., Li, L. M., & Botega, N. J. (2010). Psychiatric comorbidity and suicidal behavior in epilepsy: A community-based case–control study. *Epilepsia, 51*, 1120–1125.

Steriade, M. (2003). *Neuronal substrates of sleep and epilepsy.* Cambridge: Cambridge University Press.

Steriade, M. (2006). Grouping of brain rhythms in corticothalamic systems. *Neuroscience, 137*, 1087–1106.

Steriade, M., Amzica, F., & Contreras, D. (1996). Synchronization of fast (30–40 Hz) spontaneous cortical rhythms during brain activation. *Journal of Neuroscience, 16*, 392–417.

Steriade, M., Contreras, D., & Amzica, F. (1994). Synchronized sleep oscillations and their paroxysmal developments. *Trends in Neurosciences, 17*, 199–208.

Steriade, M., Contreras, D., Curro Dossi, R., & Nunez, A. (1993). The slow (< 1 Hz) oscillation in reticular thalamic and thalamocortical neurons: Scenario of sleep rhythm generation in interacting thalamic and neocortical networks. *Journal of Neuroscience, 13*, 3284–3299.

Steriade, M., & Deschenes, M. (1973). Cortical interneurons during sleep and waking in freely moving primates. *Brain Research, 50*, 192–199.

Stickgold, R., & Walker, M. P. (2007). Sleep-dependent memory consolidation and reconsolidation. *Sleep Medicine, 8*, 331–343.

Takeuchi, D., Hirabayashi, T., Tamura, K., & Miyashita, Y. (2011). Reversal of interlaminar signal between sensory and memory processing in monkey temporal cortex. *Science, 331*, 1443–1447.

Terzaghi, M., Sartori, I., Tassi, L., Didato, G., Rustioni, V., LoRusso, G., et al. (2009). Evidence of dissociated arousal states during NREM parasomnia from an intracerebral neurophysiological study. *Sleep, 32*, 409–412.

Tononi, G., & Cirelli, C. (2006). Sleep function and synaptic homeostasis. *Sleep Medicine Reviews, 10*, 49–62.

Valderrama, M., Crepon, B., Botella-Soler, V., Martinerie, J., Hasboun, D., Alvarado-Rojas, C., et al. (2012). Human gamma oscillations during slow wave sleep. *PLoS ONE, 7*, e33477.

van Mill, J. G., Hoogendijk, W. J., Vogelzangs, N., van Dyck, R., & Penninx, B. W. (2010). Insomnia and sleep duration in a large cohort of patients with major depressive disorder and anxiety disorders. *Journal of Clinical Psychiatry, 71*, 239–246.

Vaughn, B. V., D'Cruz, O. F., Beach, R., & Messenheimer, J. A. (1996). Improvement of epileptic seizure control with treatment of obstructive sleep apnoea. *Seizure*, *5*, 73–78.

Vignatelli, L., Bisulli, F., Naldi, I., Ferioli, S., Pittau, F., Provini, F., et al. (2006). Excessive daytime sleepiness and subjective sleep quality in patients with nocturnal frontal lobe epilepsy: A case–control study. *Epilepsia*, *47*(Suppl 5), 73–77.

Vincent, J. L., Patel, G. H., Fox, M. D., Snyder, A. Z., Baker, J. T., Van Essen, D. C., et al. (2007). Intrinsic functional architecture in the anaesthetized monkey brain. *Nature*, *447*, 83–86.

Volgushev, M., Chauvette, S., Mukovski, M., & Timofeev, I. (2006). Precise long-range synchronization of activity and silence in neocortical neurons during slow-wave oscillations [corrected]. *Journal of Neuroscience*, *26*, 5665–5672.

Vyazovskiy, V. V., Faraguna, U., Cirelli, C., & Tononi, G. (2009a). Triggering slow waves during NREM sleep in the rat by intracortical electrical stimulation: Effects of sleep/wake history and background activity. *Journal of Neurophysiology*, *101*, 1921–1931.

Vyazovskiy, V. V., Olcese, U., Hanlon, E. C., Nir, Y., Cirelli, C., & Tononi, G. (2011). Local sleep in awake rats. *Nature*, *472*, 443–447.

Vyazovskiy, V. V., Olcese, U., Lazimy, Y. M., Faraguna, U., Esser, S. K., Williams, J. C., et al. (2009b). Cortical firing and sleep homeostasis. *Neuron*, *63*, 865–878.

Wagner, T., Axmacher, N., Lehnertz, K., Elger, C. E., & Fell, J. (2010). Sleep-dependent directional coupling between human neocortex and hippocampus. *Cortex*, *46*, 256–263.

Wilson, M. A., & McNaughton, B. L. (1994). Reactivation of hippocampal ensemble memories during sleep. *Science*, *265*, 676–679.

Worrell, G. A., Parish, L., Cranstoun, S. D., Jonas, R., Baltuch, G., & Litt, B. (2004). High-frequency oscillations and seizure generation in neocortical epilepsy. *Brain*, *127*, 1496–1506.

Zeitzer, J. M., Maidment, N. T., Behnke, E. J., Ackerson, L. C., Fried, I., Engel, J., Jr., et al. (2002). Ultradian sleep-cycle variation of serotonin in the human lateral ventricle. *Neurology*, *59*, 1272–1274.

Zeitzer, J. M., Morales-Villagran, A., Maidment, N. T., Behnke, E. J., Ackerson, L. C., Lopez-Rodriguez, F., et al. (2006). Extracellular adenosine in the human brain during sleep and sleep deprivation: An in vivo microdialysis study. *Sleep*, *29*, 455–461.

11 Studying Thoughts and Deliberations Using Single Neuron Recordings in Humans

Moran Cerf, Hagar Gelbard-Sagiv, and Itzhak Fried

Much of cognitive neuroscience, investigated in humans and animals, is based on careful evaluation of neural responses to sensory stimulation or external cues. This is the ultimate product of the stimulus–response paradigm, which has been the cornerstone of neuroscience research. Yet, a considerable extent of mental life is spent in thought processes that may not arise as a direct response to external input. Such processes, which include imagery, free recall, and internal deliberations, are the very processes which are difficult to study in animal models and require the cooperation of human subjects who can express their thoughts and wishes and make them accessible to an external observer.

A basic tenet of modern cognitive neuroscience is that thoughts correspond to patterns of neuronal activity. Indeed, the idea that thoughts may be decoded from such patterns has posed a considerable challenge. It is not that we necessarily want to become "mind readers," but the prospect of such decoding presents theoretical and clinical benefits, not the least of which is the construction of neuroprosthetic devices that will enable patients that are unable to speak or move to communicate.

While numerous methods of neuroscience research, such as electroencephalography (EEG), functional magnetic resonance imaging (fMRI), positron emission tomography, magnetoencephalography, and others, offer some ways of inferring thoughts from patterns of neural activity (Kamitani & Tong, 2005; Haynes & Rees, 2006; Norman et al., 2006; Kay et al., 2008; Nishimoto et al., 2011), these methods are limited by the nature of the signals measured, including their spatial and temporal resolution of readouts from the brain. It seems that the decoding of discrete thoughts and concepts requires access to a different kind of signal—spikes of single neurons. Such recordings are feasible in humans under special clinical circumstances, thus enabling investigation of signals at the neuronal level in human subjects who are able to report their imagery, recollections, and other mental experiences directly and who are often able to elicit such processes at will when instructed to in an experiment.

In a series of studies conducted recently using single neuron recordings in humans, scientists were able to decode internal mental processes, in relatively high specificity, and in some cases prior to their actual manifestation by language or motor behavior. These pioneering studies form the scientific basis for further investigation into the possibility of decoding freely occurring thoughts and deliberations.

Looking inside Our Mind's Eye Using Imagery

Trying to direct another person to our home, or trying to recall a detail from a picture we have seen before, we often close our eyes and reconstruct an image in our mind's eye. This employs an internal process by which we visualize content that is not present on our retina, using imagery. The single neuron correlate of this internal process was studied by Kreiman and colleagues (Kreiman et al., 2000) by looking at medial temporal lobe (MTL) neurons in patients with intracranial depth electrodes.

In a preliminary stage subjects were shown a variety of images from a selected set of categories: objects, familiar faces, animal images, and so forth for 1 s each in the course of a short viewing period. Following this viewing period, the activity of a small number of brain cells was analyzed, and a set of neurons were identified, each showing a clear response to images from a particular category (e.g., a neuron showing a response to animals and another showing a response to objects). These images were then used for a subsequent experiment where subjects were asked to imagine the images seen previously. Subjects were asked to first view two images from two categories, shown repeatedly in a random order, for 1 s each, with an accompanying simultaneous tone. The tone was either a high-pitched or a low-pitched one. The subjects quickly learned to associate the high-pitched tone, for example, with the image of the animal, and the low-pitched tone with that of the object. These sets of trials were labeled "vision trials." Next, "imagery trials" followed. In these trials subjects were asked to listen to the high/low-pitched tones while closing their eyes and imagine the picture corresponding to the particular tone. The tone therefore triggered the internal imagery in the subjects' minds. The tones alternated in a sequence allowing the subject to perform alternate imagery of the two different stimuli. Sixteen percent of the recorded MTL neurons showed selectivity during vision—that is, they reliably responded with an elevated firing rate each time a specific image was presented and remained silent while other images were presented. Eight percent of the recorded neurons showed selectivity during imagery. Remarkably, almost 90% of the imagery neurons showed the same selectivity also during vision.

Comparisons of the neuronal response characteristics between vision and imagery in this experiment show various differences in latency and duration, as expected, taking longer for images to imprint via our imagination than through vision (see figure 11.1). Yet, two striking findings emerged. First was the presence of a set of neurons responding selectively to one image, which then showed the *same* selective responses during imagery. The second was that the firing rate for imagery was 85% that of vision. This suggests an only slightly diminished response in our mind to imagery with no retinal input relative to normal vision. Ultimately, these results enable an external observer an access to the internal imagery process. One could effectively parse the stimulus imagined by the subject simply by the readout of the activity of a few corresponding

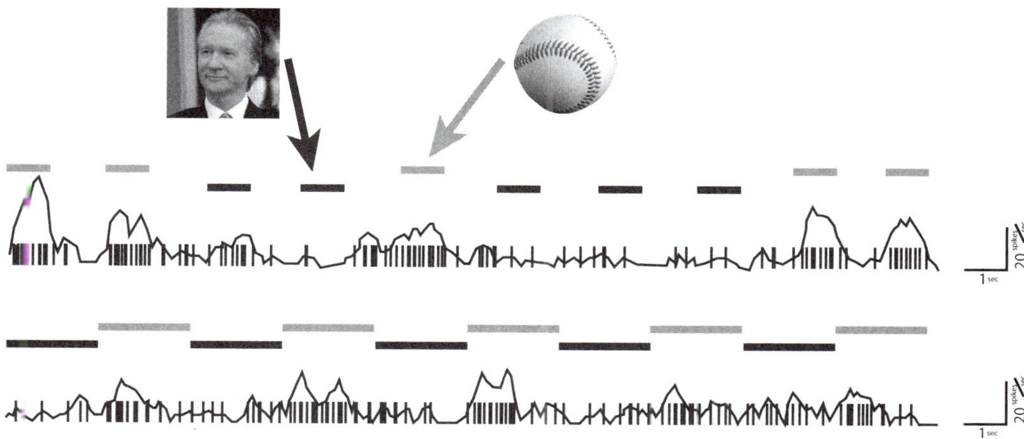

Figure 11.1
Individual responses of a single neuron during vision. Two images were shown separately for 1 s each, with five repetitions per image, indicated by horizontal black and gray bars. After each picture, subjects pressed a button to indicate whether or not the picture was a human face. The continuous thin black line shows the spike density function. After ten visual presentations, subjects closed their eyes and imagined one picture on hearing a high tone and the other picture on hearing a low tone. Tones were alternated every 3 s. Data shown are from the same neuron during visual imagery. This neuron showed a similar pattern of firing during visual presentation and visual imagery, with increased firing rate in response to the image of the ball, but not that of a human face (adapted from Rees et al., 2002).

individual neurons. Indeed, on average, in seven out of ten trials, an observer could correctly guess the stimulus category in imagery or visual trials simply by counting the number of spikes of the neurons selective to these categories. Presumably, given a larger set of neurons from the population that encodes a particular category, the prediction rate could be much higher.

Thoughts and Recollections

Recollection is another form of thought that is often triggered in our brain. Retrieving information from the past requires an internal search process that ultimately results in the recall of a certain event or a piece of knowledge. Several studies have tried to explore how similar the neuronal activity patterns during recall are to the ones evoked during the actual original event (Nyberg et al., 2000; Wheeler et al., 2000; Polyn et al., 2005). However, given the limited resolution of imaging methods, and the difficulty in accessing mental content in animals, this question has remained open.

Gelbard-Sagiv and colleagues (Gelbard-Sagiv et al., 2008) directly investigated the underlying neuronal correlates of recollection by having human subjects freely recall previously shown content while recording single neurons from their brains.

Thirteen subjects first viewed a series of short (5–10 s long) video clips. Each clip was repeated six times in random order and depicted an episode featuring famous people, characters or animals engaged in activity, or landmarks which were often familiar to the subjects. These clips included well-known scenes, for example, the famous scene of Marilyn Monroe's dress floating around her above a subway grate, or President Bush's speech after Saddam Hussein's capture. About half of the recorded neurons responded reliably and selectively during one or more of the clips. Ten percent of all responsive neurons showed sustained responses, that is, an elevated firing rate that was maintained throughout the clip, and also after the clip removal, when no content was presented. In some cases this prolonged firing was attenuated only by the onset of the following clip. Following the viewing session, and after performing a distracting task, subjects were asked to freely recall as many of the previously seen clips as possible and to verbally report immediately whenever a specific clip came to mind. Thus, the retrieval of information was triggered internally in the patient's mind and not by an external cue. Similar to the viewing session, this free-recall session was designed as a simple and natural task in order to mimic real-life recollections. Indeed, subjects performed well and, on average, recalled over 80% of the clips presented. Interestingly, cells maintained the same selectivity in the viewing and in the free-recall session. That is, a cell that was firing during each of the six presentations of the Marilyn Monroe clip fired again just before the subject reported recalling that particular clip. This result extends our understanding of the abstract nature of MTL representations: Not only do neurons in these regions respond to very different external world representations of a given episode but they also respond when retrieving the content internally. In the imagery experiment we learned that a simple tone associated with a given concept could trigger the same neuronal response during imagery whereas this study demonstrated that no external cue was required—simply "thinking of" or recalling a concept is associated with the same neuronal firing patterns evoked by sensory exposure to it (see figure 11.2 for an example). Remarkably, neurons started firing in the free-recall session 500–1500 ms *before* the onset of verbal report of the recall of their preferred clip. Although we cannot completely rule out the possibility that this was merely an issue of reaction time or that the subjects were waiting for a second or two before reporting their thoughts, the early neuronal activity observed here could represent the early stages of the formation of a thought or recollection, the stages at which we may not yet be fully aware of the content of the thought. Thus, it is possible that by monitoring the neuronal firing online, we could have told what the subjects were recalling or thinking before they were able to do so.

Decoding a Thought from Neural Activity

Ultimately, our experiences, our knowledge, and our thoughts are represented as patterns of neuronal activity in our brain. Thus, decoding a thought is analogous to figuring out the corresponding pattern of activity that underlies a particular thought.

In our recordings we can only sample a very small subset of the entire population of neurons. However, given the sparse and thus nearly binary nature of the MTL neuronal responses (i.e., a strong bursting response for a specific concept and low baseline firing during other times), decoding a particular concept from this small subset is nevertheless feasible.

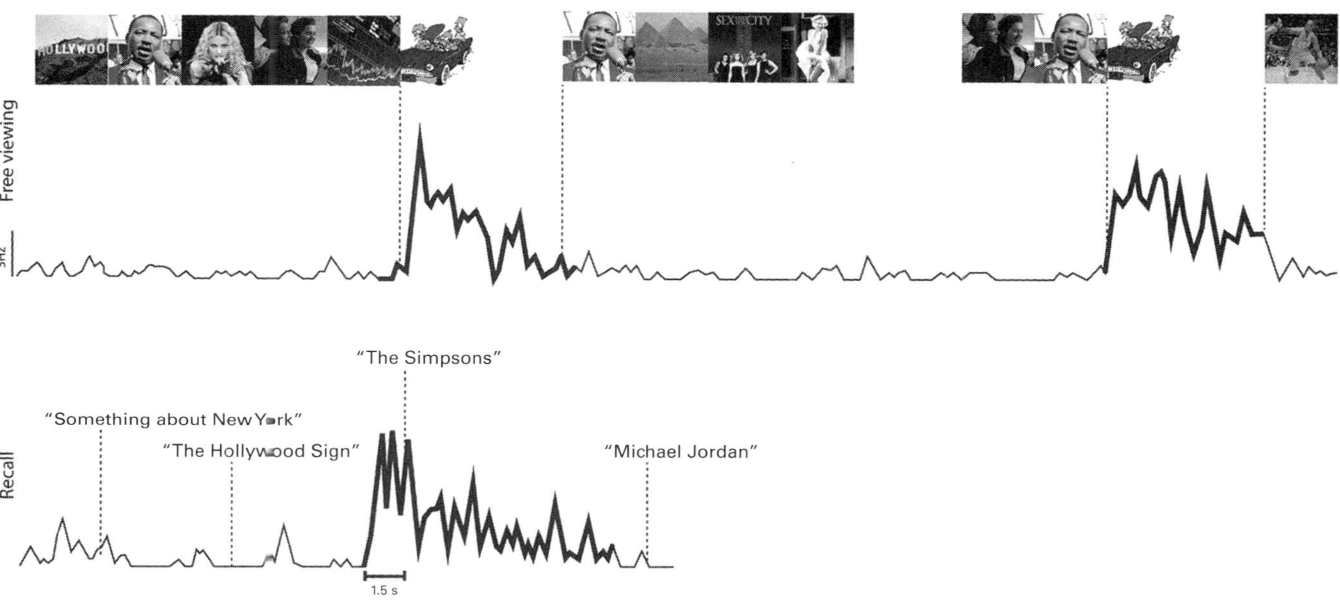

Figure 11.2
Subject freely views video clips (top panel). Each 5-s clip is depicted here by one frame; 5-s blank periods appear in between some clips. Note that the response is sustained throughout the "Simpsons" clip and also throughout the blank period even though the clip is no longer on the screen. Subject freely recalls the clips (bottom panel). The increased firing rate response starts over 1 s prior to the verbal expression of the recollection of the "Simpsons" clip.

In a recent study by Quian Quiroga and colleagues (2007), neuronal data collected while 11 patients were watching a set of images, presented for 1 s and repeated six times each, was used to quantitatively test our ability to decode the stimulus presented from the patterns of neuronal firing. Baseline firing was defined as the median firing rate from 1000 to 300 ms before stimulus onset. Similarly, the response to a specific image was defined as the median firing rate during the 300- to 1000-ms period following image onset. Neurons that elevated their firing rate to more than 5 standard deviations above the baseline firing rate for a given image were considered responsive to that image.

Using a leave-one-out decoder (using the five other image presentations to "train" the decoder), Quian Quiroga and colleagues could decode significantly above chance which picture was presented to the subject on a trial-by-trial basis. Decoding performance increased linearly with the number of units used as an input for the decoder. In accordance with the invariant nature of the MTL representations, decoding was not limited to repetition of a single photo exemplar but could also be generalized to new pictures of a given person or landmark.

Figure 11.3 depicts the decoding performance in one of the sessions in the form of a confusion matrix. Based on the activity of 19 responsive neurons, the investigators could decode which

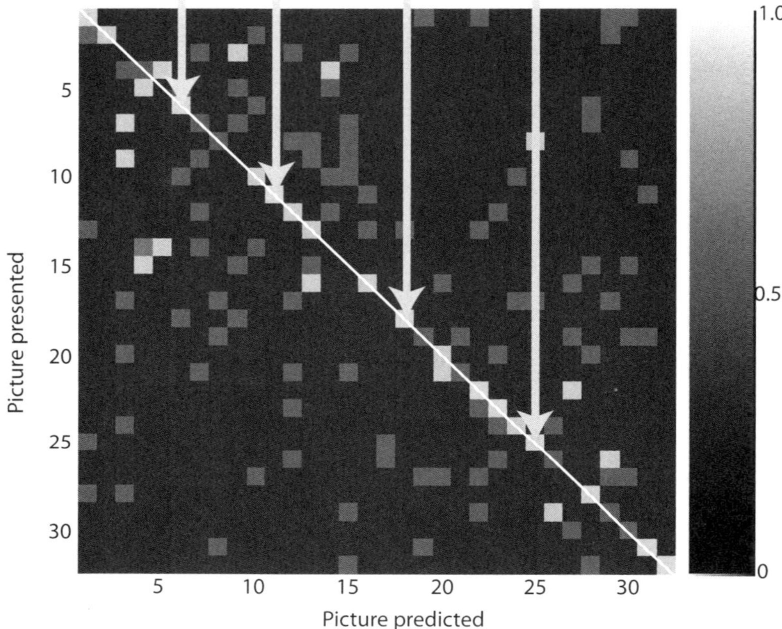

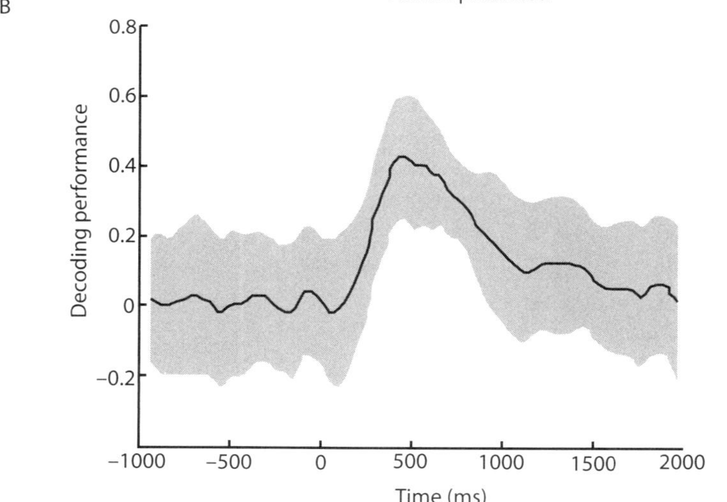

Figure 11.3
(A) Example decoding of 32 pictures that generated a significant response in any of 19 responsive units in one session. Average decoding performance for all 32 pictures was 35.4%, with a chance level of 1/32 = 3.1%. (B) Time profile of the normalized decoding performance using a (half-overlapping) moving window of 100 ms, averaged across all sessions. Band shows the 95% confidence intervals. Performance peaks between 400 and 500 ms. Values were smoothed using a 3-point moving average (adapted from Quiroga et al., 2007).

of the 32 different pictures was presented onscreen in any given moment with a probability of 35.4% (compared to 3.1% chance level).

Overall, the decoding performance first became significant 300 ms after stimulus onset with a 20% decoding performance; it then peaked between 400 and 500 ms with a performance of around 40%, followed by a gradual decay (see figure 11.3). The time window between 300 and 600 ms contained the most informative spikes, and the decoding performance in longer time windows was not significantly higher. Additionally, it was shown that using isolated putative single units resulted in superior performance relative to using multiunits (without spike sorting).

This study demonstrated the feasibility of decoding a visual stimulus presented to the subject from the activity of a few responsive neurons in the human MTL, ultimately showing that the performance relies mainly on an average of 4–5 spikes occurring between 300 and 600 ms after image onset.

Reading Thoughts in Real Time

Following the work by Quian Quiroga and colleagues, Cerf and colleagues (Cerf et al., 2010) tested the ability to use population decoding to identify a subject's thought in real time. Using a similar method, patients were first presented with a set of four images that elicited neuronal responses in an earlier screening session. Images were presented for 1 s, 12 times each, in random order. Based on the neuronal firing during image presentation, one target neuron was selected for each image such that this neuron responded strongly to that image and was indifferent to the other three images (i.e., remained at baseline firing when the other three images were presented). The firing rates of these four target neurons in the 300- to 700-ms interval after stimulus onset were represented in a four-number vector. This yielded a single data point in a four-dimensional (4D) space for each image presentation. The 12 data points for each image (12 repetitions of that image) create a cluster in this 4D space. Thus, when the responses are strong and reliable, the data points for the four images create four clearly separated exclusive clusters.

Based on these four predetermined clusters, Cerf tested the ability to decode the internal representation of the patient in real time. The subject was asked to voluntarily think of one of the four preselected images. The activity of the four target neurons during 100-ms time windows was projected to the 4D *concept space* set before, and using a distance metric from the center of each concept centroid, each momentary thought was assigned to the cluster to which it belonged. The algorithm could also assign a "none" result, meaning that the new data point could not be identified as part of any of the clusters. Based on the result of this decoder, an image corresponding to a concept cluster gradually appeared ("faded in") onscreen. The contrast of the image was increased as neuronal activity was repeatedly identified as part of the same cluster in 100-ms intervals. This served as a real-time feedback for the subject, who could control the visibility of an image by focusing his or her thoughts on a concept. This reflects a gradual transition from a thought or imagery limited to the subject's brain alone to a visual manifestation of a subject's thought.

Next, Cerf and colleagues targeted the ability to override visual input and control the competition between multiple thoughts/concepts in the brain. Similar to the "fading-in" experiment described above, subjects were instructed to make a target image more visible; however, this time a superposition of two images (the target image and a competing image) was presented on the screen. Each of the images was the preferred image of a different neuron in the set of four neurons used by the decoder. The decoder result was used to update the relative contrast of each of the two images superimposed on the screen. The starting point was 50% contrast for both images. The subject then had to not only increase the firing of the neuron (or neuronal network) coding the target image but also suppress the activity of the neuron coding the competing image while the competing image was visible onscreen. Subjects' performance on this complex task was surprisingly high, with over 70% success by the eighth trial (see figure 11.4, plate 11).

Activating External Devices Using Thoughts

The ability to decode one's thought with such a high level of accuracy has implications beyond the pure understanding of the ways by which concepts are encoded in the MTL. One obvious implication is the precise control of an external device using these signals.

To test for the feasibility of this, Cerf and colleagues conducted another experiment in which subjects selected the image that they were able to fade in on the screen in the previous experiment most easily and were playing a computer game where the activity of the corresponding neuron was used to control a spaceship (see figure 11.5). The spaceship was moving while obstacles were appearing on the screen—heading toward the spaceship. The subject's task was to try to avoid these obstacles by controlling the height at which the spaceship flies. He or she could do this by thinking of/imagining that particular image and thus increase the firing rate of the corresponding neuron and make the spaceship fly higher or, by "suppressing their thinking of that image," decrease the firing rate of the corresponding neuron and make the spaceship fly lower. Subjects were able to control the activity of an individual neuron (or a neuronal network of which this neuron was a part) to the level of roughly 70% performance on that task. When debriefed, subjects mentioned that they mainly tried to imagine the target image.

Similar to the control of the spaceship in the example above, decoding of signals from the brain with high precision can be used to control devices such as a robotic arm (Velliste et al.,

Figure 11.4 (plate 11)
(a) Recording from intracranial electrodes, neurons are identified that respond to a specific concept: in this instance, a cell responsive to the image of Marilyn Monroe. This cell increases its firing rate to the image or thought of Monroe. (b) This cell is then pitted against a different cell which was found to represent the Eiffel Tower. The two images are superimposed, and the subject is asked to bring the image of Monroe to maximum visibility. The visibility of the image is controlled by real-time decoding of the activity of each neuron relative to the other neuron and its own baseline. In this example, we show a case where the subject initially begins to fail the experiment—the firing of the Eiffel Tower neuron increases and the visibility of the tower increases, creating negative feedback. However the subject is able to exert control and, by concentrating on the internal thought of Monroe, is able to override this sensory input and increase the firing rate of the Monroe neuron and decrease that of the Eiffel Tower neuron, bringing the image of Monroe to visibility. The scans show the location of the respective electrodes within the brain.

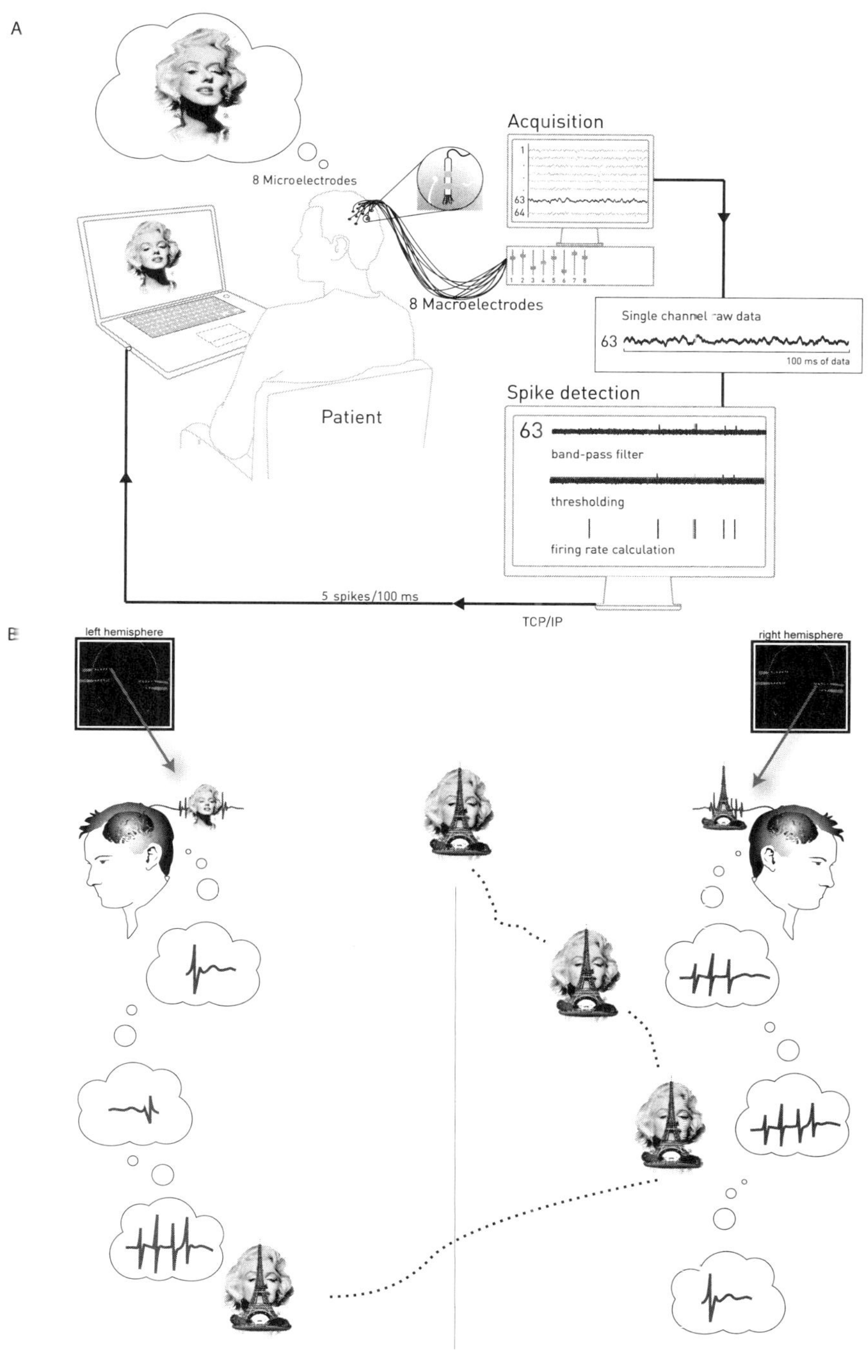

A
Acquisition
1
63
64
8 Microelectrodes
8 Macroelectrodes
Patient
Single channel raw data
63
100 ms of data
Spike detection
63
band-pass filter
thresholding
firing rate calculation
5 spikes/100 ms
TCP/IP
B
left hemisphere
right hemisphere

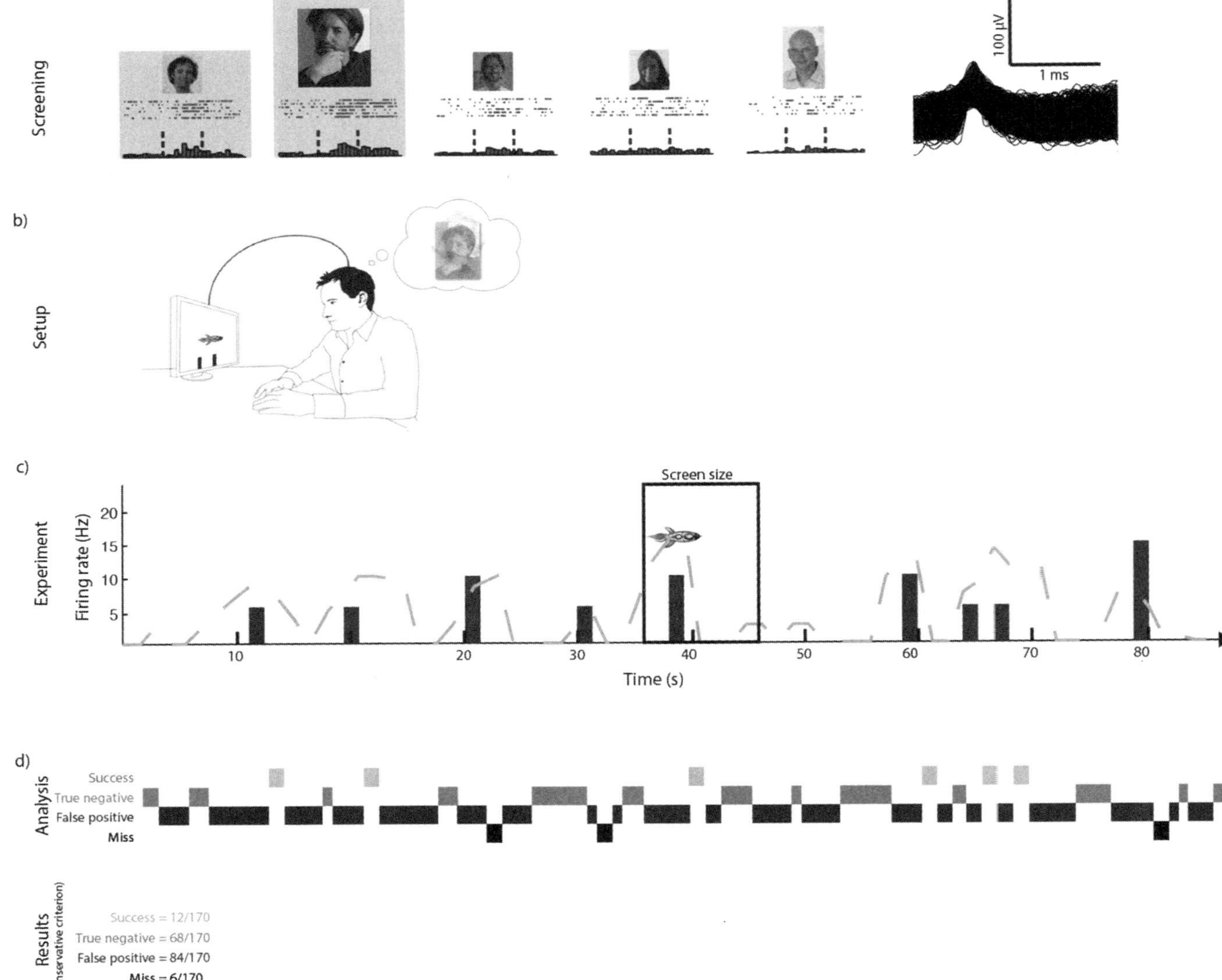

Figure 11.5
(a) Poststimulus time histogram of a single unit in the left hippocampus that responded selectively to a picture of the first author and various other people the subject was familiar with. Each image was repeated 16 times during a screening session prior to the brain–machine interface experiment. To the right are the spikes shapes. The two images with strongest response are highlighted, with that of the first author, which was used for further experiments, magnified. (b) The experimental setup. The subject was watching the game on the screen and controlled the flight of a spaceship by thinking of the first author. (c) A depiction of 90 s of the game. The black bars are the obstacles the subject was asked to avoid by elevating the firing rate of the hippocampal unit and, accordingly, the spaceship's height. The rectangle marked "Screen size" shows the part of the obstacle course the subject sees at a given window (7 s ahead and 3 s before). (d) In order to evaluate the performance, the subject was scored by the amount of hits/misses and true/false positives. A hit was considered when the spaceship hovered above an obstacle, a miss was colliding with one, a true positive required being on the ground when no obstacle is there, and a false positive involved hovering above ground when not needed. The scores match the route shown in (b). Below are the total scores based on the route shown.

2008; Hochberg et al., 2012). In this case, the decoding of neural activity serves as a brain–machine interface (BMI).

The advantage of using MTL neuronal activity for purposes of BMIs is in the accuracy of the decoding, which can outperform other methods such as external noninvasive methods to record field potentials (Lebedev & Nicolelis, 2006). Spikes from a neuron that indicates a specific concept are highly informative and thus have a high signal-to-noise ratio. Additionally, the latencies of these neurons are quite short and allow for rapid control. As the free-recall study demonstrated, neurons start increasing their firing rate well before subjects reported their recollection. Given these properties, BMI that utilizes these recording can be not only precise but also rapid and efficient.

However, BMIs activated by MTL neurons may not be practical. Apart from technical and safety concerns with long-term implants, these neurons are not ideal for the task since their representation may be dynamic and of considerable plasticity. Thus, building a BMI based on the ability of the patient to think of a specific concept will prove futile once the neuron modifies its preference. BMI research might be more efficient in regions such as motor brain areas, which may have more stable representation based on population and not sparse coding. That said, the results from single unit studies in the MTL and frontal regions can provide significantly improved methods for testing BMIs with human subjects and for the improvement of the techniques used to read out the neural signal for future BMIs based on nonmotor regions (Naci et al., 2013).

Reading Thoughts and Urges during Their Formation

Recording the activity of individual nerve cells and identifying the pattern of activity underlying a particular percept can result not only in the real-time decoding of present thoughts but also in the prediction of coming thoughts. As the results of the free-recall study (Gelbard-Sagiv et al., 2008) suggest, it may be possible to trace neuronal activity that occurs *before* the actual experienced or reported recall (see figure 11.2).

In a study inspired by the classic Libet experiment from the 1980s (Libet et al., 1983), Fried and colleagues were able to isolate a set of neurons whose firing reliably predicted volition (Fried et al., 2011). Fried asked patients to look at a one-dial clock, which was revolving rapidly (roughly 2.2 s for a revolution). Subjects were asked to let the clock complete one full revolution before choosing, at their own will, to press a key (notated "P" for "Press"). Then subjects were asked to point to the time where they "felt the *urge*" to perform the movement. Thus, subjects were asked to separate the moment of the actual action from the perceived moment of the decision or will to act (notated "W" for "Will"). Subjects timed this moment at around 200 ms before the actual button press.

Fried and colleagues were able to identify individual neurons, mostly in the supplementary and presupplementary motor areas and in anterior cingulate cortex, which became active up to 1000 ms *prior to* this estimated urge moment. The results from this study not only show that one can predict an action before it was consciously recognized but also show the mechanisms of such volition. The main finding suggests an increase in the number of neurons involved in the encoding of the movement or the intention to move as the urge moment approaches. Of the 760 medial frontal neurons recorded in 12 subjects, 55 changed their activity 1000 ms before the urge moment, and the number increased to 128 neurons in the 400-ms window before this moment. A linear decoder based on the population activity could detect significant changes in activity 500 ms before the urge moment on 90% of the trials and 1000 ms before on 70% of the trials (see figure 11.6).

The ability to decode an action before its realization on *a single trial* basis, as seen here, is unique. For comparison, most works in fMRI or EEG, including that of Libet, required the averaging of dozens of trials in order to achieve similar level of accuracy in predicting the will. These results not only shed light on one of the most intriguing questions in neuroscience and philosophy, pertaining to the notion of free will and its underlying neuronal mechanisms, but also demonstrate access to information that is not necessarily available to the subject.

The Future: Accessing Conscious and Unconscious Thoughts

What makes one pattern of neuronal activity a part of our own subjective experience while other patterns remain "locked" in our brain and never reach our conscious awareness? Single spike resolution, together with the ease of getting reports on subjective experience, makes neurophysiological recording in humans a unique and singular setup to explore this type of question.

Being able to both ask subjects about their conscious thoughts, as well as independently read their formation, directly from the brain, has fundamental implications for understanding the processes that lead to the formation of conscious percepts and thoughts. Additionally, this direct readout of brain activity can access those percepts and thoughts which are processed by the brain but remain inaccessible and unconscious. Here, methods like binocular rivalry, flash suppression, change blindness, or attentional manipulations that separate conscious perception from external stimulations can prove extremely helpful.

Similarly, another interesting aspect of the process of human recollection that one can investigate is its associative nature. Thoughts and recollections are triggered many times by related or associated stimuli, which are later used to generate predictions (Bar, 2007). The network of associations is being built throughout our life based on our experiences and is unique to each person. Working with humans who can easily provide us with an aperture to their own network of associations enables us to see this process unfold in terms of neuronal activity. As we are able to see an individual concept being reflected upon, and occasionally even identify an association related to it, we can move on to study the formation of an association or the learning of a new concept. After we create an association between concepts in the patient's mind, we can study the speed of the interaction, the timing and latency of the neuronal responses to each associated concept, and the similarity between multiple subjects in this associations buildup.

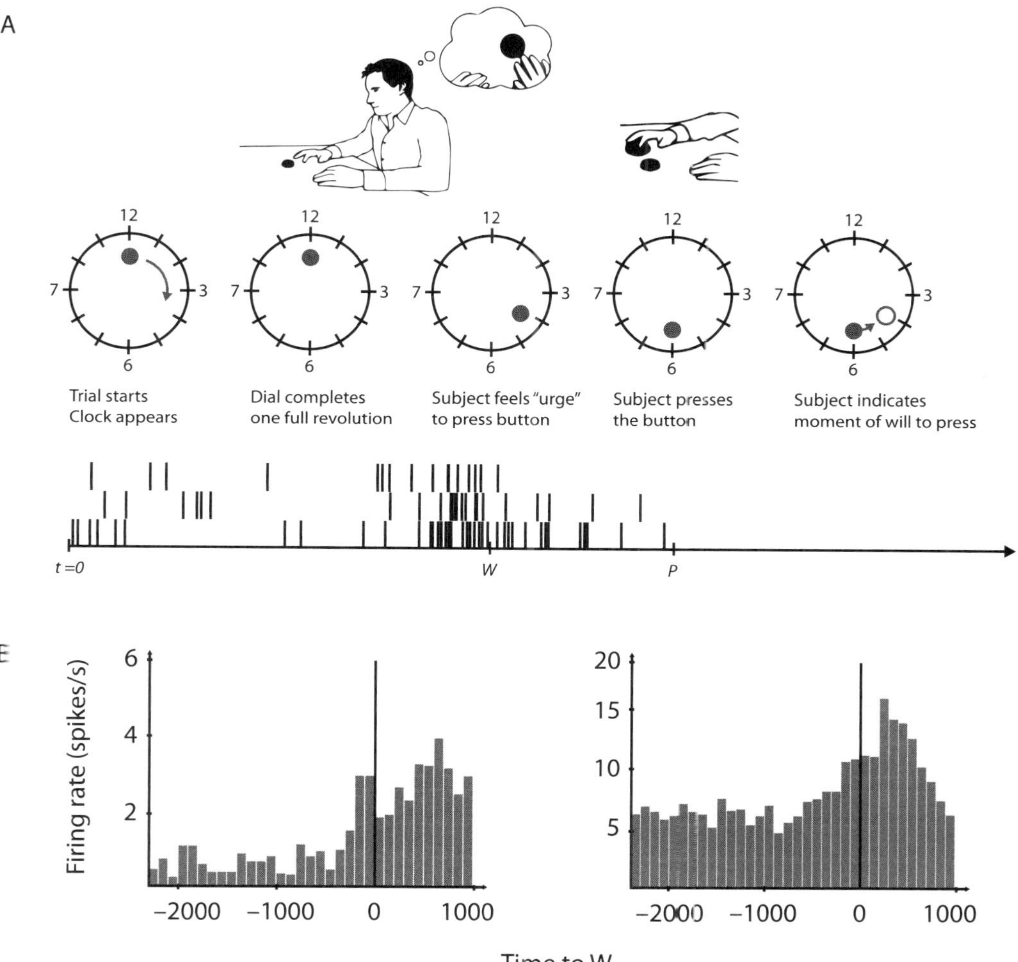

Figure 11.6
(A) Schematic diagram depicting the experimental paradigm (Libet et al., 1983). Subjects were shown an analog clock and were asked to press a key with their right index finger, at will, anytime after one rotation of the clock. After the key press event (P), the clock dial stopped and subjects were asked to indicate the time of onset of the "urge/decision" to press the key (W). (B) Histograms showing the responses of a neuron (left) in the left anterior cingulate displaying a significant response after W (rank sum test, $p < 10^{-6}$) and a neuron (right) in the left presupplementary motor area with response onset prior to W (rank sum test, $p < 10^{-3}$). All plots are aligned to W (time = 0).

Finally, the ability to read thoughts from patterns of neuronal activity may open a window to the reading of the thoughts that run through our mind when we are asleep, our dreams. Ultimately, when we study dreams, we are interested in accessing the internal states that generate the dreams—having access to the single neuron level during sleep can serve as a unique platform to conduct such studies. One way to study this will involve an earlier "learning phase" during wakefulness, in which an algorithm learns to decode percepts and imagery of a large number of concepts. Then the algorithm will use this database to try to pick up these same activity patterns during sleep. If the database is extensive enough, we can reconstruct the narrative of the dream from the recorded activity and compare it with patients' reports when they wake up. A recent study (Shibata et al., 2013) has demonstrated such mechanisms for dream decoding using fMRI in humans. The spatiotemporal nature of fMRI limited this study to the ability to decode high-level categories of dream, such as the general and gist content of the dream. The group was able to tell whether the subject was seeing a person, a landmark, or objects in a given moment but was not able to increase the resolution to the level of the specificity of the person, the landmark, or the object. They surmise that the main limitation was merely the resolution of the imaging device. If this is true, then with the high resolution provided by single neuron recordings we might be able to provide a more detailed interpretation of dream content and begin to understand how are dreams similar and how they differ from real-life experiences.

In summary, we are now in an opportune and exciting time where understanding of the way thoughts are constructed in our brain, alongside the high spatiotemporal resolution of single neuron recordings in humans, allows further exploration of some of the most intriguing questions of neuroscience pertaining to those very faculties which make us human.

References

Bar, M. (2007). The proactive brain: Using analogies and associations to generate predictions. *Trends in Cognitive Sciences, 11*, 280–289.

Cerf, M., Thiruvengadam, N., Mormann, F., Kraskov, A., Quian Quiroga, R., Koch, C., et al. (2010). On-line, voluntary control of human temporal lobe neurons. *Nature, 467*, 1104–1108.

Freud, S. (1954). The interpretation of dreams. In J. Strachey (Ed. & Trans.), *The standard edition of the complete psychological works of Sigmund Freud* (one vol. reprint of Vols. 4–5). London: Hogarth Press. (Original work published 1900)

Fried, I., Mukamel, R., & Kreiman, G. (2011). Internally generated preactivation of single neurons in human medial frontal cortex predicts volition. *Neuron, 69*, 548–562.

Gelbard-Sagiv, H., Mukamel, R., Harel, M., Malach, R., & Fried, I. (2008). Internally generated reactivation of single neurons in human hippocampus during free recall. *Science, 322*, 96–101.

Haynes, J.-D., & Rees, G. (2006). Decoding mental states from brain activity in humans. *Nature Reviews. Neuroscience, 7*, 523–534.

Hochberg, L. R., Bacher, D., Jarosiewicz, B., Masse, N. Y., Simeral, J. D., Vogel, J., et al. (2012). Reach and grasp by people with tetraplegia using a neurally controlled robotic arm. *Nature, 485*, 372–375.

Kamitani, Y., & Tong, F. (2005). Decoding the visual and subjective contents of the human brain. *Nature Neuroscience, 8*, 679–685.

Kay, K. N., Naselaris, T., Prenger, R. J., & Gallant, J. L. (2008). Identifying natural images from human brain activity. *Nature, 452*, 352–355.

Kreiman, G., Koch, C., & Fried, I. (2000). Imagery neurons in the human brain. *Nature, 408*, 357–361.

Lebedev, M. A., & Nicolelis, M. A. L. (2006). Brain–machine interfaces: Past, present and future. *Trends in Neurosciences, 29*, 536–546.

Libet, B., Gleason, C. A., Wright, E. W., & Pearl, D. K. (1983). Time of conscious intention to act in relation to onset of cerebral activity (readiness-potential): The unconscious initiation of a freely voluntary act. *Brain, 106*, 623–642.

Naci, L., Cusack, R., Jia, V. Z., & Owen, A. M. (2013). The brain's silent messenger: Using selective attention to decode human thought for brain-based communication. *Journal of Neuroscience, 33*, 9385–9393.

Nishimoto, S., Vu, A. T., Naselaris, T., Benjamini, Y., Yu, B., & Gallant, J. L. (2011). Reconstructing visual experiences from brain activity evoked by natural movies. *Current Biology, 21*, 1641–1646.

Norman, K. A., Polyn, S. M., Detre, G. J., & Haxby, J. V. (2006). Beyond mind-reading: Multi-voxel pattern analysis of fMRI data. *Trends in Cognitive Sciences, 10*, 424–430.

Nyberg, L., Habib, R., McIntosh, A. R., & Tulving, E. (2000). Reactivation of encoding-related brain activity during memory retrieval. *Proceedings of the National Academy of Sciences of the United States of America, 97*, 11120–11124.

Polyn, S. M., Natu, V. S., Cohen, J. D., & Norman, K. A. (2005). Category-specific cortical activity precedes retrieval during memory search. *Science, 310*, 1963–1966.

Quiroga, R. Q., Reddy, L., Koch, C., & Fried, I. (2007). Decoding visual inputs from multiple neurons in the human temporal lobe. *Journal of Neurophysiology, 98*, 1997–2007.

Rees, G., Kreiman, G., & Koch, C. (2002). Neural correlates of consciousness in humans. *Nature Reviews. Neuroscience, 3*, 261–270.

Shibata, K., Watanabe, T., Sasaki, Y., & Kawato, M. (2011). Perceptual learning incepted by decoded fMRI neurofeedback without stimulus presentation. *Science, 334*, 1413–1415.

Velliste, M., Perel, S., Spalding, M. C., Whitford, A. S., & Schwartz, A. B. (2008). Cortical control of a prosthetic arm for self-feeding. *Nature, 453*, 1098–1101.

Wheeler, M. E., Petersen, S. E., & Buckner, R. L. (2000). Memory's echo: Vivid remembering reactivates sensory-specific cortex. *Proceedings of the National Academy of Sciences of the United States of America, 97*, 11125–11129.

12 Human Single Neuron Reward Processing in the Basal Ganglia and Anterior Cingulate

Shaun R. Patel, Demetrio Sierra-Mercado, Clarissa Martinez-Rubio, and Emad N. Eskandar

Animals are fundamentally governed by rewards. It is a central component of our behavior and the pursuit of which represents the integration of a broad range of cognitive capacities. These functions take place in a distributed reward network and include regions such as the frontal cortex, cingulate cortex, basal ganglia, and midbrain dopaminergic systems (see figure 12.1, plate 12). Although better understanding the role of these brain regions has been the focus of many recent studies, the vast majority of studies are conducted in animal models that ultimately serve only as a proxy for human reward processing. Imaging modalities such as functional magnetic resonance imaging (fMRI) allow us to correlate whole brain activity with behavior in human subjects, however at the cost of poor temporal and spatial resolution and only indirect measures of neural activity. Within the last decade, a few pioneering studies have explored reward processes in the human brain at the single neuronal level (table 12.1), providing the first evidence on how this integral cognitive function is represented at the level of the individual neuron. In this chapter, we will highlight these studies in the context of previous animal physiology and human imaging studies. Finally, we will conclude with a section on how future studies can help to expand our understanding of human reward processing.

Basal Ganglia

The basal ganglia were classically thought of as a set of nuclei related to motor function given pathophysiological evidence from movement disorders and original anatomical evidence pointing to dense connectivity with motor cortical areas (Nauta & Mehler, 1966). However, recently, this view has changed substantially as growing evidence has unraveled a wide range of cognitive components to the basal ganglia network based on both anatomical and physiological data. Functions such as reward processing, motivation, and learning have since been attributed to basal ganglia circuits.

These discoveries originated from anatomical evidence in the early 1970s demonstrating the connectivity of the ventral striatum and the ventral pallidum with cortical areas (Heimer & Kalil, 1978). It was later discovered that many cortico–basal ganglia–thalamocortical loops (basal ganglia loops) existed, and projected to various cortical regions, including sensorimotor,

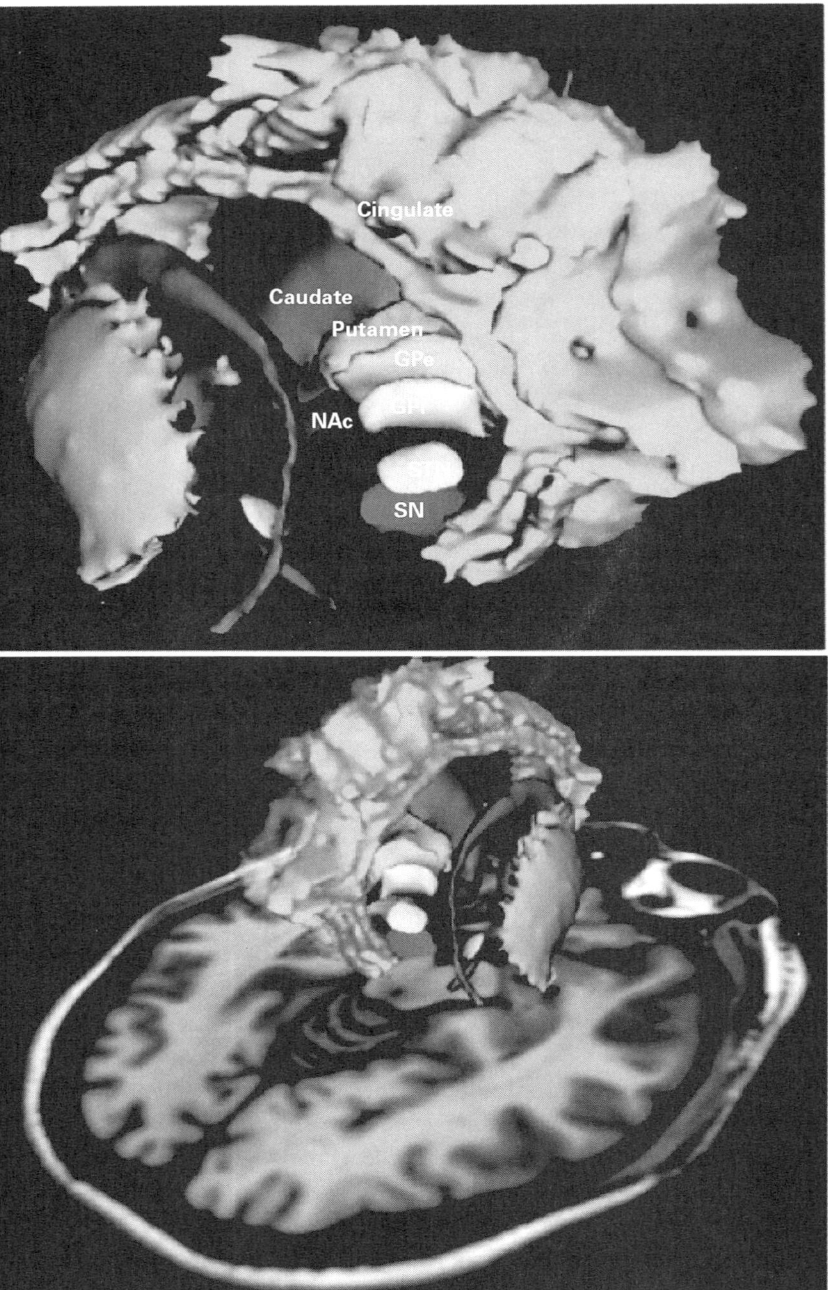

Figure 12.1 (plate 12)
Anatomical relationship of corticostriatal structures implicated in reward processing, reinforcement learning, and decision making. The cingulate cortex is thought to play an important role in cognitive functions, including reward anticipation, error detection, and decision making. The nucleus accumbens (NAc) is frequently studied for its role in addiction, but it also participates in learning and motivation through its connections with dopaminergic neurons, limbic areas, and prefrontal cortex. The substantia nigra (SN) sends dopaminergic signals to the striatum and cortex and encodes prediction error signals that drive learning. STN, subthalamic nucleus; GPe, globus pallidus externa; GPi, globus pallidus interna. Images provided by Kirk Finnis, Ph.D., and Medtronic, Inc.

Table 12.1
Table of current human single neuron studies on reward processing

Task	Brain region	Comments	Results	Reference
• War Task	Nucleus accumbens	Examined binding stimulus value to action	• Prediction of future financial decisions on a trial-by-trial basis and under conditions of uncertainty Encoding of positive/negative prediction error signals	Patel et al. (2012)
• Rewarding Stimuli Task	Nucleus accumbens	Examined processing of reward information	• Activity encoded reward anticipation Alpha oscillations are sensitive to positive feedback Beta oscillations exhibit higher power during unrewarded trials	Lega et al. (2011)
• Mental Arithmetic Counting Stroop Emotional Stroop Interference	Caudal anterior cingulate	Examined impact of emotional valance and cACC responsiveness to complex attention tasks	• cACC neurons may act as salience detectors when faced with conflict and difficult or emotional stimuli	Davis et al. (2005)
• Mental Arithmetic Counting Stroop Word Generation	Anterior cingulate	Examined responses to painful stimulus applied to skin	• First report of single neurons in humans with modified firing rates by cognitive tasks that require attention Painful stimuli did not alter the firing rate	Davis et al. (2000)
• Two-choice selection	Dorsal anterior cingulate	Examined the role of the dACC in reward-based decision making	• dACC activity increased when instructed to alter movement direction Highest response occurred when instruction also indicated an affiliated reduction in reward Level of activity predicted whether a correct choice was going to be made dACC ablation increased errors when required to change behavior based on reward reduction	Williams & Eskander (2004)
• Multi-Source Interference Task	Dorsal anterior cingulate	Examined dACC function by using functional imaging and single unit recordings in human subjects undergoing surgical cingulotomy	• dACC activity is modulated by previous trial activity producing a behavioral adaptation that accelerates reactions to cues of similar difficulty and delays reactions to cues of different difficulty These effects are abolished by ablation of dACC	Sheth et al. (2012)
• Probabilistic Learning	Substantia nigra	Examined reward properties of dopaminergic neurons using virtual financial reward	SN neurons' responses to positive and negative feedback were mainly driven by unexpected outcomes suggesting a role in reward-based learning modulated by the discrepancy between the expected and the realized outcome.	Zaghloul et al. (2009)

cACC, caudal anterior cingulate cortex; dACC, dorsal anterior cingulate cortex; SN, substantia nigra.

associative, and limbic regions (Parent & Hazrati, 1995). The notion of parallel and segregated basal ganglia loops became the dominant thinking of basal ganglia architecture. This organizational scheme was optimally suited for integration of discrete associative and limbic streams of information while connecting with motor cortical and motor output regions within the basal ganglia—requirements for adaptive behavior.

Midbrain Dopamine and Reward Signaling

Midbrain dopaminergic neurons play a critical role in moderating reward processing and adaptive behavior (Schultz, 2002). The two primary dopamine centers innervating the cerebrum are the ventral tegmental area (VTA) and the substantia nigra pars compacta (SNpc).

The VTA lies medial to the substantia nigra (SN) and ventral to the red nucleus in the midbrain. The VTA primarily receives glutamatergic input from the prefrontal cortex (PFC), lateral hypothalamus, and superior colliculus. It also receives gamma-aminobutyric acid (GABA) input from medium-spiny neurons in the ventral pallidum and nucleus accumbens (NAcc). Approximately 60% of the cell bodies are dopaminergic, of which most project densely to the ventromedial striatum, specifically the NAcc core and shell (Swanson, 1982); GABA-ergic (Carr & Sesack, 2000) and glutamatergic containing neurons are also present (Lavin et al., 2005).

The SNpc is a mesencephalic structure that lies immediately dorsal to the cerebral peduncles. Roughly, 90% of cell bodies are dopaminergic and project primarily to the dorsal striatum and PFC (Haber & Knutson, 2010). Interestingly, the SNpc receives very little reciprocal innervation from the dorsal striatum but instead receives dense projections from the ventral striatum.

The role of midbrain dopaminergic neurons in reward processing has been extensively studied (Schultz et al., 1997; Schultz, 2002). Most notably, midbrain dopaminergic neurons demonstrate a rapid phasic increase in firing rate to unexpected rewards. In other words, when there is a difference in expected and actual outcome—a prediction error signal—midbrain dopaminergic neurons rapidly fire at the onset of the unexpected reward. Furthermore, if a stimulus is consistently presented prior to the delivery of the reward, the dopamine-mediated change in firing rate transitions from the receipt of the reward to the onset of the predictive cue. This feature is thought to drive reward-based learning and adaptive behavior.

It is important that animals be able to respond to rewards, but it is in addition advantageous for them to be able to make predictions about future rewards. More recently, neuroimaging studies have found that the amygdala, orbital-frontal cortex, and ventral striatum are implicated in reward prediction (Knutson et al., 2001; Gottfried et al., 2002; O'Doherty et al., 2002). In its simplest form, this type of prediction can occur through classical conditioning. In such cases, a previously neutral stimulus which is paired with a rewarding outcome is assigned a predictive value in future instances. It is thought that phasic dopaminergic activity is the neural substrate of this type of learning (Schultz, 1998).

Human neuroimaging studies also support this notion (McClure et al., 2003; O'Doherty et al., 2003). Using classical conditioning tasks, two studies were able to identify prediction error profiles using event-related fMRI in downstream targets from midbrain dopaminergic systems. These signals were seen in the ventral striatum and orbital-frontal cortex. Although these are indirect measures of neural activity, further evidence using positron emission tomography (PET) has found dopamine release in the striatum during reward prediction errors (Zald et al., 2004).

There are many difficulties in using fMRI to measure changes in midbrain dopaminergic systems because of its proximity to tissue boundaries, endogenous motion due to its nearness to the carotid artery, and partial voluming from its small size (Haber & Knutson, 2010). Despite this, researchers have reported increases in midbrain activity to anticipation of pleasant tastes (D'Ardenne et al., 2008; O'Doherty et al., 2002) and monetary rewards (Knutson et al., 2005). Interestingly, in these studies, midbrain activity did not attenuate to expected rewards that were ommitted, in contrast with findings from nonhuman primate (NHP) work (Schultz et al., 1997). Although this may reflect an insensitivity of the imaging modality, it remains a critical point in understanding reward processing in the human brain.

The first study to explore single neuronal correlates of dopaminergic neurons in the human brain was that of Zaghloul and colleagues (Zaghloul et al., 2009). They were able to record activity from individual neurons in the human SN in patients undergoing deep brain stimulation surgery for Parkinson's disease. In this study, they asked patients to play a computerized task while surgeons performed microelectrode recordings to physiologically map the target brain region prior to implantation of the deep brain stimulation electrode (Zaghloul et al., 2009). It is worth noting that microelectrode recordings are a routine part of deep brain stimulation surgery and the only addition to the experimental design is the incorporation of the computerized behavioral task.

During surgery, a computer monitor and input device (button box) is mounted to the operating bed. After the neurosurgery team makes two small craniotomies, microelectrode wires are then inserted and lowered along the planned trajectory before the final stimulating electrode is permanently implanted. This procedure is performed to physiologically map the gray/white matter boundaries to confirm the electrode position and to functionally isolate the motor compartment of the subthalamic nucleus (STN). This is a standard component of deep brain stimulation surgery and helps to ensure optimal placement of the stimulating electrode. Although the SN is not currently an approved deep brain stimulation surgery target, it rests immediately inferior to the STN and can be reached when mapping the ventral boundary of the STN during the microelectrode recording portion of the surgery.

The task required the subjects to draw a card from one of two decks of cards. The subjects were informed that one deck probabilistically carried a higher chance of resulting in a financial reward. The researchers recorded 67 neurons from the SN and limited their analysis to 15 neurons that met stringent criteria (firing rate, waveform morphology, and feedback responsive) of dopamingeric cells. By comparing differences in observed and expected feedback, the researchers reported that 14 of the 15 dopaminergic neurons in the SN encoded both positive and negative prediction error signals, consistent with previous NHP and neuroimaging studies. This was the first concrete evidence of prediction error signals in human dopaminergic neurons.

An unavoidable confound when interpreting any human single neuronal study is the underlying disease pathology of the study population. In this case, Zaghloul et al. recorded from dopaminergic cells in the SN of subjects with severe Parkinson's disease—atrophy of which is itself the pathophysiological cause of Parkinson's disease. Although it is difficult or, in some cases, impossible to control for these effects, a discussion is necessitated to address these issues. The researchers in this case performed a very strict analysis of the physiological properties of the neurons they recorded: Through this analysis they reduced the 67 recorded neurons to 15 dopaminergic cells that were used for this analysis. The rationale is that they were able to study the remaining active dopaminergic cells and their output. The caveat is that this activity may be attenuated due to the degeneration of the cell population and therefore the healthy network.

Ventral Striatum and Reward Processing

The link between the ventral striatum and reward processing has been known since the 1950s when Olds and Milner demonstrated that rats would continually self-stimulate their reward circuit through an implanted electrode, even at the cost of maintaining basic homeostatic needs like eating and sleeping (Olds & Milner, 1954). Ever since, the ventral striatum has become a focal point in studies of reinforcement learning, reward processing, and drug addiction.

In the late 1970s, Lennart Heimer coined the term "ventral striatum." In humans and primates, the ventral striatum is the composite of the NAcc and the amorphous boundary blending caudally into the caudate and putamen in the dorsal striatum. Interestingly, neither cytoarchitectural or histochemical features demarcate a border between the ventral and dorsal striatum (Haber & Knutson, 2010); instead, anatomical connectivity provides the most consistent feature to identify the two regions.

The ventral striatum, like the dorsal striatum, receives three major inputs: a massive topographic glutamatergic input from the cerebral cortex, a large glutamatergic input from the thalamus, and a smaller dopaminergic input from midbrain dopaminergic centers (Haber & Knutson, 2010). The NAcc is classically thought of as the "motor–limbic interface" because of its rich limbic afferents from the hippocampus, amygdala, and frontal cortices and its efferent projections to motor output centers in the basal ganglia (Mogenson et al., 1980). Given this anatomical connectivity, the NAcc is in a unique position to integrate a wide range of information to modulate behavior. Furthermore, the NAcc receives widespread input from midbrain dopaminergic centers, thus placing it under the influence of reward centers. This combination makes the NAcc well suited for processing reward information, guiding goal-directed behaviors, and playing a role in drug addiction.

The role of dopamine in the striatum continues to be poorly understood. A central tenet of dopamine's influence on striatal activity (both dorsal and ventral) is that it exerts its influence via two mechanisms: tonically and phasically active neurons. Evidence suggests that tonically active neurons, thought to be cholinergic interneurons that make up about 1% of striatal neurons, may perform a gating function allowing the passage of information (Tremblay et al., 1989; Pessiglione et al., 2005) whereas phasically active neurons, thought to be the medium spiny neurons that make up about 95% of striatal neurons, are more relevant to strengthening functional circuits during instrumental conditioning and procedural learning (Graybiel, 2008; Jog et al., 1999).

In contrast, neurons in the NAcc also send reciprocal projections back to midbrain dopamine neurons. These projections are GABA-ergic and inhibitory in nature. Although the NAcc is not the only region that exerts control over midbrain dopamine centers, it provides the largest afferent control. The exact function of these projections is not well understood; however, computational approaches have led to very interesting hypotheses. One particular example is the actor–critic model of reinforcement learning (van der Meer & Redish, 2011; Montague et al., 2004). In this model, an actor signal maintains stimulus–response associations based on historically rewarding experiences while the critic signal serves as a "teaching signal" to update the actor to newly rewarding stimulus–response associations. In this sense, the striatonigral projection to midbrain dopamine centers could regulate responses to rewards, serving as a control mechanism for learning. An integral component of this system is that neurons in the NAcc/ventral striatum are sensitive to reward probability.

To interrogate neuronal activity during reward processing, Schultz and colleagues trained monkeys to perform a computer-guided task. In this task, one of three different conditioned stimuli were presented on any given trial, to which they learned the corresponding associations through brute-force training: (1) make a reaching action and receive reward, (2) refrain from performing any action and receive reward, and (3) make a reaching action followed by an auditory tone and no reward. The animals performed this task while researchers recorded single neuronal activity from the ventral striatum. Upon examination of the phasically active subgroup of neurons, the researchers found that the responses were very heterogeneous to task-related epochs. Among the findings, however, they noticed the largest and consistent categories of responses were in anticipation of a rewarding event (during the presentation of the conditioned stimulus that predicted reward) and during receipt of the reward (Hollerman & Schultz, 1998). This activity occurred irrespective of whether a movement was required, suggesting that the these neurons are not encoding a general reinforcing signal, but rather a signal for appetitive rewards.

These data were further supported by human PET and fMRI studies that have shown that the NAcc/ventral striatum is activated with receipt of both primary (food, sounds, etc.) and secondary (monetary gains) rewards (Small et al., 2001; Blood & Zatorre, 2001; Martin-Solch et al., 2001; O'Doherty et al., 2004). Further evidence using ligand-based imaging techniques have shown dopamine release in the NAcc with feelings of euphoria (Drevets et al., 2001), drug and alcohol consumption (Boileau et al., 2003), and even gambling (Koepp et al., 1998). Using event-related fMRI, researchers found that activity in the NAcc increased during the anticipation of primary and second rewards (Knutson et al., 2001; O'Doherty et al., 2002; Knutson et al., 2003). Knutson and colleagues further demonstrated that NAcc activation increases proportional to the magnitude of the anticipated monetary reward (Knutson et al., 2001).

In 2009, Cohen and colleagues performed the first study, to the best of our knowledge, to directly sample human ventral striatum activity during a reward learning and decision-making task (Cohen et al., 2009). Five patients underwent deep brain stimulation surgery for the treatment of major depressive disorder (MDD), in which a stimulating electrode was surgically implanted into the ventral striatum. Following the implant surgery and before the electrode leads were attached to the pulse generator (which occurs in a second operation after a few days), the researchers connected the leads to an acquisition system and had the subjects engage in a computer-guided task. This allowed researchers to correlate local field potential data from the NAcc to the behavior engaged in during the task. Their behavioral paradigm consisted of two tasks: a "learning" and "choosing" task. In the learning task, one of two cues was presented on any given trial. Unbeknownst to the subject, one cue was the "safe" cue, which always resulted in a small reward, while the other, "risky" cue, resulted in a high reward 75% of the time. The subjects quickly learned these associations and were able to verbally report their findings. In the "choosing" task, the subjects were now presented with both cues and asked to make a selection. Again, unbeknownst to the subjects, these cues retained the same reward contingencies as in the "learning" task. The dichotomy of these two tasks allowed researchers to interrogate the role of the NAcc in associative learning versus reward-guided behavior. The results were consistent with imaging studies, that NAcc activity reliably encodes the anticipation of reward proportional to reward magnitude. More specifically, activity increased from losses to safe to risky rewards. Interestingly, their data suggested that at the feedback period, NAcc activity does not encode a prediction error (difference between expected and actual outcome) because the activity should be zero for safe rewards (e.g., where expectation and outcome are equal). We will find that this is in contrast with single neuronal findings from the NAcc.

In 2011, Lega and colleagues presented the first single neuronal evidence from a human subject undergoing deep brain stimulation surgery (Lega et al., 2011). This subject was a 42-year-old female undergoing deep brain stimulation surgery for treatment refractory major depression. During the microelectrode recording portion of the surgery, the research team asked the subject to participate in a computer-guided visual-reward task; single neuronal and local field potential data were collected during the performance of the task. In a preoperative session, the participant was exposed to a set of 120 images from 11 categories to which she ranked her preference in a pairwise fashion. These images were then used as the rewarding stimuli during the recording session. In the operating room, at the beginning of each trial, the subject was cued to the category from which the subsequent rewarding stimuli would be presented if the trial was successfully completed. An object would then appear from either the left or right side of the screen; the subject was asked to appropriately press the left or right mouse button. If the trial was successful, the corresponding reward image from that category would be presented along with a positive reinforcing tone. If the trial was incorrect, a negative image was presented along with an aversive tone. Two neurons were identified from the recording. One neuron showed significant increase in activity during the feedback period to positively rewarding outcomes but not to negative or neutral trials. Similarly, the local field potential data showed a stark increase in alpha-band activity during the feedback of successful trials, again in comparison with negative and neutral outcomes. The authors note that their single neuronal findings most likely result from the torically active subpopulation of NAcc neurons. Although the data here are limited and must be interpreted with caution, they provided some evidence at a time where little existed.

In 2012, Patel and colleagues performed a more detailed examination of single neuronal activity in the NAcc during a financial decision-making task (Patel et al., 2012). This study was performed in patients undergoing deep brain stimulation surgery for the treatment of MDD or obsessive–compulsive disorder (OCD). The researchers implemented a computerized gambling task based on the classic card game War. In this task, the subject is presented with a card on a computer monitor. Following the presentation, the subject is given the opportunity to place a wager, either $5 or $20, via a button box. Once the wager is entered, the opponent's card is revealed—the player with the higher card wins. A feedback period follows with an image of the wager amount and the words "win" or "lose" depending on the outcome of the trial. A total of 19 neurons were examined from 8 subjects (5 MDD and 3 OCD). Similar to the results from Schultz and colleagues, the data revealed that neuronal responses in the NAcc were very heterogeneous to task-related epochs. However, there were two epochs that clearly carried most of the phasic activity: the go-cue and feedback period (the point in the trial at which the computer's card is revealed and the first point at which the subject realizes the outcome of the trial). The researchers report two main findings from the study. The first is that NAcc population activity predicts, on a trial-by-trial basis, the upcoming financial decision up to 2 s before the wager is physically manifested. This activity is time-locked to the onset of the go-cue (the first point at which the wagers appear on the screen, allowing the subject to initiate the bet). The second is that NAcc population activity encodes positive and negative prediction error signals during the feedback period. Activity in trials with expected positive and negative outcomes showed no difference during the same period.

These data provide the first single neuronal evidence from both animal and human studies that a decision signal is encoded in the NAcc (van der Meer & Redish, 2011). Although more work is needed to explore this signal, it may provide significant evidence for the role of the "actor" signal within the reinforcement learning framework. Furthermore this data provides concrete evidence that a prediction error signal is represented in the NAcc, which is more and more proving to be a ubiquitous signal represented throughout the brain.

Cingulate

Anatomy of Anterior Cingulate

Many subregions of the PFC are implicated in a greater network for reward circuitry. A major component of the reward circuitry in the PFC is the anterior cingulate cortex (ACC). The ACC covers a large region of the medial wall and is abutted to the corpus callosum (Paus, 2001). Importantly, the ACC is uniquely situated to handle an array of functions with respect to reward. Furthermore, each subregion of the ACC is predominately involved with distinct roles, such as motivation and cognition (Paus, 2001). Notably, these areas are not homogeneous and demonstrate some overlap in function. Nevertheless, there is a general agreement as to the primary role of each subregion. Two of the major parts are the dorsal ACC (dACC) and ventral ACC, which is also known as the ventral medial prefrontal cortex (vmPFC).

The portion of the ACC implicated in reward processing and cognition is the dACC. The dACC is often divided into Areas 24b and 32. Much of the dACC is located in the anterior cingulate sulcus (ACS). Each subdivision has robust anatomical connectivity to other brain regions that serve parallel functions (Haber et al., 2006). A dense set of dACC efferents are focal projections that reach a large part of the dorsal striatum, primarily from the caudate head and rostral putamen to these regions bordered by the anterior commissure. Projections to the ventral striatum are much more sparse. The area reached by focal projections of the dACC is increased by diffuse projections (Haber et al., 2006). These diffuse projections also infiltrate other parts of the striatum, such as the dorsolateral caudate and ventral putamen. Moreover, regions of the putamen that are caudal to the anterior commissure also receive afferents from the dACC. Overall, the dACC is able to interact with a large part of the dorsal striatum.

Interestingly, the ACC is also connected to motor areas of the brain. This part of the ACC is rostral and includes part of the cingulate sulcus. The rostral cingulate motor area, also known as Area 24c, has a strong influence over the primary motor cortex and other motor areas (Picard & Strick, 1996). Moreover, Area 24c is strongly connected to the striatum (Kunishio et al., 1994). It should be noted that other cortical and subcortical areas, such as the striatum, activate Area 24c. Finally, Area 24c is likely influenced by neuromodulators, such as dopamine and other monoamines. This type of connectivity may facilitate the motor outputs of behaviors and is likely influenced by cognition.

Another part of the ACC that has received attention is the ventral region, often referred to as the vmPFC. The vmPFC is strongly connected to the limbic system, which is a group of brain structures that are implicated in several functions such as emotion, motivation, and memory. Some of the main components of the vmPFC are the rostral ACC (rACC) and subgenual ACC. Like the dACC, the vmPFC also has subdivisions that are known as Area 24a and Area 25. Generally, efferents to the striatum are less robust than those of the dACC (Haber et al., 2006). However, the bulk of the focal projections from the vmPFC target the shell of the NAcc in the ventral striatum. Moreover, and in lesser amounts, the vmPFC targets some of the caudate.

Anterior Cingulate and Prediction Error Signaling

Memories linked to emotional events, such as rewarding experiences, guide behavior. The ability to regulate the expression of such memories is necessary for survival. Moreover, this ability requires a fine-tuned interaction between structures of the basal ganglia with different regions of the PFC. One region of the PFC that is implicated in reward processing is the ACC.

A wealth of evidence describes the neural signatures of prediction error related to reward in dopaminergic systems. Dopaminergic systems send robust projections to the ACC. Accordingly, neurons in the ACC also signal prediction error. Unlike the dopaminergic systems, distinct populations of neurons in the ACC signal either positive or negative events. The increased activity when an error occurs likely modifies subsequent behavior in NHPs (Michelet et al., 2009). This further supports the idea that the ACC monitors the consequences of actions and mediates subsequent changes in behavior. As described previously, signals of prediction error are represented in the NAcc region of the ventral striatum of humans (O'Doherty et al., 2004), and the behavioral choice based on predicted reward is influenced by the dorsal striatum. Both regions of the striatum are likely to become under control of the ACC, given the anatomical projections of the ACC.

One aspect of cognition that allows a species to thrive is the ability to update the value representation when expected reward is higher or lower than the level of reward received. Consequently, the magnitude and frequency by which the discrepancy occurs directly affects the update. In humans, several techniques have been used to implicate the ACC in updating. The ACC is implicated in processing errors and helps optimize behavior. Early work in humans began with noninvasive approaches, such as electroencephalography (EEG) and measured error-related negativity (ERN). ERN is thought to arise from the dACC (Ullsperger & von Cramon, 2001; Holroyd et al., 2004). Some studies focused on changes in the ERN to help understand the neural mechanisms of behavior in humans (Gehring et al., 1993).

The ACC is necessary for error detection and is involved in the selection of correct choice. Experimental data using PET in humans implicated the ACC when learning was influenced by feedback in a prelearned task (Jueptner et al., 1997a, 1997b). ERN studies suggest that activity in ACC serves as an error detection between choices (Coles et al., 2001). Data from EEG and fMRI studies demonstrate that the ACC becomes active when an error is committed (Braver et al., 2001). The increase in activity is likely involved in driving behavior to correct errors for subsequent events. Notably, fMRI data suggest that errors are processed differently depending on whether a subject's choice was based on the subject's own volition or based on a command (Walton et al., 2004; Rushworth et al., 2007).

Converging lines of evidence support the idea that the PFC is necessary for cognitive control and behavioral flexibility. One type of behavioral flexibility is predictability. Predictability requires calculations based on previous experience, risk of change, and level of reward to be obtained (Kennerley et al., 2006). More neurons respond in the ACC than other PFC subregions in NHPs (Kennerley & Wallis, 2009). In primates, early work involving lesions of the ACC rendered subjects unable to allow new information to influence subsequent choices (Kennerley et al., 2006). Several reports implicate the ACC in prediction error and the respective influence on behavioral modification in humans (Nieuwenhuis et al., 2004; Jocham et al., 2009).

Furthermore, the ACC of NHPs is thought to play a key role in assigning and reevaluating possible outcomes (Ito et al., 2003; Matsumoto et al., 2003; McCoy et al., 2003; Samejima et al., 2005). The ACC evaluates associations in the presence of varying amounts of reward. Neural activity in the ACC, the cingulate motor areas or ACS, responds when reward is reduced (Shima & Tanji, 1998). Consistent with this view, activity in the ACC increases when levels of reward change throughout a training session (Amiez et al., 2006). Interestingly, a reversible lesion blocked this effect. The ACC is also affected by temporal factors. For example, activity in the ACC of humans was further increased if a loss had recently occurred (Gehring & Willoughby, 2002). It is also likely that dopaminergic inputs drive this response (Holroyd & Coles, 2002).

Anterior Cingulate and Temporal Signaling

Future behavior is manipulated by the ACC, which signals prior experiences. Anatomical inputs to the ACC include basal ganglia, the primary motor cortex, and premotor areas. Given the convergence of various signals in the ACC, it is likely that the ACC is able to access information in a unique matter that links potential outcomes that drive behavior in NHPs (Hayden & Piatt, 2010). Activity in the ACC is necessary to adapt to a changing environment.

One behavioral situation that can influence the ACC is that of previous experiences. Previous experiences, both recent and remote, involve associative learning that influences future decisions in NHPs and humans (Samejima et al., 2005; Daw et al., 2006). This type of associative learning requires an intricate system of brain structures that can link potential outcomes with variable actions. Several brain regions, such as the midbrain in NHPs (Bayer & Glimcher, 2005) and striatum (Samejima et al., 2005), interact with the cortical system when the value of a decision is relearned and updated. There are a few components of the prefrontal cortical system that are involved, such as the ACC (Procyk et al., 2000; Walton et al., 2004).

Previous influences of the environment affect subsequent behavior. Behavioral tasks in NHPs are often designed such that levels of reward or aversive stimuli are changed to indicate that the subjects need to modify their behavior. Data from lesion studies suggest that the ACC is recruited under these conditions (Kennerley et al., 2003). However, some reports suggest that the size of the lesion to the ACC influences the ability to modify behavior (Rushworth et al., 2004; Kennerley et al., 2006). For example, subjects with large lesions were able to modify their behavior but could not maintain choice after contingencies were switched (Kennerley et al., 2006). In this study, NHPs with lesions only considered recent trials to modify their behavior whereas monkeys without lesions considered many previous trials. Together, these data suggest that the ACC is necessary for relating behavior to the outcome of an event.

The contribution of subcortical regions to the ACC depends on the amount of change in the environment, which would influence how fast the cortical system updates the expected outcome of the action (Behrens et al., 2007). Once updated, the ACC regulates activity in subcortical regions (Kunishio et al., 1994). Studies using fMRI in humans show increased activity when there is a change in current behavioral situations (Walton et al., 2004; Yoshida & Ishii, 2006). On the other hand, data from NHPs demonstrate that this increase in activity is no longer present when levels of uncertainty are diminished (Procyk et al., 2000). Integration of these distinct types of information likely occurs in the ACC. Accordingly, activity in ACC is increased when events are more likely to predict subsequent behavior (Behrens et al., 2007).

Learning a pattern that is uncertain can challenge the survival of a species. Experiments using lesions in NHPs suggest that the ACC is able to detect changes that are probabilistic (Kennerley et al., 2006). During learning, neuronal activity in ACC of NHPs is increased (Procyk et al., 2000). As with previous reports, neurons in the ACC signaled the amount of reward received (Amiez et al., 2006). Furthermore, another population of neurons responded to the likelihood that a reward would be received. That is, high reward was delivered with a high probability, and low reward delivered with a low probability. Conversely, high reward was delivered with low probability, and low reward delivered with high probability. This could only be determined and learned over many trials. In this case, increased activity is thought to influence subsequent behavior to favor a reward. The ACC is necessary for learning the reward that is presented in a probabilistic, albeit fixed, manner.

In addition to the ACC's being responsible for the history of the expected value of reward received, the ACC can survey the amount of reward received. Other reports demonstrate that activity in the dACC increases when there is a loss of reward (Holroyd & Coles, 2002; Gehring & Willoughby, 2002; Bush et al., 2002). Neurons in the ACC also become active when rewards could have been obtained (Hayden et al., 2009). This is consistent with the idea that the loss of expected reward could be aversive (Kim et al., 2006). Moreover, complete absence of reward drives the ACC to influence behavior (Chudasama et al., 2012). One report observed an increase in single unit activity of the ACC as the amount of predicted reward increased. Furthermore, it is suggested that motivation to perform the task increased with an increase in predicted reward (Shidara & Richmond, 2002). Together, these studies provide strong support that uncertain situations influence activity in the ACC.

Anterior Cingulate Influences Goals

In humans, the dACC becomes more active as the amount of reward decreases (Bush et al., 2002). Moreover, activity in ACC increased if reward loss was sequential (Gehring & Willoughby, 2002). This is probably because the effects of accumulating losses are compounded. In some cases, the implicated role of the ACC is different between species. This difference could be explained by type of reward used; different reward types may not be represented equally across species (Schultz, 2000; Bush et al., 2002). For example, human studies often use monetary rewards whereas primate studies use a liquid reward. Thus, an unanswered question in the field is the contribution of either primary or secondary rewards to ACC activity in humans. However, the influence of reward magnitude to neuronal activity in ACC of humans is unclear.

Achieving a goal requires planning and action that could lead to a rewarding event. Linking a planned action to a certain stimulus is accompanied by an anticipated reward (Hadland et al., 2003). Single cells in the ACC, likely the ACS region, respond to rewards more so than associations of stimulus to rewards (Matsumoto et al., 2003). In one study, subjects were presented with different stimuli, responses, and reward deliveries. Another role of the ACC is that it may influence the amount of effort invested in the planning of goals. The planning required to obtain a reward requires the ACC (Hadland et al., 2003). Moreover, the expectation of reward influences this (Shima & Tanji, 1998; Procyk et al., 2000).

Little is known about the behavioral significance of increased activity in ACC. To address this issue, Eskandar and colleagues (Williams & Eskandar, 2004) recorded activity from single neurons from the ACC in patients undergoing planned surgery for cingulotomy. These subjects performed a behavior task with visual cues that instructed subjects to perform specific types of actions. During parts of the task, subjects adjusted their behavior for subsequent trials and reward amounts were changed. Neurons in the ACC increased the most when the amount of expected reward was reduced. These data suggest that decreased activity in ACC is associated with reduced reward. After cingulotomy, subjects were less likely to change their behavior as instructed. This within-subject design allowed for assessment of the ACC before and after cingulotomy. Therefore, strong interpretations as to what the contributions of the ACC are to human behavior could be inferred. Thus, it is likely that the ACC is necessary for evaluating the net gain of reward and the influence on choice.

Anterior Cingulate and Cognitive Interference

Executive functioning is a conjunction of several cognitive processes and goal-directed behaviors, which require planning, decision making, and action control (Banich, 2004). Coordination of these cognitive resources requires regions of the PFC, such as the ACC. Thus, the ACC is thought to be an interface between emotion and cognition (Banich, 2004). Inherently, damage to the PFC diminishes top-down control. In said cases, subcortical and other automatic processes dominate (Banich, 2004). As described previously, the ventral division of the ACC is associated with emotional tasks whereas the dorsal division is associated with cognitive ones (Bush et al., 2000). Interestingly, these divisions can deactivate each other.

Furthermore, the ACC may monitor the presence of conflicts (Botvinick et al., 2001). Recently, an fMRI study was able to disentangle the role of ACC in monitoring and selection (Walton et al., 2004). This research used an elegant task design and demonstrated that the ACC became more active with selection. Further increase in activation was observed when subjects were required to monitor the outcome. Activity in ACC increased when subjects made a decision of their own accord. In stark contrast, activity in ACC was decreased when the a subject was influenced by an experimenter. Thus, the ACC is necessary for selection and monitoring (Holroyd et al., 2004; Rushworth et al., 2004).

Experiments focusing on conflict often involve a type of cognitive interference which occurs when characteristics of a stimulus hinder the processing of another property of that stimulus. The increase in reaction times is a result of cognitive interference and is commonly known as the "Stroop effect" (Vendrell et al., 1995; Bush et al., 1998), named after John Stroop, a forefather of interference theory (Stroop, 1935).

The two divisions of the ACC described are reliably activated in the Stroop task. As noted, examining cognitive interference can be achieved in the Stroop task. Paradigms like the Stroop task set up a situation known as a congruent stimulus response, where relevant information results in a correct response that occurs with a short reaction time (fast response). Here, a subject is required to make a correct choice as fast as possible by focusing on only a relevant characteristic of the presented stimulus while ignoring other irrelevant information that is presented. The conflict exists when the irrelevant information increases the risk of committing an incorrect response and results in long reaction times (slow response). For example, subjects may be presented with a word of a color that is printed in another color. Next, subjects identify the color of ink that the word is printed on, and not the word. This type of cognitive interference leaves a previous memory difficult to retrieve because of the presence of discordant information related to color and creates a conflict. This conflict requires that the subjects inhibit an impulse to read the color, which increases response times.

Behavioral tasks adapted from the Stroop task involved separate regions of the ACC (Bush et al., 2000, 2003). For example, the rostral ACC is thought to resolve emotional conflicts (Etkin et al., 2006). The ventral region of the cingulate is activated much more in emotional versions of the Stroop task, which can measure bias to emotional stimuli and emotional processing. In emotional versions of the Stroop task, words are charged with emotion in a similar matter as the color conflict version. However, emotional processing requires distinct mechanisms that are influenced by attention and the emotional relevance of the words (McKenna & Sharma, 2004). The type of cognitive control involved in emotional Stroop tasks is modulated by the rACC (Milham et al., 2001; MacDonald et al., 2000; Bush et al., 2000). Single unit recording experiments in the rACC/vmPFC of humans are difficult. Future studies in patients undergoing planned surgery may consider measuring neural activity from the vmPFC.

Optimizing behaviors can be achieved with increased attention. This could occur, for example, after commission of an error, reduced reward, or in the presence of a distractor (Lavie, 2005). Increased attention may serve to minimize a decline in behavioral performance. Early work using fMRI implicates the ACC in cognitive processes and attention (Wager et al., 2004). Unfortunately, the amount of information that can be extrapolated from imaging data is limited. At the turn of the century, Davis and colleagues (2000) took advantage of a rare opportunity to measure behavior in attention-demanding tasks while recording activity from single neurons in the ACC of humans undergoing planned surgery (Davis et al., 2000). They hypothesized that neurons in the ACC signal attention. To test this, Davis and colleagues (2000) administered a set of tasks, including arithmetic and Stroop tests, and compared different levels of attention. In paradigms such as the Stroop task, a behavioral marker of increased attention is an increase in time required to perform a task, which can be measured by increased reaction times. Their results demonstrated that neurons in the ACC signal attention during tasks that require high levels of cognition.

The role of single neurons in ACC of humans during other cognitive processes remained largely unexplored. Using similar techniques as in their previous study on attention, Davis and colleagues (2005) administered emotional Stroop tests while recording activity from ACC in humans (Davis et al., 2005). Their results demonstrated that activity in ACC signals attention and responds to emotionally salient stimuli. Together, these findings suggest that the ACC is necessary to disentangle situations of uncertainty while processing emotions. Additionally, the increased attention and emotion may directly influence motivation (Botvinick et al., 2001). This supports the idea that the ACC may signal cognitive processes to modify subsequent behavior.

Much of what is known about the ACC and cognitive processes stemmed from studies using imaging (Botvinick et al., 1999; Kerns et al., 2004) and event-related potential recordings in humans (Gehring & Fencsik, 2001). The behavioral relevance of neuronal activity in the ACC during cognitive processes like conflict, however, is still unclear. Indeed, the uncertainty is partly due to the differences in paradigms used and technical limitations between studies. For example, data from lesion studies in humans (Mansouri et al., 2007) have provided disparate conclusions compared to those from work using imaging.

The ability to measure the magnitude of cognitive interference is made possible by a counting version of the Stroop task, known as the Multi-Source Interference Task (MSIT). In the MSIT, subjects are presented with a picture on a screen that has three numbers. Two of the numbers are the same and are classified as "distractors," whereas one number is unique and is classified as the "target." The subjects are asked to indicate the position of the "target" by pressing the corresponding button: either one, two, or three, represented from left to right on the button box. In general, there are three levels of interference—namely, type 0, type 1, or type 2. The numbers are presented with distinct levels of interference. For example, for type 0 trials the number 1 would be presented in the leftmost position on the screen followed by two zeros. This is considered no interference, given that there is not an option for "zero" on the button box. For low-interference trials (type 1), the number 1 would be in the leftmost position on the screen followed by two other numbers of the same type, requiring that the first button on the box be pressed. For high-interference trials (type 2), on the other hand, the position of the target number on the screen did not correspond to its numerical position on the button box and was presented alongside two distractors. Accordingly, the reaction time in response to stimuli increases as the level of cognitive interference increases.

Recent reports demonstrate that the MSIT combines many sources of cognitive interference that robustly activate the dACC (Bush et al., 2003; Bush & Shin, 2006), providing a potential mechanism for how the dACC can modulate cognitive control (Fellows & Farah, 2005; van Veen et al., 2001). Increased cognitive control is required in the presence of emotional conflicts and incongruent stimuli. Indeed, increased cognitive interference with the MSIT would increase the probability of committing errors (Brown & Braver, 2005). Subsequently, there would likely be an increase in attention (Davis et al., 2000, 2005) to enhance decision making (Botvinick, 2007).

The behavioral function of signaling interference load by neurons in the dACC is unclear. Many hypotheses stemming from fields such as decision making and conflict monitoring suggest that activity in dACC from previous experience influences future levels of activity in dACC neurons. Due to technical limitations, no study has been able to provide compelling support for these hypotheses at the neuronal level. Moreover, some studies measuring single unit activity in NHPs do not support the role of ACC in monitoring conflict (Ito et al., 2003; Nakamura et al., 2005). To address this issue, Sheth and colleagues combined behavioral measures of the MSIT with imaging and single unit recording in patients undergoing planned surgery. As expected, behavioral responses and reaction times were similar to those in previous reports (Kerns et al., 2004; Botvinick, 2007; Ridderinkhof, 2002). Similarly, they reported a dose-dependent increase in fMRI and single neuronal signaling in dACC when comparing high-interference trials with low-interference trials (see figure 12.2, plate 13; Bush et al., 2003). Furthermore, Sheth and colleagues found that reaction times were faster if a previous trial was different than the current trial whereas reactions times were slower if a previous trial was the same as the current trial. Interestingly, they demonstrate that single unit activity in the dACC increased when

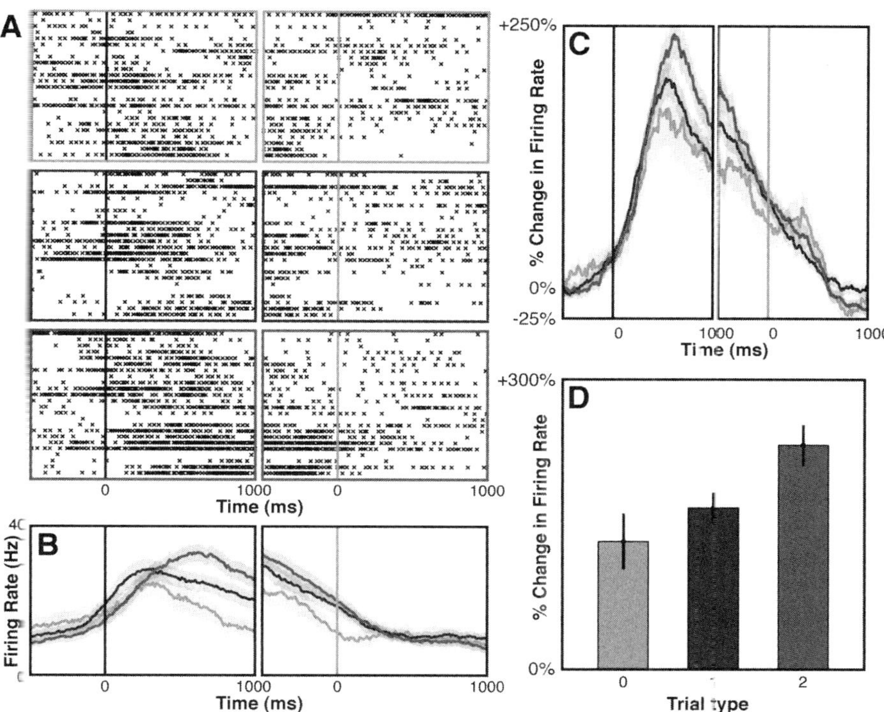

Figure 12.2 (plate 13)

Individual and population neuronal responses. (A) Example neuron showing modulation of firing based on cue-related interference. Rasters for Type 0 (green), 1 (blue), and 2 (red) trials are shown aligned to the cue (black line) and choice (gray line). (B) Average firing rates of the same neuron, demonstrating increasing firing with increasing interference. Error bars (SEM) are depicted with shading. (C) Average firing of all cue-related neurons. (D) Same as in (C), but showing activity averaged within a 200-ms-wide window centered 500 ms after the cue. Neuronal firing increased with cognitive interference (p = 0.02, analysis of variance), correlating with reaction time. From Sheth et al. (2012).

high-interference trials were preceded by high-interference trials, unlike results from previous imaging data demonstrating decreased activity in dACC when previous and current trials were the same (Kerns et al., 2004; Botvinick et al., 1999).

One possible explanation for the difference in findings between studies is the lack of temporal resolution of fMRI compared to single unit recordings (Sheth et al., 2012). Moreover, the peak signal in the dACC occurs much faster than previously thought. Interestingly, Sheth and colleagues successfully performed the planned cingulotomy and observed successful behavioral performance, albeit slower reaction between trials, in the presence of varying levels of interference.

The results of Sheth et al. showed that previous activity in the dACC influences current neuronal activity. They suggest that the dACC is necessary for behavioral adaptation, either by enhancing reactions to cues with similar levels of interference or hindering reactions to cues with varying levels of interference. If true, then removal of the dACC would impair behavioral adaptation. Interestingly, Sheth et al. noted that subjects with lesions of the dACC had similar levels of success in behavior as compared to preoperative levels. On the other hand, dACC lesions eliminated the influence of previous trials to the modulation of reaction times. These results provided empirical support that the dACC is necessary to evaluate the influence of previous experience to current behavior.

Limitations of Methodology

Human single neuronal studies provide unique data for understanding and exploring neural mechanisms of human cognition. However, fundamentally, studies of this nature are limited not only by moral and ethical responsibilities but also by technical and physical constraints. Many considerations must be taken into account in designing and executing this type of study. Importantly, all experiments must be approved by an institutional review board, and consent must be obtained from any subject participating in the study.

Conducting research studies in the operating room can be demanding, but patient care is always the foremost priority. Generally, microelectrode research studies are limited to roughly 30 minutes, and thus careful attention must be taken when designing and implementing a behavioral task. Tasks should be simple and easy to understand with little requirement for training and should provide enough statistical power to adequately explore questions of interest. It can be helpful to chunk sessions into 10-minute groups to allow for multiple neuronal isolations during a study and provide brief breaks for subjects. The underlying disease pathology of the study population must also be considered when designing the task. For example, subjects with motor impairments (e.g., tremor or dystonia) may have difficulty in performing joystick-based or other movement-demanding tasks. Similarly, laterality of cognitive function must also be considered when designing an experiment.

Neurophysiological recordings are also susceptible to many issues within the operating room environment. Ambient electrical (60-Hz) noise from surgical or anesthesia equipment often poses the most common issue. This can generally be alleviated by turning off the culprit component. Another common issue when performing human microelectrode recordings is mechanical noise induced from cardioballistic effects. This occurs when the microelectrode is in close proximity to the vasculature and occurs through the expansion/contraction of the artery/arteriole.

Inherent in these experiments are limitations on the generalizability of the results. For obvious ethical reasons, invasive studies cannot be performed in normal subjects, and thus inferences must always be considered within the context of the study population. Despite this, there are a number of ways to address these concerns. Behavioral or fMRI data can be collected and compared in a matched healthy cohort, potentially corroborating microelectrode findings. Another possibility is performing the microelectrode study in orthogonal patient populations. For example, the ventral striatum is the target for two underlying diseases, OCD and MDD. Similar findings from two separate populations can also strengthen the validity of the data.

Future Directions and Conclusions

Although there are relatively few human single neuron studies to date, and even fewer that have explored reward processing, human studies provide a unique data set that have the potential to guide both innovative basic and translational science. All of the studies presented in this review, with the exception of Sheth et al., represent correlative findings from microelectrode studies. With the advancement of invasive and noninvasive stimulation technology, such as intermittent deep brain stimulation or transcranial magnetic stimulation, future studies will explore cognitive functions through causal experimentation guided by the underlying neurophysiology. This is already a rapidly progressing avenue of research.

A fundamental limitation to these studies is that they are first and foremost guided by clinical care. With the proven effectiveness of deep brain stimulation surgery, more and more brain regions are being explored for the treatment of various neurodegenerative and neuropsychiatric disorders. As these studies progress, we will have the opportunity to investigate activity from many more regions of the widely distributed reward network. Human intraoperative studies will continue to be guided by NHP physiology and human imaging studies, which serve as a spotlight to focus these difficult and limited experiments.

In conclusion, we present evidence in this review of human single neuronal studies of reward processing in the anterior cingulate and basal ganglia. Much of the work contributed by these studies supports evidence from both NHP physiology and human neuroimaging studies. Although these studies are difficult and limited in many ways, they provide invaluable data for how we encode, represent, and process reward information within the context of the human brain. Currently, no other technique allows this type of access to the fundamental computational unit of our nervous system, and thus it will remain a gold standard for studying human behavior.

Acknowledgments

We would like to acknowledge the Sackler Programme in Psychobiology for their support.

References

Amiez, C., Joseph, J. P., & Procyk, E. (2006). Reward encoding in the monkey anterior cingulate cortex. *Cerebral Cortex*, *16*, 1040–1055.

Banich, M. T. (2004). *Cognitive neuroscience and neuropsychology* (student text; 2rd ed.). Wilmington, MA: Wadsworth Publishing.

Bayer, H. M., & Glimcher, P. W. (2005). Midbrain dopamine neurons encode a quantitative reward prediction error signal. *Neuron, 47*, 129–141.

Behrens, T. E. J., Woolrich, M. W., Walton, M. E., & Rushworth, M. F. S. (2007). Learning the value of information in an uncertain world. *Nature Neuroscience, 10*, 1214–1221.

Blood, A. J., & Zatorre, R. J. (2001). Intensely pleasurable responses to music correlate with activity in brain regions implicated in reward and emotion. *Proceedings of the National Academy of Sciences of the United States of America, 98*, 11818–11823.

Boileau, I., Assaad, J.-M., Pihl, R. O., Benkelfat, C., Leyton, M., Diksic, M., et al. (2003). Alcohol promotes dopamine release in the human nucleus accumbens. *Synapse (New York, N.Y.), 49*, 226–231.

Botvinick, M., Nystrom, L. E., Fissell, K., Carter, C. S., & Cohen, J. D. (1999). Conflict monitoring versus selection-for-action in anterior cingulate cortex. *Nature, 402*, 179–181.

Botvinick, M. M. (2007). Conflict monitoring and decision making: Reconciling two perspectives on anterior cingulate function. *Cognitive, Affective & Behavioral Neuroscience, 7*, 356–366.

Botvinick, M. M., Braver, T. S., Barch, D. M., Carter, C. S., & Cohen, J. D. (2001). Conflict monitoring and cognitive control. *Psychological Review, 108*, 624–652.

Braver, T. S., Kerns, J. G., Barch, D. M., Cohen, J. D., Gray, J. R., MacDonald, A. W., et al. (2001). Anterior cingulate cortex and response conflict: Effects of frequency, inhibition and errors. *Cerebral Cortex, 11*, 825–836.

Brown, J. W., & Braver, T. S. (2005). Learned predictions of error likelihood in the anterior cingulate cortex. *Science, 307*, 1118–1121.

Bush, G., Luu, P., & Posner, M. (2000). Cognitive and emotional influences in anterior cingulate cortex. *Trends in Cognitive Sciences, 4*, 215–222.

Bush, G., & Shin, L. M. (2006). The Multi-Source Interference Task: An fMRI task that reliably activates the cingulo–frontal–parietal cognitive/attention network. *Nature Protocols, 1*, 308–313.

Bush, G., Shin, L. M., Holmes, J., Rosen, B. R., & Vogt, B. A. (2003). The Multi-Source Interference Task: Validation study with fMRI in individual subjects. *Molecular Psychiatry, 8*, 60–70.

Bush, G., Vogt, B. A., Holmes, J., Dale, A. M., Greve, D., Jenike, M. A., et al. (2002). Dorsal anterior cingulate cortex: A role in reward-based decision making. *Proceedings of the National Academy of Sciences of the United States of America, 99*, 523–528.

Bush, G., Whalen, P. J., Rosen, B. R., Jenike, M. A., McInerney, S. C., & Rauch, S. L. (1998). The counting Stroop: An interference task specialized for functional neuroimaging—Validation study with functional MRI. *Human Brain Mapping, 6*, 270–282.

Carr, D. B., & Sesack, S. R. (2000). GABA-containing neurons in the rat ventral tegmental area project to the prefrontal cortex. *Synapse (New York, N.Y.), 38*, 114–123.

Chudasama, Y., Daniels, T. E., Gorrin, D. P., Rhodes, S. E. V., Rudebeck, P. H., & Murray, E. A. (2012, September 3). The role of the anterior cingulate cortex in choices based on reward value and reward contingency. *Cerebral Cortex* [Epub ahead of print].

Cohen, M. X., Axmacher, N., Lenartz, D., Elger, C. E., Sturm, V., & Schlaepfer, T. E. (2009). Neuroelectric signatures of reward learning and decision-making in the human nucleus accumbens. *Neuropsychopharmacology, 34*, 1649–1658.

Coles, M. G., Schultz, W. W., Scheffers, M. K., & Holroyd, C. B. (2001). Why is there an ERN/Ne on correct trials? Response representations, stimulus-related components, and the theory of error-processing. *Biological Psychology, 56*, 173–189.

D'Ardenne, K., McClure, S. M., Nystrom, L. E., & Cohen, J. D. (2008). BOLD responses reflecting dopaminergic signals in the human ventral tegmental area. *Science, 319*, 1264–1267.

Davis, K. D., Hutchison, W. D., Lozano, A. M., Tasker, R., & Dostrovsky, J. O. (2000). Human anterior cingulate cortex neurons modulated by attention-demanding tasks. *Journal of Neurophysiology, 83*, 3575–3577

Davis, K. D., Taylor, K. S., Hutchison, W. D., Dostrovsky, J. O., McAndrews, M. P., Richter, E. O., et al. (2005). Human anterior cingulate cortex neurons encode cognitive and emotional demands. *Journal of Neuroscience, 25*, 8402–8406.

Daw, N. D., O'Doherty, J. P., Dayan, P., Seymour, B., & Dolan, R. J. (2006). Cortical substrates for exploratory decisions in humans. *Nature, 441*, 876–879.

Drevets, W. C., Gautier, C., Price, J. C., Kupfer, D. J., Kinahan, P. E., Grace, A. A., et al. (2001). Amphetamine-induced dopamine release in human ventral striatum correlates with euphoria. *Biological Psychiatry, 49*, 81–96.

Etkin, A., Egner, T., Peraza, D. M., Kandel, E. R., & Hirsch, J. (2006). Resolving emotional conflict: A role for the rostral anterior cingulate cortex in modulating activity in the amygdala. *Neuron, 51*, 871–882.

Fellows, L. K., & Farah, M. J. (2005). Is anterior cingulate cortex necessary for cognitive control? *Brain, 128*, 788–796.

Gehring, W. J., & Fencsik, D. E. (2001). Functions of the medial frontal cortex in the processing of conflict and errors. *Journal of Neuroscience, 21*, 9430–9437.

Gehring, W. J., Goss, B., Coles, M. G. H., Meyer, D. E., & Donchin, E. (1993). A neural system for error detection and compensation. *Psychological Science, 4*, 385–390.

Gehring, W. J., & Willoughby, A. R. (2002). The medial frontal cortex and the rapid processing of monetary gains and losses. *Science, 295*, 2279–2282.

Gottfried, J. A., O'Doherty, J., & Dolan, R. J. (2002). Appetitive and aversive olfactory learning in humans studied using event-related functional magnetic resonance imaging. *Journal of Neuroscience, 22*, 10829–10837.

Graybiel, A. M. (2008). Habits, rituals, and the evaluative brain. *Annual Review of Neuroscience, 31*, 359–387.

Haber, S. N., Kim, K.-S., Mailly, P., & Calzavara, R. (2006). Reward-related cortical inputs define a large striatal region in primates that interface with associative cortical connections, providing a substrate for incentive-based learning. *Journal of Neuroscience, 26*, 8368–8376.

Haber, S. N., & Knutson, B. (2010). The reward circuit: Linking primate anatomy and human imaging. *Neuropsychopharmacology, 35*, 4–26.

Hadland, K. A., Rushworth, M. F. S., Gaffan, D., & Passingham, R. E. (2003). The anterior cingulate and reward-guided selection of actions. *Journal of Neurophysiology, 89*, 1161–1164.

Hayden, B. Y., Pearson, J. M., & Piatt, M. L. (2009). Fictive reward signals in the anterior cingulate cortex. *Science, 324*, 948–950.

Hayden, B. Y., & Piatt, M. L. (2010). Neurons in anterior cingulate cortex multiplex information about reward and action. *Journal of Neuroscience, 30*, 3339–3346.

Heimer, L., & Kalil, R. (1978). Rapid transneuronal degeneration and death of cortical neurons following removal of the olfactory bulb in adult rats. *Journal of Comparative Neurology, 178*, 559–609.

Hollerman, J. R., & Schultz, W. (1998). Dopamine neurons report an error in the temporal prediction of reward during learning. *Nature Neuroscience, 1*, 304–309.

Holroyd, C. B., & Coles, M. G. H. (2002). The neural basis of human error processing: Reinforcement learning, dopamine, and the error-related negativity. *Psychological Review, 109*, 679–709.

Holroyd, C. B., Larsen, J. T., & Cohen, J. D. (2004). Context dependence of the event-related brain potential associated with reward and punishment. *Psychophysiology, 41*, 245–253.

Ito, S., Stuphorn, V., Brown, J. W., & Schall, J. D. (2003). Performance monitoring by the anterior cingulate cortex during saccade countermanding. *Science, 302*, 120–122.

Jocham, G., Neumann, J., Klein, T. A., Danielmeier, C., & Ullsperger, M. (2009). Adaptive coding of action values in the human rostral cingulate zone. *Journal of Neuroscience, 29*, 7489–7496.

Jog, M. S., Kubota, Y., Connolly, C. I., Hillegaart, V., & Graybiel, A. M. (1999). Building neural representations of habits. *Science, 286*, 1745–1749.

Jueptner, M., Frith, C. D., Brooks, D. J., Frackowiak, R. S., & Passingham, R. E. (1997a). Anatomy of motor learning: II. Subcortical structures and learning by trial and error. *Journal of Neurophysiology, 77*, 1325–1337.

Jueptner, M. M., Stephan, K. M. K., Frith, C. D. C., Brooks, D. J. D., Frackowiak, R. S. R., & Passingham, R. E. R. (1997b). Anatomy of motor learning: I. Frontal cortex and attention to action. *Journal of Neurophysiology, 77*, 1313–1324.

Kennerley, S. W., Rushworth, M. F. S., Walton, M. E., Hadland, K. A., Behrens, T. E. J., Gaffan, D., et al. (2003). The effect of cingulate cortex lesions on task switching and working memory. *Journal of Cognitive Neuroscience, 15*, 338–353.

Kennerley, S. W., & Wallis, J. D. (2009). Evaluating choices by single neurons in the frontal lobe: Outcome value encoded across multiple decision variables. *European Journal of Neuroscience, 29*, 2061–2073.

Kennerley, S. W., Walton, M. E., Behrens, T. E. J., Buckley, M. J., & Rushworth, M. F. S. (2006). Optimal decision making and the anterior cingulate cortex. *Nature Neuroscience, 9*, 940–947.

Kerns, J. G., Cohen, J. D., MacDonald, A. W., Cho, R. Y., Stenger, V. A., & Carter, C. S. (2004). Anterior cingulate conflict monitoring and adjustments in control. *Science, 303*, 1023–1026.

Kim, H., Shimojo, S., & O'Doherty, J. P. (2006). Is avoiding an aversive outcome rewarding? Neural substrates of avoidance learning in the human brain. *Public Library of Science Biology, 4*, e233.

Knutson, B., Adams, C. M., Fong, G. W., & Hommer, D. (2001). Anticipation of increasing monetary reward selectively recruits nucleus accumbens. *Journal of Neuroscience, 21*, RC159.

Knutson, B., Fong, G. W., Bennett, S. M., Adams, C. M., & Hommer, D. (2003). A region of mesial prefrontal cortex tracks monetarily rewarding outcomes: Characterization with rapid event-related fMRI. *NeuroImage, 18*, 263–272.

Knutson, B., Taylor, J., Kaufman, M., Peterson, R., & Glover, G. (2005). Distributed neural representation of expected value. *Journal of Neuroscience, 25*, 4806–4812.

Koepp, M. J., Gunn, R. N., Lawrence, A. D., Cunningham, V. J., Dagher, A., Jones, T., et al. (1998). Evidence for striatal dopamine release during a video game. *Nature, 393*, 266–268.

Kunishio, K., Kunishio, K., Haber, S. N., & Haber, S. N. (1994). Primate cingulostriatal projection: Limbic striatal versus sensorimotor striatal input. *Journal of Comparative Neurology, 350*, 337–356.

Lavie, N. (2005). Distracted and confused? Selective attention under load. *Trends in Cognitive Sciences, 9*, 75–82.

Lavin, A., Nogueira, L., Lapish, C. C., Wightman, R. M., Phillips, P. E. M., & Seamans, J. K. (2005). Mesocortical dopamine neurons operate in distinct temporal domains using multimodal signaling. *Journal of Neuroscience, 25*, 5013–5023.

Lega, B. C., Kahana, M. J., Jaggi, J., Baltuch, G. H., & Zaghloul, K. (2011). Neuronal and oscillatory activity during reward processing in the human ventral striatum. *Neuroreport, 22*, 795–800.

MacDonald, A. W., Cohen, J. D., Stenger, V. A., & Carter, C. S. (2000). Dissociating the role of the dorsolateral prefrontal and anterior cingulate cortex in cognitive control. *Science, 288*, 1835–1838.

Mansouri, F. A., Buckley, M. J., & Tanaka, K. (2007). Mnemonic function of the dorsolateral prefrontal cortex in conflict-induced behavioral adjustment. *Science, 318*, 987–990.

Martin-Solch, C., Magyar, S., Kiinig, G., Missimer, J., Schultz, W., & Leenders, K. L. (2001). Changes in brain activation associated with reward processing in smokers and nonsmokers: A positron emission tomography study. *Experimental Brain Research, 139*, 278–286.

Matsumoto, K., Suzuki, W., & Tanaka, K. (2003). Neuronal correlates of goal-based motor selection in the prefrontal cortex. *Science, 301*, 229–232.

McClure, S. M., Berns, G. S., & Montague, P. R. (2003). Temporal prediction errors in a passive learning task activate human striatum. *Neuron, 38*, 339–346.

McCoy, A. N., Crowley, J. C., Haghighian, G., Dean, H. L., & Piatt, M. L. (2003). Saccade reward signals in posterior cingulate cortex. *Neuron, 40*, 1031–1040.

McKenna, F. P., & Sharma, D. (2004). Reversing the emotional Stroop effect reveals that it is not what it seems: The role of fast and slow components. *Journal of Experimental Psychology. Learning, Memory, and Cognition, 30*, 382–392.

Michelet, T., Bioulac, B., Guehl, D., Goillandeau, M., & Burbaud, P. (2009). Single medial prefrontal neurons cope with error. *PLoS ONE, 4*(7), e6240.

Milham, M. P., Banich, M. T., Webb, A., Barad, V., Cohen, N. J., Wszalek, T., et al. (2001). The relative involvement of anterior cingulate and prefrontal cortex in attentional control depends on nature of conflict. *Brain Research. Cognitive Brain Research, 12*, 467–473.

Mogenson, G. J., Jones, D. L., & Yim, C. Y. (1980). From motivation to action: Functional interface between the limbic system and the motor system. *Progress in Neurobiology, 14*(2–3), 69–97.

Montague, P. R., Hyman, S. E., & Cohen, J. D. (2004). Computational roles for dopamine in behavioural control. *Nature, 431*, 760–767.

Nakamura, K. K., Roesch, M. R., & Olson, C. R. (2005). Neuronal activity in macaque SEF and ACC during performance of tasks involving conflict. *Journal of Neurophysiology, 93*, 884–908.

Nauta, W. J., & Mehler, W. R. (1966). Projections of the lentiform nucleus in the monkey. *Brain Research, 1*, 3–12.

Nieuwenhuis, S., Holroyd, C. B., Mol, N., & Coles, M. G. H. (2004). Reinforcement-related brain potentials from medial frontal cortex: Origins and functional significance. *Neuroscience and Biobehavioral Reviews, 28*, 441–448.

O'Doherty, J., Dayan, P., Schultz, J., Deichmann, R., Friston, K., & Dolan, R. J. (2004). Dissociable roles of ventral and dorsal striatum in instrumental conditioning. *Science, 304*, 452–454.

O'Doherty, J. P., Dayan, P., Friston, K., Critchley, H., & Dolan, R. J. (2003). Temporal difference models and reward-related learning in the human brain. *Neuron, 38*, 329–337.

O'Doherty, J. P., Deichmann, R., Critchley, H. D., & Dolan, R. J. (2002). Neural responses during anticipation of a primary taste reward. *Neuron, 33*, 815–826.

Olds, J., & Milner, P. (1954). Positive reinforcement produced by electrical stimulation of septal area and other regions of rat brain. *Journal of Comparative and Physiological Psychology, 47*, 419–427.

Parent, A., & Hazrati, L. N. (1995). Functional anatomy of the basal ganglia: I. The cortico-basal ganglia-thalamo-cortical loop. *Brain Research. Brain Research Reviews, 20*, 91–127.

Patel, S. R., Sheth, S. A., Mian, M. K., Gale, J. T., Greenberg, B. D., Dougherty, D. D., et al. (2012). Single-neuron responses in the human nucleus accumbens during a financial decision-making task. *Journal of Neuroscience, 32*, 7311–7315.

Paus, T. (2001). Primate anterior cingulate cortex: Where motor control, drive and cognition interface. *Nature Reviews. Neuroscience, 2*, 417–424.

Pessiglione, M., Guehl, D., Rolland, A.-S., Francois, C., Hirsch, E. C., Feger, J., et al. (2005). Thalamic neuronal activity in dopamine-depleted primates: Evidence for a loss of functional segregation within basal ganglia circuits. *Journal of Neuroscience, 25*, 1523–1531.

Picard, N., & Strick, P. L. (1996). Motor areas of the medial wall: A review of their location and functional activation. *Cerebral Cortex, 6*, 342–353.

Procyk, E., Bush, G., Tanaka, Y. L., Vogt, B. A., Joseph, J. P., Holmes, J., et al. (2000). Anterior cingulate activity during routine and non-routine sequential behaviors in macaques. *Nature Neuroscience, 3*, 502–508.

Ridderinkhof, K. R. (2002). Micro- and macro-adjustments of task set: Activation and suppression in conflict tasks. *Psychological Research, 66*, 312–323.

Rushworth, M. F. S., Behrens, T. E. J., Rudebeck, P. H., & Walton, M. E. (2007). Contrasting roles for cingulate and orbitofrontal cortex in decisions and social behaviour. *Trends in Cognitive Sciences, 11*, 168–176.

Rushworth, M. F. S., Walton, M. E., Kennerley, S. W., & Bannerman, D. M. (2004). Action sets and decisions in the medial frontal cortex. *Trends in Cognitive Sciences, 8*, 410–417.

Samejima, K., Ueda, Y., Doya, K., & Kimura, M. (2005). Representation of action-specific reward values in the striatum. *Science, 310*, 1337–1340.

Schultz, W. (1998). Predictive reward signal of dopamine neurons. *Journal of Neurophysiology, 80*, 1–27.

Schultz, W. (2000). Multiple reward signals in the brain. *Nature Reviews. Neuroscience, 1*, 199–207.

Schultz, W. (2002). Getting formal with dopamine and reward. *Neuron, 36*, 241–263.

Schultz, W., Dayan, P., & Montague, P. R. (1997). A neural substrate of prediction and reward. *Science, 275*, 1593–1599.

Sheth, S. A., Mian, M. K., Patel, S. R., Asaad, W. E., Williams, Z. M., Dougherty, D. D., et al. (2012). Human dorsal anterior cingulate cortex neurons mediate ongoing behavioural adaptation. *Nature, 488*, 218–221.

Shidara, M., & Richmond, B. J. (2002). Anterior cingulate: Single neuronal signals related to degree of reward expectancy. *Science, 296*, 1709–1711.

Shima, K., & Tanji, J. (1998). Role for cingulate motor area cells in voluntary movement selection based on reward. *Science, 282*, 1335–1338.

Small, D. M., Zatorre, R. J., Dagher, A., Evans, A. C., & Jones-Gorman, M. (2001). Changes in brain activity related to eating chocolate: From pleasure to aversion. *Brain, 124*, 1720–1733.

Stroop, J. R. (1935). Studies of interference in serial verbal reactions. *Journal of Experimental Psychology, 18*, 643–662.

Swanson, L. W. (1982). The projections of the ventral tegmental area and adjacent regions: A combined fluorescent retrograde tracer and immunofluorescence study in the rat. *Brain Research Bulletin, 9*(1-6), 321–353.

Tremblay, L., Filion, M., & Bedard, P. J. (1989). Responses of pallidal neurons to striatal stimulation in monkeys with MPTP-induced Parkinsonism. *Brain Research, 498*, 17–33.

Ullsperger, M. M., & von Cramon, D. Y. D. (2001). Subprocesses of performance monitoring: A dissociation of error processing and response competition revealed by event-related fMRI and ERPs. *NeuroImage, 14*, 1387–1401.

van der Meer, M. A. A., & Redish, A. D. (2011). Ventral striatum: A critical look at models of learning and evaluation. *Current Opinion in Neurobiology, 21*, 387–392.

van Veen, V., Cohen, J. D., Botvinick, M. M., Stenger, V. A., & Carter, C. S. (2001). Anterior cingulate cortex, conflict monitoring, and levels of processing. *NeuroImage, 14*, 1302–1308.

Vendrell, P., Junque, C., Pujol, J., Jurado, M. A., Molet, J., & Grafman, J. (1995). The role of prefrontal regions in the Stroop task. *Neuropsychologia, 33*, 341–352.

Wager, T. D., Jonides, J., & Reading, S. (2004). Neuroimaging studies of shifting attention: A meta-analysis. *NeuroImage, 22*, 1679–1693.

Walton, M. E., Bayer, H. M., Devlin, J. T., Glimcher, P. W., & Rushworth, M. F. S. (2004). Interactions between decision making and performance monitoring within prefrontal cortex. *Nature Neuroscience, 7*, 1259–1265.

Williams, Z. M., & Eskandar, E. N. (2004). Human anterior cingulate neurons and the integration of monetary reward with motor responses. *Nature Neuroscience, 7*, 1370–1375.

Yoshida, W., & Ishii, S. (2006). Resolution of uncertainty in prefrontal cortex. *Neuron, 50*, 781–789.

Zaghloul, K. A., Blanco, J. A., Weidemann, C. T., McGill, K., Jaggi, J. L., Baltuch, G. H., et al. (2009). Human substantia nigra neurons encode unexpected financial rewards. *Science, 323*, 1496–1499.

Zald, D. H., Boileau, I., El-Dearedy, W., Gunn, R., McGlone, E., Dichter, G. S., et al. (2004). Dopamine transmission in the human striatum during monetary reward tasks. *Journal of Neuroscience, 24*, 4105–4112.

13 Electrophysiological Responses to Faces in the Human Amygdala

Ralph Adolphs, Hiroto Kawasaki, Oana Tudusciuc, Matthew Howard III, Chris Heller, William Sutherling, Linda Philpott, Ian Ross, Adam N. Mamelak, and Ueli Rutishauser

Structures in the temporal lobe have long been known to be important for object recognition, categorization, and naming. The most prominent examples of this come from face recognition. It was known for some time that lesions to ventral temporal cortex result in face agnosia, a disproportionate difficulty in naming, recognizing, and sometimes perceiving images of faces despite otherwise often largely spared visual perception, memory, and language (Damasio et al., 1990). Similarly, single unit recordings in monkeys had demonstrated regions of temporal cortex containing cells with a high selectivity for faces (Perrett et al., 1982; Hasselmo et al., 1989). The specialization of cortical sectors in the temporal lobe for face processing was further supported by the finding of the so-called "fusiform face area," a region in ventral temporal cortex showing greater blood-oxygen-level-dependent/functional magnetic resonance imaging (BOLD–fMRI) response to faces than to other classes of visual stimuli (Kanwisher et al., 1997; McCarthy et al., 1997) and, at least in monkeys, consisting of neurons that respond almost exclusively to faces (Tsao et al., 2006). These findings, together with a wealth of psychophysical behavioral data, all provided compelling evidence for specialized circuits in the temporal lobe of primates for processing information about faces (Kanwisher & Yovel, 2006; Tsao & Livingstone, 2008), data that have leveraged a veritable industry of research on the topic. It has even been shown that discrete electrical stimulation of the human ventral temporal cortex can cause highly specific, and reversible, changes in the conscious experience of how a face is seen (Parvizi et al., 2012).

Two key themes from this literature provide important background also to the present chapter, our understanding of the amygdala's role in face processing. One theme is greater anatomical and circuit-level specificity. It is now clear that there is a small set of "patches," discrete islands of cortex in the temporal cortex (Pinsk et al., 2009; Tsao et al., 2009) and frontal cortex (Tsao et al., 2008), that contain cells with high selectivity to faces and that correspond to activated regions in BOLD–fMRI studies in both monkeys and humans (Tsao et al., 2009). Most importantly, initial findings have now begun to sketch connectivity between these patches, conceiving of them as a distributed neural system for processing faces (Moeller et al., 2008). This is beginning to provide considerably more detail to the story, allowing us to understand how face representations are transformed from one face patch to another, and permitting some assignment of computational goals to different patches—for instance, a recent account of how viewpoint invariance is computed across these patches (Freiwald & Tsao, 2010). A challenge from these studies is reconciling findings from fMRI with also patchy representations of faces recorded with grids electrophysiologically (Allison et al., 1999; Kawasaki et al., 2012): They do not always match up and, of course, to some extent reflect different types of measurements.

The second key theme has been informed by diverse approaches, including electrophysiology, psychophysics, fMRI, and lesion studies: What is it about faces that the brain represents? Classic psychological models already conceived of faces as one source of input to representing a person, with associations to the person's name and semantic information about them (Bruce & Young, 2012). Some information about faces was also known to be dissociable, such as information about the identity of the person versus the person's emotional state (Bruce & Young, 1986; Tranel et al., 1988; Calder & Young, 2005). More recent work has used factor-analytic approaches with parametric variations of synthetic faces stimuli to argue for simple dimensional schemes (usually two-dimensional) that are thought to capture psychological principles of how viewers represent faces (Young et al., 1997; Calder et al., 2001; Oosterhof & Todorov, 2008, 2009), with the dimensions often aligned with aspects of social evaluation or emotion.

It is at the intersection of these two themes that we just reviewed that the amygdala comes into focus: We need to know how it is connected to the rest of the face-processing system; and we need to know what about faces it represents. Of course, these two aspects are highly related, and we need to understand both to build a comprehensive picture of human face processing in which the amygdala plays specific computational roles as one component of a distributed system. Basic evidence that the amygdala plays such a role, or set of roles, is well-known. There are neurons in the amygdala of monkeys that respond with high selectivity to faces (Leonard et al., 1985; Gothard et al., 2007; Kuraoka & Nakamura, 2007), and across fMRI studies the amygdala is reliably activated by faces (Mende-Siedlecki et al., 2013). Focal lesions of the human amygdala can result in impairments in social judgments about faces (Adolphs et al., 1998), notably in the ability to recognize facial expressions of fear (Adolphs et al., 1994; Adolphs et al., 1999), findings also consistent with fMRI studies showing differential activation in humans (Morris et al., 1996; Winston et al., 2002) as in monkeys (Hoffman et al., 2007). Especially intriguing are a handful of studies that have sought to combine lesion studies with fMRI, which have documented effects of amygdala lesions on BOLD–fMRI activation in temporal cortex during processing of emotional face stimuli in both humans (Vuilleumier et al., 2004) and monkeys (Hadj-Bouziane et al., 2012). Taken together, all these studies argue that the amygdala is indeed a component of the distributed system for face processing, as would be expected given its substantial reciprocal connectivity with temporal neocortex (Amaral et al., 1992) (see figure 13.1), but leave relatively unexplored the details of where in the network it is situated. And they argue for an important role in processing social information from faces but leave relatively unexplored the details of precisely what social information, and what features of faces, this is based on.

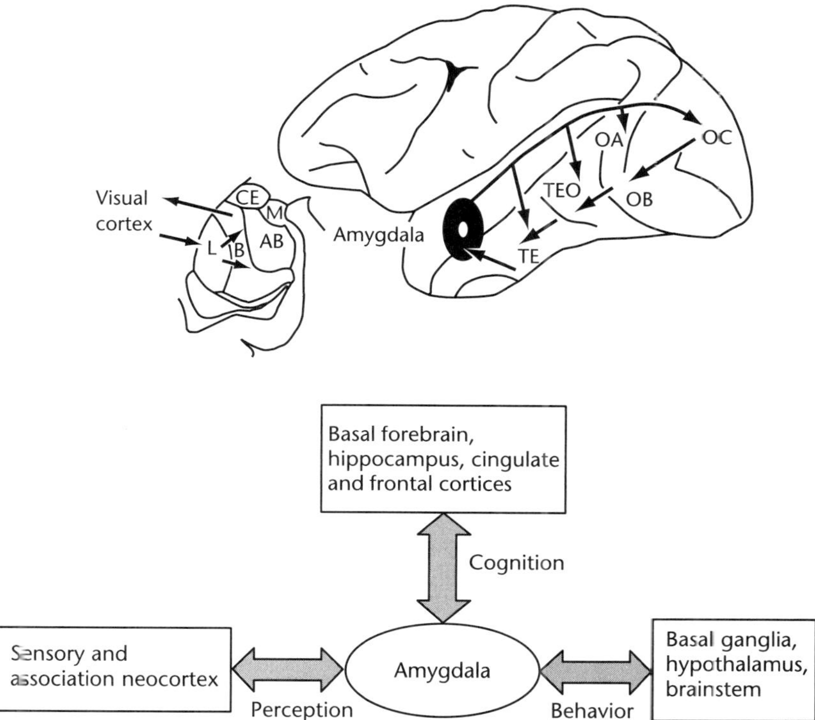

Figure 13.1
Amygdala connectivity. Top: Connections of the amygdala with visual cortex. In primates, the amygdala has both reciprocal and nonreciprocal feedback with visual cortices in the temporal lobe. In fact, it projects all the way back to V1. This may permit the amygdala to modulate visual responses in cortex based on the emotional significance of the stimuli being processed. Adapted from Amaral et al. (1992). AB, accessory basal nucleus; B, basal nucleus; CE, central nucleus; L, lateral nucleus; M, medial nucleus; OA, OB, OC, sectors of occipital cortex; TE, TEO, sectors of temporal cortex. Bottom: Schematic of systems-level connectivity between the amygdala and other brain regions, indicating the broad class of processes subserved by this connectivity. The amygdala can modulate perception, cognition, and behavior based on the emotional significance of a stimulus.

Eyes and the Amygdala

There is a long history of findings, and debate, concerning the nature of face representations in the brain (Bruce & Young, 2012). On the one hand, faces could be encoded as the sum of their constituent features. On the other hand, considerable evidence suggests that at least a simple linear version of this idea is not generally true, and that faces are encoded in a more global or "holistic" fashion that cannot be traced to individual features but depends crucially on their relationships. Nonetheless, findings in monkeys have documented responses to specific features in faces that would contribute to face perception (Freiwald et al., 2009). Similarly, fMRI adaptation studies in humans have found evidence for responses to individual features (although more so in the left than the right side of the brain; Harris & Aguirre, 2010). One feature shown to modulate electrophysiological responses in humans are the eyes in faces (McCarthy et al., 1999). In all of these studies, the focus was on regions of temporal and occipital cortex highly selective to faces. What about the amygdala?

There is plenty of evidence that the eye region of faces is the most salient and important region of the face. Viewers preferentially fixate the eyes in faces (Scheller et al., 2012), a region particularly important for recognizing fear (Smith et al., 2005) and one that is also processed early by the brain (Schyns et al., 2007). Interestingly, such processing also appears to involve, and indeed depend upon, the amygdala. A correlation between BOLD response in the amygdala and the time that subjects spent looking at the eyes in face stimuli was first noticed in a study of people with autism (Dalton et al., 2005), then extended to the first-degree relatives of people with autism (who themselves did not have a diagnosis of autism; Dalton et al., 2007), fueling a line of research that has tied abnormal amygdala structure or function to the social impairments seen in autism or the broad autism phenotype (Schumann and Amaral, 2006; Amaral et al., 2008; Kliemann et al., 2012) as well as other neurodevelopmental disorders (Schumann et al., 2011). A more detailed analysis in both typically developed participants as well as high-functioning people with autism showed that amygdala BOLD–fMRI activation was in fact correlated with the probability of making a saccade onto the eye region of faces (Gamer & Buechel, 2009; Kliemann et al., 2012), suggesting a more instrumental role in relation to eye gaze: The amygdala may encode what is salient and what should be fixated or attended, rather than simply responding to what is seen. All of these findings were also broadly consistent with a larger literature finding that the eye region of faces is particularly potent in eliciting BOLD responses from the human amygdala (Kawashima et al., 1999; Morris et al., 2002; Whalen et al. 2004; Demos et al., 2008) and that amygdala lesions impair fixations onto, and processing of information from, the eye region (Adolphs et al., 2005; Spezio et al., 2007).

Electrophysiologically, there was evidence from analysis of scalp-recorded event-related potentials (ERPs) in healthy participants that the eye region of the face was one of the earliest features processed, as we already noted above (Schyns et al., 2007), but further details in regard to the amygdala of course required depth-electrode recordings. Meletti and colleagues (Meletti et al., 2012), in a study of four patients with such depth electrodes, found strong evidence for considerably larger evoked responses from field potentials recorded in the amygdala to the eyes in faces than to the mouth region—or even to the whole face. These recordings were predominantly from the basolateral amygdala and had relatively long latencies (200–300 ms), and an event-related band power analysis showed that the largest effects appeared to be in the theta range (4–7 Hz). The long latencies observed are consistent with other reports (Mormann et al., 2008) and our own observations (Rutishauser et al., 2011a, b) when recording single unit responses, as well as reports using field potentials (Sato et al., 2012), although there are some reports of intracranial ERPs in the amygdala with shorter latencies as well (Oya et al., 2002).

In our recent work, we recorded from the amygdala in a sample of epilepsy patients with implanted depth electrodes, and used an unbiased approach in which randomly sampled pieces of faces ("bubbles") were shown to participants (see figure 13.2; Gosselin & Schyns, 2001).

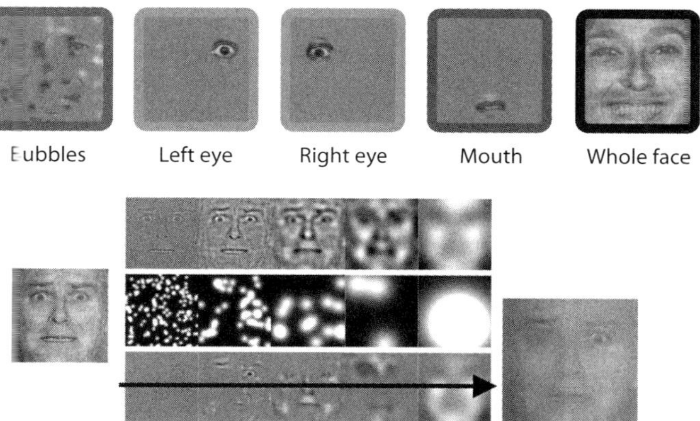

Figure 13.2
Stimuli used in the experiments. Top row: Classes of stimuli used in the experiments shown in figures 13.3 and 13.4 (plates 14 and 15). The starting point were images of whole faces (far right), from which several other stimulus categories were derived (parts, like eyes and mouth), and homogeneously sparsely sampled face stimuli, so-called "bubbles," shown at the far left. Bottom: construction of the "bubbles" stimuli from a whole face. The initial face image was filtered through Gaussian bubbles to yield a sparsely sampled face.

Subjects were asked to judge for each partially revealed face whether it was happy or fearful. The reaction time and accuracy of the subject's performance across the many trials was correlated, for each location on the face image, with whether or not a given trial revealed or occluded that pixel, yielding a two-dimensional correlation map referred to as the "classification image." Pixels with high correlations are important for distinguishing happy from fearful stimuli whereas pixels with 0 correlation will not be important. This technique has been used by numerous authors to derive behavioral classification images (Adolphs et al., 2005; Gosselin & Schyns, 2001; Smith et al., 2005). Our epilepsy patients exhibited similar behavioral classification images (see figure 1 in Rutishauser et al., 2011b), revealing that they utilized information used by the mouth and the eye to make their judgment about the emotion expressed by a face.

By analogy to deriving a behavioral classification image as above—which depicts regions of the face that are significantly associated with behavioral performance accuracy—we developed a technique to define neuronal classification images (NCI) for every single neuron recorded while patients performed this task (unpublished data and Rutishauser et al., 2011a). For this we correlated, for every trial and pixel, the number of spikes fired by a putative single unit and whether a pixel was revealed or not. If revealing a particular pixel results in an increased or decreased firing rate, the resulting correlation is expected to be significantly different from zero. We identified a significant NCI for 19% of amygdala single neurons. Significant NCIs were found to be distributed over the face to some extent, but to notably cluster around the eyes and mouth (figure 13.3, plate 14, shows an example). We thus conclude that a sizeable portion of about 20% of single neurons in the human amygdala respond specifically to certain informative parts of the face during an emotional categorization task, making these neurons also candidates for participating in the behavioral categorization. This population of neurons with responses

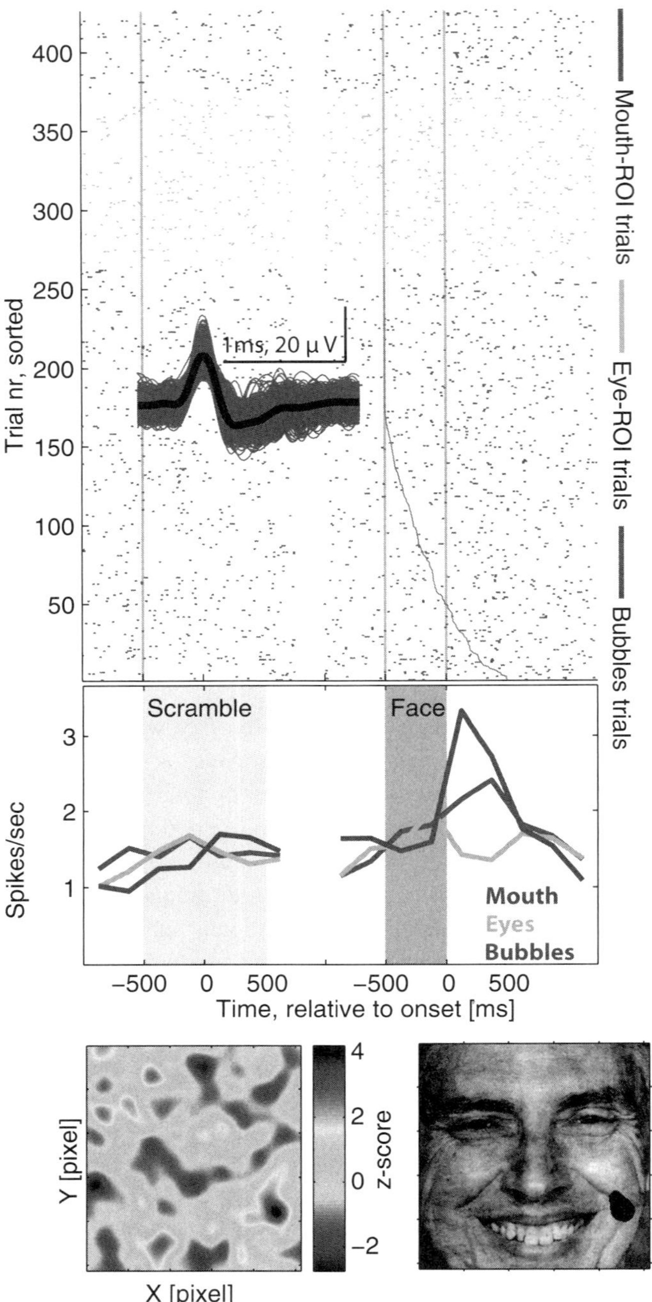

Figure 13.3 (plate 14)

Single neuron example of an amygdala neuron with a part-selective response. Shown are the raster (top) with the wave-forms superimposed, the poststimulus time histogram (middle), and the neuronal classification image (bottom). The significant portion of the classification image is superimposed on a face. The neuronal classification image is located close to the mouth, and the neuron also responds to faces where the mouth region of interest (ROI) is revealed. Nr, number.

driven by specific parts of a face stands in contrast to a likely distinct population of amygdala neurons that responds exclusively to whole faces (see below) but not to their parts.

One immediate difficulty is that the socially most salient features of faces are also typically the ones with the highest bottom-up saliency defined by contrast and/or motion: The eyes and mouth both feature large contrast changes, and both move more than other regions of the face. The theme of responses to eyes and to salient regions, whatever the source of that saliency, is consistent with emerging views on the amygdala's function in encoding anything about which an organism needs to learn more. A basic mechanism driving such learning is attention to those stimuli about which little is yet known—that is, those that are unpredicted, ambiguous, or novel. A large literature ranging from rodents through humans has documented amygdala responses that are best interpreted in this way (Herry et al., 2007; Whalen, 2007; Adolphs, 2010; Roesch et al., 2010; Rutishauser et al., 2006).

A second issue is that it is often presumed that whole-face processing that is holistic is a prerequisite for many or most of our social judgments about the faces. Indeed, in the next section we will describe a class of neurons that are at the opposite end of the spectrum in terms of their response selectivity: neurons that are exquisitely tuned to only whole faces and that show a dramatic decrement in response if only a small portion of a face is occluded. However, there is behavioral data suggesting that disruption of holistic face processing can indeed severely impair the recognition of identity (as occurs in prosopagnosia), but that this can leave intact the ability to make judgments about the emotion shown in the face, or attributes such as age and gender (Tranel et al., 1988). Similarly, even complex social judgments (such as of trustworthiness or intelligence) can be made normally in such patients (Quadflieg et al., 2012). Thus, it need not be the case that neurons with responses tuned to specific features, such as the ones we have described in this section, play no role in associating faces with social information. They may provide input to whole-face-selective neurons, but they may also perfectly well function in and of themselves, in ensemble fashion, to achieve such a computation. Nonetheless, it appears that there is also a distinct class of neuron in the amygdala that cares a lot about representing a face in its entirety, a topic we turn to next.

Whole-Face Responses in the Amygdala

Holistic facial responses are the other extreme end of face-selective responses. Such a hypothesized response would only occur if an entire face is present whereas no response would occur to individual parts such as eyes or mouths. A separate and independent question is what drives such a response—is it the nonlinear sum of responses to parts, or are whole faces detected by independent means (Tang & Kreiman, 2011), as has been suggested based on latency arguments?

We investigated this question by recording from over 200 single neurons in the amygdalae of seven neurosurgical patients with implanted depth electrodes (Rutishauser et al., 2011b). We again presented partially revealed faces using the bubbles technique (see above and figure 13.2) while patients were asked to judge for every trial whether the face presented was happy or fearful. In addition to partially revealed faces, we also presented whole faces (WFs) and selected regions of interest (ROIs; eye and mouth cutouts; see figure 13.2). Before every fully or partially revealed face, a baseline scramble was presented that contained equal statistical power but was indistinguishable from noise (figure 13.2). We analyzed a subset of, in total, 185 units that had an average firing rate of 0.2 Hz or more. Of these units 51.4% changed their firing rate in response to presentation of faces relative to the scramble baseline (42% of the responsive units responded with an increase and 58% with a decrease in firing rate). In contrast, only 1.4% of units responded to the scramble stimulus relative to blank screen. Thus, about half of the units responded selectively when facial features were revealed but not when a visual image of equal visual complexity was presented.

What features of faces did these units respond to? A subset of 19.5% (n = 36) of all units responded specifically only when the entire face was presented (Rutishauser et al., 2011b). These are WF selective neurons (figure 13.4A, plate 15, shows an example). We selected the WF-responsive neurons by comparing WF trials with cutout trials (eye or mouth revealed, defined by ROIs) and later quantified the response by using the independent bubbles trials using a whole-face index (WFI). The WFI is thus a measure that is statistically independent from how the neurons were selected in the first place. It is the percentage difference in firing rate between partially revealed (bubbled) faces and WFs, normalized to baseline. The average WFI for the WF neurons was 53% ± 7% and significantly different from the cells not classified as WF (18% ± 2%). We conclude that a subpopulation of about 20% of amygdala neurons is particularly responsive to WFs.

How did the WF neurons respond to the partially revealed faces? Some bubbled faces were almost entirely revealed faces (only a few parts removed) whereas others only revealed a small patch such as the tip of the nose. We systematically analyzed the response of the WF-selective neurons as a function of the proportion of the eye and mouth region revealed in each bubble trial. We found that the relationship between WF and non-WF responses was highly nonlinear. We separately analyzed two groups of neurons, those that increased (n = 32) and those that decreased (n = 4) their firing rate to WFs relative to baseline, respectively (see figure 13.4, plate 15). The neuronal response of the subgroup of WF cells that increase their rate (n = 32) was progressively smaller the more the face was revealed (see figure 13.4B, plate 15). Thus, the more similar to a WF the partially revealed face looked, the more different was the response of the neuron compared to when the entire face was revealed. The same pattern, but of opposite sign, was true for neurons that decrease their firing to WFs (see figure 13.4C, plate 15): The more the face was revealed, the higher the firing rate. The largest difference in neuronal response was thus between WFs and "almost" WFs, those bubbled faces where only a small proportion was hidden. We conclude that the WF neurons respond specifically to WFs only, and not to faces that reveal almost a WF but have a small portion occluded.

In conclusion, we found that over half of all neurons in the amygdala respond to face stimuli and a substantial proportion of these responded selectively only to WFs. The response latency

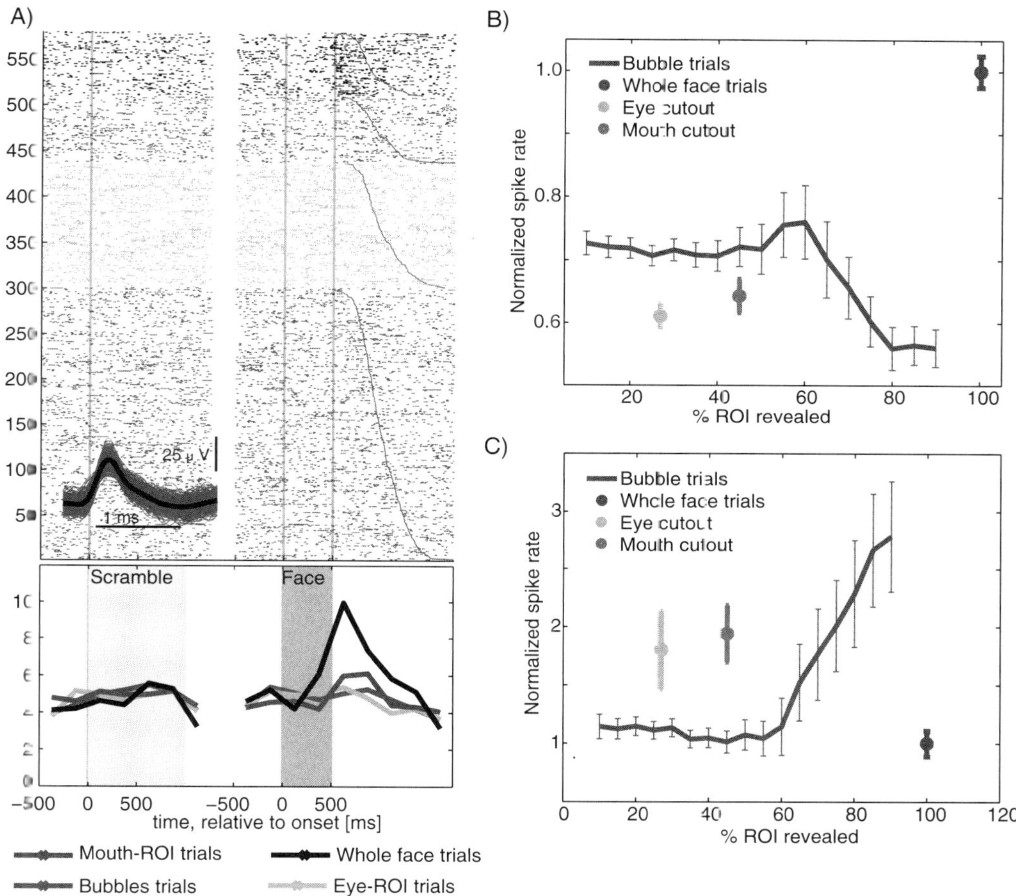

Figure 13.4 (plate 15)

Whole face selectivity of amygdala neurons. (A) Single neuron example of a whole-face selective neuron. Trials are shown ordered according to category (color code) and reaction time (magenta line). The stimulus (face) was on the screen for 500 ms. The superimposed waveforms show the waveform of each spike displayed in the raster; y-axis is in units of trial number (top) and spikes/second (bottom). (B, C) Quantification of nonlinear response to whole faces (WFs) and parts for neurons that increase their firing rate to WFs (B) and those that decrease (C). ROI, region of interest. Modified from Rutishauser et al. (2011b).

of these neurons was long, with the earliest responses occurring within 250–500 ms. The WF-selective neurons responded highly nonlinearly to partially revealed faces: Their response to WFs was inversely correlated with their response to variable amounts of partially revealed faces or facial features (eyes and mouth). Thus, the response to partially revealed faces was not predictive of the response to WFs. We conclude that these results provide strong support for the hypothesis that some amygdala neurons encode holistic information about faces beyond what a summed response to the parts would reveal.

Categorical and Dimensional Representation of Faces in the Amygdala

Early studies of the amygdala focused in large part on fear—ranging from unconditioned fear behaviors to Pavlovian fear conditioning to recognition of fear in faces. These studies prompted the hypothesis that the amygdala is a module concerned with detecting threat or danger in the environment. While this view may still be partly true, it is now clear that it cannot be the whole story. The reason is that there is a wealth of studies demonstrating amygdala responses to all kinds of stimuli that are not just threatening. For instance, appetitive sexual stimuli activate the human amygdala (Hamann et al., 2002), and single neurons in the monkey amygdala respond to stimuli with appetitive value (Nishijo et al., 1988; Paton et al., 2006).

Following an initial investigation of amygdala responses to emotions shown in faces (Fried et al., 1997), several other reports have found differential responses in the amygdala to faces, depending on the emotion expressed by the face (Krolak-Salmon et al., 2004; Rutishauser et al., 2011b; Meletti et al., 2012). This finding is also consistent with responses from single units recorded in the amygdala of monkeys, some of which show high selectivity for the identity of the face and invariance with respect to its emotion whereas others show exactly the converse pattern (high selectivity to emotion with invariance to the identity; Gothard et al., 2007). It is similarly consistent with amygdala intracranial EEG responses to the emotions depicted by, or induced by, a wide variety of complex images (Oya et al., 2002). What is less clear is how these findings are related to the amygdala's role in representing the value of stimuli, independently of their sensory properties, as has been found for both appetitive and aversive learning in monkeys (Paton et al., 2006).

One large study (Mormann et al., 2011) recorded from over 400 single neurons in 41 patients and examined this pooled data set in relation to category-selective responses in various sectors of the medial temporal lobe. One category–structure association stood out: The right amygdala, on average, had the most neurons that responded to pictures of animals, compared to other categories (people, landmarks, objects; see figure 13.5). The finding is surprising for a number of reasons. First, one might have expected pictures of faces (i.e., people) to be the predominant category, but this was not the case. Second, the particular animals to which amygdala neurons responded varied tremendously, and they showed no particular preference for animals that are threatening (snakes, spiders, predators)—cells responded to images of horses and rabbits as well. Given the large sample, the finding appears statistically robust. It is also consistent with other

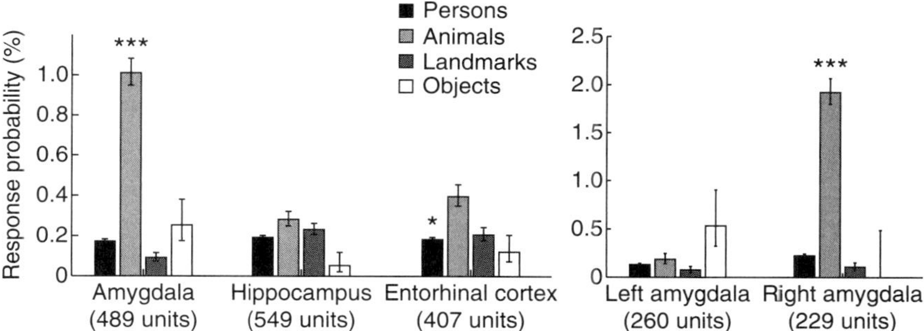

Figure 13.5
Stimulus selectivity in the amygdala. Shown is the probability of neuron responses recorded from over 400 neurons in 41 patients across various sectors of the medial temporal lobe. Error bars are 68% confidence intervals and stars denote significant difference in category tuning with respect to all responses within the same area at p < 0.001 (***) and p < 0.05 (*). Adapted from Mormann et al. (2011).

findings that animals can be detected rapidly and preattentively, arguing for a possible module for detecting animals in the environment very automatically—a module that could form the front end of a richer mechanism for detecting conspecifics and predators as well. That story is consistent with the amygdala's broader role in saliency that we mentioned earlier.

Exactly how responses encoding social information in the amygdala might be constructed remains unclear. In temporal cortex, electrophysiological signals encode emotional information at later points in time than simply detecting a face, in monkey single units (Sugase et al., 1999) as well as in human intracranial EEG (Pourtois et al., 2009), consistent with a model whereby the emotional information requires feedback from the amygdala as we reviewed above (although alternative models are possible also). Intracranial recordings in humans have suggested a spread of response through time as information about fearful faces is encoded, beginning in the amygdala and then spreading to other cortices (Krolak-Salmon et al., 2004). There is also a long-standing debate regarding the prepotent effects of emotion, with some findings that amygdala responses are driven by emotion in an automatic fashion that does not require attention (or, in some studies, conscious awareness of the stimuli; Whalen et al., 1998), contradicted by other studies claiming attention is required for emotional stimuli to be processing by the amygdala (Pessoa et al., 2006). All these studies were based on fMRI, however, and a recent reconciliation was offered from intracranial recordings in one surgical patient, which showed that fast-latency responses to emotional faces were insensitive to attention whereas later responses were modulated by attention (Pourtois et al., 2010). Possibly, the substantial temporal smoothing that occurs with BOLD–fMRI measures failed to dissociate these two components, resulting in discrepant findings in the literature.

Neuroimaging studies in humans have argued that the BOLD response in the amygdala encodes basic social attributes of faces, notably their valence and intensity, on which other social judgments can then be based (Todorov & Engell, 2008; Mende-Siedlecki et al., 2013).

Two-dimensional accounts of social evaluation have enjoyed considerable popularity, ranging from schemes whereby we evaluate other people (Fiske et al., 2007) to dimensions on which we attribute minds to any entity (Waytz et al., 2010). How do these relate to amygdala responses? And at what point in time do responses in the amygdala construct such high-level representations of people? These and other open questions for the future will require a richer and more ecologically valid set of stimuli, as well as more detailed analyses of the amygdala's connectivity with other brain structures and of its temporal profile in encoding social information from faces.

Summary

The evidence reviewed here clearly implicates the amygdala in processing faces and suggests several key points for summary. First, there are multiple populations of neurons within the amygdala. Some of these code for the identity of faces whereas others code for their emotional expression (Gothard et al., 2007). Most appear to show some sensitivity to a specific emotion or set of emotions shown in a face (Rutishauser et al., 2011b). About half of all neurons show some preferential response to faces over their scrambled versions (Rutishauser et al., 2011b) although this number may be smaller when contrasting faces with other specific object categories (Seeck et al., 2001; Mormann et al., 2011). Some amygdala neurons care about the whole face and show very intriguing nonlinear responses when even a small part of the face is occluded (Rutishauser et al., 2011b) whereas other neurons encode specific features of faces, notably eyes and mouth (Rutishauser et al., 2011a).

Taken together with other data on the amygdala and face processing, the emerging picture is one of a rather distributed nature in both space and time. Somewhat surprisingly, the amygdala appears to come into play mostly at relatively long latencies (>200 ms), suggesting prior extended processing in cortex. This finding supports models of face processing in which there is reciprocal interaction between the amygdala and temporal neocortex for processing faces rather than one in which there is an initial wave of processing to the amygdala through exclusively subcortical structures (Pessoa & Adolphs, 2010). There is evidence for considerably more rapid response to emotional faces in visual cortex (Seeck et al., 1997; Tsuchiya et al., 2008) and even in frontal cortex (Kawasaki et al., 2001), but a more complex evaluation of the social meaning of the face, together with modulation of its cortical representation through feedback, likely involves the amygdala (Vuilleumier & Pourtois, 2007). Indeed, concurrent recordings in amygdala together with other temporal lobe structures have shown that initial responses to fearful faces in the amygdala spread to cortex at later points in time (Krolak-Salmon et al., 2004).

Future Directions

Task and Context Effects

One important challenge is to better understand interactions between task and stimuli. Whereas considerable attention has been paid to the stimuli, less has been paid to the task employed, even though there is good evidence that this also modulates the electrophysiological responses seen

(Seeck et al., 2001). In part, task effects can be seen as one aspect of the broader theme of top-down modulation, for which there is substantial evidence (Puce et al., 1999). However, in terms of electrophysiological studies, most of the data come from recordings in cortex, not in the amygdala—indeed, a common mechanism for such top-down modulation, as we reviewed above, is that there is feedback to all temporal and occipital visual cortices from the amygdala, providing a top-down modulatory signal that may be especially relevant when processing highly emotional stimuli (cf. also figure 13.1).

The challenge of task and context effects is, of course, related to another challenge that is important—especially with respect to our understanding of the role of different brain structures at distinct points in time—namely, the challenge of extending our understanding of the processing of static faces to that of dynamic faces, an issue that is especially pertinent for facial emotions, which in real life are always dynamic. Little has yet been done to explore this rich and complex issue (Tsuchiya et al., 2008; Foley et al., 2011; Kawasaki et al., 2012). Another related topic is that of cognitive control and emotion regulation—a domain where one might expect particularly interesting differences between humans and nonhuman primates (or, for that matter, between adult humans and children/infants—or, between different adults, i.e., individual differences among adults). There are findings from fMRI studies of the amygdala showing that the BOLD response to emotional stimuli can be modulated by volitional reappraisal (instruction to change one's emotional response to the stimuli by reinterpreting their meaning; Ochsner et al., 2002; Schaefer et al., 2002; Koenigsberg et al., 2010). Where such modulatory signals originate remains unclear since fMRI lacks the temporal-and spatial resolution to tease this issue apart. However, one good candidate is the prefrontal cortex, which shows surprisingly short latency responses to emotionally laden stimuli in some single unit recording studies (Kawasaki et al., 2001), has anatomical connectivity suited for driving the responses of amygdala neurons (Öngür & Price, 2000; Miyashita et al., 2007), and has long been linked to cognitive and emotional control.

Three Recommendations

Three key future directions are at the top of our list for where to go next. First, we need more anatomical detail. This recommendation goes in both directions, actually: We need more precision and uniform reporting regarding the subnuclei of the amygdala from which recordings are obtained, and we also need information regarding responses in other structures with which the amygdala is connected. Both are quite feasible. Three immediate places from which to record would be temporal neocortex, hippocampus, and medial frontal cortex. All three locations are commonly targeted with depth electrodes. The ideal experiment would record from amygdala and some of these distal targets concurrently and assess relative response latencies and functional connectivity, possibly even in conjunction with electrical stimulation in one of the structures. This kind of experiment would really take our understanding of the amygdala's contribution to face processing to the network level. While appealing, it is also likely that many aspects of such a network understanding would require new conceptual and analytic tools. Some initial examples of concurrent recordings in amygdala in prefrontal cortex in monkeys, for instance, show that the dynamic interplay between responses in these two structures is very complex (Morrison et al., 2011) and will probably require a dynamical systems-level description if we want to make sense of it.

A related possibility that sometimes arises in a neurosurgical setting is the conjunction of a focal lesion and the opportunity to record in a distal target. Some patients have a lesion that is the result of prior neurosurgery, or the presumptive locus of seizure activity, and recording from other structures distal to the lesion would provide valuable information about the necessary role that the lesioned structure plays in shaping responses recorded in the distal region. Something like this has been done using the conjunction of lesions and fMRI (lesion in the amygdala while measuring BOLD activation in medial prefrontal cortex; Hampton et al., 2007).

A second important direction we would like to emphasize is the combination of eye tracking with concurrent recordings. This is technically challenging but now also entirely feasible and in fact being carried out by several groups. Just as fMRI studies have combined BOLD–fMRI with concurrent eye tracking to show how amygdala responses correlate with saccades onto features of faces (Gamer & Buechel, 2009; Kliemann et al., 2012), the same basic design (but with, of course, much greater power in terms of temporal resolution) can be translated into electrophysiological studies. It will be critical to do this for a number of reasons. One is ecological: Under normal circumstances, we are always moving our eyes, and it would seem reasonable that the representation of social stimuli in the brain is modulated by visual attention and fixations. A second is conceptual: Are amygdala responses driven primarily by the sensory input, or are they more premotor in nature and serve to allocate attention toward subsequent targets of sensory inputs (places we fixate), much as the amygdala is known to provide attentional modulation to other cognitive processes?

A third direction is a better characterization of the stimuli and their social properties. This is perhaps the most challenging of all since, in principle, faces are extremely high-dimensional stimuli and we cannot possibly search the entire space. Instead, we should pick small regions within that space, or focused hypotheses, and begin to build up our inventory of the factors and conditions that modulate neuronal responses to faces. This could perhaps be done in two directions, of which only one has received any attention. One is reductionist and aims to control and dissect the stimuli, but the downside of this approach is a possible loss of ecological validity. The second would aim to use dynamic facial expressions or, in the extreme case, even real people as the stimuli. Again, something approaching this has been done using fMRI although it is highly constrained by the artificial environment of the scanner—an issue where we would have better opportunities with intracranial recordings.

References

Adolphs, R. (2010). What does the amygdala contribute to social cognition? *Annals of the New York Academy of Sciences, 1191*, 42–61.

Adolphs, R., Gosselin, F., Buchanan, T. W., Tranel, D., Schyns, P., & Damasio, A. R. (2005). A mechanism for impaired fear recognition after amygdala damage. *Nature, 433*, 68–72.

Adolphs, R., Tranel, D., & Damasio, A. R. (1998). The human amygdala in social judgment. *Nature 393*, 470–474.

Adolphs, R., Tranel, D., Damasio, H., & Damasio, A. (1994). Impaired recognition of emotion in facial expressions following bilateral damage to the human amygdala. *Nature, 372,* 669–672.

Adolphs, R., Tranel, D., Hamann, S., Young, A., Calder, A., Anderson, A., et al. (1999). Recognition of facial emotion in nine subjects with bilateral amygdala damage. *Neuropsychologia, 37,* 1111–1117.

Allison, T., Puce, A., Spencer, D., & McCarthy, G. (1999). Electrophysiological studies of human face perception: I. Potentials generated in occipitotemporal cortex by face and non-face stimuli. *Cerebral Cortex, 9,* 415–430.

Amaral, D. G., Price, J. L., Pitkanen, A., & Carmichael, S. T. (1992). Anatomical organization of the primate amygdaloid complex. In J. P. Aggleton (Ed.), *The amygdala: Neurobiological aspects of emotion, memory, and mental dysfunction* (pp. 1–66). New York: Wiley-Liss.

Amaral, D. G., Schumann, C. M., & Nordahl, C. W. (2008). Neuroanatomy of autism. *Trends in Neurosciences, 31,* 137–145.

Bruce, V., & Young, A. (1986). Understanding face recognition. *British Journal of Psychology, 77,* 305–327.

Bruce, V., & Young, A. W. (2012). *Face perception.* New York: Psychology Press.

Calder, A. J., Burton, A. M., Miller, P., Young, A. W., & Akamatsu, S. (2001). A principal component analysis of facial expressions. *Vision Research, 41,* 1179–1208.

Calder, A., & Young, A. (2005). Understanding the recognition of facial identity and facial expression. *Nature Reviews. Neuroscience, 6,* 641–651.

Dalton, K. M., Nacewicz, B. M., Alexander, A. L., & Davidson, R. J. (2007). Gaze-fixation, brain activation, and amygdala volume in unaffected siblings of individuals with autism. *Biological Psychiatry, 61,* 512–520.

Dalton, K. M., Nacewicz, B. M., Johnstone, T., Schaefer, H. S., Gernsbacher, M. A., Goldsmith, H. H., et al. (2005). Gaze fixation and the neural circuitry of face processing in autism. *Nature Neuroscience, 8,* 519–526.

Damasio, A. R., Tranel, D., & Damasio, H. (1990). Face agnosia and the neural substrates of memory. *Annual Review of Neuroscience, 13,* 89–109.

Demos, K. E., Kelley, W. M., Ryan, S. L., Davis, F. C., & Whalen, P. J. (2008). Human amygdala sensitivity to the pupil size of others. *Cerebral Cortex, 18,* 2729–2734.

Fiske, S. T., Cuddy, A. J. C., & Glick, P. (2007). Universal dimensions of social cognition: Warmth and competence. *TICS, 11,* 78–83.

Foley, E., Rippon, G., Thai, N. J., Longe, O., & Senior, C. (2011). Dynamic facial expressions evoke distinct activation in the face perception network: A connectivity analysis study. *Journal of Cognitive Neuroscience, 24,* 507–520.

Freiwald, W. A., & Tsao, D. Y. (2010). Functional compartmentalization and viewpoint generalization within the macaque face-processing system. *Science, 330,* 845–851.

Freiwald, W. A., Tsao, D. Y., & Livingstone, M. S. (2009). A face feature space in the macaque temporal lobe. *Nature Neuroscience, 12,* 1187–1196.

Fried, I., MacDonald, K. A., & Wilson, C. L. (1997). Single neuron activity in human hippocampus and amygdala during recognition of faces and objects. *Neuron, 18,* 753–765.

Gamer, M., & Buechel, C. (2009). Amygdala activation predicts gaze toward fearful eyes. *Journal of Neuroscience, 29,* 9123–9126.

Gosselin, F. & Schyns, P. G. (2001). Bubbles: A technique to reveal the use of information in recognition. *Vision Research, 41,* 2261–2271.

Gothard, K. M., Battaglia, F. P., Erickson, C. A., Spitler, K. M., & Amaral, D. G. (2007). Neural responses to facial expression and face identity in the monkey amygdala. *Journal of Neurophysiology. 97,* 1671–1683.

Hadj-Bouziane, F., Liu, N., Bell, A. H., Gothard, K. M., Luh, W.-M., Tootell, R. B. H., et al. (2012). Amygdala lesions disrupt modulation of functional MRI activity evoked by facial expression in the monkey inferior temporal cortex. *Proceedings of the National Academy of Sciences of the United States of America, 109,* e3640–e3648.

Hamann, S. B., Ely, T. D., Hoffman, J. M., & Kilts, C. D. (2002). Ecstasy and agony: Activation of the human amygdala in positive and negative emotion. *Psychological Science, 13,* 135–141.

Hampton, A., Adolphs, R., Tyszka, J. M., & O'Doherty, J. (2007). Contributions of the amygdala to reward expectancy and choice signals in human prefrontal cortex. *Neuron, 55,* 545–555.

Harris, A., & Aguirre, G. K. (2010). Neural tuning for face wholes and parts in human fusiform gyrus revealed by fMRI adaptation. *Journal of Neurophysiology, 104,* 336–345.

Hasselmo, M. E., Rolls, E. T., & Baylis, G. C. (1989). The role of expression and identity in the face-selective responses of neurons in the temporal visual cortex of the monkey. *Behavioural Brain Research, 32*, 203–218.

Herry, C., Bach, D. R., Esposito, F., DiSalle, F., Perrig, W. J., Scheffler, K., et al. (2007). Processing of temporal unpredictability in human and animal amygdala. *Journal of Neuroscience, 27*, 5958–5966.

Hoffman, K. L., Gothard, K. M., Schmid, M. C., & Logothetis, N. K. (2007). Facial-expression and gaze-selective responses in the monkey amygdala. *Current Biology, 167*, 766–772.

Kanwisher, N., McDermott, J., & Chun, M. M. (1997). The fusiform face area: A module in human extrastriate cortex specialized for face perception. *Journal of Neuroscience, 17*, 4302–4311.

Kanwisher, N., & Yovel, G. (2006). The fusiform face area: A cortical region specialized for the perception of faces. *Philosophical Transactions of the Royal Society of London. Series B, Biological Sciences, 361*, 2109–2128.

Kawasaki, H., Adolphs, R., Kaufman, O., Damasio, H., Damasio, A. R., Granner, M., et al. (2001). Single-unit responses to emotional visual stimuli recorded in human ventral prefrontal cortex. *Nature Neuroscience, 4*, 15–16.

Kawasaki, H., Tsuchiya, N., Kovach, C. K., Nourski, K. V., Oya, H., Howard, M. A., et al. (2012). Processing of facial emotion in the human fusiform gyrus. *Journal of Cognitive Neuroscience, 24*, 1358–1370.

Kawashima, R., Sugiura, M., Kato, T., Nakamura, A., Hatano, K., Ito, K., et al. (1999). The human amygdala plays an important role in gaze monitoring: A PET study. *Brain, 122*, 779–783.

Kliemann, D., Dziobek, I., Hatri, A., Baudewig, J., & Heekeren, H. R. (2012). The role of the amygdala in atypical gaze on emotional faces in autism spectrum disorders. *Journal of Neuroscience, 32*, 9469–9476.

Koenigsberg, H. W., Fan, J., Ochsner, K., Liu, X., Guise, K., Pizzarello, S., et al. (2010). Neural correlates of using distancing to regulate emotional responses to social situations. *Neuropsychologia, 48*, 1813–1822. doi:10.1016/j.neuropsychologia.2010.03.002.

Krolak-Salmon, P., Henaff, M. A., Vighetto, A., Bertrand, O., & Mauguiere, F. (2004). Early amygdala reaction to fear spreading in occipital, temporal, and frontal cortex: A depth electrode ERP study in human. *Neuron, 42*, 665–676.

Kuraoka, K., & Nakamura, K. (2007). Responses of single neurons in monkey amygdala to facial and vocal emotions. *Journal of Neurophysiology, 97*, 1379–1387.

Leonard, C. M., Rolls, E. T., Wilson, F. A. W., & Baylis, G. C. (1985). Neurons in the amygdala of the monkey with responses selective for faces. *Behavioural Brain Research, 15*, 159–176.

McCarthy, G., Puce, A., Belger, A., & Allison, T. (1999). Electrophysiological studies of human face perception: II. Response properties of face-specific potentials generated in occipitotemporal cortex. *Cerebral Cortex, 9*, 431–444.

McCarthy, G., Puce, A., Gore, J. C., & Allison, T. (1997). Face-specific processing in the human fusiform gyrus. *Journal of Cognitive Neuroscience, 9*, 605–610.

Meletti, S., Cantalupo, G., Benuzzi, F., Mai, R., Tassi, L., Gasparini, E., et al. (2012). Fear and happiness in the eyes: An intra-cerebral event-related potential study from the human amygdala. *Neuropsychologia, 50*, 44–54.

Mende-Siedlecki, P., Said, C. P., & Todorov, A. (2013). The social evaluation of faces: A meta-analysis of functional neuroimaging studies. *Social Cognitive and Affective Neuroscience, 8*, 285–299.

Miyashita, T., Ichinohe, N., & Rockland, K. S. (2007). Differential modes of termination of amygdalothalamic and amygdalocortical projections in the monkey. *Journal of Comparative Neurology, 502*, 309–324.

Moeller, S., Freiwald, W. A., & Tsao, D. Y. (2008). Patches with links: A unified system for processing faces in the macaque temporal lobe. *Science, 320*, 1355–1359.

Mormann, F., Dubois, J., Kornblith, S., Milosavljevic, M., Cerf, M., Ison, N., et al. (2011). A category-specific response to animals in the right human amygdala. *Nature Neuroscience, 14*, 1247–1249.

Mormann, F., Kornblith, S., Quian Quiroga, R., Kraskov, A., Cerf, M., Fried, I., et al. (2008). Latency and selectivity of single neurons indicate hierarchical processing in the human medial temporal lobe. *Journal of Neuroscience, 28*, 8865–8872.

Morris, J. S., deBonis, M., & Dolan, R. J. (2002). Human amygdala responses to fearful eyes. *NeuroImage, 17*, 214–222.

Morris, J. S., Frith, C. D., Perrett, D. I., Rowland, D., Young, A. W., Calder, A. J., et al. (1996). A differential neural response in the human amygdala to fearful and happy facial expressions. *Nature, 383*, 812–815.

Morrison, S. E., Saez, A., Lau, B., & Salzman, C. D. (2011). Different time courses for learning-related changes in amygdala and orbitofrontal cortex. *Neuron, 71*, 1127–1140.

Nishijo, H., Ono, T., & Nishino, H. (1988). Single neuron responses in amygdala of alert monkey during complex sensory stimulation with affective significance. *Journal of Neuroscience, 8*, 3570–3583.

Ochsner, K., Bunge, S. A., Gross, J. J., & Gabrieli, J. D. E. (2002). Rethinking feelings: An fMRI study of the cognitive regulation of emotion. *Journal of Cognitive Neuroscience, 14,* 1215–1229.

Öngür, D., & Price, J. L. (2000). The organization of networks within the orbital and medial prefrontal cortex of rats, monkeys, and humans. *Cerebral Cortex, 10,* 206–219.

Oosterhof, N. N., & Todorov, A. (2008). The functional basis of face evaluation. *Proceedings of the National Academy of Sciences of the United States of America, 105,* 11087–11092.

Oosterhof, N. N., & Todorov, A. (2009). Shared perceptual basis of emotional expressions and trustworthiness impressions from faces. *Emotion (Washington, D.C.), 9,* 128–133.

Oya, H., Kawasaki, H., Howard, M. A., & Adolphs, R. (2002). Electrophysiological responses in the human amygdala discriminate emotion categories of complex visual stimuli. *Journal of Neuroscience, 22,* 9502–9512.

Parvizi, J., Jacques, C., Foster, B. L., Withoft, N., Rangarajan, V., Weiner, K. S., et al. (2012). Electrical stimulation of human fusiform face-selective regions distorts face perception. *Journal of Neuroscience, 32,* 14915–14920.

Paton, J. J., Belova, M. A., Morrison, S. E., & Salzman, C. D. (2006). The primate amygdala represents the positive and negative value of visual stimuli during learning. *Nature, 439,* 865–870.

Perrett, D. I., Rolls, E. T., & Caan, W. (1982). Visual neurons responsive to faces in the monkey temporal cortex. *Experimental Brain Research, 47,* 329–342.

Pessoa, L., & Adolphs, R. (2010). Emotion processing and the amygdala: From a "low road" to "many roads" of evaluating biological significance. *Nature Reviews. Neuroscience, 11,* 773–782.

Pessoa, L., Japee, S., Sturman, D., & Ungerleider, L. G. (2006). Target visibility and visual awareness modulate amygdala responses to fearful faces. *Cerebral Cortex, 16,* 366–375.

Pinsk, M. A., Arcaro, M., Weiner, K. S., Kalkus, J. F., Inati, S. J., Gross, C. G., et al. (2009). Neural representations of faces and body parts in macaque and human cortex: A comparative fMRI study *Journal of Neurophysiology, 101,* 2581–2600.

Pourtois, G., Spinelli, L., Seeck, M., & Vuilleumier, P. (2009). Modulation of face processing by emotional expression and gaze direction during intracranial recordings in right fusiform cortex. *Journal of Cognitive Neuroscience, 22,* 2086–2107.

Pourtois, G., Spinelli, L., Seeck, M., & Vuilleumier, P. (2010). Temporal precedence of emotion over attention modulations in the lateral amygdala: Intracranial ERP evidence from a patient with temporal lobe epilepsy. *Cognitive, Affective & Behavioral Neuroscience, 10,* 83–93.

Puce, A., Allison, T., & McCarthy, G. (1999). Electrophysiological studies of human face perception: III. Effects of top-down processing on face-specific potentials. *Cerebral Cortex, 9,* 445–458.

Quadflieg, S., Todorov, A., Laguesse, R., & Rossion, B. (2012). Normal face-based judgments of social characteristics despite severely impaired holistic face processing. *Visual Cognition, 20,* 865–882.

Roesch, M. R., Calu, D. J., Esber, G. R., & Schoenbaum, G. (2010). Neural correlates of variations in event processing during learning in basolateral amygdala. *Journal of Neuroscience, 30,* 2490–2495.

Rutishauser, U., Mamelak, A. N., & Schuman, E. M. (2006). Single-trial learning of novel stimuli by individual neurons of the human hippocampus–amygdala complex. *Neuron, 49,* 805–813.

Rutishauser, U., Tudusciuc, O., Neumann, D., Mamelak, A., Heller, A. C., Ross, I. B., et al. (2011a). Single-neuron correlates of abnormal face processing by the amygdala in autism. International Society for Autism Research Annual Meeting, San Diego, 2011.

Rutishauser, U., Tudusciuc, O., Neumann, D., Mamelak, A., Heller, A. C., Ross, I. B., et al. (2011b). Single-unit responses selective for whole faces in the human amygdala. *Current Biology, 21,* 1654–1660.

Sato, W., Kochiyama, T., Uono, S., Matsuda, K., Usui, K., Inoue, Y., et al. (2012). Temporal profile of amygdala gamma oscillations in response to faces. *Journal of Cognitive Neuroscience, 24,* 1420–1433.

Schaefer, S. M., Jackson, D. C., Davidson, R. J., Aguirre, G. K., Kimberg, D. Y., & Thompson-Schill, S. L. (2002). Modulation of amygdalar activity by the conscious regulation of negative emotion. *Journal of Cognitive Neuroscience, 14,* 913–921.

Scheller, E., Büchel, C., & Gamer, M. (2012). Diagnostic features of emotional expressions are processed preferentially. *PLoS ONE, 7,* e41792.

Schumann, C. M., & Amaral, D. G. (2006). Stereological analysis of amygdala neuron number in autism. *Journal of Neuroscience, 26,* 7674–7679.

Schumann, C. M., Bauman, M. D., & Amaral, D. G. (2011). Abnormal structure or function of the amygdala is a common component of neurodevelopmental disorders. *Neuropsychologia, 49*, 745–759.

Schyns, P. G., Petro, L. S., & Smith, M. L. (2007). Dynamics of visual information integration in the brain for categorization facial expressions. *Current Biology, 17*, 1580–1585.

Seeck, M., Michel, C. M., Blanke, O., Thut, G., Landis, T., & Schomer, D. L. (2001). Intracranial neurophysiological correlates related to the processing of faces. *Epilepsy & Behavior, 2*, 545–557.

Seeck, M., Michel, C. M., Mainwaring, N., Cosgrove, R., Blume, H., Ives, J., et al. (1997). Evidence for rapid face recognition from human scalp and intracranial electrodes. *Neuroreport, 8*, 2749–2754.

Smith, M. L., Cottrell, G. W., Gosselin, F., & Schyns, P. G. (2005). Transmitting and decoding facial expressions. *Psychological Science, 16*, 184–189.

Spezio, M. L., Huang, P.-Y. S., Castelli, F., & Adolphs, R. (2007). Amygadala damage impairs eye contact during conversations with real people. *Journal of Neuroscience, 27*, 3994–3997.

Sugase, Y., Yamane, S., Ueno, S., & Kawano, K. (1999). Global and fine information coded by single neurons in the temporal visual cortex. *Nature, 400*, 869–872.

Tang, H., & Kreiman, G. (2011). Face recognition: Vision and emotions beyond the bubble. *Current Biology, 21*, R888–R890.

Todorov, A., & Engell, A. D. (2008). The role of the amygdala in implicit evaluation of emotionally neutral faces. *Social Cognitive and Affective Neuroscience, 3*, 303–312.

Tranel, D., Damasio, A. R., & Damasio, H. (1988). Intact recognition of facial expression, gender, and age in patients with impaired recognition of face identity. *Neurology, 38*, 690–696.

Tsao, D. Y., Freiwald, W. A., Tootell, R. B. H., & Livingstone, M. S. (2006). A cortical region consisting entirely of face-selective cells. *Science, 311*, 670–674.

Tsao, D. Y., & Livingstone, M. S. (2008). Neural mechanisms for face perception. *Annual Review of Neuroscience, 31*, 411–438.

Tsao, D. Y., Moeller, S., & Freiwald, W. A. (2009). Comparing face patch systems in macaques and humans. *Proceedings of the National Academy of Sciences of the United States of America, 49*, 19514–19519.

Tsao, D. Y., Schweers, N., Moeller, S., & Freiwald, W. A. (2008). Patches of face-selective cortex in the macaque frontal lobe. *Nature Neuroscience, 11*, 877–879.

Tsuchiya, N., Kawasaki, H., Oya, H., Howard, M. A., & Adolphs, R. (2008). Decoding face information in time, frequency and space from direct intracranial recordings of the human brain. *PLoS ONE, 3*, e3892.

Vuilleumier, P., & Pourtois, G. (2007). Distributed and interactive brain mechanisms during emotion face perception: Evidence from functional neuroimaging. *Neuropsychologia, 45*, 174–194.

Vuilleumier, P., Richardson, M. P., Armony, J. L., Driver, J., & Dolan, R. J. (2004). Distant influences of amygdala lesion on visual cortical activation during emotional face processing. *Nature Neuroscience, 7*, 1271–1278.

Waytz, A., Gray, K., Epley, N., & Wegner, D. M. (2010). Causes and consequences of mind perception. *Trends in Cognitive Sciences, 14*, 383–388.

Whalen, P. J. (2007). The uncertainty of it all. *Trends in Cognitive Sciences, 11*, 499–500.

Whalen, P. J., Kagan, J., Cook, R. G., Davis, F. C., Kim, H., Polis, S., et al. (2004). Human amygdala responsivity to masked fearful eye whites. *Science, 306*, 2061.

Whalen, P. J., Rauch, S. L., Etcoff, N. L., McInerney, S. C., Lee, M. B., & Jenike, M. A. (1998). Masked presentations of emotional facial expressions modulate amygdala activity without explicit knowledge. *Journal of Neuroscience, 18*, 411–418.

Winston, J. S., Strange, B. A., O'Doherty, J., & Dolan, R. J. (2002). Automatic and intentional brain responses during evaluation of trustworthiness of faces. *Nature Neuroscience, 5*, 277–283.

Young, A. W., Rowland, D., Calder, A. J., Etcoff, N. L., Seth, A., & Perrett, D. I. (1997). Facial expression megamix: Tests of dimensional and category accounts of emotion recognition. *Cognition, 63*, 271–313.

14 Human Lateral Temporal Cortical Single Neuron Activity during Language, Recent Memory, and Learning

George Ojemann

This chapter reviews studies from the author's laboratory on the changes in single neuron activity in human lateral temporal cortex during measures of language, recent memory, associative learning, and related studies. These studies span the period from 1985 to 2011. All represent intraoperative microelectrode recordings from lateral temporal cortex during craniotomies for epilepsy, using a technique where the patient is awake under local anesthesia for a portion of the operation, so that physiological guides unperturbed by general anesthesia can be used to plan the surgical resection (Ojemann, 1995). This surgical technique is most often applicable to patients with temporal lobe epilepsy, thus the focus of these studies on lateral temporal cortex. At the Epilepsy Center with which the author has been associated it has been the policy to manage patients with clear unilateral ictal and interictal temporal lobe foci on preoperative scalp EEG evaluation without any type of chronic intracranial electrode evaluation. The basis for this policy is well beyond the scope of this chapter but has been discussed extensively elsewhere (Ojemann et al., 1993). In these patients, then, the investigative microelectrode studies, and all the physiological data needed to plan the resection, were acquired acutely in the operating room. In the technique used in all but the earliest studies, with the patient asleep under intravenous propofol, a local anesthetic scalp field block was placed, a craniotomy performed, and dural sensation blocked with local anesthetic. This blocks all the pain sensitive structures in the region of the craniotomy, as the surface of the brain has no pain or touch sensation. Thus the patient can be awakened and will be reasonably comfortable for up to 2 hours. During this time the information used to plan the resection was obtained, from electrocorticography and functional identification of sensorimotor cortex and in the dominant hemisphere language cortex using an electrical stimulation mapping technique (Ojemann et al., 1989), followed by the investigative microelectrode recordings providing single neuron data. All studies were carried out only after each patient had given preoperative informed consent, with the methods for obtaining that consent and the studies approved annually by the Institutional Review Boards of the University of Washington.

For the intraoperative microelectrode studies included in this chapter, the same technique of recording was used in all cases, with sharpened tungsten microelectrodes backloaded into a hydraulic adjustable microdrive through a 1-cm diameter transparent footplate. The microdrive was placed on a surface area of cortex that was free of vessels, with the positioning adjusted so that the footplate damped cortical pulsations without blanching pial vessels. In the more recent studies, pairs of microelectrodes were mounted on each microdrive with pairs of microdrives sampling different cortical regions used at each procedure. All recordings were from cortex that, based on preoperative and intraoperative evaluation, including electrocorticography and functional electrical stimulation mapping, would be subsequently resected as part of the operation. Thus no recordings were made in "eloquent" cortex, such as the focal areas identified as critical to language by electrical stimulation mapping (Ojemann et al., 1989). Regions selected were those with the least epileptiform activity in intraoperative electrocorticography. With these restrictions, recordings were most often from the lateral surface of anterior portions of superior temporal gyrus, with the posterior limit determined by the individually variable location of the sites identified as crucial for language and from anterior and middle portions of the middle temporal gyrus (see figure 14.1).

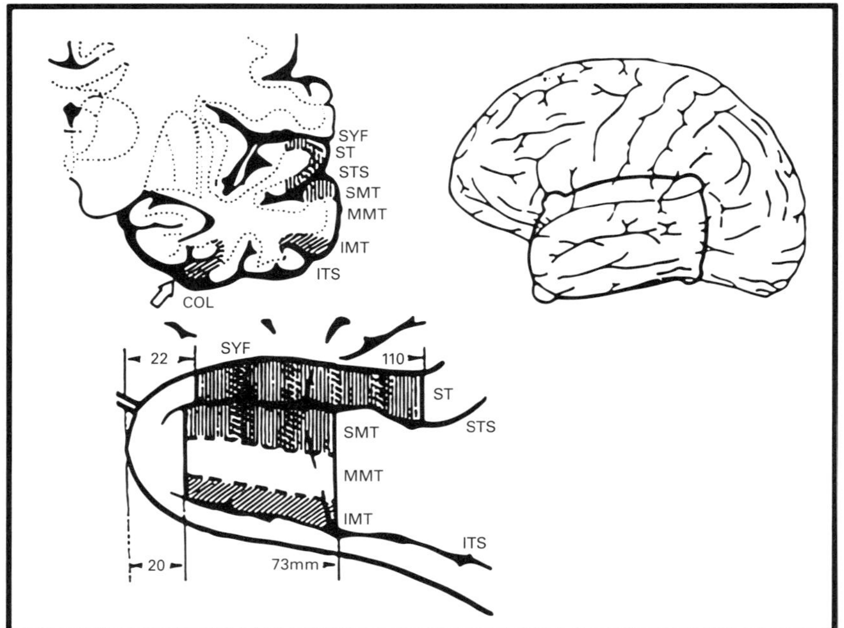

Figure 14.1
A representative sample of the location of microelectrode recordings from lateral temporal cortex reported in this chapter. This sample includes the recordings from 26 patients at 57 sites with 105 single neurons reported in Ojemann et al. (2002) and is unusual only in that the sample extends more posteriorly in superior temporal gyrus than most other series and included some from fusiform gyrus. Recordings were obtained from superior temporal gyrus (ST), superior, middle and inferior portions of middle temporal gyrus (SMT, MMT, IMT), and medial portion of basal temporal cortex in region of the collateral fissure (COL). SYF, Sylvian fissure; STS, superior temporal sulcus; ITS, inferior temporal sulcus. Dimensions are distances from the temporal tip. All recording sites have been projected onto the left hemisphere regardless of which hemisphere they were obtained from. From Ojemann et al. (2002).

Neurons at these sites selected for recording during cognitive measures were those that did not demonstrate epileptiform "bursting" (Calvin et al., 1973) or evidence of injury. Otherwise the recorded neurons represent a random sample of neurons at various known depths usually in the first 3–4 mm of the crowns of gyri. Each microdrive was adjusted, advancing the microelectrodes mounted on it until stable activity that was without epileptiform or injury changes and was stable for several minutes was present in at least one of them. The patients then engaged in a variety of behavioral and control measures as indicated below. In this setting 30–45 minutes was generally available for these measures, limiting the variety of behavioral assessments that could be performed in any one patient. Most behavioral measures were designed to provide a continuous assessment of patient performance to ensure that the patients were actually doing them. With the limited available time, standardized behavioral assessments were used that were not modified for an individual subject. Once recording started, so long as at least one microelectrode demonstrated stable activity, the recording continued uninterrupted until the end of that behavioral paradigm. Earlier recordings were archived on analog FM tape (frequency response 0.1–6.5K Hz); more recent ones were stored after digitization at 4–10K Hz, all with channels marking item presentation and patient's voice as well as those with activity from individual electrodes. Location of recordings was recorded photographically and related to a gyrus and distance from the temporal tip. Both the setting and technique for acquiring the data reported here differed considerably from the chronic microelectrode recording technique utilized in most of the other single neuron studies reported in this volume.

All further analysis was done "offline." Over the years the techniques for subdividing activity into that of individual neurons evolved from amplitude discrimination from film strips to modern programs such as that of Quiroga et al. (2004). Recordings have been obtained from both right and left temporal lobes, though in different patients. Over the years the technique of surgical resection also evolved, with less resection of superior temporal gyrus, so that more recent studies have less frequently sampled that region. Language dominance had been established for nearly all patients as part of the routine presurgical evaluation, in most cases with the intracarotid amobarbital perfusion technique (Wada & Rasmussen, 1960).

Recordings obtained with this technique had useful data from one to four microelectrodes at one to two sites in an individual patient, separated into activity from one to four single neurons/electrode. Those lateral temporal cortical neurons frequently have low firing rates, often averaging in the 3–6 Hz range during a recording, and sometimes in the absence of a task, to less than 1 Hz. These relatively low average firing rates are similar to firing rates reported for lateral temporal lobe neurons in monkey (Fuster & Jervey, 1982). Peak activity during various tasks commonly averaged around 20 Hz in well-isolated units with maximum recorded values up to 50+ Hz. Our studies have included from 9 to 34 patients each, recording from 14 to 57 lateral temporal cortical sites, yielding from 17 to 105 neurons. Table 14.1 indicates these data for the published series cited in this chapter divided by the major cognitive behaviors examined by us. In some patients activity of the same neuron(s) was assessed during several paradigms, each measuring different behaviors. Thus table 14.1 also indicates the total number of published

Table 14.1
Number of patients, sites, and single neurons included in the series of single neuron recordings from lateral temporal cortex included in those publications from the author's laboratory cited in this chapter, divided by different categories of behaviors assessed

Data category	Patients	Sites	Single neurons
Any published recording of single neuron activity during cognitive assessments	189	258	473
Any language assessment*	154	220	421
Auditory repetition	45	54	121
Object naming	66	73	115
Text word reading	72	93	157
Any recent verbal memory assessment*	86	136	243
For object names	62	99	173
For text words	75	125	223
For auditory words	26	57	105
Verbal associative learning*	36	54	139
Any assessment for nonverbal material*	35	53	73

*Each of these categories includes recordings where other categories were also assessed. Where subdivisions are present, more than one subdivision was commonly assessed in the same recording.

lateral temporal neurons, sites, and patients from which we have successfully recorded. These intraoperative microelectrode recording findings then reflect changes generalized across the patient population rather than many sites in an individual.

Language

Most of our investigations of single neuron changes with language have utilized simple tasks. Initial studies (Creutzfeldt et al., 1989a, 1989b; J. Ojemann et al., 1992) investigated auditory word perception alone and with overt production of the word. Tones at 500 or 1000 Hz, random noises of the environment, speech degraded by removing spectral components, and speech backward provided controls for activity related to auditory perception. Additional studies, each in a separate series of patients, investigated object naming, overtly and silently (Ojemann et al., 1988; Haglund et al., 1994; Schwartz et al., 1996; Ojemann & Schoenfield-McNeill, 1999). Comparison was to matching a spatial feature on the same items used for the naming task. Single word identification was studied, in conjunction with naming (Ojemann et al., 1988; Schwartz et al., 1996; Ojemann & Schoenfield-McNeill, 1999) or in other patient series as the only identification task (Weber & Ojemann, 1995; Ojemann & Schoenfield-McNeill, 1998). The control items in these last studies were the words backward, while matching of a spatial feature on the word items was the behavioral control for the other studies. The earliest auditory studies were descriptive, while the other studies established changes statistically.

Despite this variety of language and control tasks and methods of analysis, several features of the single neuron activity changes with the language tasks are evident in all the studies.

Changes for all tasks were not lateralized to the language-dominant hemisphere but rather present in about equal proportion in either hemisphere. This is in marked contrast to language changes present with lesions or electrical stimulation mapping, where the changes are confined to the dominant hemisphere. In the earliest auditory listening and repeating studies activity changes recorded from superior temporal gyrus were predominately increased activity during listening, while that from middle gyrus was more frequent during repeating and predominately decreased activity, a relative inhibition, with a variable onset sometimes preceding voice onset by up to 250 ms (Creutzfeldt et al., 1989a, 1989b). This effect was also identified during object naming when it was found during both overt and silent naming, commonly with onset immediately after item presentation. Although no lateralization of this relative inhibition was identified in the early study of auditory word listening and repetition, in the later studies of object naming early inhibition was shown to be significantly lateralized to the dominant hemisphere and late excitation to the nondominant hemisphere (Schwartz et al., 1996). Early inhibition was also a feature lateralized to the dominant hemisphere during a rhyming task (Schwartz et al., 2000). Figure 14.2 shows an example of this relative inhibition occurring during naming. This effect was less evident during reading tasks where no feature of those changes could be significantly lateralized.

Several hypotheses for the mechanism of this relative inhibition have been proposed. Initially it was suggested as a mechanism to block perception of one's own voice during overt speech, but its presence during silent naming, well before voice onset, and relative absence with overt speech evoked by word reading suggest that this is not the full explanation. It may represent an inhibitory surround, perhaps around an area of increased activity at a site crucial for language based on stimulation mapping, from which we have not recorded. Alternatively, the inhibited neurons may be that part of the network for a language modality that is active with learning of new material in that modality, such as new verbal associations, but inhibited during overlearned aspects such as those tested with object naming or word reading. We will return to this hypothesis later in this chapter when discussing the changes in activity with verbal associative learning. Relative inhibition could also reflect cortical activity that modulates subcortical circuits, such that reduced cortical activity allows the higher subcortical firing rates that have been recorded during language (Bechtereva et al., 1992). Why the early inhibition is less evident with word reading is also unclear, though stimulation mapping suggests that anterior temporal cortex may be somewhat more likely to be crucial for reading than naming (Ojemann, 1989) and thus possibly have more excitatory activity. Early inhibition may be a more general property of functional lateralization in temporal cortex, for a similar effect was lateralized to nondominant hemisphere recordings during a visuospatial task, face matching (Lucas et al., 2003).

Another feature present in multiple series of recordings during language measures is the frequent separation of activity for different aspects of language. Neurons changing activity in the same direction with naming and reading have been infrequently recorded (Schwartz et al., 1996; Ojemann and Schoenfield-McNeill, 1999). Separation of neurons changing activity with naming in one language and not another have been identified in the few recordings obtained in

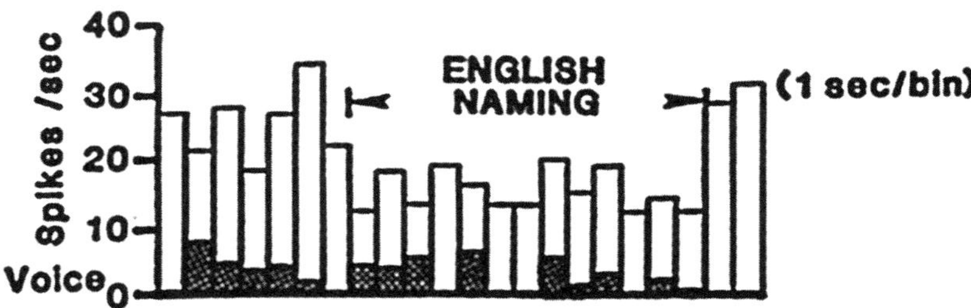

Figure 14.2
Single neuron activity from left, dominant superior temporal gyrus during overt naming of the same object pictures in two languages, English and Spanish. (Top) Activity in 2-s bins during English naming. It shifts to a higher level with the instruction to name the same pictures in Spanish. It then continued at the same high level for some minutes during English and Spanish reading and the spatial matching tasks to the naming and reading slides (Bottom) Activity, now shown in 1-s bins, then abruptly decreased again during repeat naming in English, increasing back to the higher level on its completion. This recording illustrates the relative inhibition with naming characteristic of dominant temporal cortical activity as well as the differential activity with the same linguistic task in different languages. Adapted from Ojemann (1990).

bilinguals (figure 14.2; Ojemann, 1990). These findings suggest that the networks that subserve different aspects of language are at least partially separate, a finding similar to that for crucial cortical areas for different aspects of language identified by stimulation mapping (Ojemann & Whitaker, 1978; Ojemann, 1989; Lucas et al., 2004).

In monkey, neurons have been identified that have similar changes in activity with perception of a movement and its production (Rizzolatti et al., 1988). It has been suggested that these "mirror" neurons had a role in the development of language, a concept reinforced by the finding in monkey of neurons "mirroring" sounds of a movement and its production (Kohler et al., 2002). Although stimulation mapping suggested a frequent overlap in human temporal cortical sites where changes in orofacial movements and phoneme perception were evoked (Ojemann, 1983), single neuron recording from temporal cortex during measures of language perception and production have much less clearly demonstrated this overlap, more often having significant changes with perception or production alone (Creutzfeldt et al., 1987, 1989a, 1989b; Schwartz et al., 1996; Ojemann & Schoenfield-McNeill, 1999), or as indicated above, when present for both the combination of excitation for perception and inhibition for production. Neurons with mirror properties are rare in human temporal cortex. The presence of those properties with stimulation mapping there likely reflects close proximity of separate neural networks for language perception and production.

Additional insight into the neural networks involved in language came from recordings from neurons with activity that discriminated correct from incorrect overt identification of verbal material (Ojemann et al., 2004). Neurons with that property were overrepresented in medial basal and hippocampal recordings and significantly less frequently present in lateral middle temporal gyrus recordings with none found in superior temporal gyrus. Since medial temporal structures have usually been related to memory and not language performance, this finding was unexpected but may indicate that explicit memory for specific items is a component of language in addition to the general semantic memory, so that the error-related activity may be a part of an ongoing monitoring of responses to the specific items. Although the study included identification of auditory, text, and named objects, and neurons were recorded that discriminated errors in each modality, no neuron had activity that discriminated more than one modality, further evidence of the separation of networks for different language modalities. In 69% of the neurons discriminating correct from incorrect performance, there was a relative inhibition for correct identification that was absent with errors, evidence of the importance for this inhibition in language. Activity discriminating correct from incorrect identification was present earlier in lateral temporal cortical neurons (the majority within 300 ms of item presentation) than in medial basal neurons.

The changes described above represent alterations in firing rates. However, a widely hypothesized model of the neuronal activity related to language encoding involves temporal patterning. Some evidence for such temporal patterning related to language has been suggested based on recordings from subcortical nuclei (Bechtereva et al., 1979), and a mechanism for temporal patterning of activity of individual neurons involving interactions with local field oscillations in the "gamma" range has been proposed (Fries et al., 2007). In our lateral temporal recordings during language measures, we have identified only a few examples that seem to show some element of temporal patterning. During the auditory tasks, a neuron was identified during auditory listening that seemed to have the same patterns of activity during perception of particular phonemes (see figure 14.3, Creutzfeldt, 1989a). Another neuron (in another patient) had similar selectivity to both auditory listening and overt repeating of the same word (Creutzfeldt et al., 1989b). A few other neurons seemed to have patterns related to word structure, during auditory listening responding selectively to the second but not first syllable of multisyllable words (see figure 14.4) and also only compound words (Creutzfeldt et al., 1989a). During the recent verbal memory paradigm described in the next section, another neuron in another patient had significant

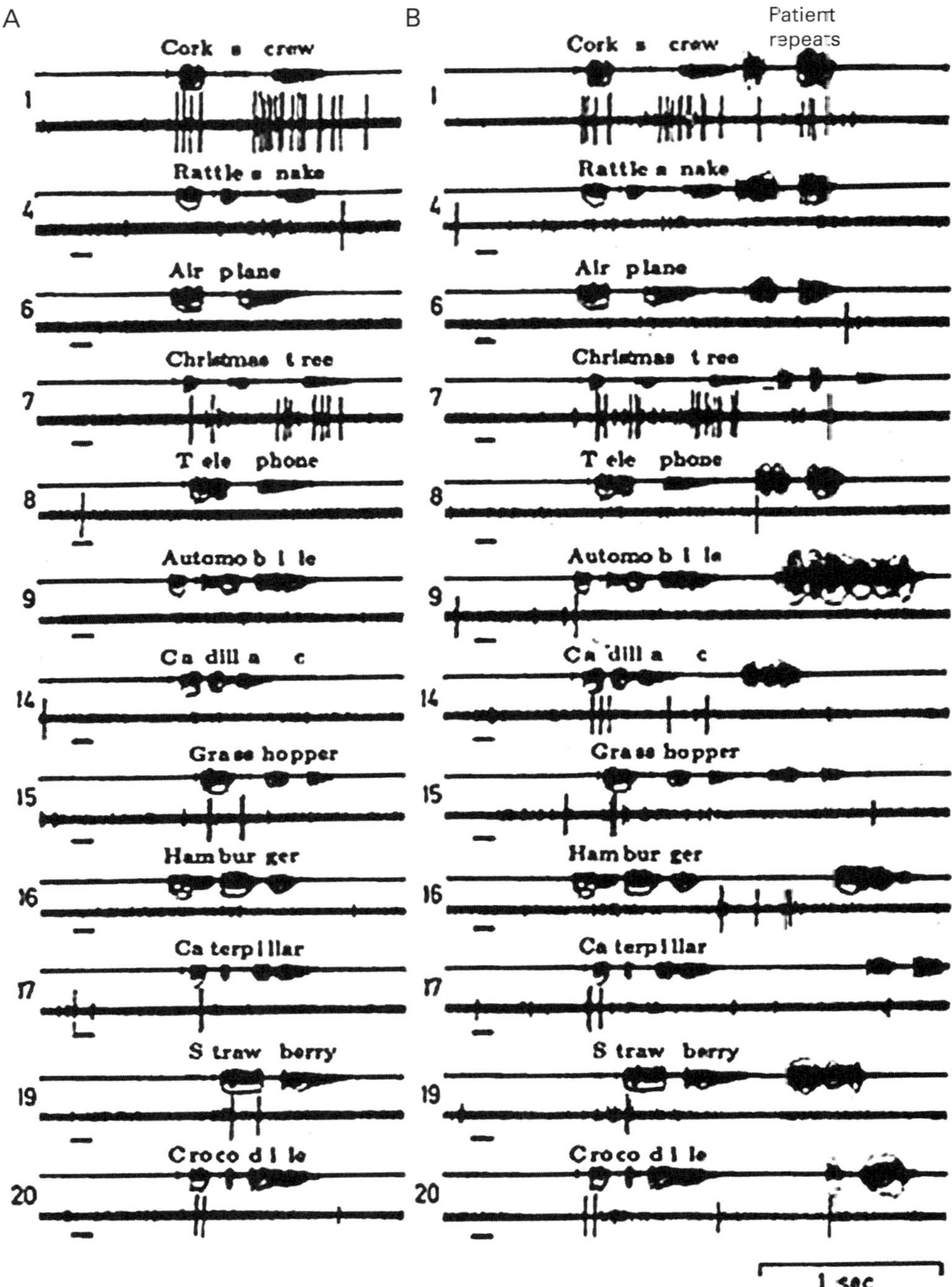

Figure 14.3
Single neuron activity recorded from right superior temporal gyrus during a sample of an auditory language task, listening (A) and listening and repeating (B) the same 20 words. The short bar to the left in each sample is a 1000-Hz tone. The patient's overt correct repeating of the word during B is indicated on the audio channel at the right. This neuron appears to have the same patterns of activity during the two samples of listening to words 1, 7, 15, 17, 19, and 20 and an absence of activity during both samples of listening to the other words. This pattern was thought to reflect responses to specific phoneme categories. There is little to no activity in response to the tone or with overt repeating. From Creutzfeldt et al. (1989a).

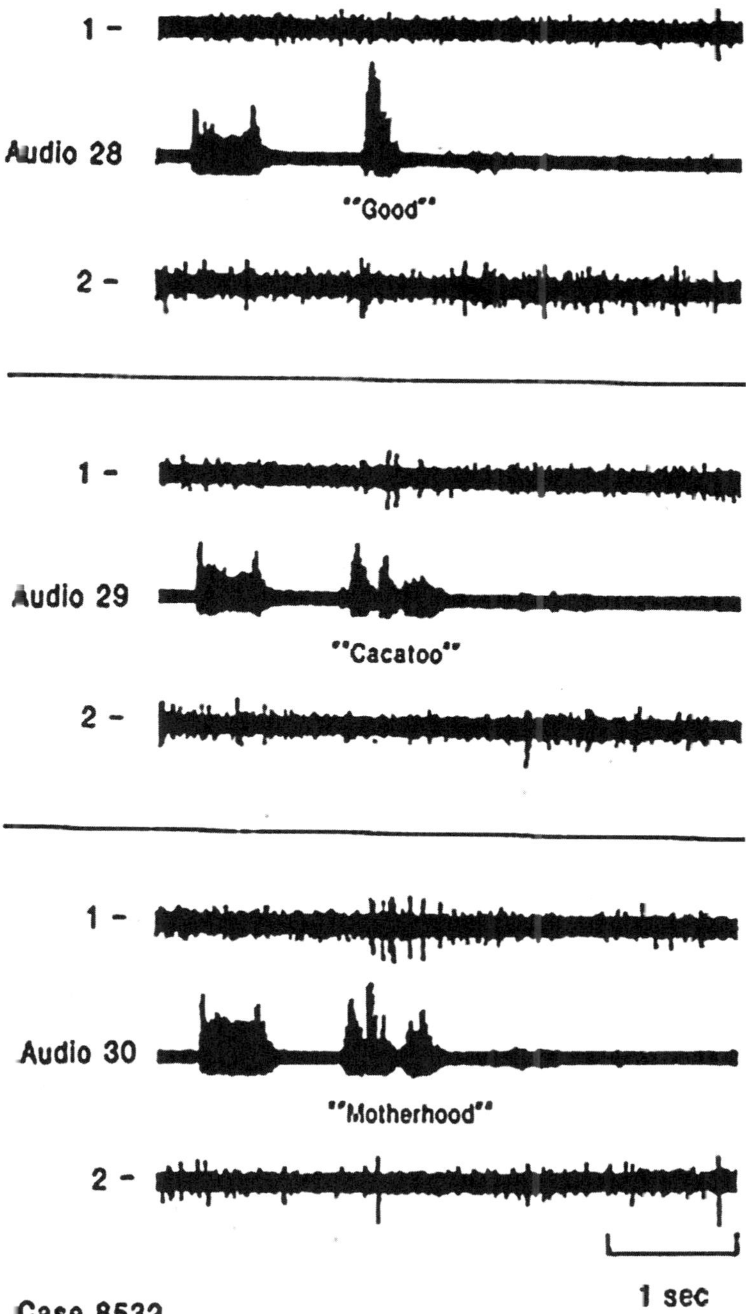

Figure 14.4
Single neuron activity related to second and subsequent syllables. Recordings from two electrodes at different depths in the same region of nondominant superior temporal gyrus during word listening. Neuron in channel 1 responds only to second and subsequent syllables of multisyllable words and not to single syllable words. Creutzfeldt and Ojemann, unpublished data (1989).

overrepresentation of the interspike intervals in the 10- to 25-ms bins at encoding and retrieval, the two tasks requiring perception or production of the same name, compared to the different names used as distractors during memory storage (Ojemann et al., 1990). These are hints, then, that there may be temporal patterns for different words, perhaps reflecting different features of the word: phonemic structure, prosody, perhaps semantics, but this question, of the role of temporal patterning compared to firing rate changes, remains an important area for future investigation.

Recent Verbal Memory

A temporal lobe role in recent memory is widely accepted, based largely on the effects of lesions. That role is now usually ascribed to medial structures, mostly to hippocampus and to some extent parahippocampal gyrus. However, there is also some evidence for a lateral temporal role in recent memory: a limited amount of evidence from lesion effects (Cheung & Chan 2003; Milner, 1967; Ojemann & Dodrill, 1987) and more from electrical stimulation mapping (Fedio & Van Buren, 1974; Ojemann, 1978; Ojemann & Dodrill, 1985, 1987; Perrine et al., 1994). Some functional magnetic resonance imaging (fMRI) memory studies have shown lateral temporal activity (Casasanto et al., 2002; Fletcher & Tyler, 2002; Kirchoff et al., 2000). We have investigated the changes in single neuron firing during recent verbal memory in a series of studies (Ojemann et al., 1988; Haglund et al., 1994; Weber & Ojemann, 1995; Ojemann & Schoenfield-McNeill, 1998, 1999; Ojemann et al., 2002). Those studies have used the same measure of recent verbal memory, multiple trials of an encoding-storage-retrieval memory paradigm. This is a modification of a standard short-term memory measure extensively studied by an earlier generation of psychologists (Peterson & Peterson, 1959) and utilized with electrical stimulation mapping of thalamus (Ojemann et al., 1971) and cortex (Ojemann, 1978). The encoding phase involved identification (usually overt) of object names or words, presented visually in all studies, and also auditorily in one, the subject having been previously instructed to remember the presented item. The storage phase usually lasted 9 s, filled with three "distractors," other items of the same modality. This was followed by the retrieval phase, the word "recall" that cued the patient to report aloud the item presented during the encoding phase. Two comparisons were made to this memory paradigm. The same items were presented in another block of trials with the instruction to just identify them overtly. Activity during this identification was compared to that during the encoding phase of the memory paradigm, that also required overt identification of the same items, differing only in the instruction to remember the item given with the memory task. The second comparison was between activity with the different phases of the memory task. For statistical analysis of most studies, activity during each phase of the memory paradigm was divided into an initial epoch related to perception and processing, lasting from item presentation to 300 ms before the onset of overt identification, a second epoch related to that overt output lasting from 300 ms before the overt output onset to 1200 ms after, and a third epoch extending from then to the next item onset. The first and second epochs were of approximately equal length, somewhat over 1 s. These relatively long epochs were used because of the low firing rate of many neurons. Except as noted, analyzed activity was only for trials with correct performance.

Various names have been applied to the type of memory assessed by this paradigm. It is a measure of *episodic, explicit*, or *declarative* memory, memory for the specific event (specific name, word) that occurs at encoding. Although encoding also requires generation of the name or word from semantic memory (and thus also assesses *verbal* memory), that is also a requirement of the identification control measure to which activity during encoding is compared. However, the finding described above that neurons discriminating correct from incorrect identification are overrepresented in medial temporal structures usually associated with explicit memory illustrates how difficult it is to make this semantic–explicit memory separation. Our memory paradigm requires storage over a short period and thus a measure of *recent* or *short-term* memory. In the functional neuroimaging literature memory for this duration is often referred to as *working* memory. Our paradigm measures a *postdistractional* memory. The presence of distractors during memory storage has been shown to adversely affect memory performance with human medial temporal lesions (Green, 1964), and the temporal cortical neural activity recorded during memory measures in monkey (Fuster, 1995), as well as performance in normal humans (and performance and neuronal activity in our subjects). Some of the "working memory" paradigms used with functional imaging have not included distractors. Our paradigm assesses cued *recall* retrieval. Recognition retrieval, often used in similar paradigms, not only is less difficult for humans but, as indicated below, is associated with different neuronal activity. The memory discussed in this section differs from long-term or consolidated memory, procedural memory (memory for purely motor tasks), implicit memory (unconscious memory processes such as priming), and the very short-term memory assessed by digit span.

Combining findings from all of our studies, lateral temporal neuronal activity has been recorded during our memory paradigm from 243 neurons at 136 sites in 86 subjects. When activity during encoding was compared to that during identification of the same material, so that the only difference in the two tasks was the requirement to retain the item in recent memory, activity changed in 135 (56%) of those neurons. Sixty-five percent of those changes were increased activity with memory encoding. No significant difference between hemispheres was present in the overall proportion of neurons changing activity with encoding. This lack of significant lateralization of neuronal activity changing with verbal memory encoding is similar to that observed with language identification. However, when encoding was assessed for different language modalities, object names, text, or auditorily presented words, neurons that changed activity with multiple modalities were significantly lateralized to the dominant hemisphere, while neurons with changes in only one modality were significantly lateralized to the nondominant (Ojemann et al., 2002). Combinations between the auditory task and one or both visual tasks were particularly likely in dominant hemisphere (Ojemann et al., 2009).

Lateral temporal cortex has not been part of most recent models of brain substrate for recent verbal memory. To establish that the changes in activity there were functionally important, the differences in activity between correct and incorrect memory performance in the presence of intact identification were examined (Ojemann et al., 2004). Significant differentiation between correct and incorrect performance was identified in 9 of the 108 neurons recorded in patients who made at least one memory error in some language modality. All these changes occurred during the encoding phase of the memory measure. They occurred in neurons that did not show any changes in activity with identification errors. They were overrepresented in recordings from superior temporal gyrus and the upper third of middle gyrus. Activity differentiating correct from incorrect performance occurred within 800 ms after item presentation in 5 of the 9 neurons, all in superior gyrus or superior third of middle gyrus. Activity in the remaining neurons in middle gyrus that differentiated correct from incorrect performance was significantly later, after 1.1 s.

The detailed timing of encoding changes during correct performance was examined in Ojemann et al. (2009). Activity was divided into that with significant changes in the first temporal epoch (item presentation to 300 ms before voice onset, the time for perception and processing), the second epoch (300 ms before voice onset to 1200 ms after, the time related to the overt output), or both. Fifty-seven percent of the neurons with any significant change had changes in different epochs for encoding of different modalities. Changes sustained through both epochs were significantly more likely in the dominant hemisphere, while those with changes in only the second epoch were significantly more likely in the nondominant. Increased activity characterized 80% of changes confined to first epoch, 68% of those confined to the second, and 56% of those sustained throughout both. Only first epoch changes were significantly more likely in superior temporal gyrus recordings, only second in middle gyrus. Changes sustained through both epochs were present throughout lateral cortex, but in dominant hemisphere, with significantly more sustained inhibition in the superior and middle thirds of the middle gyrus than in surrounding cortex. When activity was divided into 50-ms bins, 40% of significant changes for any modality were excitatory with peak activity in the first epoch. These peaks were not randomly distributed. Rather, 83% were within the first 750 ms, which accounted for only the first 52% of the first epoch. They were significantly earlier in superior temporal gyrus (average peak= 310 ms) than middle gyrus (650 ms). This parallels the timing difference between superior and middle temporal gyri in timing of activity that discriminates between correct and incorrect performance indicating that it is specifically this activity that is functionally crucial. One third of the first epoch peak excitation was within the first 300 ms and in some within the first 50 to 100 ms (see figure 14.5). These very early peaks were recorded from six different neurons from multiple subjects, usually for only one modality in each neuron but with all modalities represented. This timing in anterior superior temporal gyrus neurons is remarkably early, given that four of the seven examples involved visually presented modalities and the average peak of the visual evoked potential in occipital visual cortex is about 55 ms (Farrell et al., 2007). During encoding there appears to be a flow of excitatory activity from earliest in superior gyrus to later in middle gyrus, followed by inhibition with the overt response. Peak excitation in neurons with significant changes sustained throughout both epochs had a pattern very similar to that of neurons with only first epoch peaks (see figure 14.6) This suggested a simultaneous convergence of an input evoking a sustained increase in activity and an input related to item perception onto

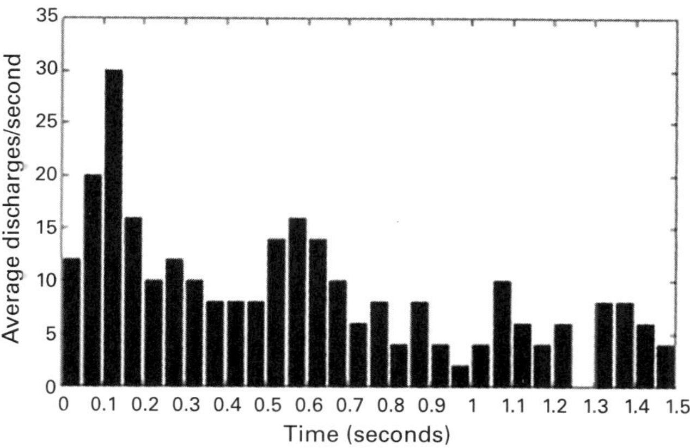
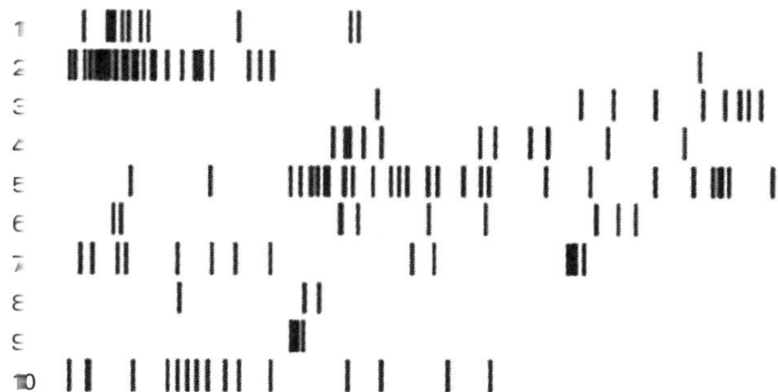

Figure 14.5
Average single neuron activity recorded from superior third of dominant middle temporal gyrus during the initial 1500 ms of ten trials of encoding of object pictures, presented in 50-ms bins. This activity differs from that with identification of these pictures without the instruction to retain the name in memory at p < 0.0001. It peaks very early, 100–150 ms after item presentation. Below is raster plot of activity during the individual trials. From Ojemann et al. (2009).

individual neurons during recent memory encoding of items of a particular language modality. This extends the Hebbian model of encoding involving potentiating effects of simultaneous presynaptic inputs (Hebb, 1949) to potentiating effects of converging "tonic" and early "phasic" inputs, an effect identified in about 10% of the sample of lateral temporal activity.

Insight into the extent of participation of an individual anterior temporal neuron in the various networks for language and memory comes from comparing activity during identification and memory encoding to the same control identification measures since both tasks required overt production of the same verbal items (Ojemann & Schoenfield-McNeill, 1999). Identification and encoding of both object names and words was assessed, using matching of a spatial feature on the objects and words backward as the respective controls. Of the 31 neurons recorded in that

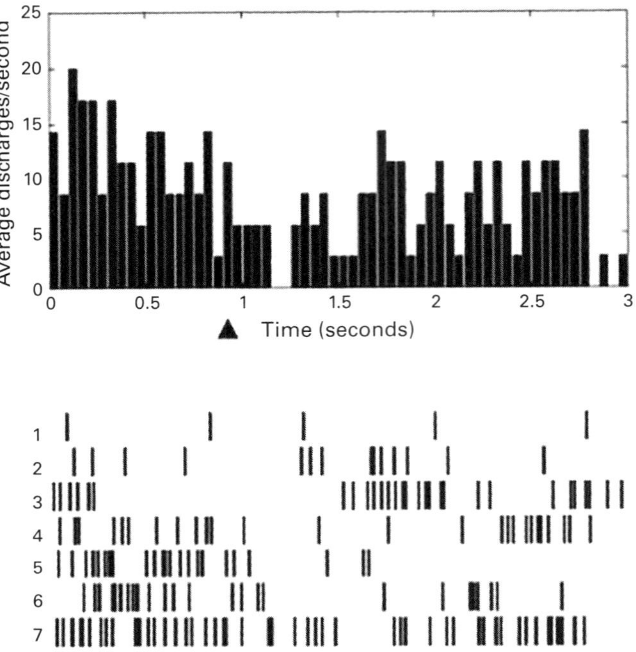

Figure 14.6
Average single neuron activity recorded from dominant superior temporal gyrus during seven trials of the initial 3 s of encoding of object pictures in 50-ms bins. The arrow marks 300 ms before the onset of the overt naming of the objects. This activity differs from that during object identification alone without the instruction to retain the name in memory at p < 0.001, for both the period before and after the arrow. This is significantly increased activity that is sustained throughout encoding that peaks very early, 100–150 ms after item presentation. Below is the raster plots for the individual trials. This neuron's activity reflects convergence of tonic and early phasic excitatory inputs. Note also the brief inhibition that appears with voice onset, 300 ms after the arrow. From Ojemann et al. (2009).

study, 8 had no significant changes on any of these identification or encoding comparisons, 5 only changes in one of the identification comparisons, and 6 only one of the encoding comparisons. Of the remaining 12 neurons, 4 had changes in both of the identification tasks, 3, in opposite directions. The remaining 8 neurons had different combinations of identification and encoding, 6 with quantitative identification-encoding differences in the same modality. Overall, then, about half the neurons that changed activity with any task participated in the network for only identification or only encoding, with most of the remainder having significant differences in the level of activity between any two tasks. In the next section, neurons with the pattern of decreased activity for identification transitioning to increased activity with memory encoding in the same modality will be shown to be important to verbal associative learning. Six neurons in five subjects showed that pattern, one for both modalities, representing 4 of 8 neurons with decreased activity during naming and 3 of 5 with decreased activity during reading. Activity

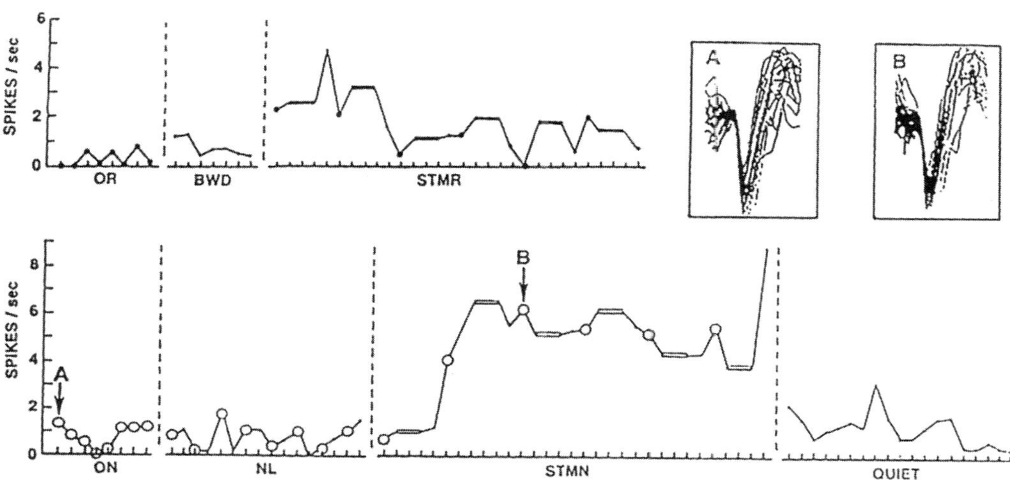

Figure 14.7
Single neuron activity from superior half of left middle temporal gyrus for individual trials of overt object word reading (OR), viewing words backwards (BWD), an encoding-storage-recall paradigm of recent memory for words (STMR), overt object naming (ON), matching line angle on the same items (NL), the same recent memory paradigm but for object names (STMN), and a quiet period at the end of the recording (Quiet). The recording is continuous except for brief periods represented by the dashed lines when instructions for the next block of trials were reviewed. (A, B) Twenty superimposed consecutive action potentials from this neuron recorded at the indicated trials of naming (A) and name encoding (B), establishing that those recordings are from the same neuron. There is significantly increased activity for memory encoding compared to identification that was sustained throughout those all phases of those blocks, is greater for name memory, and is not present in the other conditions or quiet period. From Ojemann and Schoenfield-McNeill (1999).

during memory storage and retrieval was compared to that during encoding. Significant differences in activity between phases of the memory task were infrequent. For example in one study, only 16 of the 105 temporal cortical neurons had phase differences, compared to 52 with encoding changes (Ojemann et al., 2002). The most common pattern was a change in activity with encoding that was sustained throughout storage and retrieval and lasting through all trials of the specific memory task, but absent with adjacent control tasks, as illustrated in figure 14.7. This effect has been related to task-specific thalamocortical activating circuits (Ojemann & Schoenfield McNeill, 1999), also the likely source of the "tonic" sustained input identified during encoding. The smaller number of neurons with changes in only one phase of the memory task was significantly overrepresented in the sample of activity from inferior and basal temporal gyri (Ojemann et al., 2002).

Activity during the storage phase was further assessed by manipulating the distracting task in two ways, comparing their presence or absence (Zamora et al., 2001), or lengthening the time between them (Zamora et al., 2013). Both studies have similar findings indicating activity that likely reflects an active rehearsal process maintaining the memory during storage. As expected, the presence of distractors interfered with memory performance. Distractors also changed activity in 44% of the 86 lateral cortical neurons sampled, in 69% with greater activity during the storage phase in the absence of distractors. Sixty-four percent of those neurons also had increased activity during encoding when encoding was compared to identification of the same material without having the requirement to retain it in memory, subtracting out any contribution from language. The decrease in this ongoing lateral temporal activity specific to memory processes that extended from encoding into the storage phase that occurs with the distracting task is associated with a deterioration in memory performance. In about half of the neurons with this change, that decreased activity with distractors was sustained throughout the storage phase. Since we had previously modeled such sustained changes in activity as reflecting attentional mechanisms specific to memory, this suggests that at least one mechanism for the distractor effect involved a reallocation of attention. Lengthening the time between distractors had similar findings. Performance was improved with the longer time between distractors despite the longer time the memory was stored. In the 80 neurons sampled in that study, 28% changed activity in the longer compared to the shorter period between distractors, all with increased activity concentrated in the longer period between distractors. Forty-three percent had increased activity for both encoding and during the long delay, again indicating this was activity specific to memory, not language. The longer period of increased activity during the longer period between distractors would seem to provide some additional rehearsal, accounting for the improved memory performance. The link between improved memory performance and increased neuronal activity is also further evidence of the functional importance of the lateral temporal cortex in recent memory.

It has been suggested that each retrieval of a memory involves reactivation of the processes of initial encoding (Nader & Hardt, 2009). To examine the extent to which this occurs in our recent memory paradigm, activity during recall retrieval was compared to that during encoding on the same trial. To date, this has been done only for memory for object names (Ojemann & Schoenfield-McNeill, 2002). Of the 93 neurons available for comparison, significant changes in frequency of activity were present in 23 neurons with encoding and 19 with recall, both compared to identification. Of these, 13 neurons had similar changes with both; 11 of these were activity sustained throughout both encoding and recall, in 6 sustained increases. Five neurons had changes only during recall. Some individual neurons with phasic changes during encoding show similar activity with correct recall (Ojemann et al., 2004, figure 4), others different phasic activity then (Ojemann et al., 2002, figures 3, 4; 2004, figure 5). In another series (Haglund et al., 1994) significant encoding changes were present in 11 of 20 neurons, 9 with increased activity. Similar recall changes were present in only 4. That study examined changes with serial recall after additional distracted storage periods and found activity in all those neurons to be diminished with subsequent retrievals. These data suggest that at most only some features of encoding activity are recapitulated with retrieval.

Recall retrieval activity was also compared to that during recognition retrieval (Ojemann et al., 2002). Of 35 neurons with a significant change in either recall or recognition retrieval compared to a control for motor output, only 4 had significant changes for both. Neurons related to recognition retrieval were not lateralized but were overrepresented in recordings from superior temporal gyrus and superior portion of middle gyrus than more inferior recording sites. Recognition retrieval is associated not only with more accurate performance than recall retrieval but with changes in different neural networks.

Neuronal activity in lateral temporal cortex related to recent memory, though with recognition rather than recall retrieval, has been identified in rat and monkey (Fuster & Jervey, 1982). The extent of this area is quite small in rat, confined to the homologue of parahippocampal gyrus. It is substantially larger in monkey, involving inferior temporal gyrus. The human temporal lobe is much larger than in monkey, with most of the expansion in anterior lateral and basal temporal cortex, the region related to recent memory in our single neuron recordings. On the other hand, the temporal lobe structures most often related to recent memory in rat, monkey, and humans, particularly hippocampus, have a much smaller expansion across these species, with the relative size of hippocampus compared to temporal cortex much smaller in humans. The evolutionary changes in the brain organization for recent memory involve an expansion of the cortical component with relative conservation of the medial temporal–hippocampal component.

Verbal Associative Learning

A substantial literature has established the importance of the temporal lobe in learning. We have assessed this at a neuronal level in lateral temporal cortex in two separate series of patients (Weber & Ojemann, 1995; Ojemann & Schoenfield-McNeill, 1998). Both studies used a verbal associative learning paradigm where a set of words, common concrete nouns, were visually presented under three conditions: with the instruction to read the word aloud, in the recent memory paradigm described above, and in a paired-associate learning task, where the words were presented as unrelated pairs in a series of encoding-test trials. During the encoding portion, the subject read aloud each pair. During the test portion, the first word of each pair was presented. The subject read this aloud and then provided the associated word, if known. The encoding-test trials were repeated up to seven times, with pairs in different orders between trials and encoding and test portions of each trial. Words backwards were the perceptual control. In the first study activity during silent reading of the words was also assessed. The first study recorded activity from 49 neurons in 15 patients. The second study included 21 neurons from 9 patients. Both studies included recordings from right or left brain.

In both studies, the majority of neurons had significant changes in activity with associative learning, 59% and 62%, respectively, the majority increased activity. In both studies, the proportion of neurons with associative learning changes was similar in the two hemispheres. Neurons were classified by the change in activity during identification and recent memory. In both studies, neurons with changes during word identification and memory were significantly more likely to have associative learning changes than neurons unrelated to either. In the initial study, neurons with both identification and memory changes had significantly greater activity for associations learned early compared to those learned later or never. Once learned, that activity rapidly decreased, within two further correct trials. The pool of increased neuronal activity shrinks once learning occurs, an effect that has also been observed with functional imaging (Raichle et al., 1994). Subjects who learned the associations poorly had less activity in these neurons than subjects who learned the associations rapidly, and conversely, the poor learners had greater activity in neurons with no relation to identification or memory, substantiating the functional significance of the increased activity in these neurons.

The second study compared activity on the first encoding trial for associations learned then to that for the remaining unlearned associations. Neurons with significantly *decreased* activity with identification and *increased* activity with memory encoding had relatively greater activity for associations learned on that first encoding trial than those not. Neurons with other relations to identification or memory did not show this effect (see figure 14.8). Neurons with the pattern associated with learning represented 8 of 11 neurons with decreased activity during word reading, recorded in six subjects. As noted in the previous section, in another series of patients, 3 of 5 neurons with decreased activity with word reading (recorded in three subjects) had this pattern (Ojemann & Schoenfield-McNeill 1999). In our second learning study the first encoding trial was subdivided by the temporal epochs related to pair presentation and the overt verbal identification of the pair (see figure 14.9). Both learned and unlearned pairs showed an increase in activity from the period immediately before pair presentation, to 300 ms before overt identification, but activity for learned pairs was at a significantly higher level during overt identification sustained into the interval before presentation of the next pair.

In summary, learning a new association to a verbal item involves a sustained increase in activity in those neurons that are inhibited with item identification. We hypothesized that those neurons were active with initial learning of that verbal item but, once it was learned, were rather rapidly inhibited as part of the general reduction in activity once learning occurs. These neurons are reactivated with the learning of new associations to the verbal item. That reactivation represents sustained "tonic" activity lasting beyond perception and identification of the new association, not unlike the sustained activity related to rehearsal we identified during recent memory storage. It is this dynamic change in activity in the existing network for a verbal item that seems to characterize associative learning, rather than recruitment of new neurons into the network, at least in temporal cortex.

Further Evidence on the Networks for Verbal Identification and Associative Learning

Our recent investigation of the ability of local field potentials (LFPs) to predict spike timing in lateral temporal neurons provides additional insights into the organization of the neural networks related to verbal identification and associative learning (Zanos et al., 2012). The data for this study were derived from recordings during an identification and paired-associate learning paradigm designed to be suitable for both intraoperative neuronal recording and fMRI, in an effort to identify any neuronal correlates of fMRI changes in human temporal association cortex that

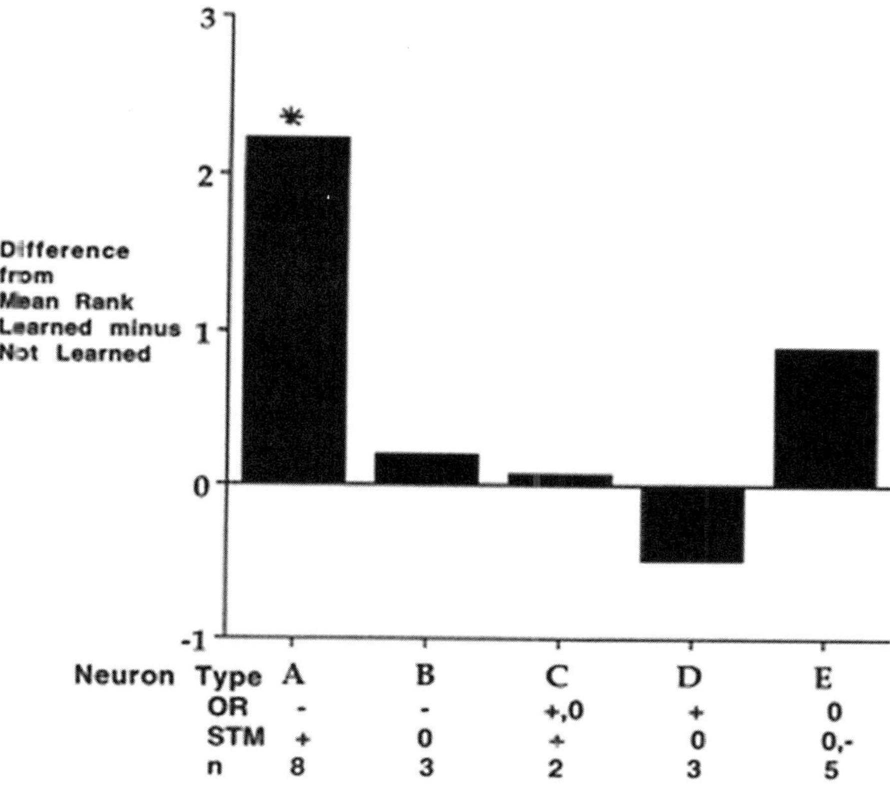

Figure 14.8
Relative difference in activity on the first encoding trial for learned compared to not-learned word pair associations in neurons grouped by the changes in activity with single word identification and recent memory encoding. A positive value indicates greater activity for learned pair associations. Neurons with inhibition during word identification and increased activity with recent memory encoding had significantly greater activity for learned pairs (Wilcoxon matched pairs sign rank, p < 0.05, two tail) and differ significantly from all other groups at p < 0.05. OR, overt word reading. STM, encoding phase of recent memory for words paradigm. From Ojemann and Schoenfield-McNeil (1998).

occurred at the same temporal lobe sites, in the same subjects, during the same behavioral task (Ojemann et al., 2010). This paradigm involved alternate blocks of trials of associate learning or identification of words, performed silently and each with a relatively large number of rapidly presented (1/s) items, but otherwise similar to the previous learning paradigm. Testing of performance after each block of associative learning demonstrated significant learning. As part of this study not only was single neuron firing assessed but also LFPs were recorded through the same electrode. In the study of Zanos et al., those LFP signals in different frequency bands from 4 to 150 Hz were convoluted with a Volterra kernel, and the timing of action potentials predicted by the output of the kernel was compared to the actual discharges. The relation of spike activity

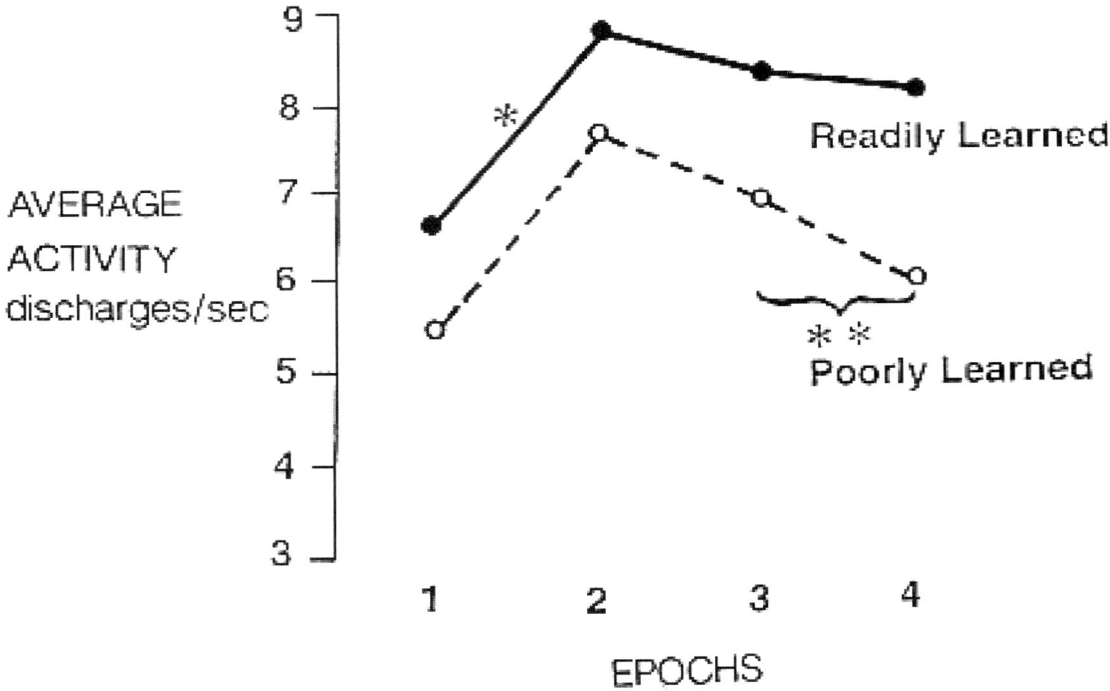

Figure 14.9
Time course for average activity during initial encoding of learned word pair associations (solid line) compared to those not (dashed line) for those neurons with inhibition during word identification and excitatior during recent memory encoding of words. Epoch 1 is the period of 700 ms before item presentation, which occurs at onset of epoch 2. It lasts until the onset of epoch 3, 300 ms before voice onset when the subject reads aloud the pictured word pair to 1200 ms after. Epoch 4 is the period from then to the next epoch 1. Significant increased activity occurs for all items between epochs 1 and 2 (*, p < 0.02). However, activity for learned pair associations is sustained throughout epochs 3 and 4 while that for unlearned pair associations declines, a difference significant (**, p < 0.02). From Ojemann and Schoenfield-McNeill (1998).

to LFP oscillations was also assessed. This study also included a number of control measures to ensure that the LFP signal was not contaminated by any signal from action potentials.

In the 69 single neurons from 14 patients included in this study, significant prediction of spike timing (to within 1 ms) was present for 25 neurons in spike-free recordings, with from 7 to 49% of spikes correctly predicted. These significant cases formed two groups. For 14 "high-frequency" neurons, spike timing was significantly more frequently predicted by high LFP frequencies (80–150 Hz) than low (4–14 Hz). In 10 other "low-frequency" neurons, it was the reverse. These two groups did not differ in power spectra of their LFPs, spike waveform, or pattern of firing. They differed in the contribution of different components of the kernel, in the relation to LFP oscillations and to the behavioral measures. High-frequency neurons had large contributions from second order kernel, low from first order. Low-frequency had more phase locking to LFP oscillations (4–10 Hz). Mean firing rate for the high-frequency neurons was significantly higher than that of low-frequency neurons for the paired-associate learning blocks, while low-frequency neurons had higher firing rates than high-frequency neurons during the identification blocks (although as in previous studies mean firing rates of all neurons were higher during associative learning than identification).

The importance of these findings to an analysis of the networks related to identification and learning is based on the analysis of these signals in experimental animals. The LFP is considered to reflect the postsynaptic inputs to a small volume of tissue around the electrode, within 250 microns diameter. The low frequency component is thought to represent activity projected from distant neuronal populations, the high frequency from local neuronal circuits. The activity with associative verbal learning then represents that of local neurons, the activity with identification effects projected from greater distances. Projecting this onto our model of the changes between relative inhibition (reduced firing rate) with identification transiting to excitation with learning of a new association to that item, the inhibition is a distant effect, perhaps favoring an inhibitory surround, the excitation a new local effect which, given its "tonic" nature, might be triggered by mechanisms such as a "thalamocortical" activating system.

Identification and Recent Memory for Visuospatial Material: Faces, Figures, Line Angles

Single neuron activity in temporal cortex has also been assessed for visuospatial materials (J. Ojemann et al., 1992; Holmes et al., 1996; Lucas et al., 2003). These separate series from the 1980s and 1990s recorded what by current standards represented small neuronal samples, most from nondominant hemisphere. Identification was assessed by overt matching of a target item to four to eight foils, in all studies for faces and complex figures with few verbal associations and in one (Holmes et al., 1996) also line angles. All three studies found neurons with significant changes for faces compared to the other conditions assessed. In the one study with recordings from both hemispheres (Lucas et al., 2003), these neurons were significantly more frequent in the right, nondominant hemisphere. That study also assessed activity with matching of individual facial features: eyes, lips, noses. One third of the neurons were "feature selective" in that activity differed between matching of intact faces or individual features. Such "feature selective" neurons have also been identified in recordings from monkey temporal cortex (Perrett et al., 1982). All three studies also identified neurons related to "matching," in that activity changed significantly with all matching conditions: faces, complex figures, and, where assessed, line angles. The direction of change differed between studies but with considerable relative early inhibition in the nondominant hemisphere during matching (J. Ojemann et al., 1992; Lucas et al., 2003). In addition to these "matching" neurons, each study also identified neurons with changes only with match for one type of item. In the two earlier studies, subjects were also asked to label facial expressions portrayed in a set of standardized actors' faces. Both studies identified neurons changing activity only to labeling of the facial expressions compared to face matching or naming.

Interference with identification of facial expressions and not face matching has been reported with electrical stimulation mapping of nondominant lateral temporal cortex (Fried et al., 1982).

Recent memory for visuospatial material was assessed in two of these studies (Holmes et al., 1996; Lucas et al., 2003), in one for line angles, the other faces. Both studies used recognition retrieval. Despite having fewer than 50 neurons in the combined studies, only 3 of which had significant changes with memory, 4 individual neurons were identified with changes in activity only during memory encoding or retrieval or both that did not show similar changes with identification of the same material, some in both studies. These were recorded from either hemisphere. However, the earlier study of memory for line angles reported relative decreased activity, the later study of memory for faces increased activity with the instruction to retain an item in memory.

Implicit Memory

The previously described identification and memory tasks all assessed conscious activities, ones the subject was aware of. However, there are also "subconscious" or "implicit" memory processes, processes of which the subject is unaware that change performance. One such process is repetition priming, a shortening of reaction time with repeated presentation of the same item. Changes in lateral temporal cortical single neuron activity related to this were assessed by comparing activity during first and repeated presentation of the same items in object naming, text reading, or auditory word repetition tasks (Ojemann et al., 2002). Significant differences were present in only 10% of the 105 neurons included in that study. This represented 38% of that study's 26 subjects. Most of these changes were for only one modality, with increased activity and a shortened reaction time (indicating priming) for the repeated item. Although that study included recordings throughout lateral temporal cortex, neurons with those significant implicit memory changes were all in superior temporal gyrus or superior portion of middle gyrus, significantly more superior and posterior than neurons with recent memory processes recorded in the same study. Moreover, recent memory encoding changes were present in only 18% of the neurons with implicit memory changes, including 4 subjects with implicit memory changes in one neuron and recent memory encoding changes in another recorded through the same microelectrode. There was also little overlap with neurons related to recent memory recognition retrieval. Implicit memory then involves neural networks in posterior superior temporal cortex, largely separate from those for recent memory, findings supported by effect of human brain lesions but not human neuroimaging or nonhuman primate neuronal recording.

Conclusions

These single neuron recordings from lateral temporal cortex have provided a number of unique insights into the neural substrate for human cognition. That those insights are functionally important has been established by relating many of them to different levels of performance. For many of the behaviors assessed, changes in neuronal activity have been widespread, without regard to lateralization, even for tasks strongly lateralized based on lesion effects. Yet within each region, these widely dispersed networks are often functionally very specific, with individual neurons often showing changes for only one function, even in only one modality. Lateralized differences, when present, have involved early and sustained changes in activity and sometimes changes in multiple modalities.

Single neuron recordings have established a major role for lateral temporal cortex in recent memory, with many of the neurons there demonstrating changes that are specific to memory. They have highlighted the importance of sustained, "tonic" changes in activity specific to the memory process including activity sustained from encoding into memory storage where it likely reflects active rehearsal to maintain the memory. Differential roles within this cortex have been identified, with superior temporal gyrus related to perception specific to memory encoding, with simultaneous convergence on some individual neurons of this encoding phasic change related to perception and a sustained "tonic" change.

These recordings have also emphasized the importance of relative inhibition, particularly in language. A potential role for these inhibited neurons has been suggested by the evidence of their activation with learning of new verbal associations, an activation characterized by sustained "tonic" activity. The dynamic nature of this network is evident by the rapid reversal of that activation once the association is learned. And the relation of LFP to timing of single neuron action potentials suggests that this activation with learning is a local network effect while the inhibition of those neurons with language reflects more distant inputs.

These findings by no means indicate the full potential for human single neuron recordings. There remain major issues that those recordings are uniquely suited to address. Perhaps the most significant is the issue of the extent to which neural networks related to cognition function with rate or temporal codes. However, the studies reviewed here give some indication of what can be accomplished when single neuron recording in the operating room is combined with standardized assessments based on neuropsychological models to investigate the neural mechanisms of human cognition.

References

Bechtereva, N. P., Abdullaev, Y. G., & Medvedev, S. V. (1992). Properties of neuronal activity in cortex and subcortical nuclei of the human brain during single-word processing. *EEG Clinical Neurophysiology, 82,* 296–301.

Bechtereva, N. P., Bundzen, P., Gogolitsin, Y., Malyshev, V., & Perepelkin, P (1979). Neurophysiological codes of words in subcortical structures of the human brain. *Brain and Language, 7,* 145–163.

Calvin, W. H., Ojemann, G. A., & Ward, A. A., Jr. (1973). Human cortical neurons in epileptogenic foci: Comparison of interictal firing patterns to those of "epileptic" neurons in animals. *EEG Clinical Neurophysiology, 34,* 337–351.

Casasanto, D., Kilgore, W., Maldjian, J., Glosser, G., Alsp, D., & Cooke, A (2002). Neural correlates of successful and unsuccessful verbal memory encoding. *Brain and Language, 80,* 287–295.

Cheung, M.-C., & Chan, A. S. (2003). Memory impairment in humans after bilateral damage to lateral temporal neocortex. *Neuroreport, 14,* 371–374.

Creutzfeldt, O., Ojemann, G., & Lettich, E. (1989a). Neuronal activity in the human lateral temporal lobe: I. Responses to speech. *Experimental Brain Research, 77*, 451–475.

Creutzfeldt, O., Ojemann, G., & Lettich, E. (1989b). Neuronal activity in the human lateral temporal lobe: II. Responses to the subjects own voice. *Experimental Brain Research, 77*, 476–489.

Creutzfeldt, O., Ojemann, G., & Lettich, E. (1987). Single neuron activity in the right and left human temporal lobe during listening and speaking. In D. Ottoson (Ed.), *Duality and unity of the brain (Wenner-Gren International Symposium Series, Vol. 47)* (pp. 295–310). Hampshire: MacMillan Press.

Farrell, D. F., Leeman, S., & Ojemann, G. A. (2007). Study of human visual cortex: Direct cortical evoked potentials and stimulation. *Journal of Clinical Neurophysiology, 55*, 1–10.

Fedio, P., & Van Buren, J. (1974). Memory deficits during electrical stimulation of speech cortex in conscious man. *Brain and Language, 1*, 29–42.

Fletcher, P., & Tyler, L. (2002). Neural correlates of memory. *Nature Neuroscience, 5*, 8–9.

Fried, I., Mateer, C., Ojemann, G., Wohns, R., & Fedio, P. (1982). Organization of visuospatial functions in human cortex: Evidence from electrical stimulation. *Brain, 105*, 349–371.

Fries, P., Nikolic, D., & Singer, W. (2007). The gamma cycle. *Trends in Neurosciences, 30*, 309–316.

Fuster, J. (1995). *Memory in the cerebral cortex: An empirical approach to neural networks in the human and nonhuman primate*. Cambridge, MA: MIT Press.

Fuster, J. M., & Jervey, J. P. (1982). Neuronal firing in the infertemporal cortex of the monkey in a visual memory task. *Journal of Neuroscience, 2*, 361–375.

Green, J. D. (1964). The hippocampus. *Physiological Reviews, 44*, 561–608.

Haglund, M. M., Ojemann, G. A., Schwartz, T. W., & Lettich, E. (1994). Neuronal activity in human lateral temporal cortex during serial retrieval from short-term verbal memory. *Journal of Neuroscience, 14*, 1507–1515.

Hebb, D. O. (1949). *The organization of behavior*. New York: Wiley.

Holmes, M., Ojemann, G. A., & Lettich, E. (1996). Neuronal activity in human right lateral temporal cortex related to visuospatial memory and perception. *Brain Research, 711*, 44–49.

Kirchoff, B., Wagner, A., Maril, A., & Stern, C. (2000). Prefrontal temporal circuitry for episodic encoding and subsequent memory. *Journal of Neuroscience, 20*, 6173–6180.

Kohler, E., Keyses, C., Umita, M. A., Fogassi, L., Gallese, V., & Rizzolatti, G. (2002). Hearing sounds, understanding actions: Action representation in mirror neurons. *Science, 297*, 846–848.

Lucas, T. H., Schoenfield-McNeill, J., Weber, P. B., & Ojemann, G. A. (2003). A direct measure of human lateral temporal lobe neurons responsive to face matching. *Brain Research. Cognitive Brain Research, 18*, 15–25.

Lucas, T. H., II, McKhann, G. M., II, & Ojemann, G. A. (2004). Functional separation of languages in the bilingual brain: A comparison of electrical stimulation language mapping in 25 bilingual patients and 117 monolingual control patients. *Journal of Neurosurgery, 101*, 449–457.

Milner, B. (1967). Brain mechanisms suggested by studies of temporal lobe. In C. Milliken & F. Darley (Eds.), *Brain mechanisms underlying speech and language* (pp. 122–1245). New York: Grune and Stratton.

Nader, K., & Hardt, O. (2009). A single standard for memory: The case for reconsolidation. *Nature Reviews. Neuroscience, 10*, 224–234.

Ojemann, G. A. (1978). Organization of short-term verbal memory in language areas of human cortex: Evidence from electrical stimulation. *Brain and Language, 5*, 331–340.

Ojemann, G. A. (1983). Brain organization for language from the perspective of electrical stimulation mapping. *Behavioral and Brain Sciences, 6*, 189–206.

Ojemann, G. A. (1989). Some brain mechanisms for reading. In C. Von Euler, I. Lundberg, & G. Lennerstrand (Eds.), *Brain and reading* (pp. 47–59). Hampshire: MacMillan Press.

Ojemann, G. A. (1990). Organization of language cortex derived from investigations during neurosurgery. *Seminars in Neuroscience, 2*, 297–305.

Ojemann, G. A. (1995). Awake operations with mapping in epilepsy. In H. H. Schmidek & W. H. Sweet (Eds.), *Operative neurosurgical techniques* (3rd ed., pp. 1317–1322). Philadelphia: W.B. Saunders.

Ojemann, G. A., Blick, K. I., & Ward, A. A., Jr. (1971). Improvement and disturbance of short-term verbal memory with human ventrolateral thalamic stimulation. *Brain and Language, 94*, 225–240.

Ojemann, G. A., Cawthon, D. F., & Lettich, E. (1990). Localization and physiological correlates of language and verbal memory in human lateral temporoparietal cortex. In A. B. Scheibel & A. F. Wechsler (Eds.), *Neurobiology of higher cognitive function* (pp. 185–202). New York: Guilford Press.

Ojemann, G. A., Corina, D. P., Corrigan, N., Schoenfield-McNeill, J., Poliakov, A., Zamora, L., et al. (2010). Neuronal correlates of functional magnetic resonance imaging in human temporal cortex. *Brain, 133*, 46–59.

Ojemann, G. A., Creutzfeldt, O. D., Lettich, E., & Haglund, M. M. (1988). Neuronal activity in human lateral temporal cortex related to short-term verbal memory, naming and reading. *Brain, 111*, 1383–1403.

Ojemann, G. A., & Dodrill, C. B. (1985). Verbal memory deficits after left temporal lobectomy for epilepsy: Mechanism and intraoperative prediction. *Journal of Neurosurgery, 62*, 101–107.

Ojemann, G. A., & Dodrill, C. B. (1987). Intraoperative techniques for reducing language and memory deficits with left temporal lobectomy. *Advances in Epileptology, 16*, 327–330.

Ojemann, G., Ojemann, J., Lettich, E., & Berger, M. (1989). Cortical language localization in left, dominant hemisphere: An electrical stimulation mapping investigation in 117 patients. *Journal of Neurosurgery, 71*, 316–326.

Ojemann, G. A., & Schoenfield-McNeill, J. (1998). Neurons in human temporal cortex active with verbal associative learning. *Brain and Language, 64*, 317–327.

Ojemann, G., & Schoenfield-McNeill, J. (1999). Activity of neurons in human temporal cortex during identification and memory for names and words. *Journal of Neuroscience, 19*, 5674–5682.

Ojemann G., & Schoenfield-McNeill, J. (2002). Neuronal activity during recall retrieval from recent verbal memory. Unpublished manuscript.

Ojemann, G. A., Schoenfield-McNeill, J., & Corina, D. (2002). Anatomic subdivisions in human temporal cortical neuronal activity related to recent verbal memory. *Nature Neuroscience, 5*, 64–71.

Ojemann, G. A., Schoenfield-McNeill, J., & Corina, D. (2004). Different neurons in different regions of human temporal lobe distinguish correct from incorrect identification or memory. *Neuropsychologia, 42*, 1383–1393.

Ojemann, G. A., Schoenfield-McNeill, J., & Corina, D. (2009). The roles of human lateral temporal cortical neuronal activity in recent verbal memory encoding. *Cerebral Cortex, 19*, 197–205.

Ojemann, G. A., Sutherling, W., Lesser, R., Dinner, D., Jayakar, P., & Saint-Hilaire, J.-M. (1993). Cortical stimulation. In J. Engel, Jr. (Ed.), *Surgical treatment of the epilepsies* (2nd ed., pp. 399–414). New York: Raven Press.

Ojemann, G. A., & Whitaker, H. A. (1978). The bilingual brain. *Archives of Neurology, 35*, 409–412.

Ojemann, J., Ojemann, G., & Lettich, E. (1992). Neuronal activity related to faces and matching in human right non-dominant temporal cortex. *Brain, 115*, 1–13.

Perrett, D., Rolls, E., & Caan, W. (1982). Visual neurons responsive to faces in monkey temporal cortex. *Experimental Brain Research, 47*, 329–342.

Perrine, K., Devinsky, O., Uysal, S., Luciano, D., & Dogali, M. (1994). Left temporal neocortex mediation of verbal memory. *Neurology, 44*, 1845–1850.

Peterson, L., & Peterson, M. (1959). Short-term retention of individual verbal items. *Journal of Experimental Psychology, 58*, 193–198.

Quiroga, R. Q., Nadasdy, Z., & Ben-Shaul, Y. (2004). Unsupervised spike detection and sorting with wavelets and superparamagnetic clustering. *Neural Computation, 16*, 1661–1687.

Raichle, M. E., Fiez, J., Videen, T., Prado, J., Fox, P., & Peterson, S. (1994). Practice-related changes in human brain functional anatomy during non-motor learning. *Cerebral Cortex, 4*, 8–26.

Rizzolatti, G., Camarda, R., Fogass, L., Gentilucci, M., Lippino, G., & Matelli, M (1988). Functional organization of inferior area 6 in macaque monkey. *Experimental Brain Research, 71*, 491–507.

Schwartz, T., Ojemann, G., Haglund, M., & Lettich, E. (1996). Cerebral lateralization of neuronal activity during naming, reading and line-matching. *Brain Research. Cognitive Brain Research, 4*, 263–273.

Schwartz, T. H., Haglund, M. M., Lettich, E., & Ojemann, G. A. (2000). Asymmetry of neuronal activity during extra-cellular microelectrode recording from left and right human temporal lobe neocortex during rhyming and line matching. *Journal of Cognitive Neuroscience, 12*, 803–812.

Wada, J., & Rasmussen, T. (1960). Intracarotid injection of sodium amytal for the lateralization of cerebral speech dominance: Experimental and clinical observations. *Journal of Neurosurgery, 17*, 266–282.

Weber, P., & Ojemann, G. A. (1995). Neuronal recordings in human lateral temporal lobe during verbal paired associates learning. *Neuroreport, 6*, 685–689.

Zamora, L., Corina, D., & Ojemann, G. (2013). Single-neuron evidence of active rehearsal during two working memory tasks from human temporal cortex. Unpublished manuscript.

Zamora, L., Corina, D., Schoenfield-McNeill, J., Lettich, E., Ojemann, G., & Holmes, M. (2001). A comparison of neuronal firing rates for distracted versus non-distracted memory. *Society for Neuroscience Abstracts, 27*, 4196.

Zanos, S., Zanos, T. P, Marmarelis, V. Z., Ojemann, G. A., & Fetz, E. E. (2012). Relationship between spike-free local field potentials and spike timing in human temporal cortex. *Journal of Neurophysiology, 107*, 1808–1821.

III CLINICAL NEUROSCIENCE

15   Microelectrode Recordings in Deep Brain Stimulation Surgery

C. Rory Goodwin, Travis S. Tierney, Frederick A. Lenz, and William S. Anderson

Several techniques have been described for the placement of deep brain stimulation (DBS) electrodes, and both radiological and physiological landmarks are usually used to accurately localize the target accurately (Bakay, DeLong, & Vitek, 1992; Garonzik, Hua, Ohara, & Lenz, 2002; Lozano, Hutchison, Kiss, Tasker, Davis, & Dostrovsky, 1996; Patel, Heywood, O'Sullivan, McCarter, Love, & Gill, 2003). These procedures are performed primarily for the treatment of movement disorders such as Parkinson's disease, essential tremor, and dystonia although indications are expanding into psychiatric and other conditions (Anderson & Lenz, 2006; Tierney, Sankar, & Lozano, 2011). In many of these procedures, surgeons use radiological localization of the anterior commissure (AC) and posterior commissure (PC) for use with indirect atlas-based targeting, or direct anatomical targeting is performed using subcortical structures identified on computed tomography (CT), magnetic resonance imaging (MRI), or a fusion of both imaging modalities (Carlson & Iacono, 1999). This radiological estimate of target location can be refined by microelectrode recordings (MERs) for physiological localization, with subsequent DBS electrode implantation or radiofrequency lesioning (Bakay et al., 1992; Mandir, Rowland, Dougherty, & Lenz, 1997; Hua, Garonzik, Lee, & Lenz, 2004). Macrostimulation techniques are also used during MER to assess clinical efficacy of the proposed treatment and to look for side effects of the stimulation (Burchiel, 1995).

Implantation of the full DBS system is usually performed as a two-stage procedure, where Stage I and II are typically separated by one week. Stage I entails the stereotactic placement of bilateral intracranial DBS leads. For essential tremor and Parkinson's disease cases, we use the Medtronic Model 3387 quadripolar electrode which consists of four 1.5-mm cylindrical contacts separated by 1.5 mm (total electrode span is 10.5 mm, Medtronic, Minneapolis, MN). The electrode is inserted through coronal burr holes while the patent is awake to assess for desired and potential adverse stimulation effects. Microelectrode recording is used to examine the electrophysiological behavior of the subcortical structures encountered during traversal to the selected target. Stage II requires general anesthesia to avoid otherwise painful tunneling of an extension lead from the scalp to the upper chest and implantation of a computerized pulse generator. We insert a dual channel Activa PC (Model 37601, Medtronic, Minneapolis, MN) into a subcutaneous pocket just inferior to the right clavicle. Between stages, we acquire a high-resolution noncontrast head CT to confirm lead placement and to assess for any intracranial bleeding.

Frame Placement

On the morning of Stage I, a Leksell G frame base (Elekta, Atlanta, GA) is affixed to the patient's head after infiltrating each pin site with 3–5 cc of a mixture of 50% bupivacaine (0.25% solution) and 50% lidocaine (1.0% solution). A number of other alternative frames and frameless systems are commercially available. The patient is informed that he or she will experience 5 to 10 minutes of pressure after the pins acquire bony purchase, but that he or she should not feel sharp pain from pin penetration of the skin. Careful reassurance that this is generally the most uncomfortable part of the day is important to maintain patient confidence and optimal intraoperative compliance. We try to position the frame base along the zygoma and parallel to the canthomeatal line, which tends to put the AC–PC line within a single axial cut. The patient is transported to the MRI machine, where thin cut axial T1- and T2-weighted images are obtained, as well as a 3-D volumetric postgadolinium T1-weighted image for identification of vascular structures when planning the trajectory with commercially available software systems. Many groups will obtain an in-frame high-resolution noncontrast head CT (0.6 mm axial slices at zero gantry angle) with subsequent coregistration with a previously acquired MRI for use in the trajectory planning station.

Radiological Localization

Both direct and indirect radiological targeting usually begin by determining the location of the AC and PC in the region of the third ventricle using standard high-resolution MRI or CT images. The AC and PC predict the locations of the different subcortical nuclei in stereotactic space when referenced to standard stereotactic atlases (Schaltenbrand & Walker, 1982; Talairach & Tournoux, 1988). For example, one method of indirect targeting for the ventral intermediate (Vim) nucleus of the thalamus is estimated with reference to the AC–PC line, implying a target of 3 mm anterior and 14 mm lateral to PC and 2 mm above the PC. Alternately, direct anatomical methods can be employed. The lateral plane of the target in Vim can be estimated as lying between the internal capsule and the T2 intense medial dorsal nucleus of the thalamus (Burchiel, 1995). In another example, the globus pallidus interna (GPi) can be estimated as 2–3 mm anterior, 3–4 mm inferior, and 20–22 mm lateral to the midcommissural point (O'Gorman et al., 2011; Vayssiere et al., 2002). Direct targeting can be carried out for the GPi by using inversion recovery and other sensitive MRI sequences (O'Gorman et al., 2011; Vayssiere et al., 2002). On an axial cut including the AC, the horizontal and anterior/posterior coordinates can be estimated from the junction of the two posterior quarters of GPi and the globus pallidus externa (GPe) (Vayssiere et al., 2002).

The location of the subthalamic nucleus (STN) can be estimated as 3–4 mm posterior and 4–5 mm inferior to the midcommissural point (MCP) and 12 mm lateral (Patel et al., 2003; Starr et al., 2002). Another indirect technique targets a point 7 mm lateral, 4 mm anterior, and axially at the position of the center of the red nucleus. The nucleus can also be directly localized on 2-mm, axial T2-weighted scans and coronal scans through the thalamus and midbrain. Fusion with other MRI series or CT can then be used to estimate the location of the STN (Patel et al., 2003). The target is then selected in the dorsal part of the posterior third of the nucleus. Another approach is to localize the STN by a fusion process based on a volumetric gradient echo T2 and a fast spin-echo sequence. Additionally, the nucleus can also be localized 7 mm lateral and 4 mm anterior to the center of the red nucleus (Starr et al., 2002). Physiological confirmation of the radiological estimate is carried out by stimulation or recording (MER) at and around the target. In our institution, a map is made to the same scale as a set of transparent atlas maps from the sagittal sections of the Schaltenbrand and Bailey atlas (Schaltenbrand & Walker, 1982; Schaltenbrand & Bailey, 1959). The 13.5-mm-lateral atlas map is used for Vim procedures, 21 mm for GPi procedures, and 11 mm for the subthalamic nucleus.

Both CT and MRI can be used for targeting, as well as merged images from the two modalities. MRI produces more distortion than CT imaging, particularly in the z-plane when acquired axially and at progressively greater distances from the center of the image. The targeting accuracy of MRI is approximately 2 mm on average and 4 mm at maximum (Kondziolka et al., 1992; Gerdes, Hitchon, Neerangun, & Torner, 1994; Holtzheimer, Roberts, & Darcey, 1999). This error is due to artifacts related to inhomogeneities in the magnetic field and nonlinearities in the gradient field—the position-dependent strength in the magnetic field (Hardy & Barnett, 1998). The artifacts can be induced by metal or magnetic susceptibility differences—produced at the interface between materials (e.g., air and bone). Attempts to decrease errors from the MRI due to these artifacts include software modifications and overlapping (fusion) of the 3-D MRI image with the CT image, which is immune to susceptibility differences (Alexander et al., 1995). Direct targeting can be assisted by computer programs which overlay atlas maps of anatomy onto the radiological section. These maps are usually deformed either to match the AC–PC line in isolation (Dostrovsky, 1998) or to match the AC–PC line and other structures, such as the margins of the third ventricle or the internal capsule (Cooper, Bergmann, & Caracalos, 1963). Several commercial targeting systems are available (FrameLink, Medtronic, Minneapolis, MD; iPlan Stereotaxy, Brainlab Inc., Westchester, IL).

DBS Lead Operative Procedure

The patient is taken to the operating room and placed on a standard adjustable operative table in a semi-Fowler's (semi-upright, 30°) position with the stereotactic frame rigidly attached to the operative table using a Mayfield adaptor. A warm blanket wrap and foam bolsters placed behind the neck and upper back provide some relief from cervical muscle strain during the operation. We position a fluoroscope for lateral skull imaging and use the arm of the scope to secure a clear plastic surgical drape. Preoperative antibiotics are given through an IV line. Blood pressure is monitored with a brachial cuff keeping systolics below 140 mm Hg, especially during electrode penetration.

Once the patient is positioned and comfortable, the scalp along the coronal suture is clipper shaved, scrubbed, and painted with povidone–iodine solution. Local anesthesia (about 20–30 cc of 1% lidocaine with epinephrine at 1:100,000) is injected along a curvilinear line marked in the coronal plane extending 3–4 cm bilaterally from the midline and just in front of bregma. Alternatively, many surgeons will make two separate parallel bilateral parasagittal incisions approximately 3.0–3.5 cm lateral to midline and straddling the coronal suture. Supraorbital and occipital nerve blocks are also often used by incorporation into the pin site injections. The scalp is then once again painted with povidone–iodine solution that is allowed to completely dry during the time it takes for preoperative computer targeting.

Using one of several standard commercially available stereotactic computer programs, preoperatively acquired inversion recovery T1 volumetric sequences (TR 4000 ms, TE 90 ms, echo train 8, FOV 270 mm^3, 256 × 256 matrix, frequency direction anterior–posterior, 2-mm axial cuts with no gap and zero gantry angle) are back-fused to the in-frame high-resolution CT scan. In this way potential MRI distortion is minimized. By convention, the posterior border of the AC and the anterior border of PC define the intercommissural line (Schaltenbrand & Bailey, 1959). These points are selected using the T1 axial sequences together with coronal and sagittal reconstructions. Most software can now eliminate pitch, yaw, and roll associated with less-than-perfect frame placement. This reformats the sagittal and axial images parallel to the intercommissural line and, together with deformable Schaltenbrand and Wahren contour overlays, assists in direct targeting. Table 15.1 lists nominal indirect targets for stereotactic coordinates that have been used in various studies to target specifically the STN, GPi, and Vim of the thalamus based on Cartesian distance in millimeters from the PC or MCP. Modification of the indirect coordinate based on direct targeting of the MRI image is necessary to avoid stimulating nearby structures that may contribute to off-target adverse effects. T2-weight coronal sequences are particularly

Table 15.1
Suggested nominal indirect stereotactic coordinates for the three deep brain stimulation targets generally used in movement disorder surgery

Target	Reference	X—lateral	Y—A/P	Z—D/V
Vim	Garonzik et al. (2002)	14 mm from PC	3 mm anterior to PC	2 mm dorsal to PC
GPi	O'Gorman et al. (2011) Vayssiere et al. (2002)	20–22 mm from MCP	2–3 mm anterior to MCP	3–4 mm ventral to MCP
STN	Patel et al. (2003) Starr et al. (2002)	12 mm from MCP	3–4 mm posterior to MCP	4–5 mm ventral to MCP

AC, anterior commissure; A/P, anterior/posterior; D/V, dorsal/ventral; MCP, midcommissural point; Vim, ventral intermediate nucleus of the thalamus; STN, subthamalmic nucleus; GPi, globus pallidus pars interna.

helpful in directly visualizing the STN in relationship to the red nucleus and other surrounding structures.

An entry point is then selected to avoid bridging veins and cortical sulci which are well vascularized. Optimal electrode trajectories also avoid ventricular transgressions because of the risk to ventricular veins and intraventricular hemorrhage. Generally, choosing entry points that are just in front of the coronal suture and 3.0 to 3.5 cm off midline will yield satisfactory trajectories for all three targets (STN, GPi, and Vim).

In the operating room (see figure 15.1), a sterile clear plastic drape is placed over the prepped scalp and base frame and secured to the arm of the fluoroscope. A short-acting light intravenous sedation (either dexmedetomidine, propofol, or selective low doses of midazolam) will decrease patient anxiety as burr hole and dural opening are performed. We make two parallel skin incisions down to bone with a #10 blade crossing just in front of the coronal suture bilaterally. The pericranium is swept aside with periosteal elevators, and small self-retaining retractors are placed to hold the scalp open. The Leksell arc is then brought into the field and attached to trunnions fixed to the base frame. The calculated X, Y, and Z coordinates are then set, as is the arc and ring angle, to allow for precise skull placement of each 14-mm burr hole. The handy self-locking lead fixation rings provided by Medtronic (Stimloc) fit into standard Codman (Raynham, MA) perforator burr holes provided that the inner rim of self-stopping bone is more than 3 mm below the outer surface. When the skull is thin, it may be necessary to use a Kerrison punch to remove this inner edge if the fixation ring does not sit flush with the skull. The dura is opened in cruciate fashion and simultaneously coagulated with an electrified dural hook, taking care to avoid any underlying cortical veins. If found, large cortical veins can usually be gently moved aside; smaller veins can be coagulated. An air embolism is possible at this time and is usually signaled by coughing or reports from the patient of a tickling sensation in the chest. Should this occur, immediately flood the burr hole with warm saline, wax the bone edge, and lower the head of the bed. It should rarely be necessary to stop the operation if this situation is immediately recognized and corrected. The underlying pia is then coagulated with bipolar cautery and sharply incised.

After checking that the systolic blood pressure is below 150 mm Hg, a 1.3-mm inner diameter brain cannula is passed through the gyrus without significantly displacing the brain surface. For placement of the microelectrode, the cannula is inserted to typically 15–25 mm above the final target, and the inner blunt stylet is removed. A Gelfoam pledget and fibrin glue seal the burr hole to prevent cerebrospinal fluid escape. Microelectrode recordings then begin from this depth to typically 5–6 mm deeper than target to fully traverse the structure of interest. Physiological testing is typically performed during these recordings to seek kinesthetic responsive cells further aiding in localization. Macrostimulation is also performed in the reverse direction as the lead is withdrawn at various depths to seek physiological and therapeutic responses as well as possible side effects.

The DBS lead is then passed to the optimally determined target position under continuous lateral fluoroscopy using the fiducial markers placed within the Leksell trunnions to position the

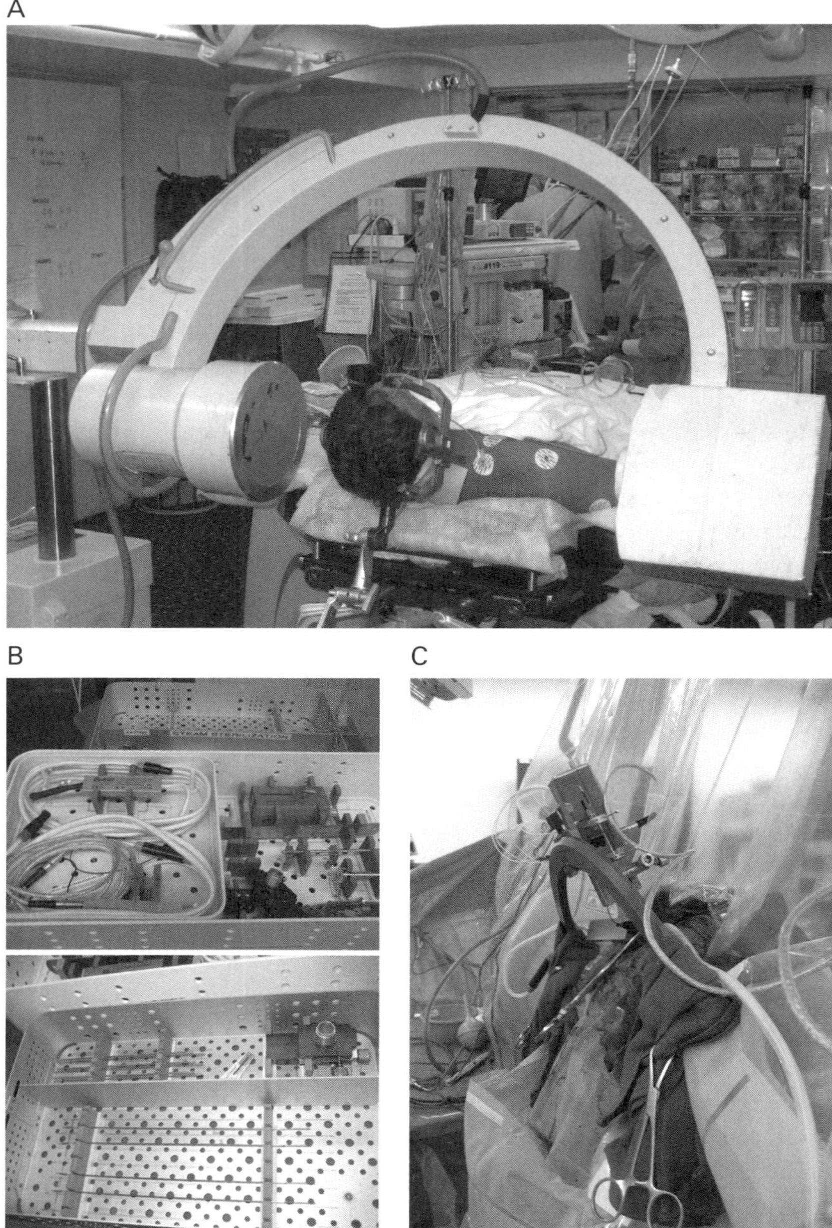

Figure 15.1

(A) Patient positioning in the operating room for deep brain stimulation placement procedure. The Leksell stereotactic frame the patient is wearing is bolted using the Mayfield head fixation system to the underlying operating room table. A fluoroscopy unit is arranged for lateral skull images; it also provides a scaffold for the operative drapes to provide a clear view of the patient from the anesthesiologist's perspective and to facilitate neurologic testing. (B) Microelectrode recording equipment (Alpha Omega Co. USA, Alpharetta, GA). The upper panel shows the tray containing the head stage amplifier (left) and the single micron resolution microdrive system (right). The lower panel demonstrates the tray containing the microelectrodes (foreground), insertion cannulae (left), and a positioning stage allowing for fine x–y trajectory adjustment (right). (C) Assembled microelectrode recording equipment, mounted on the arc of the Leksell frame.

most distal contact at the final target. Further physiological testing of the inserted lead can be then performed with a provided Medtronic externalized testing and control cable at this time, to again look for side effects and clinical efficacy with the final implanted device. The jaws of the Medtronic Stimloc system are then inserted into the previously placed burr hole ring and closed around the lead after gently raising the cannula under fluoroscopic view. Note that the lead is not completely secure at this stage, but only partially stabilized by the locking jaws. The stylet inside the macroelectrode lead is then carefully removed. Intermittent lateral fluoroscopy is used to confirm that tip migration has not occurred throughout this process. Finally, the locking cap is positioned to secure the lead firmly in place. We then radiographically document the final lead placement with a single lateral image saved to the radiology viewing system.

The distal end of the lead is covered with a dummy cap and tunneled beneath the galea with a Kelly clamp for retrieval at Stage II (the implantable pulse generator placement). The frame coordinates are then reset and confirmed for the contralateral side, and the procedure is repeated for bilateral lead placement. To avoid a wrong side placement error that may not be evident on lateral fluoroscopy, be certain that the frame coordinates are reset and confirmed. For a right-handed person in a bilateral implantation case, typically both lead cables are tunneled beneath the left-sided scalp so the patient can have an easier time placing the handheld patient programmer on the implantable pulse generator.

Implantable Pulse Generator

The implantable pulse generator can either be placed at the time of the DBS lead placement or, frequently, approximately one week later. This procedure is usually performed under general anesthesia because of the painful subcutaneous tunneling involved. The patient is prepped and draped at three sites: the location of the pocketed distal DBS lead cabling under the scalp, a retroauricular incision, and a subclavicular incision to house the pulse generator. The distal lead cabling is first exposed at the scalp site and then tunneled under the skin to the retroauricular incision. Extension lead cables (40–60 cm in length) are then tunneled from the subclavicular incision to the retroauricular incision. The extension leads are connected to the distal DBS lead cabling and to the implantable pulse generator. The connection fixture behind the ear is typically sutured to the underlying fascia and periosteum to inhibit migration. Similarly the pulse generator is provided with several suture attachment holes for securing to the underlying pectoralis fascia. Redundant extension lead cabling is placed to the sides and behind the pulse generator, but not on top of the device. Impedance checking of the leads using a Medtronic-provided transcutaneous interrogation system can be performed on the field to make sure all connection points are good. It is also helpful to mark either the right or left distal DBS lead cabling with a black silk suture to aid in connecting the two sides appropriately to the pulse generator.

Several schemes for initial programming have been investigated and published in which generally the goal is to maximize clinical efficacy in controlling the disease symptoms, minimize any stimulation-induced side effects, and (a somewhat lesser goal since rechargeable systems are now available) maximize the battery life (Volkmann, Herzog, Kopper, & Deuschl, 2002). There are four primary adjustable parameters for the charge-balanced pulses including pulse width, pulse application frequency, the identification of the contacts used to transmit the pulses (either between two of the quadripolar contacts at the end of the lead or between one contact and the pulse generator case as a distant reference), and the applied voltage between the specified contacts. Initial programming is probably best done with the patient skipping the morning dose of medication. The pulse width may be initially set at 60 μsec (Volkmann et al., 2002). The pulse frequency may also be initially set at 130 Hz; previous studies of frequency-dependent DBS effects show increasing clinical efficacy above 100 Hz (Volkmann et al., 2002). Subsequent programming then includes an assessment of each contact as the voltage is stepwise increased to seek the threshold of either clinical efficacy or side effects. The initial goal is to achieve the best clinical response with the minimum applied voltage (Volkmann et al., 2002).

In the case of Parkinson's disease, it is generally easiest to observe clinical effects on cogwheel rigidity in the upper extremities followed by tremor when tuning the stimulation parameters (Krack, Fraix, Mendes, Benabid, & Pollak, 2002). Follow-up visits may be necessary particularly in the early period after device turn-on to balance the effects of medication (particularly lowered dopaminergic agent dosing) and stimulation. Specific stimulation-induced side effects additionally will fade or change over time and may require alterations in the stimulation parameters. These include voice alterations, paresthesias, dyskinesias, and occasionally mood alterations (Krack et al., 2002). With essential tremor, the goal is tremor reduction or arrest. The treatment of dystonia with DBS and subsequent pulse generator programming is complicated by the fact that the patient's response to the therapy may be quite delayed (Kupsch et al., 2011). Because of the long delay in efficacy, many centers adjust the stimulation intensity to 10–15% below the level at which nontransient side effects occur; the intensity can then be decreased at interval visits until the dystonic symptoms recur, providing the clinician with a useful range of stimulation intensities (Kupsch et al., 2011). Other programming protocols in the case of dystonia have been published (Kupsch et al., 2011). Okun et al. have additionally published a helpful toolbox of programming adjustments to attempt in the context of troubleshooting therapeutic issues (Okun et al., 2008).

Physiological Localization

Radiological targeting is subsequently supported by identifying the different target structures (GPi, STN, and Vim) based on their electrophysiological properties. These properties are defined by recording spontaneous activity, and neuronal response to passive and active movements. Regional electrophysiological properties are also defined by microstimulation (< 100 μA), evoked sensations, motor effects such as muscular contractions, and alterations in the pathological movement disorder, for example, decreased tone or tremor. Physiological localization has been carried out by stimulation using a macroelectrode (impedance < 1000 Ω), stimulation and recording using a semi-microelectrode (impedance < 100 kΩ), and by performing MERs (impedance > 500 kΩ). High-impedance microelectrodes for physiological monitoring and recording are designed to isolate single neuron action potentials (Lenz, Dostrovsky, Kwan, et al., 1988; Jasper & Bertrand, 1966; Hubel, 1957). In addition, the electrode must be durable to withstand microstimulation. The electrode impedance is usually greater than 500 kΩ at 1000 Hz (Lozano et al., 1996; Lenz, Dostrovsky, Kwan, et al., 1988; Jasper & Bertrand, 1966). These characteristics can be achieved by constructing electrodes from a platinum–iridium alloy or from tungsten, producing a tapered tip sometimes bearing glass or lacquer insulation (Lenz, Dostrovsky, Kwan, et al., 1988; Jasper & Bertrand, 1966; Hubel, 1957; Umbach & Ehrhardt, 1965; Abe-Fessard, 1973; Ohye, 1982; Wolbarsht, MacNichol, & Wagner, 1960).

Figure 15.1 demonstrates typical patient positioning in the operating room during DBS surgery, with the commercial MER equipment in place. The assembled electrode is attached to a commercial piezoelectric or hydraulic microdrive mounted on the stereotaxic frame. Some microdrive systems incorporate a coarse drive for large individual steps so that overlying structures can be traversed quickly. The tip is then retracted into a protective cylindrical housing while the whole assembly is advanced to the starting depth above target (Vitek et al., 1998), usually 10–30 mm. The microdrive may then be employed from this new depth for detailed exploration of deeper structures. The signal from the microelectrode is amplified, band-pass filtered (300 Hz–6 kHz), and digitized for disk storage or offline analyses. Single unit action potential signals of constant shape and amplitude can be sorted with a variety of template matching or other algorithmic online sorting procedures.

Localization of Vim

Microelectrode recordings are begun with a pass made through a coronal burr hole about 25 mm from the midline. The Vim and adjacent nuclei show characteristic patterns of thalamic neural activity. Sensory cells responding to cutaneous sensory stimulation in small, well-defined receptive fields are found in the ventral caudal nucleus (Vc), posterior to Vim (see figure 15.2) (Lenz, Dostrovsky, Tasker, et al., 1988). Some have described a mediolateral somatotopy within Vc (Jasper & Bertrand, 1966; Lenz, Dostrovsky, Tasker, et al., 1988) proceeding from cutaneous or mucosal representations of oral structures medially to leg laterally (superficial sensory cells). In anterior Vc and further anterior, in Vim and the ventral oral posterior (Vop), thalamic neuronal firing is correlated with passive (deep sensory cells) joint movement (voluntary cells) (Raeva, 1986; Lenz et al., 1990).

Movement-related activity of most cells in Vim and Vop occurs during execution of particular movements with a somatotopy parallel to that of the sensory thalamus (Lenz et al., 1990). As shown in figure 15.3, some neurons respond both to active movement and to deep sensory stimulation such as joint movement (so-called combined cells; Lenz et al., 1990; Lenz et al., 1994). Many sensory, combined, and voluntary cells have activity correlated with electromyography (EMG) activity during tremor (Lenz, Tasker, et al., 1988; Hua, Lenz, Zirh, Reich, & Dougherty, 1998; Lenz et al., 1995). In addition to MER, microstimulation of subcortical

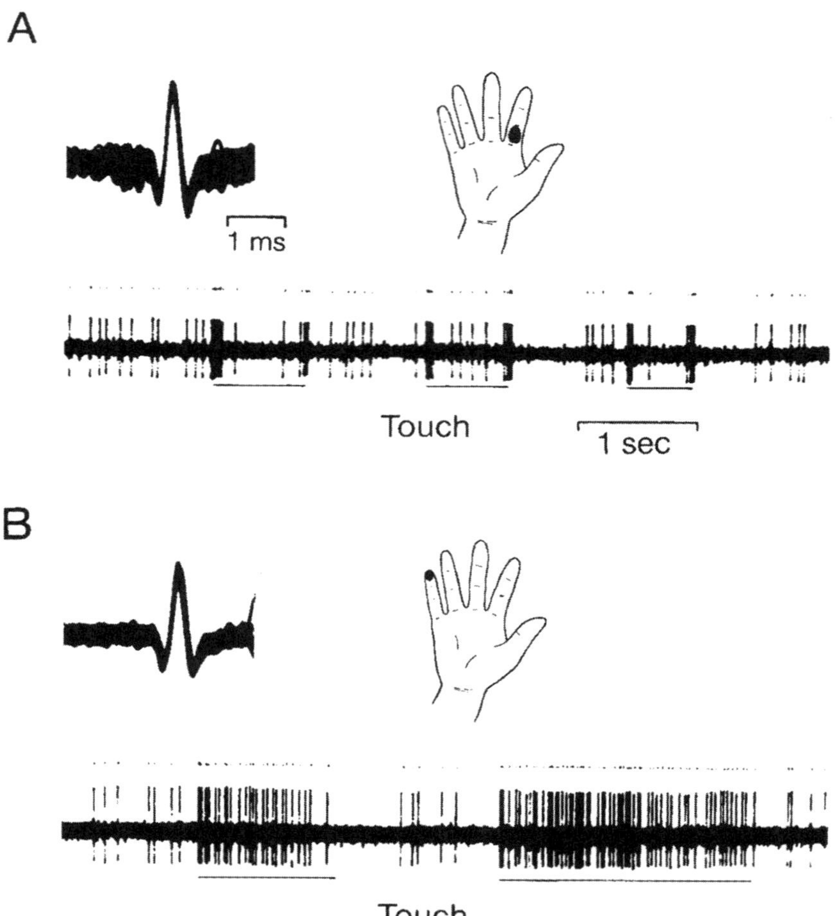

Figure 15.2
(A, B) Photographs of oscilloscope traces showing single unit activity elicited by cutaneous sensory stimulation in two neurons. The lower oscilloscope tracing in each section demonstrates the neuronal responses evoked during cutaneous sensory stimulation. The approximate duration of somatosensory stimulation is indicated by horizontal bars beneath the track. The dots above each spike train indicate the occurrence of discriminated action potentials having the shape displayed at high sweep speed in the upper trace. The receptive field for each cell is indicated in the illustrations located to the right of the upper traces. Horizontal calibrations for upper and lower traces are as marked in (A). Adapted from figure 1 in Lenz, Dostrovsky, Tasker, et al. (1988) with permission; neurons are recorded in ventral caudal nucleus of the thalamus.

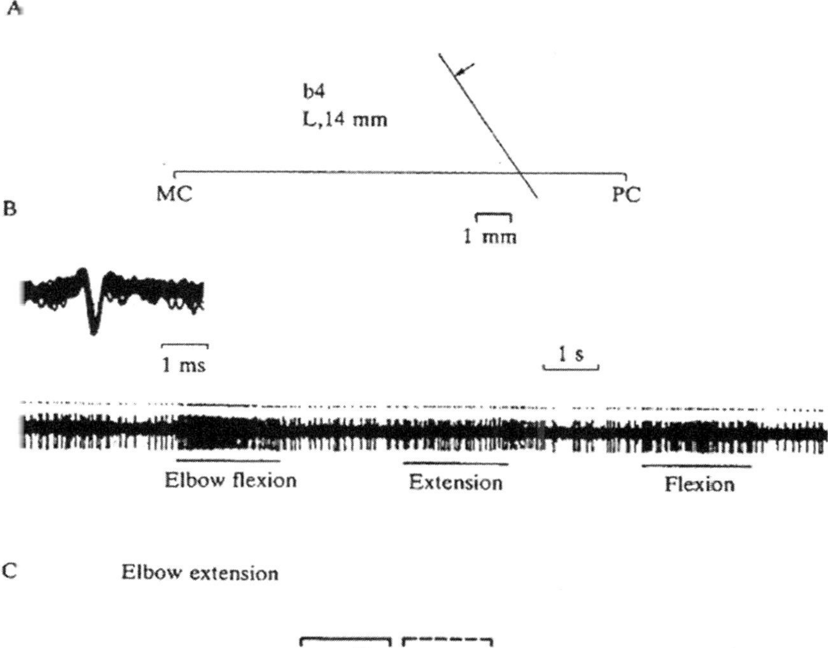

Figure 15.3
Combined cell exhibiting activity related both to active movement and somatosensory stimulation. (A) Illustrates the location of this cell (arrow) in relation to the trajectory along which it is recorded (oblique line) and the intercommissural line, indicated by the posterior commissure (PC) and the midcommissural point (MC). The patient number (b4), parasagittal plane 14 mm lateral to the midline, and the scale are indicated in this panel. (B) Illustrates the response of the cell to passive elbow movement. The timescale is as indicated. Lines below the spike train in the lower panel of (B) indicate the approximate interval of the elbow flexion and hold or extension and hold passive movements, as labeled. (C) Shows the raster and histogram for the activity of this cell during active elbow extension. Time "0" is locked to the onset of detected electromyograph activity. The continuous and dashed lines above the spike histogram demonstrate epochs with statistically higher (continuous) and lower (dashed) firing rates when compared to the mean (M) firing rate to the left. (Imp/s, impulses per second, s, second, ms, millisecond). Adapted from figure 3 in Lenz et al. (1990) with permission.

structures through the microelectrode may be employed in physiological localization. Typically, microstimulation (30–100 µA) is delivered in biphasic, square wave pulse trains of 0.1 to 0.3 ms pulses for times up to 10 s at a frequency of 300 Hz (Ranck, 1975). The current used determines the volume of tissue affected by the stimulation. Stimulation in Vc will evoke somatic sensations (Lenz, et al., 1993) while stimulation in Vim may alter the ongoing tremor or dystonia (Lenz et al., 1999).

Macrostimulation through a low-impedance macroelectrode ring built into many commercial microelectrodes can reliably identify the internal capsule by stimulation-evoked tetanic contraction of skeletal muscle at low thresholds and is also helpful for delineating the boundary between Vc and Vim (Ojemann & Ward, 1982; Tasker, Organ, & Hawrylyshyn, 1982). Stimulation of intralaminar nuclei medial to Vc or Vim may evoke the recruiting response—long latency, high voltage, negative waves occurring over much of the cortex at the frequency of stimulation (usually less than 10 Hz). Many other stimulation sites may similarly evoke the recruiting response. The target area in Vim can be identified by stimulation-evoked increase or decrease in the amplitude of tremor (Tasker, Organ, & Hawrylyshyn, 1982).

Future work in this area may focus on the possibility of controlling or altering abnormal rhythmic activity associated with tremor, perhaps using closed-loop control for the application of the stimulation pulses through the DBS implant. This requires the active detection of the underlying pathological rhythmic activity in real time (Santaniello, Fiengo, Glielmo, & Grill, 2011). Hauptmann et al. have described a clinical testing version of such a closed-loop system which can not only rely on local field potential (LFP) recordings from the implanted elements to detect abnormal rhythmicity but also can incorporate electroencephalography (EEG) and EMG input to fine-tune the control (Hauptmann et al., 2009). These closed-loop techniques will increasingly rely on novel LFP recording techniques that may additionally require recording during tonic electrode stimulation for real-time efficacy assessment (Holdefer, Cohen, & Greene, 2010; Yoshida et al., 2010). Additionally, device innovation may utilize detection of movement initiation or planning (Anderson et al., 2011).

Localization of GPi

Using atlas-based techniques, the initial coordinates of GPi can again be calculated from the midcommissural point (Laitinen, Bergenheim, & Hariz, 1992). Similar to the procedure for Vim electrode placement, the burr hole is placed approximately 25 mm from midline, through which microelectrode penetrations are made. Striatum, GPe, and GPi each have a characteristic pattern of neural activity (figure 15.4; Lozano et al., 1996; Vitek et al., 1998; . Striatal cells are characterized by broad action potentials and a slow firing rate (approximately 1 Hz), often with long silent periods. Cells in GPe have narrower action potentials and fire either at high frequency (approximately 50 Hz) with intermittent pauses (pause cells) or at lower frequency (approximately 20 Hz) with intermittent bursts (burst cells). Cells in GPi fire with tonically high rates (60–80 Hz). Cellular activity is also recorded during passive movements of upper and lower

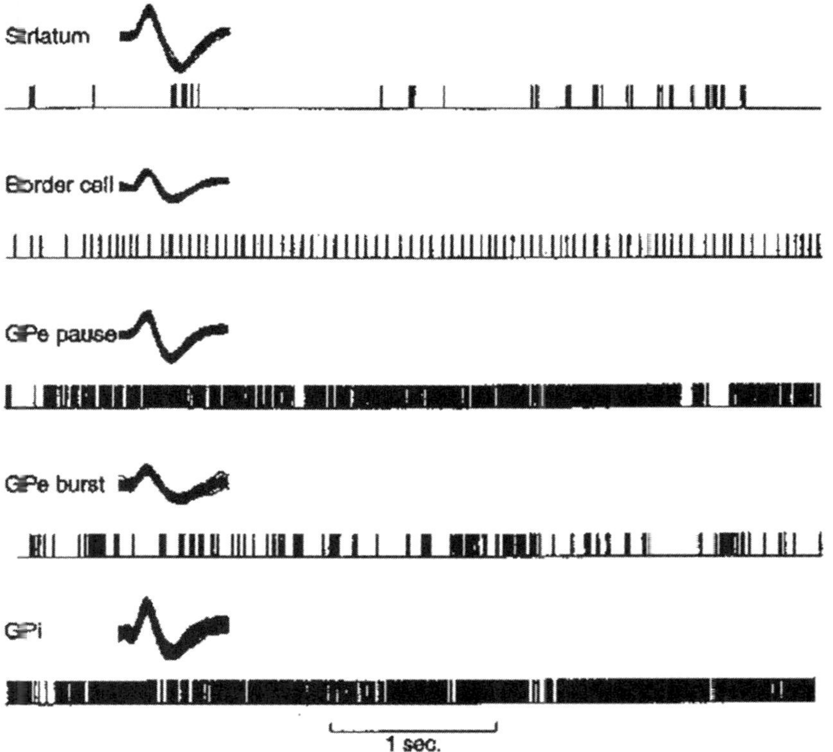

Figure 15.4
Examples of characteristic spike trains of neurons recorded in the basal ganglia of patients with Parkinson's disease. Each panel includes the label of the type of cell recorded (upper left). The shape of discriminated action potentials included in the spike train is shown to the right of the label. The spike train is shown in the line below the label. The time scale is as indicated below the lowest spike train. GPe, globus pallidus externa; GPi, globus pallidus interna. Adapted from figure 14-1 in Mandir et al. (1997) with permission.

extremities both ipsilateral and contralateral to the recording site to identify the cellular receptive field (Mandir et al., 1997).

Characteristically, the optic tract is about 1 mm below GPi. The optic tract is identified by listening for the background hissing from the firing of multiple myelinated axons. This sound increases in response to a flashing light (visual-evoked potentials), and microstimulation in the optic tract evokes multiple colored points of light. The same axonal sound and microstimulation-evoked muscle twitches can be evoked by stimulation of the internal capsule which is immediately behind this portion of GPi. These data are used to generate a functional map of the target zone in sagittal section.

Recent novel MER efforts in GPi include those of Chen et al., who have explored the synchronizing of neuronal activity with LFP oscillations at 3–12 Hz in patients with dystonia (Chen et al., 2006), implying that the disruption of this rhythm might be useful for therapy. The globus pallidus is involved in a variety of movement disorders including Parkinson's disease, dystonia, and degenerative disorders such as Huntington's disease. It is possible to compare pathological firing behavior among neurons in this nucleus between different disease states to glean the relative involvement of this structure, as has been performed by Tang et al. and other groups (Tang et al., 2005; Lee et al., 2007). Other researchers have sought pharmacological effects on neuronal firing rates in the basal ganglia, such as Steigerwald et al., who demonstrated the lack of an effect of propofol anesthesia on pallidal neuronal firing rates in the context of dystonia (Steigerwald et al., 2005).

Localization of STN

For subthalamic exploration, trajectories are made from a coronal burr hole about 30 mm from the midline or using the entry point specified by one of the commercially available planning systems. Striatum and Voa thalamus are successively encountered, and each have a characteristic pattern of neural activity (Hutchison, Allan, et al., 1998). Striatal cells are found most dorsally and are characterized by broad action potentials and a slow firing rate (approximately 1 Hz), often with long silent periods. When the thalamus is entered, the action potentials become narrower and often occur in bursts of the low threshold spike (LTS) type. LTS bursts are preceded by a silent period of 20–100 ms and consist of an interspike interval of less than 6 ms followed by interspike intervals of less than 16 ms (Zirh, Lenz, Reich, & Dougherty, 1997; Ramcharan, Gnadt, & Sherman, 2000). The zona incerta, a relatively acellular gap of 1 to 6 mm, is observed below the thalamus and above the subthalamic nucleus, depending upon the anterior–posterior location of the trajectory. The subthalamic nucleus is characterized by high-amplitude multiple spike trains recorded from the densely packed neurons, each with a mean rate of about 50/s. Microstimulation evokes paresthesias posterior to STN (lemniscal tract), muscle contractions anterior or lateral to STN (descending corticospinal tracts within the cerebral peduncle), ipsilateral internal eye deviation medial to STN (third cranial nerve fibers), and decreases in tremor or tone within STN. Ventral to the STN, substantia nigra reticulata type cells are encountered with characteristic monotonous spikes of smaller amplitude compared with the cells in the STN.

In addition to single unit recordings in the STN, recordings of the LFP as a signature of neural activity in aggregate are also being used to explore both physiological and pathophysiological signatures. For instance, Hebb et al. (Hebb, Darvas, & Miller, 2011) have recently reported modulations of beta-oscillation power (13–30 Hz) synchronous with finger movement and speech production. Reductions in abnormal levels of beta-oscillation power have been shown to occur in patients with Parkinson's disease undergoing DBS stimulation of the STN, both in the acute (Rosa et al., 2011; Bronte-Stewart et al., 2009) and chronic settings (>30 days postimplantation; Rosa et al., 2011). Similarly, transcranial magnetic stimulation of motor cortex is also able to suppress beta oscillations in patients with Parkinson's disease, as demonstrated by Gaynor et al. during concurrent LFP recordings from implanted DBS stimulators (Gaynor et al., 2008).

Tremor-Related Activity

Subsets of neurons in the thalamic nucleus Vim and in GPi and STN have been shown to fire at rates related to the pathological tremor frequency (Lenz, Tasker, et al., 1988; Lemstra, Verhagen, Lee, Dougherty, & Lenz, 1999; Hutchison, Benko, Dostrovsky, Lang, & Lozano, 1998; Rodriguez-Oroz et al., 2001). These studies indicate that thalamic cells have tremor frequency firing patterns that are linearly related to forearm EMG signals during essential tremor (tremor-related activity). Many of these cells respond to deep sensory stimulation. Three models could account for this correlation: that thalamic activity drives EMG activity, that thalamic activity is driven by sensory input generated by tremor movement, or that an oscillator outside the thalamus drives both thalamic and EMG activity. Surgical lesions in the thalamus abolish essential, parkinsonian, and cerebellar tremor (Schuurman et al., 2000; Lenz et al., 1995; Jankovic, Cardoso, Grossman, Hamilton, 1995; Goldman & Kelly, 1992).

The correlation between thalamic activity and EMG activity in tremor is found in parkinsonian, essential, and cerebellar tremor. It seems likely that thalamic activity in parkinsonian tremor is driven by sensory input which is itself generated by tremor movement in a feedback loop (Lenz et al., 1994). Thalamic oscillations in essential tremor may be facilitated by motor circuits through the thalamus that enable tremor-related thalamic activity during voluntary movement (Hua & Lenz, 2005). Cerebellar intention tremor may result from deafferentation of neurons in the cerebellar thalamus which leads to a phase lag relative to tremor that is not found in normal oscillations during active movement or in tremor-related activity in the ventral oral nucleus (Vo) (Lenz, Jaeger, Seike, Lin, & Reich, 2002). The involvement of Vim may explain the effectiveness of Vim thalamotomy and Vim DBS in these three types of tremor (Schuurman et al., 2000).

Neuronal firing rates are different in different regions of the thalamus. This may be examined in conditions as diverse as Parkinson's disease, essential tremor, and pain. Molnar et al. demonstrated differences in thalamic relay cell firing rates specifically between the pallidal (Voa and Vop) and cerebellar (Vim) receiving areas of the thalamus and found that the Voa and Vop firing rates were much lower than in essential tremor and pain (Molnar, Pilliar, Lozano, & Dostrovsky, 2005). The authors took this as further evidence demonstrating the dysfunction of the pallidal group in Parkinson's disease. Additionally, firing rates in portions of Vim were much higher in essential tremor compared to the other two groups, shedding light on the role of the cerebellum in this disease (Molnar, Pilliar, Lozano, & Dostrovsky, 2005). As in the case with the STN and GPi targets, recent efforts in electrophysiology recording in the case of the thalamus have also turned toward exploring aggregate measures (LFP) of neural activity (Holdefer et al., 2010).

Controversies and Discussion

Although widely used, the role of MER and stimulation as an aid in target localization for DBS implantation remains an area of debate. In a review that compares modern techniques with the older unguided lesioning procedures of the 1960s (Hariz, 2002), Hariz et al. argue that the technique of multiple microelectrode passes for localization results in more intracranial hemorrhages but no significant difference in outcome. This was further supported by a recent study of hemorrhages in DBS cases, which were positively correlated with the use of MER, and number of MER penetrations (Zrinzo, Foltynie, Limousin, & Hariz, 2012). Studies have shown good correspondence between the proposed electrophysiology-derived map with the MRI-defined target, indicating that the electrophysiology might not be essential (Vitek et al., 1998; Mohadjer, Goerke, Milios, Etou, & Mundinger, 1990; Hamani et al., 2005). Additionally, several series of implanted patients have been published utilizing anatomical targeting solely in an interventional MRI suite without the performance of MER (Starr et al., 2010; Nakajima et al., 2011).

In one report, MER led to a change in final location for lesion placement in 14 of 20 cases (70%; Hiner, Madden, & Neal, 1992) with an average change of 3.5 mm. In another, 75% of cases were within 3 ± 1 mm of the final lesion site, while 25% (4 of 16) were more than 5 mm away (Azizi, Goldman, & Moreledge, 1996). Still a third reported a rate of target change based upon MER which was considered too high to support pallidotomy without MER (Taha, Favre, & Burchiel, 1996). Although the accuracy of targeting without the microelectrode may be quite good in many cases, the high incidence of changes in final lesion location reported in these studies suggests that MER yields improved target localization when initial targeting is performed in an atlas-based manner. Although these reports suggest improved accuracy with microelectrode mapping, there remains no consensus regarding the effect on long-term clinical benefit. We conclude that precise mapping using the microelectrode is at present a helpful aid to define the boundaries of the target and optimize stimulator placement.

The role of MERs in the performance of emerging indications for DBS surgery is also being explored. For instance MER has been performed in the context of cingulotomy for obsessive–compulsive disorder to differentiate gray and white matter in the upper and lower banks of the cingulate gyrus (Richter et al., 2004). MER has also been used to distinguish cortical gray matter from white matter in the placement of DBS electrodes in the subgenual cingulate white matter for the treatment of depression (Mayberg et al., 2005). Similarly, stable single unit recordings have been obtained over the course of several days in patients undergoing invasive monitoring for epilepsy (Rutishauser, Ross, Mamelak, & Schuman, 2010), which may be useful for informing the placement of stimulation systems for the treatment of dementia or early cognitive impairment (Suthana et al., 2012). As new indications for stimulation of subcortical structures evolve, MER will likely continue to play a role at least in the development of the therapies.

References

Albe-Fessard, D. G. (1973). Electrophysiological methods for the identification of thalamic nuclei. *Zeitschrift fur Neurologie, 205*, 15–28.

Alexander, E., Kooy, H. M., van Herk, M., Schwartz, M., Barnes, P. D., Tarbell, N., et al. (1995). Magnetic resonance image-directed stereotactic neurosurgery: Use of image fusion with computerized tomography to enhance spatial accuracy. *Journal of Neurosurgery, 83*, 271–276.

Anderson, W. S., & Lenz, F. A. (2006). Surgery insight: Deep brain stimulation for movement disorders. *Nature Clinical Practice. Neurology, 2*, 310–320.

Anderson, W. S., Weiss, N., Lawson, H. C., Ohara, S., Rowland, L., & Lenz, F. A. (2011). Demonstration of imagined- and phantom-movement related neuronal signals in human thalamus. *Neuroreport. 22*, 78–82.

Azizi, A., Goldman, W. H., & Moreledge, D. (1996). Posteroventral pallidotomy: Comparison of the accuracy of anatomical targets defined by MR and CT imaging with physiological targets defined by microelectrode recording. *Neurology, 46*(Suppl), A199.

Bakay, R. A., DeLong, M. R., & Vitek, J. L. (1992). Posteroventral pallidotomy for Parkinson's disease. *Journal of Neurosurgery, 77*, 487–488.

Bronte-Stewart, H., Barberini, C., Koop, M. M., Hill, B. C., Henderson, J. M., & Wingeier, B. (2009). The STN beta-band profile in Parkinson's disease is stationary and shows prolonged attenuation after deep brain stimulation. *Experimental Neurology, 215*, 20–28.

Burchiel, K. J. (1995). Thalamotomy for movement disorders. In P. L. Gildenberg (Ed.), *Neurosurgery Clinics of North America* (pp. 55–71). Philadelphia: W.B. Saunders.

Carlson, J. D., & Iacono, R. P. (1999). Electrophysiological versus image-based targeting in the posteroventral pallidotomy. *Computer Aided Surgery, 4*, 93–100.

Chen, C. C., Kühn, A. A., Trottenberg, T., Kupsch, A., Schneider, G. H., & Brown, P. (2006). Neuronal activity in globus pallidus interna can be synchronized to local field potential activity over 3–12 Hz in patients with dystonia. *Experimental Neurology, 202*, 480–486.

Cooper, I. S., Bergmann, L. L., & Caracalos, A. (1963). Anatomic verification of the lesion which abolishes parkinsonian tremor and rigidity. *Neurology, 13*, 779–787.

Dostrovsky, J. O. (1998). The use of inexpensive personal computers for map generation and data analysis. In P. L. Gildenberg & R. R. Tasker (Eds.), *Textbook of stereotactic and functional neurosurgery* (pp. 2031–2036). New York: McGraw-Hill.

Garonzik, I. M., Hua, S. E., Ohara, S., & Lenz, F. A. (2002). Intraoperative microelectrode and semi-microelectrode recording during the physiological localization of the thalamic nucleus ventral intermediate. *Movement Disorders, 17*(Suppl 3), S135–S144.

Gaynor, L. M., Kühn, A. A., Dileone, M., Litvak, V., Eusebio, A., Pogosyan, A., et al. (2008). Suppression of beta oscillations in the subthalamic nucleus following cortical stimulation in humans. *European Journal of Neuroscience, 28*, 1686–1695.

Gerdes, J. S., Hitchon, P. W., Neerangun, W., & Torner, J. C. (1994). Computed tomography versus magnetic resonance imaging in stereotactic localization. *Stereotactic and Functional Neurosurgery, 63*, 124–129.

Goldman, M. S., & Kelly, P. J. (1992). Symptomatic and functional outcome of stereotactic ventralis lateralis thalamotomy for intention tremor. *Journal of Neurosurgery, 77*, 223–229.

Hamani, C., Richter, E. O., Andrade-Souza, Y., Hutchison, W., Saint-Cyr, J. A., & Lozano, A. M. (2005). Correspondence of microelectrode mapping with magnetic resonance imaging for subthalamic nucleus procedures. *Surgical Neurology, 63*, 249–253.

Hardy, P. A., & Barnett, G. H. (1998). Spatial distortion in magnetic resonance imaging: Impact on stereotactic localization. In P. L. Gildenberg & R. R. Tasker (Eds.), *Textbook of stereotactic and functional neurosurgery* (pp. 271–280). New York: McGraw-Hill.

Hariz, M. I. (2002). Safety and risk of microelectrode recording in surgery for movement disorders. *Stereotactic and Functional Neurosurgery, 78*, 146–157.

Hauptmann, C., Roulet, J. C., Niederhauser, J. J., Döll, W., Kirlangic, M. E., Lysyansky, B., et al. (2009). External trial deep brain stimulation device for the application of desynchronizing stimulation techniques. *Journal of Neural Engineering, 6*, 066003.

Hebb, A. O., Darvas, F., & Miller, K. J. (2011). Transient and state modulation of beta power in human subthalamic nucleus during speech production and finger movement. *Neuroscience.* doi:10.1016/j.neuroscience.2011.11.072.

Hiner, B., Madden, K., & Neal, J. (1992). Effect of microelectrode recording on final lesion placement in pallidotomy. *Neurology, 48*(Suppl), A251–A252.

Holdefer, R. N., Cohen, B. A., & Greene, K. A. (2010). Intraoperative local field recording for deep brain stimulation in Parkinson's disease and essential tremor. *Movement Disorders, 25*, 2067–2075.

Holtzheimer, P. E. I., Roberts, D. W., & Darcey, T. M. (1999). Magnetic resonance imaging versus computed tomography for target localization in functional stereotactic neurosurgery. *Neurosurgery, 45*, 290–298.

Hua, S. E., Garonzik, I. M., Lee, J.-I., & Lenz, F. A. (2004). Thalamotomy for tremor. In R. H. Winn (Ed.), *Youmans neurological surgery* (5th ed., pp. 2769–2784). Philadelphia: W.B. Saunders.

Hua, S. E., & Lenz, F. A. (2005). Posture-related oscillations in human cerebellar thalamus in essential tremor are enabled by voluntary motor circuits. *Journal of Neurophysiology, 93*, 117–127.

Hua, S. E., Lenz, F. A., Zirh, T. A., Reich, S. G., & Dougherty, P. M. (1998). Thalamic neuronal activity correlated with essential tremor. *Journal of Neurology, Neurosurgery, and Psychiatry, 64*, 273–276.

Hubel, D. H. (1957). Tungsten microelectrode for recording from single units. *Science, 125*, 549–550.

Hutchison, W. D., Allan, R. J., Opitz, H., Levy, R., Dostovsky, J. O., Lang, A. E., et al. (1998). Neurophysiological identification of the subthalamic nucleus in surgery for Parkinson's disease. *Annals of Neurology, 44*, 622–628.

Hutchison, W. D., Benko, R., Dostrovsky, J. O., Lang, A. E., & Lozano, A. M. (1998). Coherent relation of rest tremor to pallidal tremor cells in Parkinson's disease patients. *Movement Disorders, 13*(Suppl 2), 204.

Jankovic, J., Cardoso, F., Grossman, R. G., & Hamilton, W. J. (1995). Outcome after stereotactic thalamotomy for parkinsonian, essential and other types of tremor. *Neurosurgery, 37*, 680–687.

Jasper, H. H., & Bertrand, G. (1966). Thalamic units involved in somatic sensation and voluntary and involuntary movements in man. In D. P. Purpura & M. D. Yahr (Eds.), *The thalamus* (pp. 365–390). New York: Columbia University Press.

Kondziolka, D., Dempsey, P. K., Lunsford, L. D., Kestle, J. R. W., Dolan, E. J., Kanal, E., et al. (1992). A comparison between magnetic resonance imaging and computed tomography for stereotactic coordinate determination. *Neurosurgery, 30*, 402–407.

Krack, P., Fraix, V., Mendes, A., Benabid, A.-L., & Pollak, P. (2002). Postoperative management of subthalamic nucleus stimulation for Parkinson's disease. *Movement Disorders, 17*, S188–S197.

Kupsch, A., Tagliati, M., Vidailhet, M., Aziz, T., Krack, P., Moro, E., et al. (2011). Early postoperative management of DBS in dystonia: Programming, response to stimulation, adverse events, medication changes, evaluations, and troubleshooting. *Movement Disorders, 26*(Suppl 1), S37–S53.

Laitinen, L. V., Bergenheim, A. T., & Hariz, M. I. (1992). Leksell's posteroventral pallidotomy in the treatment of Parkinson's disease. *Journal of Neurosurgery, 76*, 53–61.

Lee, J. I., Verhagen-Metman, L., Ohara, S., Dougherty, P. M., Kim, J. H., & Lenz, F. A. (2007). Internal pallidal neuronal activity during mild drug-related dyskinesias in Parkinson's disease: Decreased firing rates and altered firing patterns. *Journal of Neurophysiology, 97*, 2627–2641.

Lemstra, A. W., Verhagen, M. L., Lee, J. I., Dougherty, P. M., & Lenz, F. A. (1999). Tremor-frequency (3–6 Hz) activity in the sensorimotor arm representation of the internal segment of the globus pallidus in patients with Parkinson's disease. *Neuroscience Letters, 267*(2), 129–132.

Lenz, F. A., Dostrovsky, J. O., Kwan, H. C., Tasker, R. R., Yamashiro, K., & Murphy, J. T. (1988). Methods for microstimulation and recording of single neurons and evoked potentials in the human central nervous system. *Journal of Neurosurgery, 68*, 630–634.

Lenz, F. A., Dostrovsky, J. O., Tasker, R. R., Yamashiro, K., Kwan, H. C., & Murphy, J. T. (1988). Single-unit analysis of the human ventral thalamic nuclear group: Somatosensory responses. *Journal of Neurophysiology, 59*, 299–316.

Lenz, F. A., Jaeger, C. J., Seike, M. S., Lin, Y. C., & Reich, S. G. (2002). Single-neuron analysis of human thalamus in patients with intention tremor and other clinical signs of cerebellar disease. *Journal of Neurophysiology, 87*, 2084–2094.

Lenz, F. A., Jaeger, C. J., Seike, M. S., Lin, Y. C., Reich, S. G., DeLong, M. R., et al. (1999). Thalamic single neuron activity in patients with dystonia: Dystonia- related activity and somatic sensory reorganization. *Journal of Neurophysiology, 82*, 2372–2392.

Lenz, F. A., Kwan, H. C., Dostrovsky, J. O., Tasker, R. R., Murphy, J. T., & Lenz, Y. E. (1990). Single unit analysis of the human ventral thalamic nuclear group: Activity correlated with movement. *Brain, 113*, 1795–1821.

Lenz, F. A., Kwan, H. C., Martin, R. L., Tasker, R. R., Dostrovsky, J. O., & Lenz, Y. E. (1994). Single neuron analysis of the human ventral thalamic nuclear group: Tremor-related activity in functionally identified cells. *Brain, 117*, 531–543.

Lenz, F. A., Normand, S. L., Kwan, H. C., Andrews, D., Rowland, L. H., Jones, M. W., et al. (1995). Statistical prediction of the optimal site for thalamotomy in parkinsonian tremor. *Movement Disorders, 10*, 318–328.

Lenz, F. A., Seike, M., Richardson, R. T., Lin, Y. C., Baker, F. H., Khoja, I., et al. (1993). Thermal and pain sensations evoked by microstimulation in the area of human ventrocaudal nucleus. *Journal of Neurophysiology*, *70*, 200–212.

Lenz, F. A., Tasker, R. R., Kwan, H. C., Schnider, S., Kwong, R., Murayama, Y., et al. (1988). Single unit analysis of the human ventral thalamic nuclear group: Correlation of thalamic "tremor cells" with the 3–6 Hz component of parkinsonian tremor. *Journal of Neuroscience*, *8*, 754–764.

Lozano, A., Hutchison, W., Kiss, Z., Tasker, R., Davis, K., & Dostrovsky, J. (1996). Methods for microelectrode-guided posteroventral pallidotomy. *Journal of Neurosurgery*, *84*, 194–202.

Mandir, A. S., Rowland, L. H., Dougherty, P. M., & Lenz, F. A. (1997). Microelectrode recording and stimulation techniques during stereotactic procedures in the thalamus and pallidum. *Advances in Neurology*, *74*, 159–165.

Mayberg, H. S., Lozano, A. M., Voon, V., McNeely, H. E., Seminowicz, D., Hamani, C., et al. (2005). Deep brain stimulation for treatment-resistant depression. *Neuron*, *45*, 651–660.

Mohadjer, M., Goerke, H., Milios, E., Etou, A., & Mundinger, F. (1990). Long term results of stereotaxy in the treatment of essential tremor. *Stereotactic and Functional Neurosurgery*, *54*, 125–129.

Molnar, G. F., Pilliar, A., Lozano, A. M., & Dostrovsky, J. O. (2005). Differences in neuronal firing rates in pallidal and cerebellar receiving areas of thalamus in patients with Parkinson's disease, essential tremor, and pain. *Journal of Neurophysiology*, *93*, 3094–3101.

Nakajima, T., Zrinzo, L., Foltynie, T., Olmos, I. A., Taylor, C., Hariz, M. I., et al. (2011). MRI-guided subthalamic nucleus deep brain stimulation without microelectrode recording: Can we dispense with surgery under local anaesthesia? *Stereotactic and Functional Neurosurgery*, *89*, 318–325.

O Gorman, R. L., Shmueli, K., Ashkan, K., Samuel, M., Lythgoe, D. J., Shahidani, A., et al. (2011). Optimal MRI methods for direct stereotactic targeting of the subthalamic nucleus and globus pallidus. *European Radiology*, *21*, 130–136.

Onye, C. (1982). Depth microelectrode studies. In G. Schaltenbrand & A. E. Walker (Eds.), *Stereotaxy of the human brain: Anatomical, physiological and clinical applications* (pp. 372–389). Stuttgart: Thieme.

Ojemann, G. A., & Ward, A. A. (1982). Abnormal movement disorders. In J. R. Youmans (Ed.), *Neurological surgery* (2nd ed., pp. 3821–3857). Philadelphia: W.B. Saunders.

Okun, M. S., Rodriguez, R. L., Foote, K. D., Sudhyadhom, A., Bova, F., Jacobson, C., et al. (2008). A case-based review of troubleshooting deep brain stimulator issues in movement and neuropsychiatric disorders. *Parkinsonism & Related Disorders*, *14*, 532–538.

Patel, N. K., Heywood, P., O'Sullivan, K., McCarter, R., Love, S., & Gill, S. S. (2003). Unilateral subthalamotomy in the treatment of Parkinson's disease. *Brain*, *126*, 1136–1145.

Raeva, S. (1986). Localization in human thalamus of units triggered during "verbal commands," voluntary movements and tremor. *Electroencephalography and Clinical Neurophysiology*, *63*, 160–173.

Ramcharan, E. J., Gnadt, J. W., & Sherman, S. M. (2000). Burst and tonic firing in thalamic cells of unanesthetized, behaving monkeys. *Visual Neuroscience*, *17*, 55–62.

Ranck, J. B. (1975). Which elements are excited in electrical stimulation of mammalian central nervous system: A review. *Brain Research*, *98*, 417–440.

Richter, E. O., Davis, K. D., Hamani, C., Hutchison, W. D., Dostrovsky, J. O., & Lozano, A. M. (2004). Cingulotomy for psychiatric disease: Microelectrode guidance, a callosal reference system for documenting lesion location, and clinical results. *Neurosurgery*, *54*, 622–628.

Rodriguez-Oroz, M. C., Rodriguez, M., Guridi, J., Mewes, K., Chockkman, V., Vitek, J., et al. (2001). The subthalamic nucleus in Parkinson's disease: Somatotopic organization and physiological characteristics. *Brain*, *124*, 1777–1790.

Rosa, M., Giannicola, G., Servello, D., Marceglia, S., Pacchetti, C., Porta, M., et al. (2011). Subthalamic local field beta oscillations during ongoing deep brain stimulation in Parkinson's disease in hyperacute and chronic phases. *NeuroSignals*, *19*, 151–162.

Rutishauser, U., Ross, I. B., Mamelak, A. N., & Schuman, E. M. (2010). Human memory strength is predicted by theta-frequency phase-locking of single neurons. *Nature Neuroscience*, *464*, 903–909.

Santaniello, S., Fiengo, G., Glielmo, L., & Grill, W. M. (2011). Closed-loop control of deep brain stimulation: A simulation study. *IEEE Transactions on Neural Systems and Rehabilitation Engineering*, *19*, 15–24.

Schaltenbrand, G., & Bailey, P. (1959). *Introduction to stereotaxis with an atlas of the human brain*. Stuttgart: Thieme.

Schaltenbrand, G., & Wahren, W. (1977). *Atlas for Stereotaxy of the Human Brain.* Stuttgart: Thieme.

Schaltenbrand, G., & Walker, A. E. (1982). *Stereotaxy of the human brain: Anatomical, physiological and clinical applications.* New York: Thieme-Stratton.

Schuurman, P. R., Bosch, D. A., Bossuyt, P. M., Bonsel, G. J., van Someren, E. J., de Bie, R. M., et al. (2000). A comparison of continuous thalamic stimulation and thalamotomy for suppression of severe tremor. *New England Journal of Medicine, 342,* 461–468.

Starr, P. A., Christine, C. W., Theodosopoulos, P. V., Lindsey, N., Byrd, D., Mosley, A., et al. (2002). Implantation of deep brain stimulators into the subthalamic nucleus: Technical approach and magnetic resonance imaging–verified lead locations. *Journal of Neurosurgery, 97,* 370–387.

Starr, P. A., Martin, A. J., Ostrem, J. L., Talke, P., Levesque, N., & Larson, P. S. (2010). Subthalamic nucleus deep brain stimulator placement using high-field interventional magnetic resonance imaging and a skull-mounted aiming device: Technique and application accuracy. *Journal of Neurosurgery, 112,* 479–490.

Steigerwald, F., Hinz, L., Pinsker, M. O., Herzog, J., Stiller, R. U., Kopper, F., et al. (2005). Effect of propofol anesthesia on pallidal neuronal discharges in generalized dystonia. *Neuroscience Letters, 386*(3), 156–159.

Suthana, N., Haneef, Z., Stern, J., Mukamel, R., Behnke, E., Knowlton, B., et al. (2012). Memory enhancement and deep-brain stimulation of the entorhinal area. *New England Journal of Medicine, 366,* 502–510.

Taha, J. M., Favre, J., & Burchiel, K. J. (1996). The value of macrostimulation in patients who underwent microrecording during pallidotomy. *Congress of Neurological Surgeons Abstract,* 46th Annual Meeting, 264–265.

Talairach, J., & Tournoux, P. (1988). *Co-planar stereotaxic atlas for the human brain: 3-D proportional system: An approach to cerebral imaging.* New York: Thieme.

Tang, J. K., Moro, E., Lozano, A. M., Lang, A. E., Hutchison, W. D., Mahant, N., et al. (2005). Firing rates of pallidal neurons are similar in Huntington's and Parkinson's disease patients. *Experimental Brain Research, 166,* 230–236.

Tasker, R. R., Organ, L. W., & Hawrylyshyn, P. (1982). *The thalamus and midbrain in man: A physiologic atlas using electrical stimulation.* Springfield, IL: Charles C Thomas.

Tierney, T. S., Sankar, T., & Lozano, A. M. (2011). Deep brain stimulation emerging indications. *Progress in Brain Research, 194,* 83–95.

Umbach, W., & Ehrhardt, K. J. (1965). Micro-electrode recording in the basal ganglia during stereotaxic operations. *Confinia Neurologica, 26,* 315–317.

Vayssiere, N., Hemm, S., Cif, L., Picot, M. C., Diakonova, N., El Fertit, H., et al. (2002). Comparison of atlas- and magnetic resonance imaging-based stereotactic targeting of the globus pallidus internus in the performance of deep brain stimulation for treatment of dystonia. *Journal of Neurosurgery, 96,* 673–679.

Vitek, J. L., Bakay, R. A., Hashimoto, T., Kaneoke, Y., Mewes, K., Zhang, J. Y., et al. (1998). Microelectrode-guided pallidotomy: Technical approach and its application in medically intractable Parkinson's disease. *Journal of Neurosurgery, 88,* 1027–1043.

Volkmann, J., Herzog, J., Kopper, F., & Deuschl, G. (2002). Introduction to the programming of deep brain stimulators. *Movement Disorders, 17*(S3), S181–S187.

Wolbarsht, M. L., MacNichol, E. F., Jr., & Wagner, H. G. (1960). Glass insulated platinum microelectrode. *Science, 132,* 1309–1310.

Yoshida, F., Martinez-Torres, I., Pogosyan, A., Holl, E., Petersen, E., Chen, C. C., et al. (2010). Value of subthalamic nucleus local field potentials recordings in predicting stimulation parameters for deep brain stimulation in Parkinson's disease. *Journal of Neurology, Neurosurgery, and Psychology, 81,* 885–889.

Zirh, A. T., Lenz, F. A., Reich, S. G., & Dougherty, P. M. (1997). Patterns of bursting occurring in thalamic cells during parkinsonian tremor. *Neuroscience, 83,* 107–121.

Zrinzo, L., Foltynie, T., Limousin, P., & Hariz, M. I. (2012). Reducing hemorrhagic complications in functional neurosurgery: A large case series and systematic literature review. *Journal of Neurosurgery, 116,* 84–94.

16 Microstimulation Effects on Thalamic Neurons

Sanjay Patra, William D. Hutchison, Clement Hamani, Mojgan Hodaie, Andres M. Lozano, and Jonathan O. Dostrovsky

Functional stereotactic surgery is a technique frequently used in procedures that target structures deep within the brain and allows verification and optimization of the target. In most cases the procedure, which is frequently performed under local anesthesia, depends on obtaining physiological information which allows identification of the target region. Usually this information comprises the neuronal activity (firing rate and pattern, and possible responses to sensory inputs [receptive field] and/or responses to passive and/or active movements) and the perceptual and/or motor effects of electrical stimulation in the target region and surrounding areas. In the past 20 years it has been used primarily for localization of the target site for deep brain stimulation (DBS) electrodes or, less frequently, a lesion. This procedure provides the unique opportunity to examine the neurophysiological properties of the target and surrounding regions in awake humans and characterize any pathophysiological features related to the specific neurological disorder of the patient.

The thalamus is a target for treating tremor and some other movement disorders, as well as chronic pain, and recently also epilepsy (Benabid et al., 1987; Kumar et al., 1997; Hodaie et al., 2002; Perlmutter & Mink, 2006). Electrical stimulation in thalamus as a treatment has largely replaced lesions which were used previously for tremor reduction and occasionally for treating pain. High-frequency stimulation (HFS) in sensory parts of thalamus elicits somatosensory (usually parasthesia), visual, or auditory percepts depending on the sensory pathway stimulated (Tasker et al., 1976; Yamashiro & Tasker, 1990). When stimulating in motor thalamus of tremor patients, in particular in the ventralis intermedius nucleus (Vim), tremor reduction or arrest is frequently induced, and this is used as a good indication that a lesion or HFS DBS at that site would produce a good result. Microelectrode recordings at such sites frequently encounter neurons, termed tremor cells, with rhythmic firing patterns in synchrony with the tremor. The association of reduction of tremor by either lesioning or HFS at such sites gave rise to the hypothesis that the mechanism underlying the effects of DBS was inhibition of neuronal firing (i.e., of the tremor cells which were part of a network driving the tremor). This hypothesis then led to trials of HFS for treating other conditions by stimulation in other regions, in particular where there was evidence that lesions of those regions could alleviate the neurological symptoms. The best examples were the globus pallidus interna (GPi) and subthalamic nucleus (STN) where it had previously been shown that lesions in humans in these regions could alleviate the symptoms of Parkinson's disease (PD; Krayenbuhl et al., 1961). HFS in thalamus, GPi, and STN has largely replaced lesions in all three regions; however, the mechanisms underlying the therapeutic benefits are still not well understood and controversial.

For the past 20 years our group has been using microstimulation associated with recordings from the neurons in the human brain during functional stereotactic surgery not only to assist in localization but also to gain a better understanding of the consequences of HFS in these regions for the neuronal activity. The methods utilized in stereotactic surgery to obtain microelectrode recordings and perform microstimulation have been described in our previous publications (Lenz et al., 1988; Hutchison et al., 1998). Furthermore, some of the results of stimulation effects in the thalamus have been previously described in abstract form by our group (Dostrovsky et al., 2002).

Methods

Patient Population and Location of the Neuronal Recordings

We examined 335 thalamic neurons in 30 patients undergoing implantation of DBS electrodes in the thalamus (n = 18) and STN (n = 12). All patients treated with STN DBS had PD and were included because microelectrode tracks often traversed the thalamus on their way to the target (Hutchison et al., 1998). Among the patients with thalamic procedures, 1 patient had PD, 4 essential tremor (ET), 1 cerebellar tremor (CB), 4 epilepsy, and 8 chronic pain (table 16.1). Sagittal maps reconstructed based on a standard atlas (Schaltenbrand & Wahren, 1977) were used to determine the locations of the neurons recorded. To adapt the maps to the morphology of the patients' brains, the images were reformatted according to the coordinates of the anterior and posterior commissures as well as key physiological landmarks. The distribution of neurons according to specific thalamic nuclei is summarized in table 16.1. In PD, ET, and CB, neurons were recorded from motor thalamic nuclei, such as the nucleus Vim, nucleus ventralis oralis anterior, and nucleus ventralis oralis posterior and possibly the reticular nucleus of the thalamus. In patients

Table 16.1
Distribution of cells according to clinical condition and thalamic nucleus

Diagnosis (target)	Number of patients	Number of cells
PD (Vim, Vop, Voa)	13	92
Pain (Vc)	8	145
ET (Vim, Vop, Voa)	4	61
Epilepsy (AN)	4	35
CB (Vim, Vop, Voa)	1	2
Total	30	335

PD, Parkinson's disease; Vim, ventralis intermedius nucleus; Vop, ventralis oralis posterior; Voa, nucleus ventralis oralis anterior; Vc, nucleus ventralis caudalis; ET, essential tremor; AN, anterior nucleus of the thalamus; CB, cerebellar tremor.

with pain, most neurons were recorded from the region of the nucleus ventralis caudalis. In the patients with epilepsy, neurons were recorded from the anterior nucleus of the thalamus.

All the procedures were approved by the University Health Network Research Ethics Board, and the patients gave free and informed consent.

Microelectrode Stimulation and Recordings

The methods for single and dual channel microelectrode mapping used in our institution during stereotactic procedures have already been published in detail (see Hutchison et al., 1998; Lozano et al., 1995; Levy et al., 2007). HFS was generally delivered through microelectrodes (tip diameter = 25 μm) and consisted of monopolar short trains of biphasic 0.15-ms pulses delivered at a frequency of 100–333 Hz at currents that ranged from 5 to 40 μA. Macroelectrodes (tip diameter = 1 mm) were used for stimulation currents between 100 μA and 800 μA. The effects of high-frequency microstimulation on neuronal activity were examined in three conditions: (1) stimulation and recording carried out from the same electrode in the vicinity of the cell; (2) stimulation and recording through electrodes that were 250 μm apart (fixed configuration); these electrodes were glued together and driven as one unit; and (3) stimulation through an independently controlled electrode 650 μm–1 mm away from the recording electrode (independent configuration).

Cell Populations and Offline Analysis

Recordings done in the operating room were initially stored on a magnetic tape and digitized for the subsequent offline analysis (CED1401 Spike 2 data acquisition system). Single units were discriminated using template-matching software (Spike2 software, Cambridge Electronic Design, Cambridge, UK).

Inhibitory or silent periods were defined by an absence of action potentials during the recordings. A cell was considered inhibited when the poststimulation silent periods were clearly longer than the baseline periods of silence. In cells that fired slowly and irregularly, poststimulus inhibitory periods were difficult to separate from random periods of silence and were not included in our analysis. In cells that were included, short spontaneous silent periods were very small compared to the periods of inhibition observed.

In this study neurons firing spontaneously in an irregular bursting firing pattern were characterized as having a bursting discharge consistent with generation from a low-threshold calcium spike (LTS) on the basis of their bursting pattern. For this purpose the activity of the neuron was characterized by graphing intraburst interspike intervals, or visually, by identifying the characteristic LTS-induced burst firing pattern (McCarley et al., 1983; Domich et al., 1986; Tsoukatos et al., 1997; Hodaie et al., 2006).

Statistical analysis comparing the percentage of particular cell responses among the different groups was performed using a Chi-square test. Confidence intervals (CIs) were used to compare mean results from distinct groups. Statistical significance was set at $p \leq 0.05$ or when CI $\geq 95\%$.

Results

Effects of High-Frequency Stimulation

The effects of HFS were examined on 335 cells. In 124 cells (37%) HFS induced long-lasting inhibition, whereas in the remaining 211 cells (63%) no observable changes in firing patterns were found. Figure 16.1 shows two examples: The top trace shows the response of a bursting neuron following the end of a 1-s, 100-Hz, 5-μA (150-μs pulse width) stimulus train to the recording electrode, and lower trace, the response following a 500-ms, 100-Hz, 10-μA (150-μs pulse width) stimulus to a second microelectrode 250 μm away (note that each spike is actually a short burst of action potentials as shown on expanded scale at right of trace). In both these cases there was a rebound bursting followed by an inhibitory period.

The incidence of inhibitory responses was highest in the fixed configuration (78%), and lowest in the independent configuration (17%) (p < 0.0001). The mean threshold current for producing an effect was significantly lower when stimulation was applied through the recording electrode

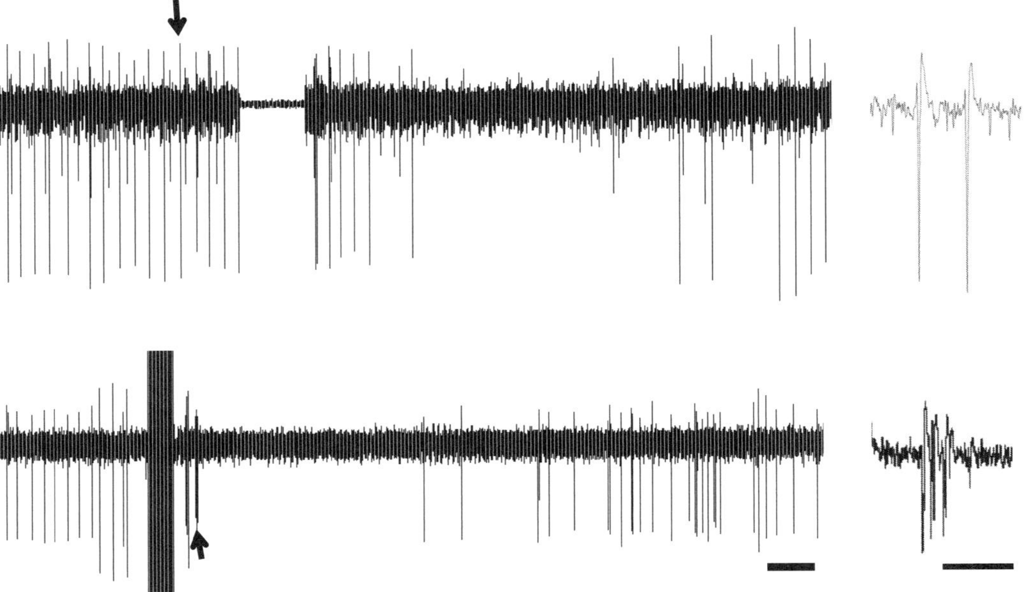

Figure 16.1
Examples of poststimulation bursting followed by a prolonged inhibitory period. The top trace shows the response of a bursting neuron following the end of a 1-s, 100-Hz, 5-μA (150-μs pulse width) stimulus train to the recording electrode, and the lower trace, the response following a 500-ms, 100-Hz, 10-μA (150-μs pulse width) stimulus to a second microelectrode 250 μm away. In both cases there was a rebound bursting followed by an inhibitory period. Note that each spike is a burst of action potentials (usually two in these examples) as can be seen in the enlarged segment to the right of each trace. The arrow shows the burst that is enlarged on the right.

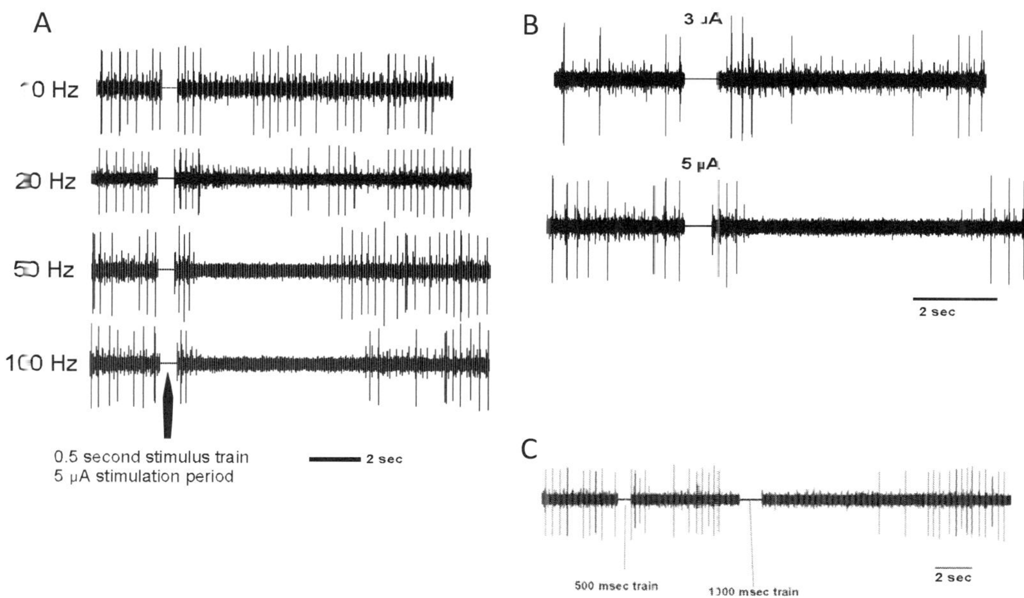

Figure 16.2
Examples of the effects of train frequency, stimulus intensity, and train duration on the inhibitory period. In all cases the stimulus was delivered through the recording electrode. In (A) the train intensity was 5 µA and duration 500ms, in (B) the train was at 50 Hz and 500 ms in duration, and in (C) it was at 200 Hz and at 3 µA intensity.

or when the electrodes were in the fixed configuration, compared to the independent configuration (CI = 95%). The mean duration of the inhibitory period was 3.9 ± 0.3 s (ranging from 0.4 to 20 s). The mean length of inhibition was greatest when stimulation was delivered through the fixed configuration or directly from the recording electrode, compared to the independent configuration. In a few cases effects of stimulation frequency, train duration, and stimulus intensity were examined and revealed that increasing the frequency (in the range from 10 to 100 Hz), duration, or intensity of the stimulus train resulted in increased duration of inhibition (see figure 16.2). The incidence of inhibition was higher in cells that had bursting activity prior to stimulation compared to nonbursting cells (63% vs. 23%; p < 0.0001). The mean duration of inhibition did not vary according to prestimulation firing pattern.

Poststimulation bursts of action potentials preceding the inhibitory period were recorded in 69% of the 124 cells examined. The pattern of firing within these bursts was similar to the one observed in thalamic bursting cells with LTS properties. The latency between the stimulation artifact and the first burst of action potentials was 179 ± 34 ms (n = 19 cells). The number of spikes per burst usually ranged from two to seven, and the interburst intervals were between 100 and 200 ms. Also of interest was the finding that the number of spikes per burst decreased with succeeding bursts. The mean duration of the posttrain bursting period was 346 ms ± 54 ms and ranged from 152 ms to 894 ms (n = 13). The mean duration of inhibition was longer in cells that had posttrain bursts (CI = 95%). In bursting cells, the number of spikes per burst in the first burst following the termination of the train was almost always greater than the average number of spikes per burst preceding the stimulus. Also, there was a higher incidence of preinhibitory period bursting in bursting cells.

The effects of HFS applied though macroelectrodes were tested on 7 cells. Four of these had inhibitory responses preceded by bursting activity.

Effects of Single Pulse Stimulation

The effects of single stimuli (generally applied at frequencies of 10 Hz or lower) were examined in 76 out of the 335 neurons tested with HFS. Excitatory responses were recorded in 16 of the 76 neurons (21%). The latency of the excitatory responses was generally <10 ms. Two of the seven cells tested with high-frequency macrostimulation were also tested with low-frequency stimulation (LFS), and both cells showed excitation similar to that seen following low-frequency microstimulation. An example can be seen in figure 16.3, bottom left. LFS applied though macroelectrodes was tested in 2 cells and evoked excitatory responses in both.

Discussion

This study has shown that following a high-frequency stimulus train in thalamus, there is a long period of inhibition in many neurons. Typically, the inhibitory periods were longer than 2 s and in many cases above 5 s. These effects were predominantly seen following high-frequency (>100 Hz) stimulation in close proximity (~250 μm) to the cell. Furthermore, the long-duration inhibition was seen primarily in bursting cells (81% in bursting cells compared to 24% in nonbursting cells). The length of long-duration inhibition was found to be dependent on current intensity, train length, and frequency of stimulation. In many cells the inhibitory period was preceded by one or a few LTS bursts. At low stimulation rates we observed excitatory responses in some thalamic neurons directly following the stimulus. These excitatory responses generally increased with each successive stimulus for the first 5 to 10 stimuli in the train at stimulation rates of around 10 Hz, but this phenomenon was not studied in detail.

Similar methods have been applied by our group in GPi, substantia nigra pars reticulata (SNr), and STN (see figure 16.3; Dostrovsky et al., 2000; Filali et al., 2004; Lafreniere-Roula et al., 2010; Liu et al., 2012). Although the effects in GPi and SNr were similar, they were quite different from those observed in STN, and both regions (GPi/SNr and STN) were different from thalamus as can be seen in figure 16.3. High-frequency microstimulation in GPi and SNr typically caused short-duration inhibition (i.e., a few hundred milliseconds) following termination of the stimulus train, and even very low intensities were generally effective (1–3 μA for stimulation through the recording electrode). Furthermore, stimulation even at low intensities with single pulses and low stimulation rates resulted in short-duration inhibition following each stimulus (~10–30 ms) in almost all cells. In contrast, in the STN, HFS (50–200 Hz) resulted in inhibition

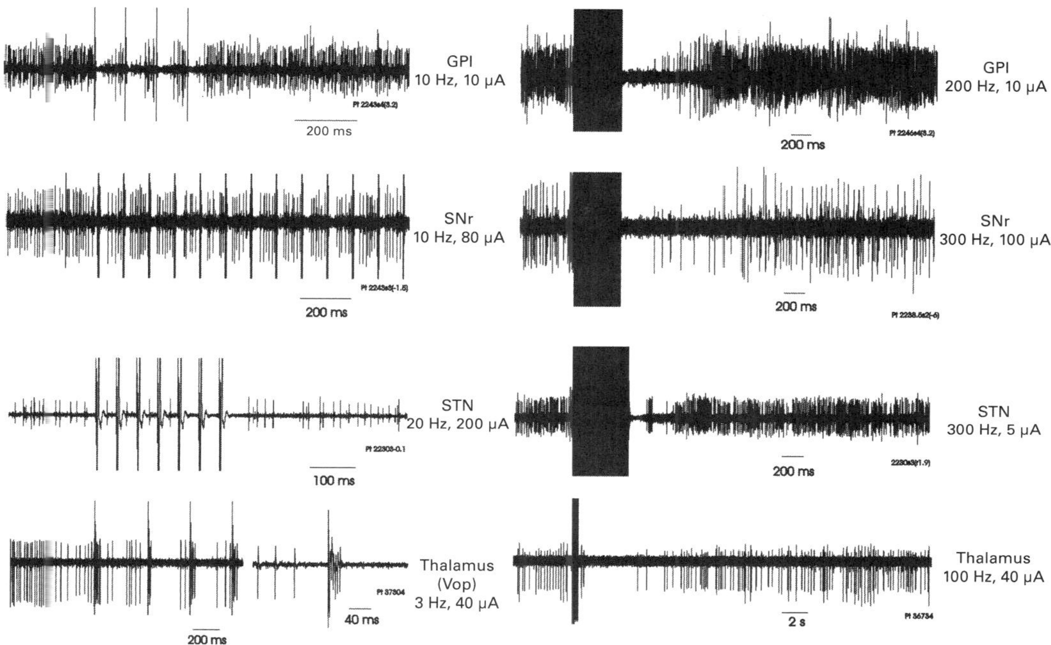

Figure 16.3
Examples of the effects of microstimulation through a nearby electrode on the firing of neurons in the globus pallidus interna (GPI), substantia nigra pars reticulate (SNr), subthalamic nucleus (STN), and thalamus. The left side of the figure shows the responses to individual stimuli at low rates of stimulation, and the right side shows the effects following high-frequency stimulation trains. Vop, ventral oralis posterior.

following the stimulus train, but stimulation currents generally had to be considerably higher than those in GPi and SNr. This inhibition ranged from 50 to over 500 ms, which was considerably shorter than the inhibitory periods observed for thalamic stimulation. For some of the STN cells the pattern of inhibition consists of inhibition followed by rebound excitation and subsequently followed by more inhibition, but of a different pattern from the rebound LTS bursting seen in thalamus. Thus, although HFS in all these regions generally results in inhibitory aftereffects, the detailed responses are quite different and suggest that the therapeutic effects of DBS in different regions may involve different mechanisms.

Stimulation in thalamus will preferentially activate axons although cell bodies close to the electrode tip may be excited directly (Nowak & Bullier, 1998; Ranck, 1975). Thus a major effect of stimulation is likely to be activation of glutamatergic thalamocortical efferents, as well as ascending thalamic afferents (e.g., lemniscal, cerebellothalamic) and descending corticothalamic afferents. The excitatory responses observed following single stimuli at lower frequencies are probably due to activation of ascending thalamic afferents. Although one might expect to also observe some antidromic activation, the occurrence of such spikes would most likely be obscured by the stimulus artifact. The prominent inhibitory responses are most likely due to activation of GABAergic inputs from the thalamic reticular nucleus and inhibitory interneurons. These inhibitory inputs may be elicited not only directly but also by the activation of the thalamic reticular nucleus neurons by the excitatory collaterals of the thalamocortical neurons. The short period of rebound-enhanced LTS bursting which frequently occurs following termination of the train strongly suggests that during stimulation, or at least toward the end of the train, the membrane potential of the neuron is hyperpolarized as LTS bursts only occur following a period (at least 100 ms) of hyperpolarization (McCormick & Huguenard, 1992).

Studies in rat thalamic slices by Kiss's group found that HFS primarily induced synaptically mediated glutamatergic membrane depolarization, which usually occurred only at the onset of the train (Anderson et al., 2004) and was followed by a period of silence. However, since the rat does not have GABAergic interneurons and the circuit via the thalamic reticular nucleus is probably destroyed by sectioning, it is difficult to assess to what extent they adequately model the situation in the human thalamus.

A recent study by Birdno et al. (2009) employed computational modeling of the stimulation effects on thalamocortical neuron responses to elucidate the possible mechanisms underlying the long inhibition observed in our studies described above. The model included a population of biophysically realistic thalamocortical relay neurons and the presynaptic axons terminating on these neurons and included the effects of neuromodulators which may have been released by stimulated terminals, including adenosine, histamine, acetylcholine, serotonin, and noradrenaline. The model was able to replicate the post-HFS responses we had observed in patients and indicated that the prolonged inhibition was associated most strongly with changes in a pertussis toxin–sensitive K+ current (I_{KG}). Because I_{KG} conductances are activated by GABAB, muscarinic, and A1 adenosine receptors, these receptors are the most likely sources of poststimulus inhibition observed in human subjects. Importantly, the model also showed that the axons of the thalamocortical relay neurons were excited by suprathreshold stimulation in the period during stimulation, and activity in these axons was regularized during HFS.

In DBS for treatment of tremor, the most common application of thalamic DBS, the effects of HFS are manifested during the stimulation and tremor generally resumes shortly after cessation of stimulation. Thus, the findings of the present study do not directly address the question of what occurs during the HFS stimulation. Nevertheless, the findings provide some insights into underlying mechanisms. The repetitive burst discharge in synchrony with the tremor of thalamic tremor cells is likely to be a major component of the circuit driving the tremor, and therefore disruption of this rhythmic activity would be expected to stop the tremor. HFS in the region of these tremor cells is thus likely to disrupt the tremor activity and lead to cessation of the tremor. This is consistent with the observed effects of microstimulation and macrostimulation (with DBS electrodes) on tremor and also with the fact that lesions to this area or inactivation by microinjection of lidocaine also block the tremor (Dostrovsky et al. 1993). The findings reported are all consistent with disruption of tremor activity, and it is possible in fact that the final outcome is a combination of several different mechanisms, for example, inhibition of cell firing and excitation of the afferents and efferents with a regular high-frequency discharge. Furthermore with prolonged high-frequency discharge of afferents and efferents it is likely that there is synaptic fatigue and failure in thalamus and motor cortex (thalamocortical efferents).

References

Anderson, T., Hu, B., Pittman, Q., & Kiss, Z. H. (2004). Mechanisms of deep brain stimulation: An intracellular study in rat thalamus. *Journal of Physiology, 559*, 301–313.

Benabid, A. L., Pollak, P., Louveau, A., Henry, S., & de Rougemont, J. (1987). Combined (thalamotomy and stimulation) stereotactic surgery of the VIM thalamic nucleus for bilateral Parkinson's disease. *Applied Neurophysiology, 50*, 344–346.

Birdno, M., Tang, W., Dostrovsky, J. O., Hutchison, W. D., & Grill, W. M. (2009). Characteristics and mechanisms of the post-stimulus responses of thalamic neurons to high frequency stimulation (HFS). *Society for Neuroscience Abstracts,* 41.12.

Domich, L., Oakson, G., & Steriade, M. (1986). Thalamic burst patterns in the naturally sleeping cat: A comparison between cortically projecting and reticularis neurones. *Journal of Physiology, 379*, 429–449.

Dostrovsky, J. O., Levy, R., Wu, J. P., Hutchison, W. D., Tasker, R. R., & Lozano, A. M. (2000). Microstimulation-induced inhibition of neuronal firing in human globus pallidus. *Journal of Neurophysiology, 84*, 570–574.

Dostrovsky, J. O., Patra, S., Hutchison, W. D., Palter, V. N., Filali, M., & Lozano, A. M. (2002). Effects of stimulation in human thalamus on activity of nearby thalamic neurons. *Society for Neuroscience Abstracts,* 62.14.

Dostrovsky, J. O., Sher, G. D., Davis, K. D., Parrent, A. G., Hutchison, W. D., & Tasker, R. R. (1993). Microinjection of lidocaine into human thalamus: A useful tool in stereotactic surgery. *Stereotactic and Functional Neurosurgery, 60*, 168–174.

Filali, M., Hutchison, W. D., Palter, V. N., Lozano, A. M., & Dostrovsky, J. O. (2004). Stimulation-induced inhibition of neuronal firing in human subthalamic nucleus. *Experimental Brain Research, 156*, 274–281.

Hodaie, M., Cordella, R., Lozano, A. M., Wennberg, R., & Dostrovsky, J. O. (2006). Bursting activity of neurons in the human anterior thalamic nucleus. *Brain Research, 1115*, 1–8.

Hodaie, M., Wennberg, R. A., Dostrovsky, J. O., & Lozano, A. M. (2002). Chronic anterior thalamus stimulation for intractable epilepsy. *Epilepsia, 43*, 603–608.

Hutchison, W. D., Allan, R. J., Opitz, H., Levy, R., Dostrovsky, J. O., Lang, A. E., et al. (1998). Neurophysiological identification of the subthalamic nucleus in surgery for Parkinson's disease. *Annals of Neurology, 44*, 622–628.

Krayenbuhl, H., Wyss, O. A., & Yasargil, M. G. (1961). Bilateral thalamotomy and pallidotomy as treatment for bilateral parkinsonism. *Journal of Neurosurgery, 18*, 429–444.

Kumar, K., Toth, C., & Nath, R. K. (1997). Deep brain stimulation for intractable pain: A 15-year experience. *Neurosurgery, 40*, 736–746.

Lafreniere-Roula, M., Kim, E., Hutchison, W. D., Lozano, A. M., Hodaie, M., & Dostrovsky, J. O. (2010). High-frequency microstimulation in human globus pallidus and substantia nigra. *Experimental Brain Research, 205*, 251–261.

Lenz, F. A., Dostrovsky, J. O., Kwan, H. C., Tasker, R. R., Yamashiro, K., & Murphy, J. T. (1988). Methods for microstimulation and recording of single neurons and evoked potentials in the human central nervous system. *Journal of Neurosurgery, 68*, 630–634.

Levy, R., Lozano, A. M., Hutchison, W. D., & Dostrovsky, J. O. (2007). Dual microelectrode technique for deep brain stereotactic surgery in humans. *Neurosurgery, 60*(4 Suppl 2), 277–283 (discussion 283–284).

Lu, L. D., Prescott, I. A., Dostrovsky, J. O., Hodaie, M., Lozano, A. M., & Hutchison, W. D. (2012). Frequency-dependent effects of electrical stimulation in the globus pallidus of dystonia patients. *Journal of Neurophysiology, 108*, 5–17.

Lozano, A. M., Hutchison, W. D., & Dostrovsky, J. O. (1995). Microelectrode monitoring of cortical and subcortical structures during stereotactic surgery. *Acta Neurochirurgica, 64*(Suppl), 30–34.

McCarley, R. W., Benoit, O., & Barrionuevo, G. (1983). Lateral geniculate nucleus unitary discharge in sleep and waking: State- and rate-specific aspects. *Journal of Neurophysiology, 50*, 798–818.

McCormick, D. A., & Huguenard, J. R. (1992). A model of the electrophysiological properties of thalamocortical relay neurons. *Journal of Neurophysiology, 68,* 1384–1400.

Nowak, L. G., & Bullier, J. (1998). Axons, but not cell bodies, are activated by electrical stimulation in cortical gray matter: II. Evidence from selective inactivation of cell bodies and axon initial segments. *Experimental Brain Research, 118,* 489–500.

Perlmutter, J. S., & Mink, J. W. (2006). Deep brain stimulation. *Annual Review of Neuroscience, 29,* 229–257.

Ranck, J. B. J. (1975). Which elements are excited in electrical stimulation of mammalian central nervous system: A review. *Brain Research, 98,* 417–440.

Schaltenbrand, G., & Wahren, W. (1977). *Atlas for stereotaxy of the human brain.* Stuttgart: Georg Thieme.

Tasker, R. R., Organ, L. W., & Hawrylyshyn, P. (1976). Sensory organization of the human thalamus. *Applied Neurophysiology, 39,* 139–153.

Tsoukatos, J., Kiss, Z. H., Davis, K. D., Tasker, R. R., & Dostrovsky, J. O. (1997). Patterns of neuronal firing in the human lateral thalamus during sleep and wakefulness. *Experimental Brain Research, 113,* 273–282.

Yamashiro, K., & Tasker, R. R. (1990). Microstimulation for stereotactic neurosurgery. *Stereotactic and Functional Neurosurgery, 54,* 168–171.

17 Human Single Unit Activity for Reach and Grasp Motor Prostheses

Arjun K. Bansal

There are over 5 million patients suffering from paralysis in the United States alone due to traumatic accidents and diseases (Christopher & Dana Reeve Paralysis Foundation). Paralysis due to spinal cord injury, amyotrophic lateral sclerosis (ALS), or stroke sometimes leads to patients becoming "locked-in," wherein the patient is cognitively intact but is unable to move or communicate with the outside world (Bauby, 1998). To restore some degree of movement control and communication ability, motor prostheses systems have attempted to tap into intact brain signals and decipher locked-in patients' intentions (see figure 17.1). Although early prostheses used noninvasive approaches such as electroencephalography (EEG), the signal-to-noise ratios of these approaches have been somewhat limited (but see Birbaumer, 2006) compared to that of using single unit activity (SUA) recorded from many neurons. Even though the extraction of SUA requires an invasive procedure, the successful use of invasive electrodes in cochlear implants and deep brain stimulation (DBS) electrodes to help cure deafness and Parkinson's disease, respectively, suggested that invasive approaches could hold promise in helping cure paralysis (Donoghue, 2008; Hatsopoulos & Donoghue, 2009). In fact, recent clinical trials have had positive results in enabling paralyzed patients to control computer cursors and robotic arms with SUAs recorded from multielectrode arrays placed in the motor cortices of paralyzed individuals. Two noteworthy efforts are the BrainGate clinical trials at Brown University (Hochberg et al., 2006; Hochberg et al., 2012) and a separate clinical trial at the University of Pittsburgh (Collinger et al., 2012). In parallel, researchers are also working on lower limb prostheses for restoring walking (He et al., 2008; Harkema et al., 2011) including electrochemical options (van den Brand et al., 2012) although this latter research has not reached clinical trial stage yet. A market research study has suggested that motor prostheses that achieve paraplegic functionality for quadriplegic patients, and thereby give them greater independence from caregivers, for at least seven consecutive years can be economically viable for insurance companies (Bansal et al., 2005). Here, we review the progress toward the development of reach and grasp prostheses that may one day achieve this goal.

Monkey electrophysiology has helped guide the understanding of the neural code underlying reach and grasp movement generation as well as helped quantify the amount of information extractable from various signals. With the recent advances in brain–machine interfaces (BMIs)

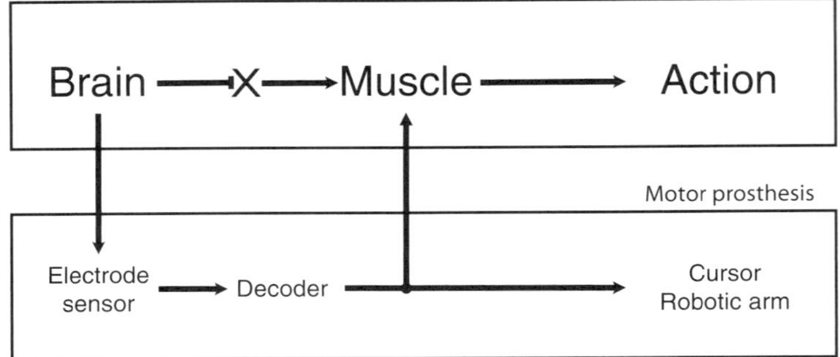

Figure 17.1
Conceptualization of a motor prosthesis or brain–machine interface (adapted from figure 1 in Donoghue, 2008). In a paralyzed subject the normal connection between brain and muscles is severed due to disease or injury. A motor prosthesis records the brain activity using electrode sensors and decodes this activity to infer the subject's motor intentions. The decoded activity is used to drive the subject's muscles using functional electrical stimulation or to control a cursor or robotic arm.

to restore motor control in paralyzed human patients (specifically the BrainGate and Pittsburgh clinical trials mentioned earlier) there is tremendous opportunity to not just apply the motor coding theories built on monkey experiments but also to compare and contrast these with human motor cortical control. These studies will help humans with paralysis and simultaneously advance our understanding of human motor neurophysiology (Donoghue, 2008; Mukamel & Fried, 2012).

This chapter is divided into three sections. In the first section, we will review the neurophysiology of motor coding based primarily on single unit recordings in monkeys and humans and its applications toward reach and grasp prostheses. In the second section, we will describe the technical considerations of researchers when building motor prostheses systems. Finally, we end with future directions.

Motor Coding

A motor prosthesis can strive to replicate various parameters of a reach and grasp movement such as the arm's or digit's end-point position, direction and velocity of movement, force, trajectory, or higher-level goals.[1] Through the work of monkey neurophysiologists over the last century, we now have a better understanding of the encoding of these parameters in neuronal populations across motor cortical areas. Feedback and plasticity also play a crucial role in shaping movements on shorter and longer timescales, respectively.

The first detailed account of motor cortical organization in monkeys (that were lightly anesthetized) came from stimulation, lesion, and cooling experiments by Leyton and Sherrington (1917). They identified several sites anterior to the central sulcus (see figure 17.2, plate 16, for a simplified schematic of cortical regions and connected pathways involved in motor

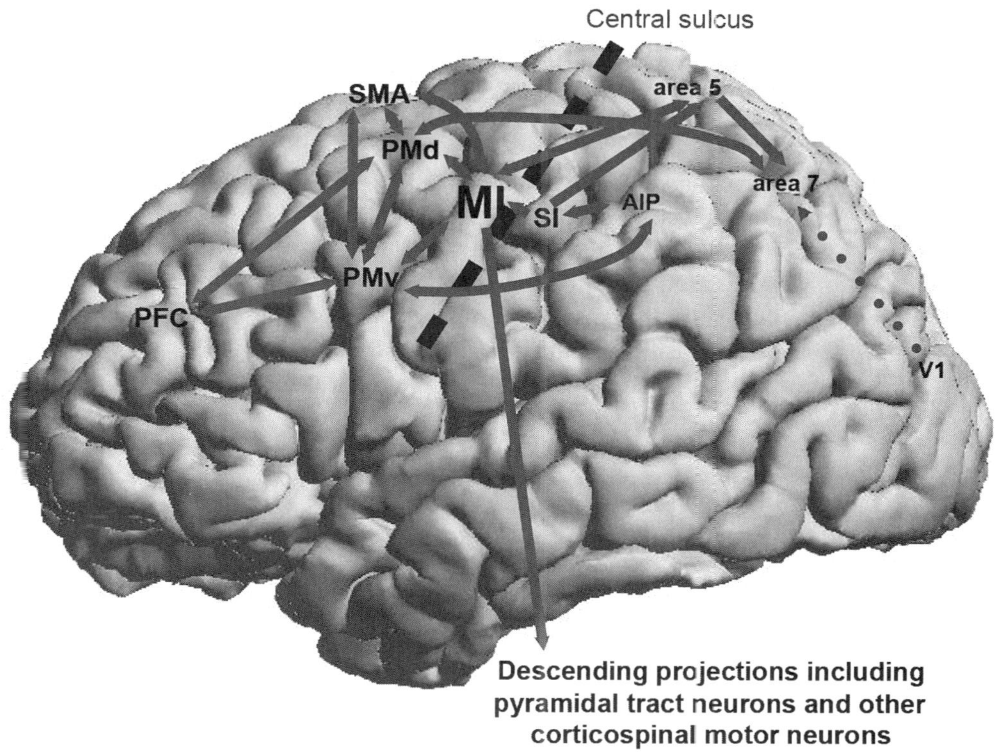

Figure 17.2 (plate 16)
A simplified schematic of cortical regions and connected pathways involved in control of reaching and grasping actions; see text. SMA, supplementary motor area; PMd, dorsal premotor cortex; PMv, ventral premotor cortex; MI, primary motor cortex; SI, primary somatosensory cortex; AIP, anterior intraparietal region; PFC, prefrontal cortex; V1 primary visual cortex. Adapted from figures in Martin (2003), and Scott (2004), and from connectivity results in monkeys from Vogt and Pandya (1978), Matelli et al. (1986), Pandya and Yeterian (1996), and Dum and Strick (2005).

control), which, when stimulated, resulted in relatively stereotyped muscle movements of particular parts of the body such as fingers, arms, neck, hip, and so forth. Such movements were not elicited by stimulating sites posterior to the central sulcus. Penfield and Boldrey (1937) reported that within primary motor cortex (MI; see figure 17.2, plate 16) there is somatotopy in the organization of neurons, with nearby neurons generally coding for the movement of nearby muscles on the body. Subsequent work related the firing rates and patterns of motor cortical neurons to kinematic parameters such as position and velocity, and dynamic parameters such as force and the rate of change of force, but found no single parameter that was best controlled by these neurons (Evarts, 1968; Humphrey et al., 1970). A key concept in our understanding of motor coding and its application in human motor prostheses is that of "directional tuning" of motor cortical cells.

Visual information about the object to be reached and grasped arrives into the brain from the eyes, and through the lateral geniculate nuclei of the thalamus (LGN), to the primary visual cortex (see figure 17.2, plate 16). The location and shape of the object to be reached and grasped are thought to be extracted via regions in the parietal cortex such as area 7 cf the posterior parietal cortex and anterior intraparietal region (AIP). Area 5 estimates the current configuration of the arm, and nearby regions compute the transformations required for the arm and hand to perform the reach and grasp. Parietal regions are also reciprocally connected with premotor regions. AIP is reciprocally connected with ventral premotor cortex (PMv) while area 7 is reciprocally connected with dorsal premotor cortex (PMd). These circuits are thought to be preferentially involved in grasp and reach, respectively (but see Vargas-Irwin, 2010, and Bansal et al., 2012a, for a discussion about reach and grasp in PMv). Premotor regions also receive higher order goal information from the prefrontal cortex. MI) is reciprocally connected with premotor cortex, area 5, and receives input from the primary somatosensory cortex. MI computes motor command signals that are transmitted to the spinal cord and brainstem structures via descending projection such as the pyramidal tract neurons and other corticospinal motor neurons. Note that the dotted arrow in figure 17.2 (plate 16) indicates indirect connections, and the dashed line indicates central sulcus. Connectivity is based on anatomical results in monkeys of homologous brain regions and is not meant to be exhaustive. Note that here we do not present other important structures important for motor control such as the basal ganglia, cerebellum, cranial nerves, and the vestibular and oculomotor systems.

Directional Tuning

Individual neurons' firing rates in monkey MI exhibit cosine-shaped tuning to a range of preferred directions of reaching arm movements in two (Georgopoulos et al., 1982) and three (Schwartz et al., 1988) dimensions. This property is analogous to the orientation tuning displayed by cells in primary visual cortex when subjects are presented with stimuli of various orientations. Remarkably, by using the weighted combination of individual neuron responses or "population vector," the direction of the monkey's arm movement was determined (Georgopoulos et al., 1986 Georgopoulos et al., 1988). Thus, each cell did not code for a unique movement, but groups of cells acted together to perform a movement. From Georgopoulos et al. (1999), we write the mathematical expression for the population vector as

$$P_j = \sum_{i=1}^{N} w_{ij} C_i,$$

where P_j is the population vector for the jth finger or wrist movement, C_i is the preferred direction of the ith cell (N total cells), and w_{ij} is a weighting function:

$$w_{ij} = d_{ij} - \overline{d_i},$$

where

$$\overline{d}_i = \frac{1}{M} \sum_{j=1}^{M} d_{ij} \, .$$

M equals the number of movements, and d_{ij} is the mean firing rate of the ith cell for the jth movement.

In principle, the population vector could then be used to control a motor prosthesis. From the perspective of motor prostheses, it is important to note that these early studies were investigating fundamental questions of motor coding and therefore reconstructed the direction of arm movements *offline*, which means after the experiment when the monkey actually performed the action. More recently, several studies motivated by building motor prostheses showed that monkeys could control a cursor *online* in real time with just their neuronal activity. Moreover, earlier experiments were considered *open loop* since the monkey did not receive any feedback about its decoded intentions, but more recent experiments are termed *closed loop* as the monkey controls an effector with its brain signals and receives sensory (typically visual) feedback during brain control (open-loop robotic arm in 1-D and 3-D, Wessberg et al., 2000; instant cursor control, Serruya et al., 2002; 2-D end point, Musallam et al., 2004; 3-D cursor, Taylor et al., 2002). It is also noteworthy that neuronal tuning properties often change during closed-loop brain-control trials, resulting in improved motor control over time (see the "Plasticity" subsection later in this section). Decoding algorithms that account for changes in tuning are an active area of research (see the "Decoding Algorithm Design" subsection of the "Technical Considerations" section below).

The invasiveness of the microelectrodes required for single neuron recordings have precluded the validation of the monkey physiology results in healthy human MI. Initial work with human ALS patients implanted with neurotrophic[2] electrodes demonstrated that they could modulate the activity of a single neuron in their MI that drove an on/off switch and controlled a cursor or a speech synthesizer (Kennedy & Bakay, 1998; Kennedy et al., 2000). Recently populations of direction-tuned neurons were found in paralyzed human patients in the BrainGate clinical trials, and these were used in an online, closed-loop prosthesis to control a cursor and the opening and closing of a robotic fist (Hochberg et al., 2006). Truccolo et al. (2008) found that more than 80% of the cells in MI were tuned to observed position and velocity of a target in a pursuit-tracking task. In addition, Truccolo et al. (2008) reported that the intended target was decoded with an accuracy of 80–95% in a 2-D center–out task, consistent with previous research in monkeys (Paninski et al., 2004). These results were striking because even though the patients in the BrainGate trials had severe loss of voluntary control of their limbs and had not used them in years,[3] they were still able to modulate their MI neuronal activity in response to intended movements. The modulation of MI neurons not just during action performance but also during action observation is a significant property for motor prostheses and has also been demonstrated in monkeys (Tkach et al., 2007; Dushanova & Donoghue, 2010). Further information about the properties of single neurons in human motor cortical areas was obtained from patients with Parkinson's disease undergoing surgery for the implantation of DBS electrodes. During DBS surgeries, premotor (area 6) cortical neurons were demonstrated to exhibit directional tuning, and contain movement intent (to move or not to move) information (Ojakangas et al., 2006). Current work regarding the selection of *area* and subregion of implantation for a prosthetic device is summarized later in this chapter.

Grasp Coding

Although direction tuning can be exploited to get an effector to a desired location, manipulating objects often requires grasping them. For motor prostheses, self-feeding is a crucial step toward conferring independence to paralyzed subjects. Ventral premotor (PMv, or area F5; see figure 17.2, plate 16) cortical neurons are involved in finger movements and encode grasp types and grasp aperture in monkeys (Kurata & Tanji, 1986; Rizzolatti et al., 1988; Umiltà et al., 2007; Vargas-Irwin, 2010; Bansal et al., 2012a). Similar grasp types are encoded in similar neuronal firing patterns in PMv (Carpaneto et al., 2011). Although many studies have examined PMv for a specific role in grasp coding, recent work has reported equivalent representation for continuous grasp in MI, and intermixed reach and grasp populations within both MI and PMv (Vargas-Irwin, 2010; Vargas-Irwin et al., 2010; Bansal et al., 2011; Bansal et al., 2012a).

A mouse click may be regarded as a simple grasp manipulation. Initial work in the BrainGate trials simulated a mouse click by asking subjects to imagine squeezing closed or opening their fists (Kim et al., 2011). In more complex applications, monkeys fed themselves by reaching and grasping for food using a robotic arm driven by signals from neuronal populations in MI in real time (Velliste et al., 2008). Paralyzed humans in the BrainGate clinical trial were recently able to reach, grasp, and drink from a coffee cup using a robotic arm driven by MI neurons (Hochberg et al., 2012). A recent clinical trial at the University of Pittsburgh demonstrated brain control of a seven-degrees-of-freedom robotic arm using SUAs from two 96-electrode arrays in MI of a tetraplegic patient (Collinger et al., 2012).

Force Coding

Even as perfect decoding of direction and grasp kinematics would allow a subject to hold an object such as an egg between his or her fingers, applying too much force at the wrong time will crush the egg and create a mess. Therefore, understanding the relationship between SUA and force generation could provide crucial signals for paralyzed patients. Despite this fact, most neurophysiology studies with human patients have so far focused on kinematics, with algorithmic control of force. In monkeys, force is known to modulate firing rates of pyramidal tract neurons (Evarts, 1968; see figure 17.2, plate 16). This property can be used to predict end point or grasping force from MI neurons (Gupta & Ashe, 2009; Ethier et al., 2012). In human patients undergoing DBS surgeries, Patil et al. (2004) demonstrated that neurons in subcortical structures such as the subthalamic nucleus and thalamic motor areas predict gripping force. In applications such as BrainGate, however, the modulation of neuronal responses with varying levels of imagined force is still unreported. Computing the appropriate force to apply is complex and depends on proprioceptive feedback that may be diminished in paralyzed patients.[4] The issue of *feedback* and how it may be conveyed to movement or force-generating neurons will be discussed later in this chapter. In addition, imagining moving a heavy load versus a lighter load may provide a gain control mechanism by which otherwise quiescent neurons bolster their firing rates and are then read out by a neural implant.

Trajectories in Space and Time

So far we have described a static view of the neuronal encoding of reach and grasp parameters such as end-point position, grasp aperture, and force. Reach and grasp movements, however, occur not just in 3-D space, but also in time. Thus, improved understanding of how neurons code for trajectories of movements may enable prostheses with better performance. Recent work has found that the activity of monkey motor cortical neurons is better explained by preferred "pathlets" or trajectories for reach and grasp rather than by preferred directions that are independent in space and time (Hatsopoulos et al., 2007; Saleh et al., 2010; Saleh et al., 2012). This is a remarkable validation of a concept proposed by Leyton & Sherrington (1917): "The motor cortex may be regarded as a synthetic organ for compounding and recompounding in varied ways movements of varied kinds of scope from comparatively small, though in themselves well coordinated, fractional movements." Furthermore, these trajectories might be generated by the coordinated activation of related muscles or *muscle synergies* (d'Avella et al., 2003; Overduin et al., 2012) by pathlet coding neurons.

For motor prostheses, there has been progress with the application of direction and velocity tuning based models in controlling a robotic arm driven by neuronal signals (Hochberg et al., 2012). Hochberg et al. (2012) used a Kalman filter based decoder (Wu et al., 2006), which is a linear state–space dynamical system model. The decoder is trained with the neuronal activity when the patient imagines moving a robotic arm to certain targets placed along the cardinal axes. Such a decoder learns an implicit knowledge of kinematic trajectories in the form of the state–space transition matrix but does not explicitly model the activity of each neuron as contributing toward a pathlet. Future work may explore whether pathlets are a better approach to building an encoding/decoding model for motor prostheses. Recent work, however, suggests that the predictive power of pathlet models is weaker than that of models using spiking history of other simultaneously recorded neurons (Truccolo et al., 2010). Thus, including spiking history in the model may be another way of improving decoding performance (Truccolo et al., 2005).

Areal Organization of Motor Coding Regions, and Sensorimotor Computation

Understanding the areal organization of motor coding for prostheses is motivated by targeting the recording electrodes toward the region(s) that will maximize the information content of the relevant motor parameter. MI is the main contributor to corticospinal motor neurons pathways and therefore a natural area for the placement of electrodes for motor prostheses. The BrainGate clinical trials report considerable success with placing the electrode array in the "knob"-shaped motor hand area (Yousry et al., 1997) of MI. Outside of MI many areas are known to code for motor parameters that may not just be useful in cases of damage to MI but also potentially offer additional information beyond that in MI. Anterior to MI, dorsal premotor cortex PMd and PMv neurons are interconnected with MI (Pandya & Yeterian, 1996; Dum & Strick, 2005; see figure 17.2, plate 16) and encode reach and grasp parameters often earlier than MI in delay paradigms (Kurata & Tanji, 1986; Rizzolatti et al., 1988; Fu et al., 1993; Messier & Kalaska, 2000; Umiltà et al., 2007; Stark & Abeles, 2007; Stark et al., 2007). Premotor areas could serve as sources of additional information for prostheses (Bansal et al., 2012a). Medial frontal cortical regions such as the supplementary motor area (SMA; figure 17.2, plate 16) are key to planning sequential movements in monkeys (Tanji and Shima, 1994) and related to movement intention in humans (Fried et al., 1991; Fried et al., 2011). SMA and PMd neurons have been used to decode two target sequences in real time (Shanechi et al., 2012). Prefrontal and frontopolar cortices are reported to encode increasingly abstract parameters related to movements such as higher order goals (Tsujimoto et al., 2011).

Although movements are not evoked by stimulating sites posterior to postcentral gyrus (Leyton & Sherrington, 1917), the posterior parietal cortex (PPC) serves a crucial role in sensorimotor computations and has been demonstrated as a useful source of reach end-point, trajectory, and grasp information (Musallam et al., 2004; Mulliken et al., 2008; Townsend et al., 2011). The PPC is composed of many regions in monkeys such as the lateral intraparietal area, ventral intraparietal area, central intraparietal area, AIP (figure 17.2, plate 16), area 7 (figure 17.2, plate 16), and the medial superior temporal area (Andersen et al., 1997), with corresponding human counterparts (Grefkes & Fink, 2005). These areas compute movement plans in diverse (but not exclusive) frames of reference such as in eye, head, hand, body, world, or object-centered coordinates using multimodal information (including posture) as inputs. The diversity of reference frames enables the computation of coordinated movements of body parts such as the neck, arms, and eyes to achieve a final goal. Furthermore, the responses of neurons in these areas are considered to be intermediate between purely sensory or motor representations and modulated by cognitive signals such as attention, intention, reward, and decision making (Andersen et al., 1997; Glimcher, 2004). On the one hand, these representations could provide additional information to drive a motor prosthesis compared to those in MI. On the other hand, the influence of these cognitive variables may make it harder to disentangle the precise motor intentions of a paralyzed subject. Complicating this simple assessment are the observations that MI neurons also exhibit postural modulation (Ajemian et al., 2008) and cognitive features such as serial order (Carpenter et al., 1999), and they may not just direct muscles but also participate in coordinate transformations (Kakei et al., 1999). Despite these complications, both the BrainGate and Pittsburgh clinical trials have targeted MI, and the use of other areas remains to be explored in human prostheses.

Feedback

Visual and proprioceptive feedback play a significant role in generating smooth movements (Scott, 2004) by providing, for example, limb or effector position information to motor control areas such as area 5 (see figure 17.2, plate 16) in parietal cortex (Kalaska et al., 1983) and MI (Goldring & Ratcheson, 1972). While visual feedback might be typically unaffected, proprioceptive feedback processing may be impaired to various degrees in patients with paralysis. While proprioceptive feedback is not strictly necessary to generate motor control signals, as demonstrated in the clinical trials mentioned earlier, providing proprioceptive feedback could enhance BMI performance (Suminski et al., 2010). In paralyzed patients with impaired feedback afferents, electrical or optical stimulation may provide an approach toward "writing in" proprioceptive information (Diester et al., 2011; Gilja et al., 2011). In healthy motor control, the process of converting motor commands to movements of the limb, as well as the estimation of limb position, is subject to errors. The proprioceptive feedback in a prosthesis could thus provide information about limb or effector position although, even in its absence, visual feedback might compensate for it to some extent. The nature and extent of this compensatory ability remain to be quantified (Scheidt et al., 2005).

State–Space Models

Alternative motor coding proposals have suggested that MI neurons are not directly coding for parameters such as arm/hand position or velocity but are coding for some state variables intrinsic to the system generating movements such as muscle length and velocity (e.g., Oby et al., 2012). There have been proposals that the motor system learns optimal feedback control laws to act on these state variables (Scott, 2004). Recent work has modeled the high-dimensional neuronal population activity (that can be quite noisy for each neuron from trial to trial: e.g., Nawrot et al., 2008) at any instant as a point on a *low-dimensional manifold*. In this view, the neuronal activity over time related to generating a movement trajectory describes a "neural trajectory" in this low-dimensional space that stays more similar compared to the noisy individual neurons and that may be indicative of a dynamical control system for generating actions (Santhanam et al., 2009; Yu et al., 2009; Shenoy et al., 2011; Churchland et al., 2012). Santhanam et al. (2009) reported significant improvements (~75%) using a factor analysis based decoding approach, which exploited the correlated trial-to-trial variability. These approaches remain to be tested in prostheses applications.

Plasticity[5]

For a motor prosthesis to work over an extended period of time, it will need to adapt to changing neuronal properties. MI is known to change its properties following traumatic injury or during skill-learning and everyday actions (Sanes & Donoghue, 2000). Many BMI studies with monkeys have shown an improvement in decoding performance over days suggesting that the monkeys' neurons learn to control the effectors better over time (Carmena et al., 2003; Taylor et al., 2002;

Musallam et al., 2004; Ganguly & Carmena, 2009; Jarosiewicz et al., 2008). A similar improvement was seen in human clinical trials (Collinger et al., 2012). Thus, neuronal plasticity has critical implications for motor prostheses. Typically, decoding algorithms initially strive to tap into the natural tuning properties of the neurons as determined by imagined or observed movements. To achieve skilled control, the subject can engage plasticity mechanisms as they modulate their neuronal responses in trying to *transmit* their intentions to a relatively stable decoding algorithm. Plasticity (or noise), however, may alter the responses of neurons involuntarily (Rokni et al., 2007), requiring a change in the algorithm to infer the subjects' true intentions. The specific balance between tuning the decoding algorithm and allowing the brain to adapt to a fixed algorithm remains to be established.

Technical Considerations

Signal Selection: Single Unit Activity versus Other Signals for Motor Prostheses

An extracellular microelectrode placed intracortically records a field potential (voltage) signal, which can be filtered into many frequency bands from 0.1 to about 5000 Hz. At the higher range of frequencies (300–5000 Hz), action potentials are detected and then sorted. Action potentials are the only signals corresponding directly to the activity of single neurons. Unsorted activity of many single units, and the band-pass filtered field potential signal at frequencies greater than 100 Hz (thought to reflect the spiking of many single units) are both confusingly referred to as multiunit activity (MUA). In our previous work (Bansal et al., 2012a) and here we refer to the high-frequency band-pass filtered signal as MUA, and "unsorted spikes" are referred to explicitly. Lower frequency bands (<100 Hz) of the field potential (FP) are called local field potentials (LFPs) when recorded using microwires, as their activity is thought to reflect the averaged synaptic inputs (and outputs) in a local brain region. Low-frequency LFPs (<2 Hz; *lf*-LFPs) including the movement-event-related potential, MUAs, and SUAs have all been demonstrated to contain information about movement kinematics. The middle-frequency bands such as the alpha (8–12 Hz) and beta (12–30 Hz) bands have relatively weaker kinematic representation (Zhuang et al., 2010) but contain go/no-go state information (Hwang and Andersen, 2009). EEG and electrocorticography (ECoG) also measure FP signals (using electrodes placed, respectively, on the surface of the scalp or brain), but on a relatively coarser spatial scale than those measured using intracortical microelectrodes (Waldert et al., 2009). EEG and ECoG also contain information related to reaching and grasping movements (Wolpaw and McFarland, 2004; Schalk et al., 2007; Kubánek et al., 2009; Bradberry et al., 2010; Pistohl et al., 2011; Milekovic et al., 2012).

If SUA in motor cortex corresponds to the output that ultimately drives muscles and generates movement, then it would seem to be the most informative signal for acquiring movement intentions for a prosthetic device. Despite this intuition, initial results suggested that *lf*-LFPs or MUAs are more informative than SUAs (Mehring et al., 2003; Stark and Abeles, 2007) in regimes with a handful of simultaneously recorded neurons and average-selection-based decoding algorithms. More recent work, however, with 96-channel multielectrode arrays and computationally intensive greedy-selection algorithms, has suggested that SUAs contain more information than MUAs and *lf*-LFPs (Bansal et al., 2012a) for 3-D reach and grasp in both MI and PMv. Following similar reasoning, human clinical trials have mostly used SUA for motor prostheses (Hochberg et al., 2006; Hochberg et al., 2012; Collinger et al., 2012) although a recent study has used ECoG (Wang et al., 2013). The FP signals may provide other advantages such as stability, invasiveness trade-offs, and simpler signal processing. We briefly review some of the trade-offs next.

Speed and Accuracy Santhanam et al. (2006) used an information theoretic measure to quantify the rate of end-point information extracted from SUA in monkey PMd. They reported obtaining up to 6.5 bits per second of information, allowing for 3.5 brain-controlled trials per second. Although not a direct comparison, this rate appears to be superior to information extracted from EEG, ECoG, and magnetoencephalography (<1 bit), and LFP (<2 bits) based methods (Waldert et al., 2009). The superior performance of SUA is probably related to the lower spatial correlation in that signal compared to the FP based signals (Bansal et al., 2012a). In the average case, however, *lf*-LFPs can outperform SUA (Mehring et al., 2003; Bansal et al., 2011). Furthermore, ECoG and EEG may contain more information than previously thought as recent studies have successfully reconstructed 3-D reach parameters offline (see table 17.1) from ECoG (Chao et al., 2010) and EEG (Bradberry et al., 2010). However, recent work directly comparing ECoG with intracortical spikes and LFPs has reported much worse performance with epidural ECoG compared to spikes or LFPs (Flint et al., 2012). Further work is needed to test the precision of online, closed-loop 3-D control that can be achieved using these techniques.

Ease of Control A distinct advantage of SUA based prostheses is the relative ease of control. Subjects imagine moving their arm, and the corresponding signals are directly interpreted to control a robotic arm (Hochberg et al., 2006; Hochberg et al., 2012; Collinger et al., 2012). On the other end of the recording spectrum, EEG based methods typically rely on the subject's performing a mental exercise that is not directly related to the desired action. For example, an EEG based 2-D cursor control prosthesis was designed based on biofeedback (Wolpaw & McFarland, 2004). Subjects controlled the two dimensions by modulating the power of mu

Table 17.1
Comparison of continuous reach (and grasp) offline decoding performance across three recording techniques in monkeys

Technique	Mean decoding performance (r)
Intracortical microelectrode arrays (Bansal et al., 2012a: spikes + LFPs)	0.76 (3-D endpoint position, velocity, and grasp aperture)
Electrocorticography (Chao et al., 2010)	0.72 (3-D endpoint position)
Scalp EEG (Bradberry et al., 2010)	0.35 (endpoint y and z velocity)

Decoding performance reports the mean Pearson's correlation coefficient (r) between original and reconstructed kinematic parameters obtained from several studies. EEG, electroencephalography; LFP, local field potential.

(alpha) and beta rhythms. The subjects took several days to learn basic cursor control because of the indirect controlling methods. In contrast, within-session control of cursors and robotic arms was achieved using SUAs (Hochberg et al., 2006; Hochberg et al., 2012). Unassisted 2-D and 3-D control within a few days has recently been achieved with ECoG in a paraplegic subject (Wang et al., 2013). LFPs have been used to control a switch by a paralyzed subject, but higher-dimensional control remains unexplored (Kennedy et al., 2004).

Invasiveness Despite the above-mentioned limitations, scalp EEG has the unique advantage that it requires neither invasive surgery nor the subsequent placement of electrodes that penetrate cortex. ECoG requires a craniotomy, and electrodes are placed epidurally or subdurally. SUA (and LFP) recordings are most informative and provide ease of control but require both a craniotomy and the placement of penetrating electrodes, although anecdotal evidence suggests that depth electrodes in epilepsy patients are tolerated more readily than subdural electrodes. The size of the craniotomy, however, may be reduced to a small burr hole (slightly larger than the 4 × 4 mm 96-microelectrode array, which is roughly the size of one ECoG electrode) that is targeted over the electrode placement location. The relative trade-offs of these approaches in terms of pain and long-term infection rates remain to be quantified.

Signal Stability (Unit Yield) and Tuning Stability A significant issue with SUA based prostheses is the number of neurons from which the electrode array can measure signals. As mentioned earlier, the power of using SUAs lies in the several independent degrees of freedom that many neurons recorded across multiple electrodes encode, compared to a relatively correlated signal measured by the FP channels. Nevertheless, if the recording quality deteriorates over time (such as because of drastic impedance changes), and the number of neurons falls, then the prosthesis designer may consider alternative approaches such as using LFP bands as supplemental signals and/or inserting multiple arrays in one or more cortical areas for redundancy (Bansal et al., 2012a). The BrainGate and Pittsburgh clinical trials have used the Utah array (manufactured by Bionics, Cyberkinetics, and Blackrock Microsystems over the past decade), which is a ~4 × 4 mm microelectrode array with 96 recording channels that floats over the brain and can record from approximately 100 neurons in motor areas. The numbers of recorded units trends upward in the first 100 days (Collinger et al., 2012), and the electrodes can record SUA for over 3–5 years (Simeral et al., 2011; Hochberg et al., 2012) despite possible initial vascular damage, bleeding, and inflammatory response. Spike shape stays stable during a session (~1 hour), but the underlying population changes slightly over time (Suner et al., 2005). In addition, spike-tuning properties can stay stable over at least a two-day period (Chestek et al., 2007). The amplitudes of recorded units may trend downward over time, but this trend is uncorrelated with decoding performance (Chestek et al., 2011). The number of recorded units may eventually decrease over time as the signal degrades over the lifetime of the electrodes (Schwartz et al., 2006). In addition, the Utah array incorporates a fixed-length electrode design that does not allow for moving the electrodes toward neurons with potentially more information (Andersen et al., 2004). Current work has also tried to ascertain the best layer to target to extract the most information and found greater information in superficial layers within 0.5 mm of the cortical surface compared to deeper layers >1.0 mm (Markowitz et al., 2011).

Reach and grasp information may also be obtained from unsorted spiking activity. Unsorted spikes based decoders may confer greater stability and the advantage of simpler (and less energy demanding) computation compared to a prosthesis system that requires online spike sorting (Ventura, 2008; Chestek et al., 2011). Finally, ECoG based approaches have reported both significant signal and tuning stability over days (visual system in human epilepsy patients: Bansal et al., 2012b) and months (motor system in monkeys: Chao et al., 2010). On account of their potentially greater tuning stability, ECoG based prostheses may require less calibration on a daily basis compared to SUA based prostheses.

Owing to the invasiveness of both SUA and ECoG based prostheses, they would need to last several years while recording useful signals, with minimal risk of infection, and minimal technician support (for recalibration of filters etc.) to make them appealing for a greater number of paralyzed patients. The exact cost–benefit calculation may be have to be performed on a case-by-case basis depending on each patient's residual motor abilities.

Decoding Algorithm Design

Successful applications of BMIs have adopted a three-step decoding process (Velliste et al., 2008; Hochberg et al., 2012; Collinger et al., 2012). In step 1, an initial model is trained using movements that are imagined or observed by the subject. In step 2, this initial model is used to guide an effector, but the actual movements are corrected toward a most direct path toward the target. The data during this step are used to refine the initial model. In step 3, the effector is allowed to completely run in brain-control mode with no assistance from the technician or knowledge of target in the algorithm.

A recent study has improved the decoding performance and doubled the speed with which monkeys acquire targets using real-time brain control (Gilja et al., 2012). The novelty of this approach was the use of brain-control data to fit the Kalman filter model, combined with including position and velocity in the same model (the latter has demonstrated improvement in Brain-Gate trials; see Kim et al., 2008). The use of brain-control data in filter training supports an optimal feedback controller view of motor and premotor cortex. Instead of building a static filter using data from previous trials with imagined movements, in this approach the patient's brain is assumed to generate neuronal firing that directs the cursor toward the target at each step along the trajectory, thus incorporating a continuous visual signal about the current cursor position and effectively minimizing an error between cursor and target location (see figure 17.3, plate 17). This process may be qualitatively similar to how a nonparalyzed brain might incorporate visual information and continuously adjust the motor commands that direct muscles toward targets. Once the cursor reaches the target, the neuronal activity is set to correspond to zero velocity,

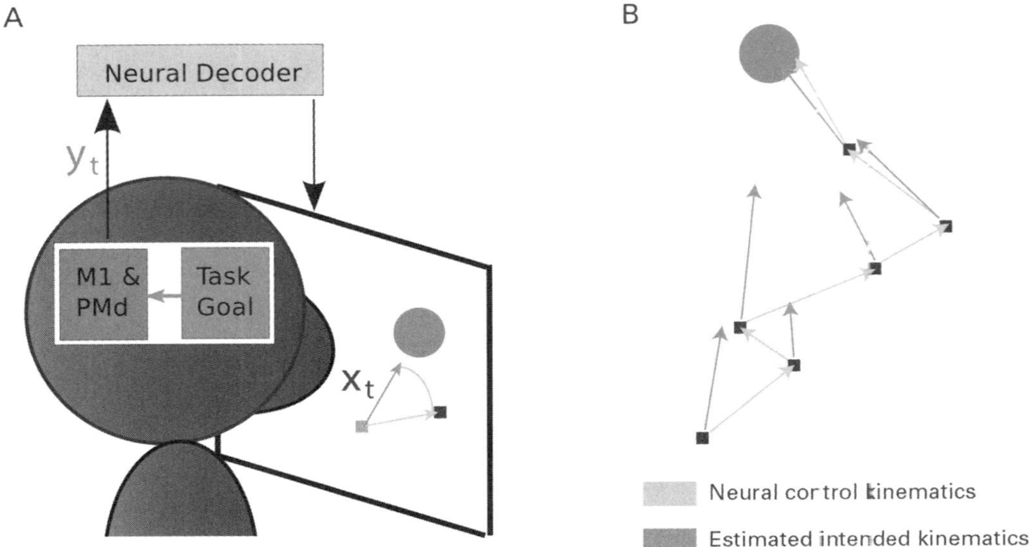

Figure 17.3 (plate 17)
Generating an "intention-based" kinematic training set. (A) The user is engaged in online control with a neural cursor. During each moment in the session, the neural decoder drives the cursor with a velocity shown as a red vector. Gilja et al. assumed that the monkey intended the cursor to generate a velocity towards the target in that moment, so following data collection the researchers rotate this vector to generate an estimate of intended velocity, shown as a blue vector. Note that this blue vector was not rendered on the screen as part of the experiment but is drawn there just to aid in explanation. This new set of kinematics is the training set used to train the control algorithm. M1, primary motor cortex; PMd, dorsal premotor cortex. (B) An example of this transformation applied to successive cursor updates. Figure reproduced and legend modified from Gilja et al. (2012) with permission from *Nature Neuroscience*.

mimicking target hold periods, and providing for a training signal that during brain-control trials achieves controlled movements that are similarly able to acquire and hold targets without overshooting them. The improvement due to the combined use of position and velocity information speaks to the postural or position effects on motor cortex neuronal tuning (Ajemian et al., 2008).

Future Direction

The vision of motor prostheses is one toward an electrode array that encapsulates recording, amplification, analog-to-digital conversion, power supply, and wireless transmission in a compact, implantable unit that runs at body temperature (Donoghue, 2008; Gilja et al., 2010). A separate, cell-phone-sized computer worn by the subject may then process the wirelessly transmitted signals to guide an effector such as a robotic arm or the subject's muscles via a functional electrical stimulation system. Together, these systems will aim to provide an *untethered*, *free-running* prosthesis that will confer to a quadriplegic user paraplegic levels of function and, thus, independence from technicians or nurses. Wireless interfaces will minimize the risk of infection that may be carried into the brain via cables that are usually connected in wired systems to the intracortical electrode array. Algorithms such as those described above are working toward requiring minimal calibration (Gilja et al., 2012). To facilitate a free-running prosthesis, a critical addition to current algorithm designs that focus on the kinematics of movements will be the ability to decode the LFP or spiking signals related to when the subject wants to move (Hwang & Andersen, 2009; Fried et al., 2011). Although single units may provide the most information related to continuous, complex movements (Bansal et al., 2012a), FP based approaches such as EEG may provide less *invasive* prostheses for subjects with less severe impairments. The exact relationship between level of impairment or injury and the invasiveness of prosthesis or signal selection remains to be established. In addition, prostheses that use SUA, LFPs, and FPs from ECoG or EEG, and from multiple cortical regions, may provide more robust, fault-tolerant performance (Bansal et al., 2012a). Another useful addition to current designs could be the extraction of a reward or error signal related to how well the subject's intention is being interpreted by the decoding algorithm. When this error signal exceeds some threshold, the prosthesis would then try to recalibrate itself. This signal may provide a solution to the problem of when to *recalibrate* the decoding algorithm versus allowing the patient to adapt to a stable but imperfect algorithm.

Perhaps, the immediate next set of improvements in prostheses may arrive in the form of the decoding of *force* information from neuronal ensembles, and some form of proprioceptive *feedback* conveyed back to the patient. In parallel, prosthesis designers may explore more *abstract approaches* where the patient's neurons provide higher order goal information such as "Turn on the light" instead of just the intermediate kinematic information of controlling an arm. *Decoding algorithms* that incorporate dimensionality reduction approaches and spiking history of ensemble neurons may improve the encoding and decoding models. More generally, improvements can be expected in the *number of simultaneous neurons* an electrode array records from. Although early studies suggested that thousands of neurons might be necessary to decode movements accurately, recent clinical studies have obtained impressive performance with tens of neurons to a few hundred neurons (Hochberg et al., 2012; Collinger et al., 2012). In laboratory studies with monkeys, decoding performance of complex 3-D reach and grasp movements saturates with the best 30 neurons (Vargas-Irwin et al., 2010; Bansal et al., 2012a). Still, just as reliably recording from ~100 neurons compared to ~10 neurons changed the conclusions about the best signal for decoding reach and grasp from LFP and MUA to SUA (Bansal et al., 2012a), recording from thousands of neurons may bring an improved understanding of collective neuronal dynamics (Truccolo et al., 2010) and new, unexpected insights (Stevenson & Kording, 2011).

Further out in the future, one might expect advances in *synthetic biology* to produce something akin to the science-fiction vision of swallowable pills that grow into electrodes, reach the right location in the brain, and are powered by the brain's glucose. These would alleviate the need for invasive surgical procedures to implant the electrode arrays and the engineering challenges of delivering power to an implanted, wireless, and power-hungry device. Without these challenges, one can also foresee BMIs becoming more commonplace for healthy individuals in an augmenting role, which could stretch beyond motor function to enhanced sensory, mnemonic, and cognitive functions (Donoghue, 2002; Serruya & Kahana, 2008).

Notes

1. It might be argued that for an effective motor prosthesis, achieving the patient's desired end goal might be good enough, and the details of kinematics and dynamics of the movement do not matter. Nevertheless, knowing the full kinematics and dynamics could avoid the complexity of inferring joint kinematics from end goal and also help provide control signals for a functional electrical stimulation system that might one day restore movement by activating a paralyzed patient's muscles (Peckham & Knutson, 2005; Moritz et al., 2008; Chadwick et al., 2011; Ethier et al., 2012).

2. Neurotrophic electrodes are filled with neurites or growth factors that facilitate the growth of neuronal processes into them.

3. One of these patients had suffered a spinal cord injury, and another patient had suffered a pontine stroke.

4. See http://www.cbsnews.com/video/watch/?id=50137987n for a video clip demonstrating the results related to Collinger et al. (2012) and an illustration of this issue (at around 11 min, 30 s).

5. The separate technical issue of signal stability will be discussed in a later section.

References

Andersen, A., Burdick, W., Musallam, S., Pesaran, B., & Cham, J. G. (2004). Cognitive neural prosthetics. *Trends in Cognitive Sciences, 8*, 486–493.

Ajemian, R., Green, A., Bullock, D., Sergio, L., Kalaska, J., & Grossberg, S. (2008). Assessing the function of motor cortex: Single-neuron models of how neural response is modulated by limb biomechanics. *Neuron, 58*, 414–428.

Andersen, R. A., Snyder, L. H., Bradley, D. C., & Xing, J. (1997). Multimodal representation of space in the posterior parietal cortex and its use in planning movements. *Annual Review of Neuroscience, 20*, 303–330.

Bansal, A. K., Jiron, F., Lin, D., & Rizzuto, D. (2005). Neural Prosthetics Technology and Market Assessment Report. https://sites.google.com/site/mindscience/home/publications/E103%20Final%20Submitted%20Report%20-%20Brainiacs.pdf?attredirects=0

Bansal, A. K., Singer, J. M., Anderson, W. S., Golby, A., Madsen, J. R., & Kreiman, G. (2012b). Temporal stability of visually selective responses in intracranial field potentials recorded from human occipital and temporal lobes. *Journal of Neurophysiology, 108*, 3073–3086.

Bansal, A. K., Truccolo, W., Vargas-Irwin, C. E., & Donoghue, J. P. (2011). Decoding 3-D reach and grasp from hybrid signals in motor and premotor cortices: Spikes, multiunit activity and local field potentials. *Journal of Neurophysiology, 105*, 1603–1619.

Bansal, A. K., Vargas-Irwin, C. E., Truccolo, W., & Donoghue, J. P. (2012a). Relationships among low-frequency local field potentials, spiking activity, and 3-D reach and grasp kinematics in primary motor and ventral premotor cortices. *Journal of Neurophysiology, 107*, 1337–1355.

Bauby, J. D. (1998). *The diving bell and the butterfly: A memoir of life in death.* New York: Vintage.

Birbaumer, N. (2006). Breaking the silence: Brain–computer interfaces (BCI) for communication and motor control. *Psychophysiology, 43*, 517–532.

Bradberry, T. J., Gentili, R. J., & Contreras-Vidal, J. L. (2010). Reconstructing three-dimensional hand movements from noninvasive electroencephalographic signals. *Journal of Neuroscience, 30*, 3432–3437.

Carmena, J. M., Lebedev, M. A., Crist, R. E., O'Doherty, J. E., Santucci, D. M., Dimitrov, D. F., et al. (2003). Learning to control a brain–machine interface for reaching and grasping by primates. *PLoS Biology, 1*(E42).

Carpaneto, J., Umiltà, M. A., Fogassi, L., Murata, A., Gallese, V., Micera, S., et al. (2011). Decoding the activity of grasping neurons recorded from the ventral premotor area F5 of the macaque monkey. *Neuroscience 188*, 80–94.

Carpenter, A. F., Georgopoulos, A. P., & Pellizzer, G. (1999). Motor cortical encoding of serial order in a context-recall task. *Science, 283*, 1752–1757.

Chadwick, E. K., Blana, D., Simeral, J. D., Lambrecht, J., Kim, S. P., Cornwell, A. S., et al. (2011). Continuous neuronal ensemble control of simulated arm reaching by a human with tetraplegia. *Journal of Neural Engineering, 8*, 034003.

Chao, Z. C., Nagasaka, Y., & Fujii, N. (2010). Long-term asynchronous decoding of arm motion using electrocorticographic signals in monkeys. *Frontiers in Neuroengineering, 3*, 3.

Chestek, C. A., Batista, A. P., Santhanam, G., Yu, B. M., Afshar, A., Cunningham, J. P., et al. (2007). Single-neuron stability during repeated reaching in macaque premotor cortex. *Journal of Neuroscience, 27*, 10742–10750.

Chestek, C. A., Gilja, V., Nuyujukian, P., Foster, J. D., Fan, J. M., Kaufman, M. T., et al. (2011). Long-term stability of neural prosthetic control signals from silicon cortical arrays in rhesus macaque motor cortex. *Journal of Neural Engineering, 8*, 045005.

Christopher & Dana Reeve Paralysis Foundation. http://www.christopherreeve.org/site/c.mtKZKgMWKwG/b.5184189/k.3587/Paralysis_Facts__Figures.htm.

Churchland, M. M., Cunningham, J. P., Kaufman, M. T., Foster, J. D., Nuyujukian, P., Ryu, S. I., et al. (2012). Neural population dynamics during reaching. *Nature Neuroscience, 15*, 1752–1757.

Collinger, J. L., Wodlinger, B., Downey, J. E., Wang, W., Tyler-Kabara, E. C., Weber, D. J., et al. (2012). High-performance neuroprosthetic control by an individual with tetraplegia. [Epub ahead of print]. *Lancet*. doi:10.1016/S0140-6736(12)61816-9.

d'Avella, A., Saltiel, P., & Bizzi, E. (2003). Combinations of muscle synergies in the construction of a natural motor behavior. *Nature Neuroscience, 6*, 300–308.

Diester, I., Kaufman, M. T., Mogri, M., Pashaie, R., Goo, W., Yizhar, O., et al. (2011). An optogenetic toolbox designed for primates. *Nature Neuroscience, 14*, 387–397.

Donoghue, J. P. (2002). Connecting cortex to machines: Recent advances in brain interfaces. *Nature Neuroscience, 5*, 1085–1088.

Donoghue, J. P. (2008). Bridging the brain to the world: A perspective on neural interface systems. *Neuron, 60*, 511–521.

Dum, R. P., & Strick, P. L. (2005). Frontal lobe inputs to the digit representations of the motor areas on the lateral surface of the hemisphere. *Journal of Neuroscience, 25*, 1375–1386.

Dushanova, J., & Donoghue, J. (2010). Neurons in primary motor cortex engaged during action observation. *European Journal of Neuroscience, 31*, 386–398.

Ethier, C., Oby, E. R., Bauman, M. J., & Miller, L. E. (2012). Restoration of grasp following paralysis through brain-controlled stimulation of muscles. *Nature Neuroscience, 485*, 368–371.

Evarts, E. V. (1968). Relation of pyramidal tract activity to force exerted during voluntary movement. *Journal of Neurophysiology, 31*, 14–27.

Flint, R. D., Lindberg, E. W., Jordan, L. R., Miller, L. E., & Slutzky, M. W. (2012). Accurate decoding of reaching movements from field potentials in the absence of spikes. *Journal of Neural Engineering, 9*, 046006.

Fried, I., Katz, A., McCarthy, G., Sass, K. J., Williamson, P., Spencer, S. S., et al. (1991). Functional organization of human supplementary motor cortex studied by electrical stimulation. *Journal of Neuroscience, 11*, 3656–3666.

Fried, I., Mukamel, R., & Kreiman, G. (2011). Internally generated preactivation of single neurons in human medial frontal cortex predicts volition. *Neuron, 69*, 548–562.

Fu, Q. G., Suarez, J. I., & Ebner, T. J. (1993). Neuronal specification of direction and distance during reaching movements in the superior precentral premotor area and primary motor cortex of monkeys. *Journal of Neurophysiology, 70*, 2097–2116.

Ganguly, K., & Carmena, J. M. (2009). Emergence of a stable cortical map for neuroprosthetic control. *PLoS Biology, 7*, E1000153.

Georgopoulos, A. P., Kalaska, J. F., Caminiti, R., & Massey, J. T. (1982). On the relations between the direction of two-dimensional arm movements and cell discharge in primate motor cortex. *Journal of Neuroscience, 2*, 1527–1537.

Georgopoulos, A. P., Kettner, R. E., & Schwartz, A. B. (1988). Primate motor cortex and free arm movements to visual targets in three-dimensional space: II. Coding of the direction of movement by a neuronal population. *Journal of Neuroscience, 8*, 2928–2937.

Georgopoulos, A. P., Pellizzer, G., Poliakov, A. V., & Schieber, M. H. (1999). Neural coding of finger and wrist movements. *Journal of Computational Neuroscience, 6*, 279–288.

Georgopoulos, A., Schwartz, A., & Kettner, R. (1986). Neuronal population coding of movement direction. *Science, 233*, 1416–1419.

Gilja, V., Chestek, C. A., Diester, I., Henderson, J. M., Deisseroth, K., & Shenoy, K. V. (2011). Challenges and opportunities for next-generation intracortically based neural prostheses. *IEEE Transactions on Bio-Medical Engineering, 58*, 1891–1899.

Gilja, V., Chestek, C. A., Nuyujukian, P., Foster, J., & Shenoy, K. V. (2010). Autonomous head-mounted electrophysiology systems for freely behaving primates. *Current Opinion in Neurobiology, 20*, 676–686.

Gilja, V., Nuyujukian, P., Chestek, C. A., Cunningham, J. P., Byron, M. Y., & Fan, J. M., et al. (2012). A high-performance neural prosthesis enabled by control algorithm design. *Nature Neuroscience*. Epub 2012 Nov 18. doi:10.1038/nn.3265.

Glimcher, P. W. (2004). *Decisions, uncertainty, and the brain: The science of neuroeconomics*. Cambridge, MA: MIT Press.

Goldring, S., & Ratcheson, R. (1972). Human motor cortex: Sensory input data from single neuron recordings. *Science, 175*, 1493–1495.

Grefkes, C., & Fink, G. R. (2005). The functional organization of the intraparietal sulcus in humans and monkeys. *Journal of Anatomy, 207*, 3–17.

Gupta, R., & Ashe, J. (2009). Offline decoding of end-point forces using neural ensembles: Application to a brain–machine interface. *IEEE Transactions on Neural Systems and Rehabilitation Engineering, 17*, 254–262.

Harkema, S., Gerasimenko, Y., Hodes, J., Burdick, J., Angeli, C., Chen, Y., et al. (2011). Effect of epidural stimulation of the lumbosacral spinal cord on voluntary movement, standing, and assisted stepping after motor complete paraplegia: A case study. *Lancet, 377*, 1938–1947.

Hatsopoulos, N. G., & Donoghue, J. P. (2009). The science of neural interface systems. *Annual Review of Neuroscience, 32*, 249–266.

Hatsopoulos, N. G., Xu, Q., & Amit, Y. (2007). Encoding of movement fragments in the motor cortex. *Journal of Neuroscience, 27*, 5105–5114.

He, J., Ma, C., & Herman, R. (2008). Engineering neural interfaces for rehabilitation of lower limb function in spinal cord injured. *Proceedings of the IEEE, 96*, 1152–1166.

Hochberg, L. R., Bacher, D., Jarosiewicz, B., Masse, N. Y., Simeral, J. D., Vogel, J., et al. (2012). Reach and grasp by people with tetraplegia using a neurally controlled robotic arm. *Nature Neuroscience, 485*, 372–375.

Hochberg, L. R., Serruya, M. D., Friehs, G. M., Mukand, J. A., Saleh, M., Caplan, A. H., et al. (2006). Neuronal ensemble control of prosthetic devices by a human with tetraplegia. *Nature Neuroscience, 442*, 164–171.

Humphrey, D. R., Schmidt, E. M., & Thompson, W. D. (1970). Predicting measures of motor performance from multiple cortical spike trains. *Science, 170*, 758.

Hwang, E. J., & Andersen, R. A. (2009). Brain control of movement execution onset using local field potentials in posterior parietal cortex. *Journal of Neuroscience, 29*, 14363–14370.

Jarosiewicz, B., Chase, S. M., Fraser, G. W., Velliste, M., Kass, R. E., & Schwartz, A. B. (2008). Functional network reorganization during learning in a brain–computer interface paradigm. *Proceedings of the National Academy of Sciences of the United States of America, 105*, 19486–19491.

Kakei, S., Hoffman, D. S., & Strick, P. L. (1999). Muscle and movement representations in the primary motor cortex. *Science, 285*, 2136–2139.

Kalaska, J. F., Caminiti, R., & Georgopoulos, A. P. (1983). Cortical mechanisms related to the direction of two-dimensional arm movements: Relations in parietal area 5 and comparison with motor cortex. *Experimental Brain Research, 51*, 247–260.

Kennedy, P., Andreasen, D., Ehirim, P., King, B., Kirby, T., Mao, H., et al. (2004). Using human extra-cortical local field potentials to control a switch. *Journal of Neural Engineering, 1*, 72–77.

Kennedy, P. R., & Bakay, R. A. E. (1998). Restoration of neural output from a paralyzed patient by a direct brain connection. *Neuroreport, 9*, 1707–1711.

Kennedy, P. R., Bakay, R. A., Moore, M. M., Adams, K., & Goldwaithe, J. (2000). Direct control of a computer from the human central nervous system. *IEEE Transactions on Rehabilitation Engineering, 8*, 198–202.

Kim, S. P., Simeral, J. D., Hochberg, L. R., Donoghue, J. P., & Black, M. J. (2008). Neural control of computer cursor velocity by decoding motor cortical spiking activity in humans with tetraplegia. *Journal of Neural Engineering, 5*, 455–476.

Kim, S. P., Simeral, J., Hochberg, L., Donoghue, J., Friehs, G., & Black, M. (2011). Point-and-click cursor control with an intracortical neural interface system in humans with tetraplegia. *IEEE Transactions on Neural Systems and Rehabilitation Engineering*, *19*, 193–203.

Kubánek, J., Miller, K. J., Ojemann, J. G., Wolpaw, J. R., & Schalk, G. (2009). Decoding flexion of individual fingers using electrocorticographic signals in humans. *Journal of Neural Engineering*, *6*, 066001.

Kurata, K., & Tanji, J. (1986). Premotor cortex neurons in macaques: Activity before distal and proximal forelimb movements. *Journal of Neuroscience*, *6*, 403–411.

Leyton, A. S. F., & Sherrington, C. S. (1917). Observations on the excitable cortex of the chimpanzee, orangutan, and gorilla. *Experimental Physiology*, *11*, 135–222.

Markowitz, A., Wong, T., Gray, M., & Pesaran, B. (2011). Optimizing the decoding of movement goals from local field potentials in macaque cortex. *Journal of Neuroscience*, *31*, 18412–18422.

Martin, J. H. (2003). *Neuroanatomy: Text and atlas* (3rd ed.). New York: McGraw-Hill.

Matelli, M., Camarda, R., Glickstein, M., & Rizzolatti, G. (1986). Afferent and efferent projections of the inferior area 6 in the macaque monkey. *Journal of Comparative Neurology*, *251*, 281–298.

Mehring, C., Rickert, J., Vaadia, E., Cardosa de Oliveira, S., Aertsen, A., & Rotter, S. (2003). Inference of hand movements from local field potentials in monkey motor cortex. *Nature Neuroscience*, *6*, 1253–1254.

Messier, J., & Kalaska, J. F. (2000). Covariation of primate dorsal premotor cell activity with direction and amplitude during a memorized-delay reaching task. *Journal of Neurophysiology*, *84*, 152–165.

Milekovic, T., Fischer, J., Pistohl, T., Ruescher, J., Schulze-Bonhage, A., Aertsen, A., Rickert, J., Ball, T., & Mehring, C. (2012). An online brain-machine interface using decoding of movement direction from the human electrocorticogram. *Journal of Neural Engineering,* 9:046003.

Moritz, C. T., Perlmutter, S. I., & Fetz, E. E. (2008). Direct control of paralysed muscles by cortical neurons. *Nature Neuroscience*, *456*, 639–642.

Mukamel, R., & Fried, I. (2012). Human intracranial recordings and cognitive neuroscience. *Annual Review of Psychology*, *63*, 511–537.

Mulliken, G. H., Musallam, S., & Andersen, R. A. (2008). Decoding trajectories from posterior parietal cortex ensembles. *Journal of Neuroscience*, *28*, 12913–12926.

Musallam, S., Corneil, B. D., Greger, B., Scherberger, H., & Andersen, R. A (2004). Cognitive control signals for neural prosthetics. *Science*, *305*, 258–262.

Nawrot, M. P., Boucsein, C., Rodriguez Molina, V., Riehle, A., Aertsen, A., & Rotter, S. (2008). Measurement of variability dynamics in cortical spike trains. *Journal of Neuroscience Methods*, *169*, 374–390.

Oby, E. R., Ethier, C., & Miller, L. E. (2012). Movement representation in primary motor cortex and its contribution to generalizable EMG predictions. *Journal of Neurophysiology* [Epub]. doi:10.1152/jn.00331.2012.

Ojakangas, C. L., Shaikhouni, A., Friehs, G. M., Caplan, A. H., Serruya, M. D., Saleh, M., et al. (2006). Decoding movement intent from human premotor cortex neurons for neural prosthetic applications. *Journal of Clinical Neurophysiology*, *23*, 577–584.

Overduin, S. A., d'Avella, A., Carmena, J. M., & Bizzi, E. (2012). Microstimulation activates a handful of muscle synergies. *Neuron*, *76*, 1071–1077.

Pandya, D. N., & Yeterian, E. H. (1996). Comparison of prefrontal architecture and connections. *Philosophical Transactions of the Royal Society of London. Series B, Biological Sciences*, *351*, 1423–1432.

Paninski, L., Fellows, M. R., Hatsopoulos, N. G., & Donoghue, J. P. (2004). Spatiotemporal tuning of motor cortical neurons for hand position and velocity. *Journal of Neurophysiology*, *91*, 515–532.

Patil, P. G., Carmena, J. M., Nicolelis, M. A., & Turner, D. A. (2004). Ensemble recordings of human subcortical neurons as a source of motor control signals for a brain–machine interface. *Neurosurgery*, *55*, 27–35.

Peckham, P. H., & Knutson, J. S. (2005). Functional electrical stimulation for neuromuscular applications. *Annual Review of Biomedical Engineering*, *7*, 327–360.

Penfield, W., & Boldrey, E. (1937). Somatic motor and sensory representation in the cerebral cortex of man as studied by electrical stimulation. *Brain*, *60*, 389–443.

Pistohl, T., Schulze-Bonhage, A., Aertsen, A., Mehring, C., & Ball, T. (2011). Decoding natural grasp types from human ECoG. *NeuroImage*, *167*, 105–114.

Rizzolatti, G., Camarda, R., Fogassi, L., Gentilucci, M., Luppino, G., & Matelli, M. (1988). Functional organization of inferior area 6 in the macaque monkey. *Experimental Brain Research, 71*, 491–507.

Rokni, U., Richardson, A. G., Bizzi, E., & Seung, H. S. (2007). Motor learning with unstable neural representations. *Neuron, 54*, 653–666.

Saleh, M., Takahashi, K., Amit, Y., & Hatsopoulos, N. G. (2010). Encoding of coordinated grasp trajectories in primary motor cortex. *Journal of Neuroscience, 30*, 17079–17090.

Saleh, M., Takahashi, K., & Hatsopoulos, N. G. (2012). Encoding of coordinated reach and grasp trajectories in primary motor cortex. *Journal of Neuroscience, 32*, 1220–1232.

Sanes, J. N., & Donoghue, J. P. (2000). Plasticity and primary motor cortex. *Annual Review of Neuroscience, 23*, 393–415.

Santhanam, G., Ryu, S. I., Yu, B. M., Afshar, A., & Shenoy, K. V. (2006). A high-performance brain–computer interface. *Nature Neuroscience, 442*, 195–198.

Santhanam, G., Yu, B. M., Gilja, V., Ryu, S. I., Afshar, A., Sahani, M., et al. (2009). Factor-analysis methods for higher-performance neural prostheses. *Journal of Neurophysiology, 102*, 1315–1330.

Schalk, G., Kubánek, J., Miller, K. J., Anderson, N. R., Leuthardt, E. C., Ojemann, J. G., et al. (2007). Decoding two-dimensional movement trajectories using electrocorticographic signals in humans. *Journal of Neural Engineering, 4*, 264–275.

Scheidt, R. A., Conditt, M. A., Secco, E. L., & Mussa-Ivaldi, F. A. (2005). Interaction of visual and proprioceptive feedback during adaptation of human reaching movements. *Journal of Neurophysiology, 93*, 3200–3213.

Schwartz, A. B., Kettner, R. E., & Georgopoulos, A. P. (1988). Primate motor cortex and free arm movements to visual targets in three-dimensional space: I. Relations between single cell discharge and direction of movement. *Journal of Neuroscience, 8*, 2913–2927.

Schwartz, A. B., Cui, X., Weber, D., & Moran, D. (2006). Brain-controlled interfaces: Movement restoration with neural prosthetics. *Neuron, 52*, 205–220.

Scott, S. H. (2004). Optimal feedback control and the neural basis of volitional motor control. *Nature Reviews. Neuroscience, 5*, 532–546.

Serruya, M. D., Hatsopoulos, N. G., Paninski, L., Fellows, M. R., & Donoghue, J. P. (2002). Instant neural control of a movement signal. *Nature Neuroscience, 416*, 141–142.

Serruya, M. D., & Kahana, M. J. (2008). Techniques and devices to restore cognition. *Behavioural Brain Research, 192*, 149–165.

Shanechi, M. M., Hu, R. C., Powers, M., Wornell, G. W., Brown, E. N., & Williams, Z. M. (2012). Neural population partitioning and a concurrent brain–machine interface for sequential motor function. [Epub 2012 Nov 11]. *Nature Neuroscience*. doi:10.1038/nn.3250.

Shenoy, K. V., Kaufman, M. T., Sahani, M., & Churchland, M. M. (2011). A dynamical systems view of motor preparation: Implications for neural prosthetic system design. *Progress in Brain Research, 192*, 33–58.

Simeral, J. D., Kim, S. P., Black, M. J., Donoghue, J. P., & Hochberg, L. R. (2011). Neural control of cursor trajectory and click by a human with tetraplegia 1000 days after implant of an intracortical microelectrode array. *Journal of Neural Engineering, 8*, 025027.

Stark, E., & Abeles, M. (2007). Predicting movement from multiunit activity. *Journal of Neuroscience, 27*, 8387–8394.

Stark, E., Asher, I., & Abeles, M. (2007). Encoding of reach and grasp by single neurons in premotor cortex is independent of recording site. *Journal of Neurophysiology, 97*, 3351–3364.

Stevenson, I. H., & Kording, K. P. (2011). How advances in neural recording affect data analysis. *Nature Neuroscience, 14*, 139–142.

Suminski, A. J., Tkach, D. C., Fagg, A. H., & Hatsopoulos, N. G. (2010). Incorporating feedback from multiple sensory modalities enhances brain–machine interface control. *Journal of Neuroscience, 30*, 16777–16787.

Suner, S., Fellows, M. R., Vargas-Irwin, C., Nakata, G. K., & Donoghue, J. P. (2005). Reliability of signals from a chronically implanted, silicon-based electrode array in non-human primate primary motor cortex. *IEEE Transactions on Neural Systems and Rehabilitation Engineering, 13*, 524–541.

Tanji, J., & Shima, K. (1994). Role for supplementary motor area cells in planning several movements ahead. *Nature Neuroscience, 371*, 413–416.

Taylor, D. M., Tillery, S. I., & Schwartz, A. B. (2002). Direct cortical control of 3D neuroprosthetic devices. *Science*, *296*, 1829–1832.

Thach, D., Reimer, J., & Hatsopoulos, N. G. (2007). Congruent activity during action and action observation in motor cortex. *Journal of Neuroscience*, *27*, 13241–13250.

Townsend, B. R., Subasi, E., & Scherberger, H. (2011). Grasp movement decoding from premotor and parietal cortex. *Journal of Neuroscience*, *31*, 14386–14398.

Truccolo, W., Eden, U. T., Fellows, M. R., Donoghue, J. P., & Brown, E. N. (2005). A point process framework for relating neural spiking activity to spiking history, neural ensemble, and extrinsic covariate effects. *Journal of Neurophysiology*, *93*, 1074–1089.

Truccolo, W., Friehs, G. M., Donoghue, J. P., & Hochberg, L. R. (2008). Primary motor cortex tuning to intended movement kinematics in humans with tetraplegia. *Journal of Neuroscience*, *28*, 1163–1178.

Truccolo, W., Hochberg, L. R., & Donoghue, J. P. (2010). Collective dynamics in human and monkey sensorimotor cortex: Predicting single neuron spikes. *Nature Neuroscience*, *13*, 105–111.

Tsujimoto, S., Genovesio, A., & Wise, S. P. (2011). Frontal pole cortex: Encoding ends at the end of the endbrain. *Trends in Cognitive Sciences*, *15*, 169–176.

Umiltà, M. A., Brochier, T., Spinks, R. L., & Lemon, R. N. (2007). Simultaneous recording of macaque premotor and primary motor cortex neuronal populations reveals different functional contributions to visuomotor grasp. *Journal of Neurophysiology*, *98*, 488–501.

van den Brand, R., Heutschi, J., Barraud, Q., DiGiovanna, J., Bartholdi, K., Huerlimann, M., et al. (2012). Restoring voluntary control of locomotion after paralyzing spinal cord injury. *Science*, *336*, 1182–1185.

Vargas-Irwin, C. E. (2010). Motor cortical control of naturalistic reaching and grasping actions. Brown University Doctoral Thesis.

Vargas-Irwin, C. E., Shakhnarovich, G., Yadollahpour, P., Mislow, J. M., Black, M. J., & Donoghue, J. P. (2010). Decoding complete reach and grasp actions from local primary motor cortex populations. *Journal of Neuroscience*, *30*, 9659–9669.

Velliste, M., Perel, S., Spalding, M. C., Whitford, A. S., & Schwartz, A. B. (2008). Cortical control of a prosthetic arm for self-feeding. *Nature*, *453*, 1098–1101.

Ventura, V. (2008). Spike train decoding without spike sorting. *Neural Computation*, *20*, 923–963.

Vogt, B. A., & Pandya, D. N. (1978). Cortico-cortical connections of somatic sensory cortex (areas 3, 1 and 2) in the rhesus monkey. *Journal of Comparative Neurology*, *177*, 179–191.

Waldert, S., Pistohl, T., Braun, C., Ball, T., Aertsen, A., & Mehring, C. (2009). A review on directional information in neural signals for brain–machine interfaces. *Journal of Physiology, Paris*, *103*, 244–254.

Wang, W., Collinger, J. L., Degenhart, A. D., Tyler-Kabara, E. C., Schwartz, A. B., Moran, D. W., et al. (2013). An electrocorticographic brain interface in an individual with tetraplegia. *PLOS*, *8*, E55344.

Wessberg, J., Stambaugh, C. R., Kralik, J. D., Beck, P. D., Laubach, M., Chapin, J. K., et al. (2000). Real-time prediction of hand trajectory by ensembles of cortical neurons in primates. *Nature Neuroscience*, *408*, 361–365.

Wolpaw, J. R., & McFarland, D. J. (2004). Control of a two-dimensional movement signal by a noninvasive brain–computer interface in humans. *Proceedings of the National Academy of Sciences of the United States of America*, *101*, 17849–17854.

Wu, W., Gao, Y., Bienenstock, E., Donoghue, J. P., & Black, M. J. (2006). Bayesian population decoding of motor cortical activity using a Kalman filter. *Neural Computation*, *18*, 80–118.

Yousry, T. A., Schmid, U. D., Alkadhi, H., Schmidt, D., Peraud, A., Buettner, A., et al. (1997). Localization of the motor hand area to a knob on the precentral gyrus: A new landmark. *Brain*, *120*, 141–157.

Yu, B. M., Cunningham, J. P., Santhanam, G., Ryu, S. I., Shenoy, K. V., & Sahani, M. (2009). Gaussian-process factor analysis for low-dimensional single-trial analysis of neural population activity. *Journal of Neurophysiology*, *102*, 614–635.

Zhuang, J., Truccolo, W., Vargas-Irwin, C., & Donoghue, J. P. (2010). Decoding 3-D reach and grasp kinematics from high-frequency local field potentials in primate primary motor cortex. *IEEE Transactions on Bio-Medical Engineering*, *57*, 1774–1784.

18 Human Single Neuron Recording as an Approach to Understand the Neurophysiology of Seizure Generation

Andreas Schulze-Bonhage and Rüdiger Köhling

Extracellular recordings capable of measuring not only field potentials representing the sum of postsynaptic potentials of large neuronal agglomerations but also action potentials of individual neurons have been used in patients with movement disorders and with pharmacoresistant focal epilepsy undergoing intracranial recordings to define the epileptogenic zone since the 1950s. Whereas presently single neuron recordings in the human are mostly used to study the behavior of individual neurons during various types of physiological behavior (e.g., during alterations in vigilance or during particular cognitive tasks like memory or sensory processing; Andrillon et al., 2011; Quiroga, 2012; Miller et al., 2013; chapters 7–14 in this volume), already the first in vivo recordings during surgical interventions in epilepsy patients were targeted toward understanding the physiology of epileptic discharges and comparing human recordings to electrophysiological findings obtained in experimental models of epilepsy.

In patients with epilepsy, "seizures" with altered perception, cognition, or motor control occur due to abnormally synchronized discharges in focal or extended brain networks. During these seizures, field potential (electroencephalogram; EEG) recordings performed intracranially or from the scalp show high-frequency discharges or rhythmic activity in specific frequency bands with variable spread over the cortex. Between seizures, in the "interictal" period, epileptic "spikes" or "sharp waves" are found, steep field potentials with subsequent slow waves which pop out from the background activity. These interictal discharges occur during wakefulness and sleep, when patients appear to be asymptomatic, and provide a diagnostic hallmark for the disease; in intracranial recordings, such spiking is often frequent in areas close to where seizures are generated (Schulze-Bonhage, 2011). In the following, the term "spike" will be used for these epileptic discharges whereas the behavior of individual neurons will be termed "action potential generation," "unit activity," "firing," or "bursting."

Early approaches to recording human neurons during the brief periods of intraoperative recordings date back to the 1950s and were extended and refined by the inclusion of semi-chronic recordings in the 1960s–1970 (see chapter 1). For such prolonged recordings, two approaches have been used: (1) unit recordings from microwires protruding from the tip of depth electrodes and (2) recordings from intracortically placed microelectrode arrays with high numbers of penetrating electrodes.

Methodological requirements to obtain information about the behavior of individual neurons in vivo are similar to recordings during cognitive studies as long as interictal periods are analyzed. For analyses of ictal neuronal activities, however, prolonged recording times are required, ranging from days to weeks. Progress in storage capacities of present-day recording systems render a continuous recording of single units possible over many days despite the high sampling rates needed to correctly characterize action potentials. Here, an overview is given on approaches to analyzing neuronal behavior during interictal and ictal epileptic discharges in the human in vivo as reported in the literature published so far.

Insights from Animal Studies

The problem of relating field potential or multiunit activity to the activity of single neurons has been addressed in a number of animal studies both in vivo and in vitro. Analyses were performed on the role of interictal spiking in the process of epileptogenesis and ictiogenesis, on cellular mechanisms of spike generation, and on changes occurring during the transition from interictal state to seizure and with spatial propagation of epileptic activity. Epileptic spikes assessed at the levels of multiunits and field potentials (EEG) are synaptic and/or population action potential events with a temporal variation, lasting from ~100 to ~3000 ms (cf. Avoli et al., 1996) depending on the definition used by the respective authors, in contrast to action potentials at the cellular and single unit extracellular levels with a duration of 1–3 ms, which are sometimes also called "spikes." During field-potential spikes, so-called paroxysmal depolarization shifts (PDSs) were found to occur at the level of individual neurons, which essentially consist of bursts of action potentials riding on a prolonged 10- to 3000-ms depolarizing envelope (Goldensohn & Purpura, 1963; Matsumoto & Marsan, 1964a, 1964b), albeit not in all cells recorded from in a focus. There has been an ongoing debate on the synaptic versus intrinsic nature of these PDSs; whereas a single neuron can generate PDSs via endogenous bursts of action potentials even without synaptic input (Segal, 1991, 1994), to generate an EEG spike, a minimal volume of synaptically/ephaptically connected neurons (220-μm radius; equivalent to ~7500 neurons) is required (Schwindt et al., 1997).

Studies in hippocampal–entorhinal cortex tissue slices have revealed that, beyond a large temporal variation, interictal field potential spikes appear to arise as different entities, depending on both location and timing as related to seizure-activity onset (Avoli et al., 1996, 2012): One type consisted of brief (80–150 ms) and relatively frequent (~1/s) interictal spikes originating in hippocampal subfields, and another of long-lasting (1–3 s), infrequent (~0.02/s) spikes appearing in all areas recorded from. In addition, a third type is observed in studies on more complex preparations (whole isolated guinea pig brain in vitro), consisting of fast oscillatory spikes at 20–60/s, with short durations, which strictly speaking are not interictal but early ictal discharges (de Curtis & Gnatkovsky, 2009). Spike morphology and types vary with their reported function: While in acute studies based on penicillin application in vivo and in cell cultures interictal spike frequency buildup correlates with seizure initiation, suggesting pro-epileptic functions (Dichter et al., 1972; Dyhrfjeld-Johnsen et al., 2010; for reviews, also see de Curtis et al., 2012), investigations in chronic models show a decrease or no change in interictal spike occurrence before seizures (reviewed in de Curtis & Avanzini, 2001). The latter suggests that interictal spikes are unrelated to seizures or actually interfere with the generation of ictal epileptic activity (Avoli, 2001; Jensen & Yaari, 1988; Librizzi & de Curtis, 2003).

Animal studies addressing the activities of single units during interictal and pre-ictal EEG spikes and during interictal–ictal transitions allow for intracellular recordings and for the assessment of physiological neuronal properties like input resistance, membrane potential, and subthreshold behavior. In summary, in epilepsy models involving the entorhinal cortex, hippocampal pyramidal cell activity during fast, interictal spikes appears to be based on glutamatergic input, leading to classical PDS (Avoli et al., 2012; de Curtis et al., 2012; Lopantsev & Avoli, 1998). These seem to curtail ictogenesis (Barbarosie & Avoli, 1997; Librizzi and de Curtis, 2003). Other spikes can be termed "pre-ictal," promoting ictogenesis. One type of "pre-ictal" spike involves slow interictal, GABAergically mediated depolarizations which, when becoming suprathreshold aided by extracellular K^+ rises, initiate seizure activity (Lopantsev & Avoli, 1998). A second type, characterized in the isolated guinea-pig brain, consists of high-frequency extracellular discharges reflecting prolonged bursting of interneurons, while principal cells remain silent. In a third type of "pre-ictal" spike, in neocortical seizures in feline models in vivo, principal cell discharges are driven by GABAergic input turned excitable (Timofeev et al., 2002). In all these cases, interneuronal activity thus plays a major role, either by generating depolarizations and eventually discharges in principal neurons or directly underlying the spikes.

In other models (isolated hippocampus or neocortical slices), the transition to the ictal state is supposed to be associated with a loss of inhibitory restraint and emergence of predominantly glutamatergic "pre-ictal" spikes, essentially indiscernible from the previous GABAergic ones (Trevelyan & Schevon, 2012), a finding which is corroborated also in human neocortical tissue (Schevon et al., 2012; Huberfeld et al., 2011). A closer look at the coherence of these spikes, but also of correlation of intracellular firing and extracellular spikes, reveals that the essence of "pre-ictalness" is the sudden increase in phase coherence among spikes, as well as intracellular–extracellular cross-correlation with the transition to seizure both in chronic or acute animal models (Cymerblit-Sabba & Schiller, 2010; Jiruska et al., 2013) and human tissue (Schevon et al., 2012; Huberfeld et al., 2011).

Animal studies thus raise some caveats relevant for human in vivo studies: Extracelullar spikes are not per se a marker of epileptogenicity but can play dual roles: controlling or initiating a seizure. Even in "pre-ictal" spikes, accompanying intracellular activity is not uniform but can in principle originate from principal cells firing due to (1) glutamatergic drive, (2) GABAergic depolarizing and thus paradoxically excitatory drive or (3) interneuronal firing. Extracellular unit recordings during epileptic spiking are not able to detect subthreshold changes; when aiming at a differentiation of these different underlying types of pathophysiology, they have to be complemented by an analysis of the cell types they originate from.

Intraoperative Recordings of Human Neuronal Units

In 1955, Ward and Thomas for the first time recorded neocortical human neuronal units in vivo in Seattle during an operative resection of a temporo–occipital epileptic focus performed under Pentothal anesthesia and muscle relaxation (Ward & Thomas, 1955). An area of interest was defined by the occurrence of afterdischarges following epicortical electrical stimulation. Recordings were performed with glass pipettes held by a hydraulic micromanipulator attached to the patient´s skull which were inserted in an oblique angle to minimize relative movements of electrode tip and cortical tissue. In this first study, neuronal action potentials could be recorded extracellularly, at this time not showing particular differences between presumed epileptogenic region and adjacent cortex.

This group extended their investigations to intraoperative analyses of epileptic activity in patients undergoing surgery during local anesthesia only. Under these conditions, autonomous high-frequency burst discharges of neurons were found in the interictal period (Ward et al., 1956; Ward, 1961; Rayport & Waller, 1967). An enduring dendritic depolarization was discussed as a possible cause underlying the bursting behavior. On rare occasions, it was also possible to record propagated ictal activity with "gross hyperactivity" of the recorded neurons

In a series of recordings obtained from 10 patients, tungsten microelectrodes were also introduced for intraoperative recordings. This allowed, for the first time, recording of multiple units with a single electrode contact, differentiated based on visual criteria like action potential amplitude and absence or presence of action potentials during the refractory phase of a bursting neuron. Evidence for nonsynchronous bursting of different units was observed, which suggested the involvement of mechanisms of burst generation independent from recruitment by massive synchronization (Calvin et al., 1973). Furthermore, bursts showed stereotyped patterns of decrease of the action potential amplitude similar to paroxysmal depolarizations in experimental epilepsy models. Sometimes, frequency buildup with decreasing intervals between subsequent action potentials was noted, suggestive of an increasing depolarization of the unit recorded from (Matsumoto, 1964).

In one of the 10 patients reported, a seizure arose during surgery; pre-ictal tonic firing of a large portion of units sampled was recorded 3.5 minutes prior to the onset of the seizure; recordings had to be terminated, however, with the first appearance of clinical seizure correlates.

Wyler et al. (1982) investigated relationships between recorded activities of individual neurons and epileptic interictal spikes as typically recorded using field potential recordings from the cortical surface of the temporal lobe. In patients undergoing focus resection under local anesthesia, cortical sites with maximal spontaneous epileptiform spiking were identified, followed by 1–2 hours of combined recordings with silver ball electrodes placed epicortically and penetrating microelectrodes at the same sites, provided that they were to be resected. For 90 units recorded from 17 patients, epicortically recorded epileptic spikes and action potentials of individual recorded units were aligned.

Results indicated that there is a remarkable variability of neuronal activity patterns even at times when highly synchronized neuronal discharges had been hypothesized based on field potential recordings. Of 90 units, 46 units did not show any correlation to epileptic surface spikes, even though 9 of these were considered epileptic due to spontaneous high-frequency burst firing. Only 40 of the 90 units were found to discharge during a spike whereas in 4 units action potential generation was inhibited during spikes.

In three cases pre-ictal and early ictal phase could be recorded during spontaneous intraoperative ictal events. Some units changed their firing behavior from individual action potentials to bursts; with onset of the behavioral seizure, multiple units recorded from the same microelectrode fired in closer synchrony, and units fired in closer temporal relation to locally recorded field potentials and to electrocorticography (ECoG) spikes.

The authors pointed out that the absence of a predictable relationship between individual units and epicortically recorded epileptic spikes differs from acute models of focal epilepsy, as in spikes arising from topical application of penicillin, and more resembles chronic in vivo models of an epileptic focus, for example, following application of subpial alumina gel in monkeys. A more distributed epileptogenesis in human epileptogenic cortex compared to a more focal and concentrated epileptogenic neuronal aggregate in experimental models was considered a possible reason for these differences (Wyler et al., 1982).

Although the findings reported above are clearly of considerable interest, all intraoperative investigations are severely limited in the time domain and in the number of microelectrodes applicable, face principal limitations in recording ictal events, and are confronted with considerable technical problems including a noisy electrical environment. This led to the development of techniques applied in the setting of invasive recordings for long-term monitoring of seizures in the presurgical evaluation of epilepsy patients.

Prolonged Recordings of Human Neuronal Units Using Depth Electrodes with Protruding Microwires

For chronic recordings of single units, both movable stiff wires and flexible wires were introduced in the late 1960s. UCLA researchers greatly advanced intraoperative recordings using semi-chronic recordings obtained during presurgical evaluation of epilepsy patients with multi-contact depth electrodes. For this purpose, fine-wire bundles which had been established for animal experiments (Harper & McGinty, 1973) were modified for human applications. Conventional blunt macroelectrodes which had been in use for recordings of the mesial temporal lobe were modified to introduce additional microwires (Crandall et al., 1963; Verzeano et al., 1971; Babb & Crandall, 1976). This new electrode consisted of a hollow stainless-steel cannula serving as a conduit for the introduction of a bundle of seven fine wires in its core which protrude for 5 mm beyond the end of the cannula. Microwires had a diameter of 30–62.5 μm, were made from tungsten or platinum alloy, and were insulated except for their tip, which was cut with sharp scissors. They were inserted after the macroelectrode had been fixed in the skull and were sealed by a cap and acrylic cement. X-ray controls of the positioning of microwires suggested that they can remain in a bundle; bending of the ends resulted in a spray-like divergence and increased the number and duration of successful unit recordings.

With this technique, stable extracellular recordings from the same neuron were reported for several days despite seizure occurrence (Babb et al., 1973). With microwire impedances of 100–300 kΩ, action potentials of negative polarity and with amplitudes of up 300 µV could be recorded. A decrease in action potential amplitude over the course of days was, however, noted, which was interpreted as possibly related to a tissue reaction. This effect was lesser with microwires consisting of platinum alloy compared to iron alloy. Both losses and new appearance of neurons over time were noted as probable consequences of minor wire movements.

The first report of Verzeano et al. (1971) on such extraoperative recordings using stereotactically implanted electrodes was based on the introduction of microwires into one amygdalar electrode used for long-term monitoring. Recordings showed variable relations between firing patterns of individual neurons and epileptic field potentials, often consisting in clustering of action potential generation. Similar to reports of Ward et al. (1956) and Ward (1961), changes in the neuronal discharges of units recorded from the amygdala were seen minutes prior to the onset of a clinical seizure, again with increased clustering of neuronal discharges comparable to pharmacologically induced seizures in in vivo animal models. In the phase preceding clinical seizure manifestations, there were also changes in the distribution of intervals between action potentials; individual neurons showed different behaviors with increases of firing rates in some neurons and decreases in others. For variable involvement of individual neurons in seizure onset see figure 18.1.

Using the simultaneous implantation of multiple depth electrodes with microwires, Babb's group identified region-dependent different neuronal discharge behaviors for amygdala, hippocampus, and adjacent neocortical regions. The majority of neurons recorded from had low discharge rates (6–15.8/minute). In contrast, neurons in the hippocampal gyrus showed higher median firing rates (136.2/minute) due to frequent bursting with interburst intervals of 16 ms. Characteristic decreasing action potential amplitudes during bursts were found in these neurons, suggesting that bursts were generated by pyramidal neurons. Furthermore, in patients with bilateral hippocampal implantation, the likelihood of recording neuronal units was lower in the epileptic hippocampus, possibly related to the loss of pyramidal cells on the epileptic side, and the degree of bursting behavior correlated with the degree of histologically ascertained hippocampal sclerosis. Interestingly, bursting behavior was *not* indicative of the predominant ictal onset side in those patients in whom bihippocampal seizure generation occurred (Babb & Crandall, 1976), a finding confirmed in a later analysis by Colder et al. (1996).

Staba et al. (2002) investigated interictal firing patterns during overnight recordings in 17 patients undergoing temporal microwire recordings. The study confirmed higher mean single neuron firing rates and burst rates within epileptic areas; differences were more marked during rapid eye movement (REM) and non-REM (NREM) sleep than during wakefulness. Burst firing

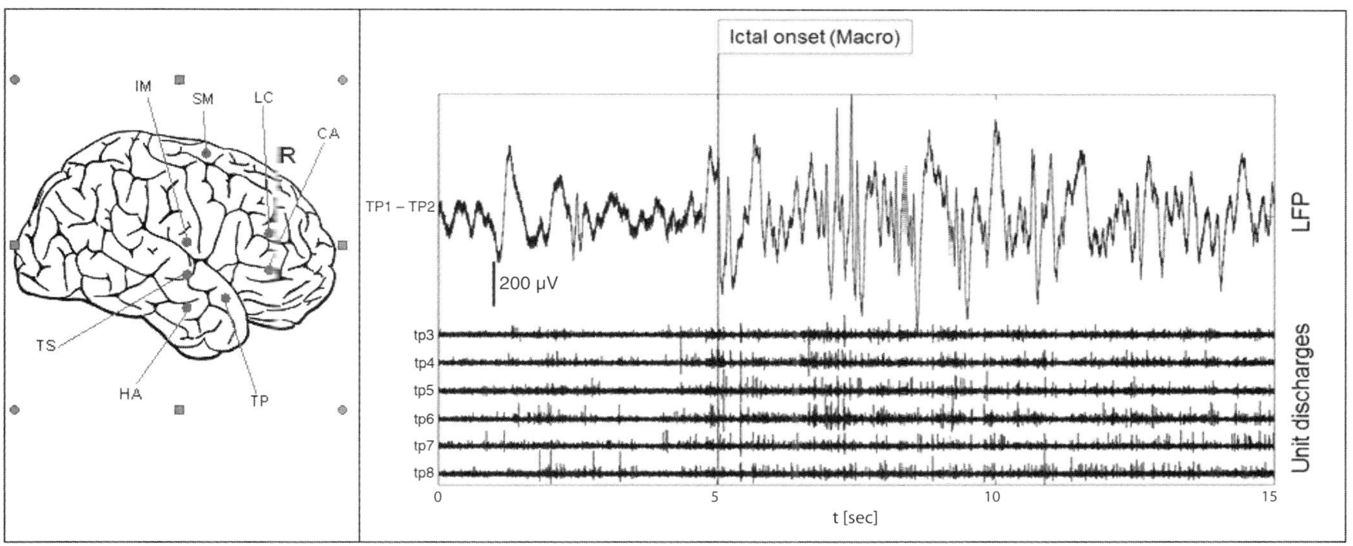

Figure 18.1
Comparison of local field potentials (LFPs) recorded from the macroelectrode at the electrode tip of a right temporopolar depth electrode (TP; see scheme on the left for localization of this electrode) and of neuronal activity recorded from six microwires protruding from the tip of this electrode during the first 10 s of the electrographic onset of a partial seizure. Note the variability of degree of increase in spiking rate and of spiking patterns within the small area recorded (note scale bar for amplitudes of LFP and unit recordings). IM, SM, LC, CA, TS, HA, indicate different recording locations.

was most frequent during NREM sleep. The inclusion of sleep recordings made it possible to show significant differences in synchronicity of discharges and coincident firing between epileptic and nonepileptic hippocampi which were not evident from analyses of the awake state alone; bursting behavior was indicative of epileptogenicity only in the hippocampus, not when recordings from the subiculum or entorhinal cortex were analyzed.

A recent study of Valdez et al. (2013) included high unit numbers from a series of 17 patients undergoing interictal microwire recordings during cognitive studies. Here, a higher yield of recorded neurons and a higher burst frequency (assessed as burst interval ratio, the fraction of interburst intervals <10 ms divided by bursts with longer interburst intervals) were statistically indicative of the seizure onset zone as assessed by intracranial EEG recordings. The authors do not provide information on how the exact comparisons of unit yields between areas were performed; if they were based on comparisons within a hippocampus, the former finding could correspond to experimental evidence that seizure generation does not start in the most heavily damaged hippocampal regions but rather in its surround (e.g., Häussler et al., 2012). More frequent interictal bursting in the seizure onset zone again is in concordance with the above-mentioned findings of the Babb and Staba groups.

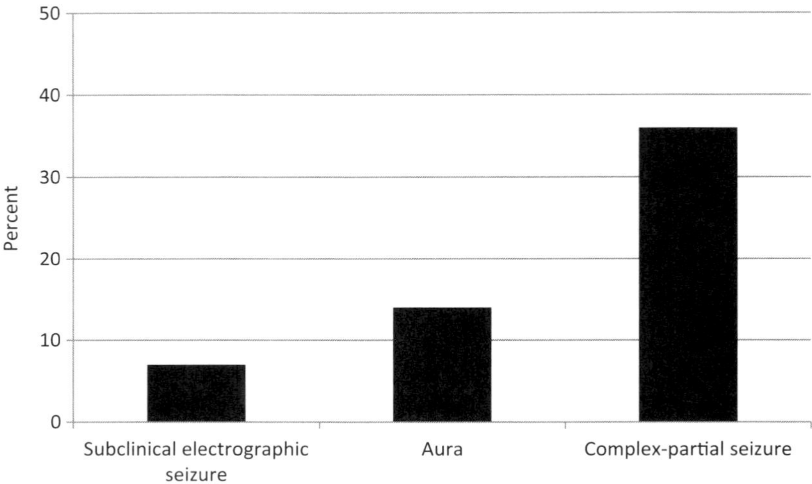

Figure 18.2
Ictal human in vivo recording of neuronal units and multiunits using multiple mesiotemporal depth electrodes with
protruding microwire bundles. The percentage of units increasing the fire rate during an ictal event (y-axis) correlates
with the severity of the seizure: purely electrographic ictal patterns show lesser involvement than seizures with subjective
symptoms only; the highest percentage of neurons is involved in seizures with objective signs (loss of consciousness
and motor phenomena). Data from Babb et al. (1987).

These investigations of spontaneous behavior of mesiotemporal neuronal units constituted the
background for studies on the ictal behavior of neurons (Babb & Crandall, 1976; Babb et al.,
1987). Based on multisite recordings with up to eight depth electrodes containing multiple
microwires, involvement of neuronal units located in the ictal onset zone and in areas of spread
was analyzed during the ictal period and related to the semiological seizure type. As a result,
only a minority of neuronal units recorded from were activated during a seizure, even if neurons
were located in the seizure focus. The percentage of single and multiunits showing an increase
in firing rate was 7% (6 of 90) during subclinical electrographic seizures, 14% (4 of 29) during
auras, and 36% (12 of 33) in seizures with loss of consciousness (see figure 18.2).

Furthermore, increases in firing rates of individual neurons did not reflect the much more
sustained field potential alterations but were often limited to a few seconds. Based on these
findings, the authors comment "the population of neurons needed to fire for seizure propagation
is surprisingly small and the rate and duration of single cell firing are not great ... brief synchronous firing of neurons in the epileptic focus is sufficient to sequentially activate neurons
one or several synapses away" (Babb et al., 1987).

In subclinical and psychomotor seizures, continuous recordings also allowed the assessment
of the postictal phase, with suppression of unit firing lasting only briefly and a return to pre-ictal
firing patterns within one minute (Babb & Crandall, 1976).

Recording of Epileptic Activity of Individual Human Neurons Using Epicortically Placed Microelectrode Arrays with Penetrating Electrodes

Microelectrode arrays with multiple electrodes penetrating the cortical layers have become an important tool in the field of brain–computer interfacing, aiming at a readout of information processing in the motor cortex to steer devices, for example, in patients with high-grade paresis (Hochberg et al., 2006). Whereas these applications use devices like the Utah array with the intention of offering new treatments, microelectrode arrays have also been implanted for scientific reasons to elucidate neurophysiological aspects of seizure generation in the human in addition to subdural recordings performed for the identification of the epileptogenic zone.

In a study aiming at an elucidation of layer-specific aspects of interictal spike generation, Ulbert et al. (2004) used specially designed penetrating microelectrodes with 24 recording sites spaced at distances of 150 or 175 µm which were placed below a subdural grid. Using simultaneous recordings of local field potentials for current source density analyses and unit recordings, this group found evidence for "de novo" spike generation in layer V only in a microelectrode positioned in a region which was considered part of the seizure onset zone. Regions outside the seizure onset zone showed only neuronal discharges which had some evidence of being propagated. These spikes could originate either in layer IV or in supragranular layers, corresponding to transsynaptic activation via thalamic or intracortical inputs. These results only partially confirmed prior investigations from neocortical slices from epilepsy patients by Köhling et al. (1999), who had found generators of epileptic spikes both in layer V and in superficial cortical layers.

House et al. (2006) reported histological data following the pneumatic insertion of Utah microelectrode arrays with 10×10 contacts with distances of 400 µm and a length of 1.5 mm; in all 6 patients minor bleeding and some shearing considered related to cardiac and respiratory pulsations was found in histological stains following surgical resection. Although the functional relevance of these findings is unclear, these histological alterations may limit the application of penetrating microelectrode arrays for purely scientific purposes.

In a study by Waziri et al. (2009), electrophysiological data from 7 epilepsy patients implanted with microelectrode arrays for the period of presurgical evaluation in addition to standard subdural grid electrodes were reported. The microelectrode arrays used consist of 96 platinum microelectrodes arranged in a 10×10 order with a size of 4 mm^2 and a recording tip 1.5 mm below the gyral crown level. Placement was performed in noneloquent and accessible cortex which was considered likely to be subsequently resected. It was reported that the quality of recording improved during the first 48 hours following implantation but could worsen again prior to explantation in individual patients.

From these arrays, "microepileptiform discharges" and "microseizures" with very limited spatial extension (involving isolated areas of the array) were recorded which remained undetected by adjacent subdural standard grids. The authors came to the conclusion that epileptogenic areas restricted to isolated submillimeter cortical domains exist. Recordings remained stable for up to 1 month. Despite these interesting findings, the authors conclude that "the potential contribution of the NeuroPort MEA to current protocols for surgical epilepsy evaluation remains unclear."

In a study by Truccolo et al. (2011) performed in Boston, similar microarrays of a size of 4×4 mm with 96 recording channels were implanted in patients undergoing subdural grid recordings to determine seizure onset. Also these microarrays were implanted additionally in close proximity to the subdural grids for purely scientific reasons, again preferring areas believed to become part of the resection zone based on prior surface EEG recordings. Individual penetrating contacts had a length of 1 mm, suggesting the majority of neurons recorded from were located in layers 3 and 4 of the neocortex. Placement outside the grid-covered region implied, however, that the expected seizure onset zone could not be implanted but that only spread of ictal epileptic activity to the cortex underlying the electrode array could be analyzed.

Visual inspection showed differences in the behavior of discharging neurons in the small areas covered by the array, for example, showing increases or decreases of discharge rates in individual neurons. Heterogeneity of discharge rates obtained from individual recording channels was in contrast to an overall trend to have an increase in generated action potentials over the whole array and to have an almost complete suppression of action potential generation in the immediate postictal phase.

Quantitative analyses were included using the Fano factor (variance divided by the mean discharge rate) as a measure to characterize the heterogeneity of discharge patterns across channels. Interestingly, there was an increase in the Fano factor both early during an ictal event and in its course. This strongly argues against the notion that during a seizure a steadily increasing proportion of neurons in the area of seizure spread is recruited and discharges synchronously with field potential alterations.

In contrast, the recordings—in concordance with microwire recordings referred to above—suggest that there are neuronal subpopulations which behave in different, sometimes opposite ways. So far, there has been no analysis of which factors contribute to this behavior: Spatial factors, connectivity patterns, or cell types recorded from might play a role. It is speculated that multicluster synchronization may be critical for the transformation of the locally different discharge patterns into a manifest seizure.

In a further step, spike sorting was employed to study the behavior of individual neuronal units. Similarly, variable time courses of changes in the discharge rates were found, often with marked changes in the ictal discharge rates from behavior which would be predicted based on prior interictal discharge rates.

In two cases in which two consecutive seizures were recorded in the same patient within a period of 1 hour, activity patterns recorded at the same electrodes appeared similar both visually and when analyzed regarding the variance of discharges. The Pearson correlation between the behaviors of individual neurons in consecutive seizures was 0.72 and 0.82, respectively.

Analyses were also performed on unit behavior during the pre- and postictal phase. Pre-ictally, changes in the action potential generation behavior of individual neurons were found in 1–29%.

Changes could consist in either an increase or a decrease of firing as compared to the interictal period, which again appeared to be consistent when two consecutive seizures were analyzed.

In the postictal phase, uniform suppression of action potential generation in the vast majority of units recorded was found. There was no tendency for reduced action potential amplitudes preceding this suppression at the end of the ictal period that might suggest a depolarization block; rather, inhibitory mechanisms were suggested as mechanisms underlying the almost complete cessation of pyramidal cell action potential generation.

A similar study was performed at Columbia University in New York; here, however, microarrays were implanted directly in the seizure onset zone (Schevon et al., 2012). The results confirmed stereotyped firing patterns of neuronal firing during ictal events and an increased variance in the firing patterns of neurons when seizures started. Stereotyped patterns of ictal neuronal involvement were depicted as "firing trajectories" which remained stable during up to five recorded sequential focal seizures recorded in the same patient. In contrast to the report of Truccolo, synchronization greatly increased during the late course of the seizure, and the Fano factor dropped to values below the interictal period. In addition, this study showed that even in the area considered to be the seizure onset zone as defined by conventional ECoG recording criteria, there were small areas which were under heavy bombardment by excitatory input but were not recruited into a firing pattern in synchrony with the field potentials, termed "penumbra" in this article. The high variability in discharge patterns of neurons thus remains a constant finding not only for secondarily involved areas but also in seizure onset zones as assessed by either microwire recordings or electrode microarrays, and in neocortical as well as archicortical areas.

Discussion

In a variety of technical approaches, using intraoperative recordings, recordings using microwires, and intracortical electrode microarrays, the behavior of individual neurons has been investigated in relation to interictal discharges and during ictal activity. Results from several groups using different methodological approaches have consistently shown complex patterns of neuronal participation in epileptic activity, in particular of involvement in interictal spiking and recruitment during ictal discharges, showing how limited the degree of synchronization is during seizures when analyzed at the unit level. Despite a number of methodological constraints limiting the interpretation of results, there is plenty of room to address interesting new questions using the recording approaches presently available.

Methodological Comparison and Limitations

Intraoperative recordings as obtained initially during the 1950s and 1960s served as pioneering work mostly confined to the proof of the principle that it is possible to record individual human neurons in vivo and to show different discharge patterns during normal physiological and epileptic interictal activity. Basically, recordings were constrained to a few neurons recorded for some minutes. Recordings were first performed under general anesthesia; when only local anesthesia was applied, recordings were endangered by ictal events. Nevertheless, these recordings were the basis for later progress applying microwires integrated into depth electrodes used for semi-chronic recordings. Similarly, implantation techniques for epicortical microarrays with penetrating electrodes were first studied intraoperatively and later used for longer recording periods.

The use of depth electrodes with protruding microelectrodes offers the option to simultaneously record from multiple brain areas, including areas suspected to be close to the seizure origin and distant areas involved or not involved in seizure spread, at times even in the opposite hemisphere. Thus, depending on the number of depth electrodes used, heterogeneous areas can be recorded from. At each individual site, the number of units recorded from is, however, low, and it may be considered impossible to obtain a representative sample of neurons, particularly when the results of recent microarray recordings are taken into account which show a wide diversity of neuronal behaviors up to the level of cortical columns.

Microelectrode arrays, on the other hand, offer the opportunity to record high numbers of units from an area of interest. This allows a comparison of firing patterns of multiple neurons in a neighborhood at the time of particular field potential changes. In the future, spike sorting applied to these recordings may allow assessment of different types of neurons simultaneously and characterization of their interactions with each other. On the other hand, penetrating microelectrode arrays pose more damage to the brain, which can limit their application, cover only a tiny area of the brain surface, and do not allow simultaneous recording from areas involved in early and late epileptic activity as usually defined using ECoG recordings. The limited spatial coverage may limit evaluations to across-subjects comparisons of various brain areas.

Recording Stability

Whereas intraoperative recordings initially were limited to a few minutes, subchronic recordings have increased this period considerably, even if not to the degree of field potential recordings. As stated by Truccolo et al. (2011), "the consistent sorting of single units over time periods longer than a few hours proved challenging" as "waveforms of extracellularly recorded action potentials could change and units appear and disappear from recordings, owing perhaps to micromotion, changes in brain states and other factors." When different seizures are to be analyzed, units were accordingly characterized for the individual periods (in Truccolo's work in a 2-hour time bin centered on the ictal onset time). On the other hand, the recent analysis of Schevon et al. (2012) gives evidence of similar neuronal recruitment patterns during the occurrence of up to five seizures, suggesting a degree of recording stability which is sufficient for analyses scheduled for the usual period of intracranial presurgical evaluation in focal epilepsy. Our own experiences with microwire recordings (Hefft et al., 2013) suggest that—unlike local field potential recordings—the stability of unit recordings is sensitive to movements as they may occur during seizures with major motor components, thus limiting repeated ictal analyses to less severe seizure types.

Have Human in vivo Unit Recordings Produced Any Insight into Seizure Generation?

In 1986, Wyler and Ward stated that "results from human recordings are confusing and have not yet yielded a great deal of insight toward understanding the basic mechanisms of the epilepsies." They state that drug effects, a lack of sufficient normative control data, and insecurity as to whether the actual focus or distant areas are recorded from may have contributed to limited scientific results, which was considered in part due to limited imaging quality available at that time.

On the other hand, studies of the behavior of neuronal units during interictal epileptic discharges and during seizures in human epilepsy are, in principle, of high relevance for comparisons with in vitro studies and in vivo experimental animal models on which our present-day understanding of epilepsy is based. This can serve to identify similarities which may make a given experimental epilepsy model suitable to give valid answers to questions of epileptogenesis, ictogenesis, and treatment response and to find out critical differences which limit the applicability of certain models and analyses.

A number of relevant insights have been obtained from human in vivo recordings, including the following:

• Evidence for bursting behavior of human neurons found particularly in the vicinity of the epileptic focus (Calvin et al., 1973) and showing evidence for synchronized occurrence in the focal brain area (Ishijima et al., 1975), validating the experimental evidence of PDS as a hallmark of epileptic activity at the level of individual neurons (Matsumoto, 1964) also in humans.
• Firing characteristics of neurons allowed conclusions to be drawn on the mechanisms involved in postictal depression of activity (Truccolo et al., 2011).
• Evidence for highly localized pre-ictal changes in discharge patterns in individual neurons (Verzeano et al., 1971) supports the concept of focal alterations in pre-ictal dynamics preceding ictal patterns as evidenced by field potential recordings.
• Interictal bursting was found to correlate with the degree of neuropathology but not to indicate the side of seizure onset in patients with bilateral temporomesial recordings (Babb & Crandall, 1976; Colder et al., 1996). Thus, interictal behavior of individual neurons, even if typical for an epileptogenic area, does not predict seizure occurrence, in striking similarity to a lack of temporal correlation between interictal spiking and temporal seizure generation found in intracranial field potential recordings (Gotman & Marciani, 1985) and with the limited value of interictal spiking for delineation of the epileptogenic zone in the spatial domain (Alarcon et al., 1997). This finding points to an important role of network properties critical for the synchronization process leading to a seizure.

- Laminar analyses of spontaneous spike generation showed differences between deep de novo spike generation (in layer V) and granular and supragranular generation of spikes presumed to be propagated via thalamic and corticocortical inputs (Ulbert et al. 2004). Due to the preserved connectivity, in vivo studies add significant information relevant to network effects on spike generation to in vitro studies using human tissue slices (Köhling et al., 1999).
- Evidence for a high diversity of neuronal discharge patterns even during ictal epileptic discharges was consistently found in studies applying different recording techniques. The absence of high degrees of synchronization of neurons during seizure onset contrasts with the pathophysiology of some acute experimental seizure models and questions simplistic views of epileptic activity as characterized by abnormal synchronization only. Human in vivo studies showing a mosaic of microdomains which may or may not participate in epileptic synchronization (Schevon et al., 2012; see also Schevon et al., 2008) thus support the hypothesis of networks of pathologically interconnected neuron clusters put forward by Bragin et al. (2000).

Future Prospects

Alterations in the behavior of neuronal units may also be relevant for a better understanding of the dynamics of interictal–ictal transitions. Interestingly, putative interneurons (with higher discharge rates) more frequently showed pre-ictal and ictal alterations in their action potential generation compared to putative principal neurons. Understanding the behavior of inhibitory and excitatory neurons in pre-ictal and early ictal phases may offer new options for pharmacological and nonpharmacological interventions both in providing a rational for acute and chronic treatment and in offering a quantifiable efficacy parameter for a given intervention. This could include mechanisms involved in the synchronization of multiple clusters of independently active neurons suggested to be critical for seizure manifestation.

Similar to validating experimental models of epilepsy, in vivo unit recordings have also to confirm findings from analyses based on in vitro studies of brain slices obtained from epilepsy surgery, for example, regarding the behavior of single neurons in relation to other biomarkers of epileptic activity like high-frequency oscillations (Roopun et al., 2010). This may also include to some degree the analysis of pharmacological effects modulating excitability as expressed in neuronal interactions.

Is There a Potential Clinical Use?

Limited spatial sampling in studies addressing the activity of single neurons poses even more limitations for representative recordings and coverage of clinically relevant areas than analyses of intracranial field potentials. There is little reason to see a role of unit recordings in the delineation of an epileptogenic area unless the interpretation is limited to a whole anatomical structure, like the amygdala. In addition, limited recording stability may render chronic analyses, for example, as an input into interventional devices, difficult if not impossible, a problem that may be even more pronounced with microwire recordings than in recordings based on multielectrode arrays, where a decline of neurons over periods of months is common (Nicolelis et al., 2003).

Thus, changes in the discharge patterns of individual neurons which could, in principle, offer chances both for diagnostic and for therapeutic applications may be difficult to use with sufficient robustness. Pre-ictal alterations in unit behavior as reported in several studies may support the concept of gradual and highly localized pre-ictal changes in EEG dynamics without offering a direct technical solution to seizure prediction. It remains to be demonstrated whether the differences in neuronal behavior found in areas of primary seizure generation and secondary spread can serve as a new biomarker of the epileptogenic zone, and to what degree high resolution field potential recordings may replace single unit recordings in this respect.

With the given methodological constraints, single unit recordings may thus primarily contribute to the formation and validation of concepts relevant for ictogenesis, rather than serving as a readily available diagnostic tool, and may influence new strategies of EEG analysis and EEG-based intervention mostly indirectly.

Acknowledgments

Single unit recordings of the author were supported by National Institute of Health grant 2R01MH061975-07A2 and BMBF grant 01GQ1010.

References

Alarcon, G., Garcia Seoane, J. J., Binnie, C. D., Martin Miguel, M. C., Juler, J., Polkey, C. E., et al. (1997). Origin and propagation of interictal discharges in the acute electrocorticogram: Implications for pathophysiology and surgical treatment of temporal lobe epilepsy. *Brain*, *120*, 2259–2282.

Andrillon, T., Nir, Y., Staba, R. J., Ferrarelli, F., Cirelli, C., Tononi, G., et al. (2011). Sleep spindles in humans: Insights from intracranial EEG and unit recordings. *Journal of Neuroscience*, *31*, 17821–17823.

Avoli, M. (2001). Do interictal discharges promote or control seizures? Experimental evidence from an in vitro model of epileptiform discharge. *Epilepsia*, *42*(Suppl 3), 2–4.

Avoli, M., Barbarosie, M., Lücke, A., Nagao, T., Lopantsev, V., & Köhling, R. (1996). Synchronous GABA-mediated potentials and epileptiform discharges in the rat limbic system in vitro. *Journal of Neuroscience*, *16*, 3912–3924.

Avoli, M., de Curtis, M., & Köhling, R. (2012). *Does interictal synchronization influence ictogenesis? Neuropharmacology*, *69*, 37–44. doi:pii: S0028–3908(12)00302–4.10.1016/j.neuropharm.2012.06.044.

Babb, T. L., Carr, E., & Crandall, P. H. (1973). Analysis of extracellular firing patterns of deep temporal lobe structures in man. *Electroencephalography and Clinical Neurophysiology*, *34*, 247–257.

Babb, T. L., & Crandall, P. H. (1976). Epileptogenesis of human limbic neurons in psychomotor epileptics. *Electroencephalography and Clinical Neurophysiology*, *40*, 225–243.

Babb, T. L., Wilson, C. L., & Isokawa-Akesson, M. (1987). Firing patterns of human limbic neurons during stereoencephalography (SEEG) and clinical temporal lobe seizures. *Electroencephalography and Clinical Neurophysiology*, *66*, 467–482.

Barbarosie, M., & Avoli, M. (1997). CA3-driven hippocampal–entorhinal loop controls rather than sustains in vitro limbic seizures. *Journal of Neuroscience*, *17*, 9308–9314.

Bragin, A., Wilson, C. L., & Engel, J., Jr. (2000). Chronic epileptogenesis requires development of a network of pathologically interconnected neuron clusters: A hypothesis. *Epilepsia*, *41*(S6), S144–S152.

Calvin, W. H., Ojemann, G. A., & Ward, A. A. (1973). Human cortical neurons in epileptogenic foci: Comparison of interictal firing patterns to those of "epileptic" neurons in animals. *Electroencephalography and Clinical Neurophysiology, 34*, 337–351.

Colder, B. W., Frysinger, R. C., Wilson, C. L., Harper, R. M., & Engel, J., Jr. (1996). Decreased neuronal burst discharge near site of seizure onset in epileptic human temporal lobes. *Epilepsia, 37*, 113–121.

Crandall, P. H., Walter, R. D., & Rand, R. W. (1963). Clinical applications of studies on stereotaxically implanted electrodes in temporal lobe epilepsy. *Journal of Neurosurgery, 20*, 827–840.

Cymerblit-Sabba, A., & Schiller, Y. (2010). Network dynamics during development of pharmacologically induced epileptic seizures in rats in vivo. *Journal of Neuroscience, 30*, 1619–1630.

de Curtis, M., & Avanzini, G. (2001). Interictal spikes in focal epileptogenesis. *Progress in Neurobiology, 63*, 541–567.

de Curtis, M., & Gnatkovsky, V. (2009). Reevaluating the mechanisms of focal ictogenesis: The role of low-voltage fast activity. *Epilepsia, 50*, 2514–2525.

de Curtis, M., Jefferys, J. G. R., & Avoli, M. (2012). Interictal epileptiform discharges in partial epilepsy: Complex neurobiological mechanisms based on experimental and clinical evidence. In J. L. Noebels, M. Avoli, M. A. Rogawski, R. W. Olsen, & A. V. Delgado-Escueta (Eds.), *Jasper's basic mechanisms of the epilepsies.* [Internet] 4th ed., pp. 213–227). Bethesda, MD: NCBI.

Dichter, M. A., Herman, C. J., & Selzer, M. (1972). Silent cells during interictal discharges and seizures in hippocampal penicillin foci: Evidence for the role of extracellular K+ in the transition from the interictal state to seizures. *Brain Research, 48*, 173–183.

Dyhrfjeld-Johnsen, J., Berdichevsky, Y., Swiercz, W., Sabolek, H., & Staley, K. J. (2010). Interictal spikes precede ictal discharges in an organotypic hippocampal slice culture model of epileptogenesis. *Journal of Clinical Neurophysiology, 27*, 418–424.

Goldensohn, E. S., & Purpura, D. P. (1963). Intracellular potentials of cortical neurons during focal epileptogenic discharges. *Science, 139*, 840–842.

Gotman, J., & Marciani, M. G. (1985). Electroencephalographic spiking activity, drug levels, and seizure occurrence in epileptic patients. *Annals of Neurology, 17*, 597–603.

Harper, R. M., & McGinty, D. J. (1973). A technique for recording single neurons from unrestrained animals. In I. Phillips (Ed.), *Brain unit activity and behavior* (pp. 80–104). Springfield, IL: Charles C Thomas Press.

Häussler, U., Bielefeld, L., Froriep, U. P., Wolfart, J., & Haas, C. A. (2012). Septotemporal position in the hippocampal formation determines epileptic and neurogenic activity in temporal lobe epilepsy. *Cerebral Cortex, 22*, 26–36.

Hefft, S., Brandt, A., Zwick, S., von Elverfeldt, D, Mader, I., Cordeiro, J., Trippel, M., & Schulze-Bonhage, A. (2013). Safety of hybrid electrodes for single neuron recordings in humans. *Neurosurgery, 73*, 78–85.

Hefft, S., Trippel, M., Brandt, A., Eble, I., Zwick, S., von Elverfeldt, D, …Schulze-Bonhage, A (under revision). Neurosurgical techniques to improve micro-wire recordings of human single neurons in vivo.

Hochberg, L. R., Serruya, M. D., Friehs, G. M., Mukand, J. A., Saleh, M., Caplan, A. H., et al. (2006). Neuronal ensemble control of prosthetic devices by a human with tetraplegia. *Nature Neuroscience, 13*, 164–171.

House, P. A., MacDonald, J. D., Tresco, P. A., & Normann, P. A. (2006). Acute microelectrode array implantation into human neocortex: Preliminary technique and histological considerations. *Neurosurgical Focus, 20*, E4.

Huberfeld, G., Menendez de la Prida, L., Pallud, J., Cohen, I., Le Van Quyen, M., Adam, C., et al. (2011). Glutamatergic pre-ictal discharges emerge at the transition to seizure in human epilepsy. *Nature Neuroscience 14*, 627–634.

Ishijima, B., Hori, T., Yoshimasu, N., Fukushima, T., Hirakawa, K., & Sekino, H. (1975). Neuronal activities in human epileptic foci and surrounding areas. *Electroencephalography and Clinical Neurophysiology, 39*, 643–650.

Jensen, M. S., & Yaari, Y. (1988). The relationship between interictal and ictal paroxysms in an in vitro model of focal hippocampal epilepsy. *Annals of Neurology, 24*, 591–598.

Jiruska, P., de Curtis, M., Jefferys, J. G., Schevon, C. A., Schiff, S. J., & Schindler, K. (2013). Synchronization and desynchronization in epilepsy: Controversies and hypotheses. *Journal of Physiology, 591*, 787–797.

Köhling, R., Qü, M., Zilles, K., & Speckmann, E. J. (1999). Current-source-density profiles associated with sharp waves in human epileptic neocortical tissue. *Neuroscience, 94*, 1039–1050.

Librizzi, L., & de Curtis, M. (2003). Epileptiform ictal discharges are prevented by periodic interictal spiking in the olfactory cortex. *Annals of Neurology, 53*, 382–389.

Lopantsev, V., & Avoli, M. (1998). Participation of GABAA-mediated inhibition in ictallike discharges in the rat entorhinal cortex. *Journal of Neurophysiology, 79*, 352–360.

Matsumoto, H. (1964). Interictal events during the activation of cortical epileptiform discharges. *Electroencephalography and Clinical Neurophysiology, 17*, 294–307.

Matsumoto, H., & Marsan, C. A. (1964a). Cortical cellular phenomena in experimental epilepsy: Interictal manifestations. *Experimental Neurology, 9*, 286–304.

Matsumoto, H., & Marsan, C. A. (1964b). Cortical cellular phenomena in experimental epilepsy: Ictal manifestations. *Experimental Neurology, 9*, 305–306.

Miller, J. F., Neufang, M., Solway, A., Brandt, A., Trippel, M., Mader, I., Hefft, S., Merkow, M., Poly, S. M., Jacobs, J., Kahana, M. J., & Schulze-Bonhage, A. (2013). Neural activity in human hippocampal formation reveals the spatial context of retrieved memories. *Science, 341*, 1111–1114.

Nicolelis, M. A., Dimitrov, D., Carmena, J. M., Crist, R., Lehew, G., Kralik, J. D., & Wise, S. P. (2003). Chronic, multisite, multielectrode recordings in macaque monkeys. *Proceedings of the National Academy of Sciences of the United States of America, 100*, 11041–11046.

Quiroga, R. Q. (2012). Concept cells: The building blocks of declarative memory functions. *Nature Reviews Neuroscience, 13*, 587–597.

Rayport, M., & Waller, H. J. (1967). Technique and results of micro-electrode recording in human epileptogenic foci. *Electroencephalography and Clinical Neurophysiology, 25*, 143–151.

Roopun, A. K., Simonotto, J. D., Pierce, M. L., Jenkins, A., Nicholson, C., Schofield, I. S., et al. (2010). A nonsynaptic mechanism underlying interictal discharges in human epileptic neocortex. *Proceedings of the National Academy of Sciences of the United States of America, 107*, 338–343.

Schevon, C. A., Ng, S. K., Cappell, J., Goodman, R. R., McKhann, G., Jr., Waziri, A., et al. (2008). Microphysiology of epileptiform activity in human neocortex. *Journal of Clinical Neurophysiology, 25*, 321–330.

Schevon, C. A., Weiss, S. A., McKhann, G., Jr., Goodman, R. R., Yuste, R., Emerson, R. G., et al. (2012). Evidence of an inhibitory restraint of seizure activity in humans. *National Communication Journal, 3*, 1060. doi:10.1038/ncomms2056.

Schulze-Bonhage, A. (2011). An introduction to epileptiform activities and seizure patterns obtained by scalp and invasive EEG recordings. In I. Osorio, H. P. Zaveri, M. G. Frei, & S. Arthurs (Eds.), *Epilepsy: The intersection of neurosciences, biology, mathematics, engineering and physics* (pp. 51–64). Boca Raton: CRP Press.

Schwindt, W., Nicholson, C., & Lehmenkühler, A. (1997). Critical volume of rat cortex and extracellular threshold concentration for a pentylenetetrazol-induced epileptic focus. *Brain Research, 753*, 86–97.

Segal, M. M. (1991). Epileptiform activity in microcultures containing one excitatory hippocampal neuron. *Journal of Neurophysiology, 65*, 761–770.

Segal, M. M. (1994). Endogenous bursts underlie seizurelike activity in solitary excitatory hippocampal neurons in microcultures. *Journal of Neurophysiology, 72*, 1874–1884.

Staba, R. J., Wilson, C. L., Bragin, A., Fried, I., & Engel, J. Jr. (2002). Sleep states differentiate single neuron activity recorded from human epileptic hippocampus, entorhinal cortex, and subiculum. *Journal of Neuroscience, 22*, 5694–5704.

Timofeev, I., Grenier, F., & Steriade, M. (2002). The role of chloride-dependent inhibition and the activity of fast-spiking neurons during cortical spike-wave electrographic seizures. *Neuroscience, 114*, 1115–1132.

Trevelyan, A. J., & Schevon, C. A. (2012). How inhibition influences seizure propagation. *Neuropharmacology, 69*, 45–54.

Truccolo, W., Donoghue, J. A., Hochberg, L. R., Eskandar, E. N., Madsen, J. R., Anderson, W. S., et al. (2011). Single-neuron dynamics in human focal epilepsy. *Nature Neuroscience, 14*, 635–641.

Ulbert, I., Heit, G., Madsen, J., Karmos, G., & Halgren, E. (2004). Laminar analysis of human neocortical interictal spike generation and propagation: Current source density and multiunit analysis in vivo. *Epilepsia, 45*(S4), 48–56.

Valdez, A. B., Hickman, E. N., Treiman, D. M., Smith, K. A., & Steinmetz, P. N. (2013). A statistical method for predicting seizure onset zones from human single-neuron recordings. *Journal of Neural Engineering, 10*, 016001.

Verzeano, M., Crandall, P. H., & Dymond, A. (1971). Neuronal activity in the amygdala inpatients with psychomotor epilepsy. *Neuropsychologia, 9*, 331–344.

Ward, A. A. (1961). The epileptic neurone. *Epilepsia, 2*, 70–80.

Ward, A. A., & Thomas, L. B. (1955). The electrical activity of single units in the cerebral cortex of man. *Electroencephalography and Clinical Neurophysiology, 7*, 135–136.

Ward, A. A., Thomas, L. N., & Schmidt, R. P. (1956). Some properties of single epileptic neurons. *Transactions of the American Neurological Association, 81*, 41–43.

Waziri, A., Schevon, C. A., Cappell, J., Emerson, R. G., McKhann, G. M., II, & Goodman, R. R. (2009). Initial surgical experience with a dense cortical microarray in epileptic patients undergoing craniotomy for subdural electrode implantation. *Neurosurgery, 64*, 540–545.

Wyler, A. R., Ojemann, G. A., & Ward, A. A. (1982). Neurons in human epileptic cortex: Correlation between unit and EEG activity. *Annals of Neurology, 11*, 301–308.

Wyler, A. R., & Ward, A. A. (1986). Neuronal firing patterns from epileptogenic foci of monkey and human. In A. V. Delgado-Escueta, A. A. Ward, Jr., D. M. Woodbury, & R. J. Porter (Eds.), *Advances in neurology* (Vol. 44, pp. 967–989). New York: Raven Press.

IV CONCLUSIONS

19   The Next Ten Years and Beyond

Ueli Rutishauser, Itzhak Fried, Moran Cerf, and Gabriel Kreiman

The brain is arguably one of the most complex systems ever studied by science. Major insights and constraints about brain function can be obtained from lesion studies as well as behavioral studies (Finger, 2000). Yet, ultimately, a mechanistic understanding of the computations that underlie cognition requires speaking and interpreting the language of the brain: spikes (Rieke et al., 1997; Kreiman, 2004; Koch, 2005). To understand how our cognition is implemented, we need to investigate the human brain from the inside.

Significant advances in our characterization of neurons and neuronal circuits came about with the advent of techniques to "listen to" spikes (Adrian, 1926; Hubel & Wiesel, 1998). Most of the efforts to examine the activity of neurons have focused on studies in animal models given the difficulties inherent in invasive studies of the human brain. Yet, even from the early days of neurophysiology, neurosurgeons became interested in exploring neuronal function in the human brain (see chapters 2 and 3).

Earlier chapters have documented the potential and insights derived from recording neuronal activity in the human brain in a variety of domains. Invasive studies of the human brain can transform our understanding of human cognition at a mechanistic level and can also help us better understand and treat neurological disorders including epilepsy, Parkinson's disease, Alzheimer's disease, and other cognitive and motor disorders.

Here we would like to suggest possible areas of investigation in the field that may provide significant insights during the next decade and beyond. While the discussion here will take a speculative tone, we hope that these notes will help inspire the next generation of researchers to push the frontiers of knowledge. The discussion here does not aim to be exhaustive in any way. We hope to be at least partly wrong and be surprised by exciting new unforeseen discoveries. We somewhat arbitrarily divide these future directions into ten different but overlapping themes.

Learning and Memory, Time Travel

Many of the depth electrode cases in epileptic patients target the medial temporal lobe (MTL) including the hippocampus and surrounding structures. These regions provide a unique opportunity to further our understanding of the neuronal circuits and mechanisms that underlie learning and the conversion of short-term to long-term memories (see chapters 7–9).

Studying the human brain can provide glimpses into aspects of episodic memory that are not easily amenable to exploration in animal models. It has been argued that sophisticated forms of episodic memory are unique to humans and can thus only be studied in humans (Tulving, 2002; Hampton & Schwartz, 2004). Different timescales of memory formation have been described at the neuronal level; it will be interesting and important to examine whether and how these different temporal scales are encoded in MTL neurons. Work in rodents has largely focused on spatial learning. It is conceivable that these spatial navigation studies constitute an example of a more generic mechanism for associations. Further strengthening the links between rodent/ macaque neurophysiological studies and human studies can prove fruitful to investigators working with all models.

It has remained very difficult to directly link the mechanisms of synaptic plasticity to memories (Martin et al., 2000). This is in part due to the difficulty of testing for the presence of a memory without being able to directly query subjects on the content of their memory. Human experiments offer the unique opportunity to make a contribution toward our understanding of which neuronal mechanisms support memories. This is particularly the case for remote memories (Frankland & Bontempi, 2005; Squire & Bayley, 2007). We are able to recall memories that were established years or even decades ago, but the processes by which such remote call works remain largely unknown. Furthermore, our ability to imagine ourselves in a future scenario and time travel into an imagined future may also be dependent on MTL machinery (Nyberg et al., 2010; Addis et al., 2011). Thus, working at the single neuron level with human patients who can time travel from their distant past into a potential future and report their experiences can provide rare insights into these uniquely human conditions.

Work with human patients offers the opportunity to directly query subjects about memories from many years ago. Additionally, emotions are likely to play a key role in memory formation (chapter 13; Cahill et al., 1995; Fanselow & Gale, 2003; Phelps, 2004). The relationship between emotional processing and memory formation has been only poorly studied in the human brain (see discussion below).

Recently, a few studies have suggested that it may be possible to enhance memory formation through electrical stimulation of sites in the MTL or sites outside the temporal lobe sites connected to it (Laxton et al., 2010; Suthana et al., 2012) or through pharmacological interventions (Bentley et al., 2011). Future studies are needed to further our understanding of how such interventions work to enhance learning and memory.

Emotions

Another area within the MTL that is often targeted in depth electrode implantation cases is the amygdala. Significant evidence from lesion studies and animal studies suggests that this large structure plays a critical role in processing emotions (see chapter 13). Certain aspects of emotional processing have been examined in animal models, particularly using behavioral paradigms such as fear conditioning. The repertoire and depth of emotions that can be studied in animal models is limited. The study of emotions constitutes a prime example where single neuron recordings in the human brain can lead us to uncover a representation that is extremely difficult to investigate with other methods and models. Important initial steps in this direction were described in chapter 13.

The amygdala is a very large structure; at the anatomical level it is composed of multiple different nuclei. Some of these nuclei are distinct in terms of their input and output. It is tempting to speculate that these different nuclei will manifest differential responses to different aspects of emotional processing, a topic that has received little attention. The large repertoire of emotions that humans are sensitive to has seen only preliminary investigations and deserves further scrutiny. Perhaps there exists a functional subdivision of different types of emotions encoded in different substructures, although this is an admittedly oversimplified hypothesis.

There is a significant overlap in the patient population between autism spectrum disorders (ASDs) and epilepsy. A behavioral hallmark of ASD involves difficulties in social interactions including interpreting and expressing emotions. It has proven remarkably difficult to identify animal models that express the behavioral anomalies of ASD patients. Investigating the neurophysiological properties of neurons in the amygdala in ASD patients can yield significant advances toward a mechanistic understanding of the elusive nature of ASD.

Additionally the role of the amygdala in posttraumatic stress disorder has been a topic of considerable investigation (Ursano et al., 2010). Studies in the human amygdala using recording and stimulation paradigms may add invaluable data with potential implications for the treatment of this disorder.

Language and Communication

Beyond memory and emotions, several aspects of cognitive science lend themselves to investigation in humans. One such example is language (e.g., see chapter 14). Language, at least in the forms typically discussed within the cognitive science literature, is specific to humans. Temporal dynamics play a central role in language generation and understanding, making it difficult to study with techniques that have poor spatial and/or temporal resolution.

Uncovering the neural code underlying language is particularly difficult as there is no good animal neurophysiology to build upon. Recent work by Tankus et al. (2012) reports on a highly structured neuronal encoding of vowel articulation. In medial–frontal neurons, highly specific tuning to individual vowels was found, whereas in superior temporal gyrus, neurons had nonspecific, sinusoidally modulated tuning (analogous to motor cortical directional tuning). At the neuronal population level, a decoding analysis revealed that the underlying structure of vowel encoding corresponded to the anatomical basis of articulatory movements. This structured encoding enabled accurate decoding of volitional speech segments and could thus potentially be applied in the development of brain–machine interfaces for restoring speech in paralyzed individuals (Tankus et al., 2012).

Another example involves the extrapolation of mirror neurons from nonhuman primates to humans. Indeed, single neuron work may shed light on this system, which appears to be quite spread out in human cortex (Mukamel et al., 2010).

Free Will

A particularly intriguing aspect of decision making involves those situations where choices are made volitionally as opposed to being triggered by external inputs. We have a strong subjective feeling that we are free agents owning our destinies. The notion of free will has been debated for millennia and is at the heart of our very essence of *self* (Haggard, 2008). Our legal system is based on the assumption of the existence of free will (Cashmore, 2010). Our daily actions depend on how we interpret free will. Yet, ultimately, our volitional decisions must be encoded by neurons in our brains. When, how, and with which neurons volitional decisions are orchestrated remains a daunting and elusive problem (see initial steps and discussion in chapter 8). Intriguing and heroic efforts have been made to characterize neuronal signatures preceding volitional movements in animal models, but it remains quite difficult to interpret the animals' notion of self and will. How free will relates to different brain structures has received significant attention in neurological studies as well as noninvasive scalp EEG studies (e.g., the *Bereitschaftspotential* and its modulations). In comparison, there has been minimal work at the neuronal circuit level in the human brain. Experiments in humans (Fried et al., 1991; Fried et al., 2011) offer a unique opportunity to interrogate the neuronal circuits and mechanisms underlying free will because tasks can be designed in which subjects make voluntary decisions and because we can control external variables and have some degree of access to subjective decisions. An intriguing question involves whether there is a "point of no return" in voluntary decisions. In a rather simplistic model, one could conceive of a hierarchical chain of commands ranging from the initial will all the way to the implementation at the level of motor neurons and muscles. What aspects or processes within this dynamical chain can be interrupted and which ones cannot be interrupted may shed light on the elusive and fascinating circuitry that can implement the most capricious of cognitive phenomena. Experimental paradigms that utilize real-time processing to visualize internal thought processes are powerful tools that will likely be instrumental in teasing apart the architecture of voluntary decisions (see chapters 6 and 8).

Consciousness

Free will is but one aspect of the general problem of finding the neuronal correlates of consciousness (chapters 8 and 11 in this volume; Koch, 2005). The study of animal models has made tremendous contributions to help formulate questions about how the contents of consciousness are represented in a rigorous fashion amenable to scientific investigation. At the same time, the lack of subjective reports leads to restricting ourselves to experimental paradigms that require careful quantification of behavior and extensive training. While we certainly need to continue with experiments in animal models, there is always the lingering question about the extent to which animals solve the tasks in the same way that we do or experience the same sensations upon presentation of the same stimulus.

Human neurophysiological recordings can provide a bridge between noninvasive measurements in humans and more invasive studies in animal models. So far, this field has been largely focusing on experiments that parallel research efforts in animal models. Experiments have been conducted in which the same stimulus in some cases leads to conscious experience and in other instances doesn't (e.g., binocular rivalry; see chapter 8). Investigators then correlate single neuron activity with conscious experience. Multiple (speculative) discussions about the representation of consciousness have argued that consciousness is the consequence of interactions between multiple brain areas. These putative interactions have been poorly explored in human neurophysiology. While in some cases, the small number of electrodes and limited sampling might make it difficult to systematically examine such interactions, it still seems that examining interactions across units, or coherence between unit activity and local field potential (LFP) activity in different areas (e.g., Womelsdorf et al., 2007) and coupled with subjective report of conscious human subjects, may provide unique insights.

Sleep

Overnight recordings offer a rich data set that allows investigators to delve into the fascinating and often-mysterious patterns of brain activity during sleep (see chapter 10). Many efforts have centered on characterizing firing rates, synchronization, and other activity patterns during each of the different sleep stages and during events defined by large-scale recordings such as spindles or K-complexes.

A fascinating and largely unexplored aspect of sleep involves dreams. Recently, a functional imaging study took initial steps in an effort to decode brain activity during dreams (Horikawa et al., 2013). While the elusive nature of dreams has always made them difficult to study, one would hope that the resolution of single unit studies may shed light on how neurons represent the contents of dreams. Furthering our understanding of brain activity during dreams may have important implications not only for addressing this age-old dilemma but also for elucidating the neural correlates of consciousness and memory recall.

There is ample evidence that multiple tasks show sleep-dependent enhancement (Stickgold, 2005). The interactions between memory formation and sleep have been investigated in rodents, where it has been shown that hippocampal neuron ensembles replay activity patterns that occurred during the awake experience (see Wilson & McNaughton, 1994, and multiple more recent efforts that have extended those discoveries and reported a wide variety of replay phenomena). Replay of neuronal activity patterns still remains to be demonstrated in the human hippocampus. Additionally, the relative easiness with which behavior can be examined in humans may pave the ways to a systematic investigation of the relationship between MTL activity during sleep and learning.

Epilepsy

The need to understand the mystery of the human brain is in itself one of the ultimate challenges of science. Yet, this need is largely driven by the goal of providing therapies for the variety of devastating neurological diseases including epilepsy, stroke, movement disorders such as Parkinson's disease, and Alzheimer's disease and other dementias. This goal is present and clear especially in the clinical situations that provide unique data, such as in single neuron recordings in humans, which are performed only in settings of neurological disorders.

In spite of significant progress in our understanding of the origins and mechanisms that give rise to epilepsy, 20–40% of patients with epilepsy remain medically refractory (Engel et al., 2012). Advances in drug development might help reduce, or hopefully one day even eliminate, the number of patients with pharmacologically intractable seizures. In the meantime, two ongoing efforts might make significant strides to reduce or eliminate seizures.

One of these involves using invasive devices to detect or even predict seizures in real time and then using electrical or pharmaceutical stimulation methods to stop the seizures (Sun et al., 2008; Morrell and the RNS System in Epilepsy Study Group, 2011). While there has been significant progress in seizure detection algorithms, seizure prediction remains a daunting problem (Mormann et al., 2007). Recently, encouraging evidence from a prospective long-term trial has been reported (Cook et al., 2013).

The higher spatial resolution of single neuron recordings could potentially provide key missing circuit-level constraints to improve seizure prediction. We expect to see major advances in this field in the next several years.

In refractory cases where invasive intervention is an option (Engel et al., 2012), we expect that the higher resolution of microelectrode recordings (single unit or localized field potentials) may help guide and improve the surgical approaches (Schevon et al., 2008; Stead et al., 2010; Truccolo et al., 2011; Alarcon et al., 2012; Bower et al., 2012; Schevon et al., 2012; Valdez et al., 2012; Mormann & Jefferys, 2013). While these and similar studies demonstrate correlations between single cell properties and later focal seizure onset location, whether these methods are useful to predict the seizure onset zone has not been tested so far in a rigorous blinded study (see discussions in chapter 18). It may be possible in the future to further delimit the boundaries of the epileptogenic areas, thus leading the way to smaller, more focal surgical resections in treatment of pharmacologically resistant epilepsy. This may help reduce potential cognitive deficits that could be associated with larger resections.

Single neuron studies could potentially be useful not only in delineating where seizures originate but also in determining whether nonepileptogenic tissue is sufficiently functional. This is of high clinical interest, as the functionality of resected areas needs to be supported by the tissue that remains. Most often this is assessed with limited spatial resolution using the Wada test (also called the intracarotid sodium amobarbital procedure), but alternatives or additional supporting tests are an active research topic and include functional neuroimaging mapping techniques. As an example for a different possible approach using neuronal recordings, novelty-sensitive neurons within the MTL exhibit differential sensitivity in putative epileptogenic versus nonepileptogenic areas (see chapter 7 and figure 11 in Rutishauser et al., 2008). This suggests that the extent of novelty sensitivity of MTL neurons might serve as an indicator for which part of the MTL is functional, a hypothesis that remains to be tested directly but that is supported by lesion studies (Knight, 1996).

Additionally, the combination of single neuron recordings and field potential recordings may shed light on the elusive nature of interictal discharges and high-frequency oscillations (Engel & da Silva, 2012; Jacobs et al., 2012; see also chapter 18). Understanding the nature and origin of these discharges might help localize epileptic activity as well as its propagation in the brain.

Brain tissue that results from removal of the putative epileptic focus can be utilized to perform in vitro experiments that permit intracellular or molecular work not possible in vivo (Williamson et al., 1993; Kohling et al., 1999; Kohling & Avoli, 2006). Such tissue can exhibit epileptic phenomena such as interictal sharp waves (see chapter 18). This tissue offers the unique opportunity to use standard laboratory techniques such as the whole-cell patch clamp to record from human neurons. This can be used to test anti-epileptic drugs as well as to test hypothesis on abnormal network organization. While rarely utilized at present, we expect significance advances in the future that will utilize such tissue.

Deep Brain Stimulation

Deep brain stimulation (DBS) has become part of the tool arsenal to treat a variety of motor disorders (see chapters 17 and 18). The placement of DBS electrodes for therapeutic ends has also been a source of single and multiunit activity data, as recordings of these units has often been an integral part of the identification of optimal therapeutic targets (Engel et al., 2005). Additionally, a number of intriguing and potentially transformative applications of DBS are currently under intense investigation. One such application involves attempts to enhance learning and memory formation in Alzheimer's patients via bilateral fornix DBS (Laxton et al., 2010; Laxton and Lozano, 2012). Other work has shown that DBS applied to entorhinal cortex at the time of learning can enhance spatial memory (Suthana et al., 2012). Other interesting applications of DBS involve the treatment of epilepsy as well as several psychiatric disorders such as obsessive–compulsive disorder and major depression (Anderson & Lenz, 2009), also part of an ongoing clinical trial.

Treatment of movement disorders—principally Parkinson's disease (PD), dystonia, and essential tremor—with implantation of chronic DBS electrodes is often highly effective and is performed routinely (Hariz, 2012; Lozano & Lipsman, 2013). However, the mechanism by which stimulation acts is poorly understood, and it is unclear why in some cases DBS works almost instantaneously (such as subthalamic nucleus stimulation for PD) whereas in others the latency to effectiveness can be very long (such as globus pallidus interna stimulation for dystonia). Also, there is considerable debate about the appropriate target structures as well as whether the primary treatment effect is due to stimulation or inhibition of cell bodies or fibers of passage. While effective, DBS treatment can also result in severe side effects. Better understanding of the mechanism and targeting of DBS is expected to significantly enhance treatment effectiveness while at the same time minimizing side effects.

Target structures are frequently identified by intraoperative microelectrode recordings (see chapters 15 and 16). These recordings can be performed anywhere along the track to the target structure, thus providing access to a number of areas within the basal ganglia, striatum, and thalamus. This technique has already revealed numerous features of both normal and abnormal function (see chapters 12, 15, and 16). Microelectrode recordings, potentially combined with subdural intracranial field potential recordings during DBS placement (Crowell et al., 2012), offer an exciting opportunity to record from brain areas impacted in movement disorders. Different movement disorders have fundamentally different underlying disease mechanisms (i.e., PD vs. dystonia), meaning that the different patient populations can serve as controls for each other. This enables investigation of disease-specific effects otherwise not possible (see de Hemptinne et al., 2013, for an example of this approach). Additionally, it should be possible to combine neuronal recordings in closed-loop systems, with real-time data analyses and electrical stimulation in different locations, timings, and frequencies dictated by actual physiological variables.

Motor Prostheses

The last decade has seen astounding progress toward developing neural prosthetic devices that can interact with the human brain to restore motor capabilities to quadriplegic patients (e.g., Carmena et al., 2003; Musallam et al., 2004; Schwartz, 2004; Hochberg et al., 2012, reviewed in chapter 17). The synergistic work across investigations in monkeys and humans can lead to important progress in these devices. Despite the advances, these devices remain under investigational inquiry and have not yet reached clinical use. Widespread adoption is held back due to factors such as limited repertoire of motor capabilities achieved, limited electrode lifetimes, clunky wires and connections, neurosurgical risk and complexity, and so forth.

Researchers are working diligently to address each of these limitations. It is conceivable that a combination of more electrodes, a better understanding of the encoding of motor signals, and exploitation of the remarkable plasticity of brain circuits may lead to an increased repertoire of motor abilities. Increasing the longevity of implanted devices with more robust electrode designs or with algorithms that use LFPs in addition to spiking activity could minimize and eventually obviate the need for periodic invasive surgery to replace the electrodes. The development of wireless and subcutaneous fiber optic technologies to transmit the signals to end effectors could minimize infection risk and further provide the patient with an outward appearance indistinguishable from that of a healthy control subject. Emerging neurotechnologies might dramatically reduce the cost and complexity of surgically placing the electrodes.

While considerable work has been done toward upper limb prostheses, we might also see clinical trials of lower limb prostheses in the form of brain-controlled exoskeleton devices. Such devices may come to incorporate sensory feedback to improve the capabilities of the exoskeleton. Thus, hitherto fantastical science-fiction creations such as Luke Skywalker's prosthetic arm in *Star Wars* may become reality for the benefit of amputees or paralyzed patients.

Neurotechnology

New technologies often open doors to examine phenomena in a different way. The rapid pace of progress in the development of neurotechnologies to examine brain function can have tremendous impact in how we invasively investigate the human brain. We foresee progress on several fronts here, including the following:

1. *Further development of techniques to process and decode data in real time* This was briefly discussed in chapter 6. Real-time experiments may permit a level of interaction with patients not possible with offline studies. Real-time data processing may transform experiments as well as provide new solutions to clinical challenges as noted above (in the "Epilepsy," "Deep Brain Simulation," and "Motor Prostheses" sections).

2. *Wireless transmission* Wireless transmission of signals may help alleviate cumbersome cables, provide more mobility to the patients, and reduce the risk of infections.

3. *Better signal isolation* Single unit isolation is always challenging, and overcoming noise is a perennial theme in human neurophysiological recordings.

4. *Novel types of electrodes* While it is often difficult to evaluate new electrodes in human recordings, several new types of technologies are being examined in animal models, and the knowledge from animal neurophysiology may be translated to human neurophysiology.

5. *Microdialysis combined with neurophysiology* Intriguing and promising initial observations were made upon combining microdialysis and electrode recordings (Fried et al., 1999; Blouin et al., 2013). While these efforts are not devoid of challenges, they may open the opportunity for direct interrogation of neuronal activity in the context of the surrounding chemical milieu and eventually, in the future, local application of different drugs.

6. *Brain–machine interfaces* Prosthetic devices for motor applications have been at the center of investigational efforts as these have a very clear and attainable use in neurological patients inflicted with paralysis such as in spinal cord injury, amyotrophic lateral sclerosis, and traumatic injuries (see the "Motor Prostheses" section above). Yet, the ability to directly interface the brain with the environment may be utilized to enhance other brain functions in different patient populations. Speech neuroprosthetic devices may be developed based on deciphering the neural code governing speech. For example, single neuron studies in humans uncovered a neural code for vowels involving different coding scheme in medial frontal and superior temporal regions

(Tankus et al., 2012). Yet, this is only a first step, and more complex schemes may yet be discovered by direct recordings from the human brain. For example, a neural code for images of specific individuals was used by Cerf et al. (2010) to construct a four-neuron brain–machine interface (see chapter 11). Initial experiments have already shown that electrical stimulation with DBS is capable of enhancing human memory (Laxton et al., 2010; Suthana et al., 2012) in some situations. Going even further, others have proposed to replace entire brain structures such as the hippocampus with electronic chips (Berger et al., 2005). The intention is to replicate the function of a brain structure such that, for example, a hippocampus damaged by seizures can be replaced to reestablish normal function.

References

Addis, D. R., Cheng, T., Roberts, R. P., & Schacter, D. L. (2011). Hippocampal contributions to the episodic simulation of specific and general future events. *Hippocampus*, *21*, 1045–1052.

Adrian, E. (1926). The impulses produced by sensory nerve endings: II. The response of a single end-organ. *Journal of Physiology*, *61*, 151–171.

Alarcon, G., Martinez, J., Kerai, S. V., Lacruz, M. E., Quiroga, R. Q., Selway, R. P., et al. (2012). In vivo neuronal firing patterns during human epileptiform discharges replicated by electrical stimulation. *Clinical Neurophysiology*, *123*, 1736–1744.

Anderson, W., & Lenz, F. A. (2009). Lesioning and stimulation as surgical treatments for psychiatric disorders. *Neurosurgery Quarterly*, *19*, 132–143.

Bentley, P., Driver, J., & Dolan, R. J. (2011). Cholinergic modulation of cognition: Insights from human pharmacological functional neuroimaging. *Progress in Neurobiology*, *94*, 360–388.

Berger, T. W., Ahuja, A., Courellis, S. H., Deadwyler, S. A., Erinjippurath, G., Gerhardt, G. A., Gholmieh, G., Granacki, J. J., Hampson, R., et al. (2005). Restoring lost cognitive function. *IEEE Engineering in Medicine and Biology*, *24*, 30–44.

Blouin, A. M., Fried, I., Wilson, C. L., Staba, R. J., Behnke, E. J., Lam, H. A., et al. (2013). Human hypocretin and melanin-concentrating hormone levels are linked to emotion and social interaction. *Nature Communications*, *4*, 1547.

Bower, M. R., Stead, M., Meyer, F. B., Marsh, W. R., & Worrell, G. A. (2012). Spatiotemporal neuronal correlates of seizure generation in focal epilepsy. *Epilepsia*, *53*, 807–816.

Cahill, L., Babinsky, R., Markowitsch, H. J., & McGaugh, J. L. (1995). The amygdala and emotional memory. *Nature Neuroscience*, *377*, 295–296.

Carmena, J. M., Lebedev, M. A., Crist, R. E., O'Doherty, J. E., Santucci, D. M., Dimitrov, D. F., et al (2003). Learning to control a brain–machine interface for reaching and grasping by primates. *PLoS Biology*, *1*, E42.

Cashmore, A. (2010). The Lucretian swere: The biological basis of human behavior and the criminal justice system. *Proceedings of the National Academy of Sciences of the United States of America*, *107*, 4499–4504.

Cerf, M., Thiruvengadam, N., Mormann, F., Kraskov, A., Quiroga, R. Q., Koch, C., et al. (2010). On-line, voluntary control of human temporal lobe neurons. *Nature Neuroscience*, *467*, 1104–1108.

Cook, M. J., O'Brien, T. J., Berkovic, S. F., Murphy, M., Morokoff, A., Fabinyi, G., et al. (2013). Prediction of seizure likelihood with a long-term, implanted seizure advisory system in patients with drug-resistant epilepsy: A first-in-man study. *Lancet Neurology*, *12*, 563–571.

Crowell, A. L., Ryapolova-Webb, E. S., Ostrem, J. L., Galifianakis, N. B., Shimamoto, S., Lim, D. A., et al. (2012). Oscillations in sensorimotor cortex in movement disorders: An electrocorticography study. *Brain*, *135*, 615–630.

de Hemptinne, C., Ryapolova-Webb, E. S., Air, E. L., Garcia, P. A., Miller, K. J., Ojemann, J. G., et al. (2013). Exaggerated phase–amplitude coupling in the primary motor cortex in Parkinson disease. *Proceedings of the National Academy of Sciences of the United States of America*, *110*, 4780–4785.

Engel, A. K., Moll, C. K., Fried, I., & Ojemann, G. A. (2005). Invasive recordings from the human brain: Clinical insights and beyond. *Nature Reviews. Neuroscience, 6,* 35–47.

Engel, J., Jr., & da Silva, F. L. (2012). High-frequency oscillations—Where we are and where we need to go. *Progress in Neurobiology, 98,* 316–318.

Engel, J., Jr., McDermott, M. P., Wiebe, S., Langfitt, J. T., Stern, J. M., Dewar, S., et al. (2012). Early surgical therapy for drug-resistant temporal lobe epilepsy: A randomized trial. *Journal of the American Medical Association, 307,* 922–930.

Fanselow, M. S., & Gale, G. D. (2003). The amygdala, fear, and memory. *Annals of the New York Academy of Sciences, 985,* 125–134.

Finger, S. (2000). *Minds behind the brain: A history of the pioneers and their discoveries.* New York: Oxford University Press.

Frankland, P. W., & Bontempi, B. (2005). The organization of recent and remote memories. *Nature Reviews. Neuroscience, 6,* 119–130.

Fried, I., Katz, A., McCarthy, G., Sass, K. J., Williamson, P., Spencer, S. S., et al. (1991). Functional organization of human supplementary motor cortex studied by electrical stimulation. *Journal of Neuroscience, 11,* 3656–3666.

Fried, I., Mukamel, R., & Kreiman, G. (2011). Internally generated preactivation of single neurons in the human brain predicts volition. *Neuron, 69,* 548–562.

Fried, I., Wilson, C. L., Maidment, N. T., Engel, J., Behnke, E., Fields, T. A., et al. (1999). Cerebral microdialysis combined with single-neuron and electroencephalographic recording in neurosurgical patients. *Journal of Neurosurgery, 91,* 697–705.

Haggard, P. (2008). Human volition: Towards a neuroscience of will. *Nature Reviews. Neuroscience, 9,* 934–946.

Hampton, R. R., & Schwartz, B. L. (2004). Episodic memory in nonhumans: What, and where, is when? *Current Opinion in Neurobiology, 14,* 192–197.

Hariz, M. (2012). Twenty-five years of deep brain stimulation: Celebrations and apprehensions. *Movement Disorders, 27,* 930–933.

Hochberg, L. R., Bacher, D., Jarosiewicz, B., Masse, N. Y., Simeral, J. D., Vogel, J., et al. (2012). Reach and grasp by people with tetraplegia using a neurally controlled robotic arm. *Nature Neuroscience, 485,* 372–375.

Horikawa, T., Tamaki, M., Miyawaki, Y., & Kamitani, Y. (2013). Neural decoding of visual imagery during sleep. *Science, 340,* 639–642.

Hubel, D. H., & Wiesel, T. N. (1998). Early exploration of the visual cortex. *Neuron, 20,* 401–412.

Jacobs, J., Staba, R., Asano, E., Otsubo, H., Wu, J. Y., Zijlmans, M., et al. (2012). High-frequency oscillations (HFOs) in clinical epilepsy. *Progress in Neurobiology, 98,* 302–315.

Knight, R. (1996). Contribution of human hippocampal region to novelty detection. *Nature Neuroscience, 383,* 256–259.

Koch, C. (2005). *The quest for consciousness* (1st ed.). Los Angeles: Roberts & Company Publishers.

Kohling, R., & Avoli, M. (2006). Methodological approaches to exploring epileptic disorders in the human brain in vitro. *Journal of Neuroscience Methods, 155,* 1–19.

Kohling, R., Qu, M., Zilles, K., & Speckmann, E. J. (1999). Current-source-density profiles associated with sharp waves in human epileptic neocortical tissue. *Neuroscience, 94,* 1039–1050.

Kreiman, G. (2004). Neural coding: Computational and biophysical perspectives. *Physics of Life Reviews, 1,* 71–102.

Laxton, A. W., & Lozano, A. M. (2012). *Deep brain stimulation for the treatment of Alzheimer disease and dementias.* World Neurosurgery.

Laxton, A. W., Tang-Wai, D. F., McAndrews, M. P., Zumsteg, D., Wennberg, R., Keren, R., et al. (2010). A phase I trial of deep brain stimulation of memory circuits in Alzheimer's disease. *Annals of Neurology, 68,* 521–534.

Lozano, A. M., & Lipsman, N. (2013). Probing and regulating dysfunctional circuits using deep brain stimulation. *Neuron, 77,* 406–424.

Martin, S. J., Grimwood, P. D., & Morris, R. G. (2000). Synaptic plasticity and memory: An evaluation of the hypothesis. *Annual Review of Neuroscience, 23,* 649–711.

Mormann, F., Andrzejak, R. G., Elger, C. E., & Lehnertz, K. (2007). Seizure prediction: The long and winding road. *Brain*, *130*, 314–333.

Mormann, F., & Jefferys, J. G. (2013). Neuronal firing in human epileptic cortex: The ins and outs of synchrony during seizures. *Epilepsy Currents*, *13*, 100–102.

Morrell, M. J., & the RNS System in Epilepsy Study Group. (2011). Responsive cortical stimulation for the treatment of medically intractable partial epilepsy. *Neurology*, *77*, 1295–1304.

Mukamel, R., Ekstrom, A. D., Kaplan, J., Iacoboni, M., & Fried, I. (2010). Single-neuron responses in humans during execution and observation of actions. *Current Biology, CB*, *20*, 750–756.

Musallam, S., Corneil, B., Greger, B., Scherberger, H., & Andersen, R. (2004). Cognitive control signals for neural prosthetics. *Science*, *305*, 258–261.

Nyberg, L., Kim, A. S., Habib, R., Levine, B., & Tulving, E. (2010). Consciousness of subjective time in the brain. *Proceedings of the National Academy of Sciences of the United States of America*, *107*, 22356–22359.

Phelps, E. A. (2004). Human emotion and memory: Interactions of the amygdala and hippocampal complex. *Current Opinion in Neurobiology*, *14*, 198–202.

Rieke, F., Warland, D., van Steveninck, R., & Bialek, W. (1997). *Spikes*. Cambridge, MA: MIT Press.

Rutishauser, U., Schuman, E. M., & Mamelak, A. N. (2008). Activity of human hippocampal and amygdala neurons during retrieval of declarative memories. *Proceedings of the National Academy of Sciences of the United States of America*, *105*, 329–334.

Schevon, C. A., Ng, S. K., Cappell, J., Goodman, R. R., McKhann, G., Jr., Waziri, A., et al. (2008). Microphysiology of epileptiform activity in human neocortex. *Journal of Clinical Neurophysiology*, *25*, 321–330.

Schevon, C. A., Weiss, S. A., McKhann, G., Jr., Goodman, R. R., Yuste, R., Emerson, R. G., et al. (2012). Evidence of an inhibitory restraint of seizure activity in humans. *Nature Communications*, *3*, 1060.

Schwartz, A. B. (2004). Cortical neural prosthetics. *Annual Review of Neuroscience*, *27*, 487–507.

Squire, L. R., & Bayley, P. J. (2007). The neuroscience of remote memory. *Current Opinion in Neurobiology*, *17*, 185–196.

Stead, M., Bower, M., Brinkmann, B. H., Lee, K., Marsh, W. R., Meyer, F. B., et al. (2010). Microseizures and the spatiotemporal scales of human partial epilepsy. *Brain*, *133*, 2789–2797.

Stickgold, R. (2005). Sleep-dependent memory consolidation. *Nature Neuroscience*, *437*, 1272–1278

Sun, F. T., Morrell, M. J., & Wharen, R. E., Jr. (2008). Responsive cortical stimulation for the treatment of epilepsy. *Neurotherapeutics; the Journal of the American Society for Experimental NeuroTherapeutics*, *5*, 68–74.

Suthana, N., Haneef, Z., Stern, J., Mukamel, R., Behnke, E., Knowlton, B., et al. (2012). Memory enhancement and deep-brain stimulation of the entorhinal area. *New England Journal of Medicine*, *366*, 502–510.

Tankus, A., Fried, I., & Shoham, S. (2012). Structured neuronal encoding and decoding of human speech features. *Nature Communications*, *3*, 1015.

Truccolo, W., Donoghue, J. A., Hochberg, L. R., Eskandar, E. N., Madsen, J. R., Anderson, W. S., et al. (2011). Single-neuron dynamics in human focal epilepsy. *Nature Neuroscience*, *14*, 635–641.

Tulving, E. (2002). Episodic memory: From mind to brain. *Annual Review of Psychology*, *53*, 1–25.

Ursano, R. J., Goldenberg, M., Zhang, L., Carlton, J., Fullerton, C. S., Li, H., et al. (2010). Posttraumatic stress disorder and traumatic stress: From bench to bedside, from war to disaster. *Annals of the New York Academy of Sciences*, *1208*, 72–81.

Valdez, A. B., Hickman, E. N., Treiman, D. M., Smith, K. A., & Steinmetz, P. N. (2012). A statistical method for predicting seizure onset zones from human single-neuron recordings. *Journal of Neural Engineering*, *10*, 016001.

Williamson, A., Spencer, D. D., & Shepherd, G. M. (1993). Comparison between the membrane and synaptic properties of human and rodent dentate granule cells. *Brain Research*, *622*, 194–202.

Wilson, M. A., & McNaughton, B. L. (1994). Reactivation of hippocampal ensemble memories during sleep. *Science*, *265*, 676–679.

Womelsdorf, T., Schoffelen, J. M., Oostenveld, R., Singer, W., Desimone, R., Engel, A. K., et al. (2007). Modulation of neuronal interactions through neuronal synchronization. *Science*, *316*, 1609–1612.

Contributors

Ralph Adolphs
Division of Humanities and Social Sciences, California Institute of Technology

William S. Anderson
Department of Neurosurgery, The Johns Hopkins University School of Medicine

Arjun K. Bansal
Children's Hospital, Harvard Medical School

Eric J. Behnke
Departments of Neurology and Neurosurgery, David Geffen School of Medicine at UCLA

Moran Cerf
Department of Neurosurgery, University of California, Los Angeles; Stern School of Business, New York University; and Kellogg School of Management, Northwestern University

Jonathan O. Dostrovsky
Department of Physiology, University of Toronto

Emad N. Eskandar
Department of Neurosurgery, Massachusetts General Hospital

Tony A. Fields
Department of Neurology, David Geffen School of Medicine at UCLA

Itzhak Fried
Department of Neurosurgery, David Geffen School of Medicine and Semel Institute for Neuroscience and Human Behavior, University of California, Los Angeles; Functional Neurosurgery Unit, Tel Aviv Medical Center; and Sackler School of Medicine, Tel Aviv University

Hagar Gelbard-Sagiv
Department of Neurosurgery, University of California, Los Angeles; and Division of Biology, Caltech

C. Rory Goodwin
Department of Neurosurgery, The Johns Hopkins University School of Medicine

Clement Hamani
Department of Surgery, Division of Neurosurgery, Toronto Western Hospital

Chris Heller
Neurosurgery and Spine Center, Rockwood Clinic

Mojgan Hodaie
Department of Surgery, Division of Neurosurgery, Toronto Western Hospital

Matthew Howard III
Department of Neurosurgery, University of Iowa

William D. Hutchison
Department of Physiology, University of Toronto; and Department of Surgery, Division of Neurosurgery, Toronto Western Hospital

Matias J. Ison
Department of Engineering and Center for Systems Neuroscience, University of Leicester

Hiroto Kawasaki
Department of Neurosurgery, University of Iowa

Christof Koch
Division of Biology, California Institute of Technology; and Allen Institute for Brain Science, Seattle

Rüdiger Köhling
Oscar-Langendorff-Institut for Physiology, Universitätsmedizin Rostock, University Rostock

Gabriel Kreiman
Children's Hospital, Harvard Medical School

Michel Le Van Quyen
Centre de Recherche de l'Institut du Cerveau et de la Moelle épinière (ICM), INSERM UMRS
975 and CNRS UMR 7225, Université Pierre et Marie Curie, Hôpital de la Pitié-Salpêtrière,
Paris

Frederick A. Lenz
Department of Neurosurgery, The Johns Hopkins University School of Medicine

Andres M. Lozano
Department of Surgery, Division of Neurosurgery, Toronto Western Hospital

Adam N. Mamelak
Department of Neurosurgery, Cedars-Sinai Medical Center

Carissa Martinez-Rubio
Department of Neurosurgery, Massachusetts General Hospital

Florian Mormann
Department of Epileptology, University of Bonn

Yuval Nir
Department of Physiology and Pharmacology, Sackler School of Medicine and Sagol School
of Neuroscience, Tel Aviv University, Israel; Department of Psychiatry, School of Medicine,
University of Wisconsin—Madison

George Ojemann
Department of Neurological Surgery, University of Washington School of Medicine

Shaun R. Patel
Department of Neurosurgery, Massachusetts General Hospital; and Department of Anatomy and
Neurobiology, Boston University School of Medicine

Sanjay Patra
Department of Clinical Neuroscience, Division of Neurosurgery, Spectrum Health System,
Grand Rapids, Michigan

Linda Philpott
Epilepsy and Brain Mapping Program, Huntington Memorial Hospital

Rodrigo Quian Quiroga
Department of Engineering and Center for Systems Neuroscience, University of Leicester

Ian Ross
Department of Neurosurgery, Huntington Memorial Hospital

Ueli Rutishauser
Department of Neurosurgery and Neurology, Cedars-Sinai Medical Center; and Division of Biology, California Institute of Technology

Andreas Schulze-Bonhage
Epilepsy Center, University Hospital Freiburg

Erin M. Schuman
Max Planck Institute for Brain Research, Frankfurt am Main, Germany

Demetrio Sierra-Mercado
Department of Neurosurgery, Massachusetts General Hospital

Richard J. Staba
Department of Neurology, David Geffen School of Medicine at UCLA

Nanthia Suthana
Department of Neurosurgery, David Geffen School of Medicine, and Semel Institute for Neuroscience and Human Behavior, University of California, Los Angeles; and Department of Psychology, University of California, Los Angeles

William Sutherling
Epilepsy and Brain Mapping Program, Huntington Memorial Hospital

Travis S. Tierney
The Brigham and Women's Hospital; and Harvard Medical School

Giulio Tononi
Department of Psychiatry, School of Medicine, University of Wisconsin—Madison

Oana Tudusciuc
Division of Humanities and Social Sciences, California Institute of Technology

Charles L. Wilson
Department of Neurology, David Geffen School of Medicine at UCLA

Index